DYNAMIC METEOROLOGY

DYNAMIC METEOROLOGY

LECTURES DELIVERED AT THE SUMMER SCHOOL OF SPACE PHYSICS OF THE CENTRE NATIONAL D'ETUDES SPATIALES, HELD AT LANNION, FRANCE, 7 AUGUST–12 SEPTEMBER 1970

Edited by

P. MOREL

Université de Paris, Laboratoire de Météorologie Dynamique

D. REIDEL PUBLISHING COMPANY

DORDRECHT-HOLLAND / BOSTON-U.S.A.

DOI 10.1007/978-94-010-2599-7

Published by D. Reidel Publishing Company,
P.O. Box 17, Dordrecht, Holland

CONTENTS

PREFACE

The development of numerical integration techniques and the pioneering efforts of Von Neumann and his associates at the Institute for Advanced Studies (Princeton) have spurred the renewed interest of many leading fluid dynamicists and meteorologists in the theory and numerical simulation of planetary atmosphere and oceans circulations. Their work during the last 15 years, now culminating in the Global Atmospheric Research Program, has led to the possibility of vastly improved weather forecasts as well as the development of a full fledged branch of the physical sciences : geophysical fluid dynamics.

Simultaneously, great strides have been made in developing powerful new instruments, operating from earth orbiting satellites, to observe the meteorological phenomena and to determine the state of motion of the atmosphere. Centre National d'Etudes Spatiales (CNES) of France has very significantly contributed to this effort by developing the EOLE navigation and data collection satellite, launched on 16 August 1971 to interrogate 500 instrumented platforms measuring meteorological parameters. It is fitting then, that CNES should have brought together leading scientists in the field of dynamic meteorology, to participate in its 1970 Summer School on Space Physics.

The CNES Summer School sessions are held every second year and intended for postgraduate students and junior scientists. The Summer School is organized to cover one field of space physics or geophysics with up-to-date tutorial reviews of the latest scientific advances. Five world known scientists from the United States : Professors Norman PHILLIPS, Jule CHARNEY, Vern SUOMI and Dr. Douglas LILLY, and from the U.S.S.R. : Professor Andreï MONIN, agreed to participate and to deliver, together with Professors Paul QUENEY and Pierre MOREL of the University of Paris, a total of 70 lectures during a period of five weeks.

This book is an edited account of these lectures, expanded with extensive bibliographic references and some new developments which were painstakingly compiled during the best part of two years. As it stands now, the book gives a nearly complete and searching presentation of the theoretical basis of dynamic meteorology. We hope that it will be a useful addition to the relatively few printed works in this new field of science.

The editor would like to extend his thanks to Centre National d'Etudes Spatiales and the many other institutions which made the school possible, and particularly to Mr. James HIEBLOT, Director of the Groupe de Recherches Ionosphériques du C.N.R.S. and Dr. André LEBEAU, Director of Programs and Planning of CNES, who organized this third CNES Summer School on Space Physics. He wishes to express also his deep appreciation to the CNES staff who contributed to the success of the School and the publication of this book.

Pierre MOREL

Directors of the School

J. HIEBLOT

Director
of the Groupe de Recherches Ionosphériques
du Centre National de la Recherche Scientifique
45 - ORLEANS 02 (France)

A. LEBEAU

Director of Programs and Planning
of the Centre National d'Etudes Spatiales
91 - BRETIGNY-SUR-ORGE (France)

P. MOREL

Professor
Laboratoire de Météorologie Dynamique
Ecole Normale Supérieure
75 - PARIS (France)

Professors

J.G. CHARNEY

Massachusetts Institute of Technology
Department of Meteorology
CAMBRIDGE, Massachusetts (U.S.A.)

D.K. LILLY

National Center for Atmospheric Research
BOULDER, Colorado (U.S.A.)

A.S. MONIN

Institute of Oceanology
Academy of Sciences
MOSCOW (U.S.S.R.)

P. MOREL

Laboratoire de Météorologie Dynamique
Ecole Normale Supérieure
75 - PARIS (France)

N.A. PHILLIPS

Massachusetts Institute of Technology
Department of Meteorology
CAMBRIDGE, Massachusetts (U.S.A.)

P. QUENEY

University of Paris
Laboratoire de Météorologie
75 - PARIS (France)

PRINCIPLES OF LARGE SCALE NUMERICAL WEATHER PREDICTION

N.A. PHILLIPS

Massachusetts Institute of Technology
Department of Meteorology
CAMBRIDGE, Massachusetts (U.S.A.)

CONTENTS

PRINCIPLES OF LARGE SCALE NUMERICAL WEATHER PREDICTION

by

N.A. PHILLIPS

MASSACHUSETTS INSTITUTE OF TECHNOLOGY

Department of Meteorology

CAMBRIDGE, Massachusetts, U.S.A.

1. EQUATIONS OF MOTION FOR THE ROTATING ATMOSPHERE

The motion of the atmosphere, when treated from the viewpoint of continuum mechanics, is governed by Newton's law of motion relating the acceleration to the forces, the thermodynamic law relating the rate of change of internal energy to the rate of heating, the principle of conservation of mass, the thermodynamic state relations, detailed mathematical formulations of the forces and rates of heating, and, finally, appropriate conditions at the boundaries of the particular mathematical model of the atmosphere being considered. Implied in this statement already is a recognition that we limit our attention to an approximate model of the real "atmosphere". For example, at very high altitudes the mean free path and time interval between molecular collisions becomes large enough that the quasi-equilibrium assumptions of continuum mechanics and thermodynamics break down. As other examples of the limitation of our meteorological viewpoint we may cite on the one hand our neglect of the interaction between the "atmosphere" and its extension to interplanetary space, and on the other hand our ignoring of the exchange of dry air across the air-ground and air-ocean interface, which we shall treat as impervious to the flow of dry air. These somewhat trite examples are cited only to show immediately that we admit to a simplified atmospheric model ; in fact, however, the practical analysis of large-scale atmospheric motions at the present state of hydrodynamical theory forces approximations upon dynamical meteorologists which have a much greater effect on the accuracy of the results than do the somewhat esoteric examples mentioned above. Chief among these approximations are the crude representations of the effects of very small scale motions upon the large scale motions. (The observational network available to define the initial state of the atmosphere will only define motions on space and time scales larger than the intervals between observations.) These effects of the small-scale motions on the large-scale momentum and temperature must be absorbed into our definition of "friction" and "heating".

We begin by formulating Newton's laws of motion for the rotating atmosphere. From elementary textbooks, we know that from the viewpoint of a coordinate system centered in the earth and rotating with the earth (Ω = 7.292 x 10^{-5} radian sec^{-1}), the "relative" or apparent acceleration $d\vec{v}/dt$ is equal to

$$\frac{d\vec{v}}{dt} = -\frac{1}{\rho}\nabla p + \vec{F} - \nabla\phi_e - \nabla\phi_t - \vec{\Omega}\times(\vec{\Omega}\times\vec{r}) - 2\vec{\Omega}\times\vec{v} \qquad (1.1)$$

where

$\vec{v}$ = velocity relative to rotating earth.

p = pressure

ρ = density

ϕ_e = Newtonian potential due to earth (the self-gravity of atmosphere is negligible).

ϕ_t = tidal potential due to moon and sun (and possibly the variable distribution of mass in oceans and solid earth). See Lamb (1932, pp. 358-362).

$\vec{\Omega}$ = angular velocity vector of earth (assumed constant).

$\vec{r}$ = position vector from point on axis of earth

$\vec{F}$ = all forces (per unit mass) other than pressure and Newtonian gravitational attraction. "Friction" is the only effect included in numerical weather prediction and in studies of the general circulation of the lowest 100 km of the atmosphere. Electromagnetic forces begin to be important above about

100 km.

(The tidal force $-\nabla\phi_t$ is the difference between the acceleration of the earth in its orbit and the combined attraction of solar and lunar gravity on the atmosphere). The operator d/dt is a time rate of change following the motion :

$$\frac{d}{dt} = \frac{\partial}{\partial t} + \vec{v}\cdot\nabla \tag{1.2}$$

$$\begin{aligned}\frac{d\vec{v}}{dt} &= \frac{\partial\vec{v}}{\partial t} + \vec{v}\cdot\nabla\vec{v} \\ &= \frac{\partial\vec{v}}{\partial t} + \nabla\left(\frac{1}{2}\vec{v}^2\right) + (\nabla\times\vec{v})\times\vec{v}\end{aligned}$$

(The symbol ∇ is the three-dimensional gradient operator in sections 1-3. Later, after the hydrostatic pressure system is introduced, it will denote a "horizontal" operator only).

We will ignore from now on the tidal potential ϕ_t since the motions produced by it seem to be small enough (below 100 km) to be simply added to whatever motions result from the remaining forces. Noting that $-\vec{\Omega}\times(\vec{\Omega}\times\vec{r})$ is equal to $\frac{1}{2}\nabla(\Omega^2\ell^2)$, where ℓ is the distance to the rotation axis, we combine this term with ϕ_e to produce the geopotential, ϕ :

$$\phi = \phi_e - \frac{1}{2}\Omega^2\ell^2$$

$$-\nabla\phi = \vec{g} = \text{"gravity"} \tag{1.3}$$

Because of inhomogeneities in the crust and other slight deviations, a reference surface of constant ϕ (the <u>geoid</u>) is not exactly a spheroidal surface of revolution but undulates slightly above and below the <u>reference spheroid</u>. The latter is a theoretically and empirically defined ellipsoid of revolution whose shape and mass is such that the field of geopotential associated with it gives a good approximation to observed gravity. The corresponding value of $\vec{g}$ is called <u>normal gravity</u>, and it is enough for our purposes to neglect any deviation of $\vec{g}$ from normal gravity. The geographic latitude, θ , is the angle between $-\vec{g}$ and the equatorial plane, in distinction to the <u>geocentric latitude</u> θ_{gc} determined by the equatorial plane and the radius vector from the center of the spheroid. On the surface of the spheroid the two latitudes are almost equal, differing by a maximum amount of 10 minutes in middle latitudes. At large distances from the earth, they differ greatly.

On the surface of the reference spheroid the magnitude of normal gravity varies only slightly with latitude,

$$g = 9.78049\,(1 + 0.0052884 \sin^2\theta - 0.0000059 \sin^2 2\theta)\ \text{m sec}^{-2}$$

differing by only 1/2 percent from pole to equator. Near this surface its variation with elevation (z) is small $(\partial g/\partial z \sim -3 \times 10^{-5} \sec^{-2})$ and g is reduced by only 3 percent at z = 100 km.

Our equation of motion

$$\frac{d\vec{v}}{dt} = -\frac{1}{\rho}\nabla p + \vec{F} - \nabla\phi - 2\vec{\Omega}\times\vec{v} \qquad (1.4)$$

(where ϕ is now the normal geopotential) must be expressed in a convenient coordinate system for computational purposes. We first take an orthogonal curvilinear system ξ_1 ξ_2 ξ_3 in which

$$\begin{aligned} \xi_1 &= \lambda && = \text{east longitude} \\ \xi_3 &= \phi && = \text{normal geopotential surfaces of revolution} \qquad (1.5) \\ & & \xi_2 & = \text{surfaces of revolution normal to the geopotential surfaces.} \end{aligned}$$

Let ϕ_0 be the value of ϕ at <u>sea-level</u> (i.e., at the surface of the reference spheroid). Each ξ_2 surface can be labelled with the value of latitude θ_0 at the point where it intersects the ϕ_0 surface. This choice for ξ_1, ξ_2 and ξ_3 is dictated by our desire to have sea-level a coordinate surface and to have gravity ($-\nabla\phi$) appear in only one component (the <u>vertical</u> component) of (1.4). However, the metric scale factors h_1, h_2, h_3 in the distance increments ds

$$ds_i = h_i\, d\xi_i \qquad (1.6)$$

are now complicated functions of ξ_2 and ξ_3 .

To simplify matters, we first define a <u>reference sphere</u> of radius equal to the mean radius of the earth

$$a = 6{,}371 \text{ km}$$

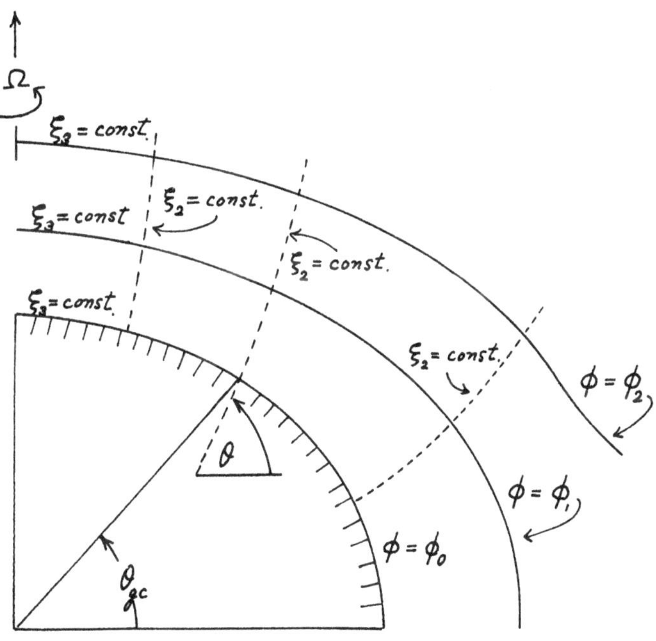

Surfaces of constant ξ_2 and constant ξ_3

Then to a point P in the atmosphere with coordinates ξ_1, ξ_2, and ξ_3, we assign spherical coordinates (λ_s, θ_s, r_s) with respect to the reference sphere by the following rules :

$$\begin{aligned} \lambda_s &= \lambda = \lambda_s(\xi_1) \\ \theta_s &= \theta_0 = \theta_s(\xi_2) \\ r_s &= a + \int_{\phi_0}^{\phi} \frac{d\phi}{\bar{g}(\phi)} = r_s(\xi_3) \end{aligned} \tag{1.7}$$

The integral is performed at constant ξ_2 and $\bar{g}(\phi)$ is an average value of g on a surface of constant ϕ. Then (1.7) is only a relabelling of the ξ_2 and ξ_3 surfaces, and λ_s, θ_s, r_s are still orthogonal coordinates. Due to the uniformity of g, $r_s - a$ is very closely equal to the geometric height of the point P above sea-level, as long as point P is not too far away from sea-level.

The obvious final step is to introduce an approximation by recognizing that, at points reasonably close to the earth, the uniformity of g and the small ellipticity imply that the scale factors for λ_s, θ_s, r_s

differ only slightly from those appropriate to true spherical coordinates, i.e., we now make the approximation

$$h_1 = r_s \cos\theta_s \qquad h_2 = r_s \qquad h_3 = 1 \tag{1.8}$$

Two important features for (1.4) have been retained by this technique, however. These are that $-\nabla\phi$ has a component only in the "radial" direction, and that spatial derivatives at constant r_s (as in ∇p) do not mean derivation at a constant distance from the center of the earth but instead mean derivation along the "horizontal" surfaces of constant geopotential.

At this point we can drop the s subscript. If we introduce the velocity component notation,

$$\begin{aligned} u &= \text{eastward} = h_\lambda \frac{d\lambda}{dt} = r\cos\theta \frac{d\lambda}{dt} \\ v &= \text{northward} = h_\theta \frac{d\theta}{dt} = r\frac{d\theta}{dt} \\ w &= \text{radial} = h_r \frac{dr}{dt} = \frac{dr}{dt} \end{aligned} \tag{1.9}$$

equation (1.4) can be expressed in components by using the last part of (1.2). The result is well-known :

$$\frac{du}{dt} = -\frac{1}{\rho r\cos\theta}\frac{\partial p}{\partial\lambda} + F_\lambda + \left(2\Omega + \frac{u}{r\cos\theta}\right)(v\sin\theta - w\cos\theta) \tag{1.10}$$

$$\frac{dv}{dt} = -\frac{1}{\rho r}\frac{\partial p}{\partial\theta} + F_\theta - \left(2\Omega + \frac{u}{r\cos\theta}\right)u\sin\theta - \frac{vw}{r} \tag{1.11}$$

$$\frac{dw}{dt} = -\frac{1}{\rho}\frac{\partial p}{\partial r} - g + F_r + \left(2\Omega + \frac{u}{r\cos\theta}\right)u\cos\theta + \frac{v^2}{r} \tag{1.12}$$

g is taken as constant, in conformity with the approximations (1.8) :

$$g = 9.81 \ m\,sec^{-2}$$

The terms proportional to 2Ω are the Coriolis acceleration (actually derived earlier by Laplace), whereas the quadratic velocity terms are a result only of our use of a non-cartesian system. Both classes of terms are orthogonal to $\vec{v}$, so that

$$\frac{d}{dt}\left(\frac{1}{2}\vec{v}^2\right) = \vec{v}\cdot\left(-\frac{1}{\rho}\nabla p + \vec{F} - \nabla\phi\right) \tag{1.13}$$

i.e., the rate of change of kinetic energy is equal to the rate at which the forces of pressure, friction, and apparent gravity do work. Our approximation (1.8) has not interfered with this physical principle. It has also not interfered with the angular momentum principle ; (1.9) - (1.12), taken together, include the relation

$$\frac{d}{dt}\left[(u + \Omega r\cos\theta)r\cos\theta\right] = r\cos\theta\left(-\frac{1}{\rho r\cos\theta}\frac{\partial p}{\partial\lambda} + F_\lambda\right) \tag{1.14}$$

Because (1.8) is an approximation, $u + \Omega r\cos\theta$ is not equal to the true eastward velocity as seen by a non-rotating observer. Therefore $(u + \Omega r\cos\theta)r\cos\theta$ is not the true absolute angular momentum (per unit mass) of a parcel of air about the axis of the earth, and (1.14) is only apparently equivalent to the principle that the time rate of change of the absolute angular momentum should be equal to the torque exerted by the longitudinal forces. Nevertheless it is very important that equations (1.9) - (1.12) contain such a statement ; an important approach in the study of the general circulation concerns the question of how the distribution of zonal velocity u , with its concentration into strong jet streams, is brought about and maintained against the destructive effect of friction. This approach would be much more difficult to follow if (1.14) did not exist.

Several further approximations are often made in equations (1.8) - (1.12)in their application to large-scale motions. These are

a) The replacement of r by a constant value, a .

b) The Coriolis and metric terms proportional to $\cos\theta$ are ignored in (1.10) and (1.12).

c) Equation (1.12) is replaced by the hydrostatic relation

$$-\frac{1}{\rho}\frac{\partial p}{\partial r} = g \tag{1.15}$$

The direct replacement of r in (1.8) - (1.12) by the constant a results in equations which do not have an angular momentum principle like (1.14). Instead, let us return momentarily to Newton's law for a non-rotating system in the vector-invariant form

$$\frac{\partial\vec{v}_a}{\partial t} + \nabla(\tfrac{1}{2}\vec{v}_a^{\,2}) + (\nabla\times\vec{v}_a)\times\vec{v}_a = -\frac{1}{\rho}\nabla p + \vec{F} - \nabla\phi_e \tag{1.16}$$

where $\vec{v}_a$ is the absolute velocity,

$$\vec{v}_a = \Omega h_\lambda\hat{e}_\lambda + \vec{v} \tag{1.17}$$

$$\left(\frac{\partial}{\partial t}\right)_a = \frac{\partial}{\partial t} - \Omega\frac{\partial}{\partial\lambda} \tag{1.18}$$

We now introduce the symbol z for the integral in (1.7) and the following simplified scale factors in place of the more accurate forms (1.8)

$$z \equiv \int_{\phi_0}^{\phi} \frac{d\phi}{\bar{g}} \approx \frac{\phi - \phi_0}{\bar{g}_0}$$

$$h_\lambda = a\cos\theta, \quad h_\theta = a, \quad h_z = 1 \tag{1.19}$$

Use of (1.19) in the vector formulas for $\vec{v}$, ∇ , and curl, now produces the following simplified versions of (1.9) - (1.12) :

$$\begin{aligned} u &= h_\lambda \frac{d\lambda}{dt} = a\cos\theta \frac{d\lambda}{dt} \\ v &= h_\theta \frac{d\theta}{dt} = a\frac{d\theta}{dt} \\ w &= h_z \frac{dz}{dt} = \frac{dz}{dt} \end{aligned} \tag{1.20}$$

$$\frac{du}{dt} = -\frac{1}{\rho a\cos\theta}\frac{\partial p}{\partial\lambda} + F_\lambda + \left(2\Omega + \frac{u}{a\cos\theta}\right) v\sin\theta \tag{1.21}$$

$$\frac{dv}{dt} = -\frac{1}{\rho a}\frac{\partial p}{\partial\theta} + F_\theta - \left(2\Omega + \frac{u}{a\cos\theta}\right) u\sin\theta \tag{1.22}$$

$$\frac{dw}{dt} = -\frac{1}{\rho}\frac{\partial p}{\partial z} - g + F_z \tag{1.23}$$

These possess the "angular momentum principle"

$$\frac{d}{dt}\left[(u + \Omega a\cos\theta) a\cos\theta\right] = a\cos\theta\left(-\frac{1}{\rho a\cos\theta}\frac{\partial p}{\partial\lambda} + F_\lambda\right) \tag{1.24}$$

in which the "angular momentum" of a parcel is evaluated as if it were located at sea-level.

The Coriolis terms proportional to $\cos\theta$ have disappeared formally because h_λ in (1.17) did not depend on z . The validity of omitting $2\Omega\cos\theta$ terms but retaining $2\Omega\sin\theta$ terms is not obvious, however. (Eckart, 1960, refers to this omission as the "traditional approximation".) A more detailed analysis (see Phillips (1968) for a discussion and references to other discussions) shows that for the motions of most interest to meteorologists, the $2\Omega\cos\theta$ terms can generally be neglected if the atmosphere has a sufficiently stable stratification. The appropriate measure of this is the buoyancy frequency N. For a perfect gas this is given by

$$N^2 = \frac{g}{c_p}\frac{\partial s}{\partial z} = \frac{g}{T}\left(\frac{g}{c_p} + \frac{\partial T}{\partial z}\right) \tag{1.25}$$

(s = entropy, T = temperature). N^2 gives the Archimedean restoring force per unit mass on a particle displaced vertically downward a unit distance without heating. The analysis referred to above suggests that $2\Omega\cos\theta$ terms can be ignored if

$$N^2 \gg \Omega^2 \tag{1.26}$$

(Roughly speaking, we may consider that large N^2 forces the fluid to move in horizontal planes, thereby allowing the vertical component $2\Omega \sin\theta$ of the Coriolis vector to be more important than the horizontal component $2\Omega\cos\theta$. It must be pointed out, however, that the arguments leading to (1.26) rest on simple models of atmospheric motions and therefore do not prove conclusively that $2\Omega\cos\theta$ plays no role in atmospheric dynamics).

For a normal temperature lapse rate of $\partial T/\partial z \sim -6^\circ \mathrm{deg/km}$, N^2 is about 10^{-4} sec^{-2}, whereas $\Omega^2 = (7.29 \times 10^{-5})^2 \sim 5 \times 10^{-9}$ sec^{-2}. The inequality (1.26) is well satisfied and we may therefore feel reasonably confident in accepting the simplified system (1.19) - (1.24). (When the atmosphere happens to have a neutral or unstable stratification, small-scale convection is present, and the frequencies of this motion are large enough that both Coriolis terms are unimportant).

Postponing discussion of the hydrostatic approximation (1.15) we now collect for reference the other physical laws governing the atmosphere. These are the conservation of mass ("continuity equation")

$$\frac{1}{\rho}\frac{d\rho}{dt} = -\nabla\cdot\vec{v} = -\frac{\partial w}{\partial z} - \frac{1}{a\cos\theta}\left[\frac{\partial u}{\partial\lambda} + \frac{\partial}{\partial\theta}(v\cos\theta)\right] \tag{1.27}$$

and the thermodynamic law

$$\frac{ds}{dt} = \frac{q}{T} \tag{1.28}$$

where s is the specific entropy and q the rate of heating per unit mass. (q should include the conversion of kinetic energy to heat by viscosity, but for all purposes except the study of the exact entropy balance of this atmosphere this heating is negligible compared to that from conduction and radiation. Equation (1.28) should also includes changes of entropy due to mixing. These too are negligible since the composition of air is almost uniform. To these we must add the equations of state. For dry air these are the ideal gas relations

$$\begin{aligned} ds &= C_p\, d\ln\Theta = C_p\, d\ln T - R\, d\ln p \\ \Theta &= \text{potential temperature} = T\left(\frac{p_{00}}{p}\right)^{R/C_p} \end{aligned} \tag{1.29}$$

$$\rho = \frac{p}{RT} \tag{1.30}$$

p_{00} = 1000 mb is a convenient reference pressure. The exact density relation for moist air is

$$\rho = (1 + r)\frac{p_d}{RT} \tag{1.31}$$

where r is the mixing ratio (vapor plus liquid) and p_d is the partial pressure of the dry air. In large-scale dynamics the difference between (1.31) and (1.30) is usually ignored.

Boundary conditions at the bottom of the atmosphere include the requirement that the normal velocity equal that of the air-ground or air-water interface. For a rigid surface $z = z_0(\lambda, \theta)$ this is

$$w = \vec{v}\cdot\nabla z_0 = \frac{u}{a\cos\theta}\frac{\partial z_0}{\partial\lambda} + \frac{v}{a}\frac{\partial z_0}{\partial\theta} \quad \text{at} \quad z = z_0 \tag{1.32}$$

When the force $\vec{F}$ depends upon a frictional stress $\vec{\tau}$ acting across horizontal surfaces, e.g.

$$F_\lambda = \frac{1}{\rho}\frac{\partial\tau_\lambda}{\partial z} \quad , \quad F_\theta = \frac{1}{\rho}\frac{\partial\tau_\theta}{\partial z} \tag{1.33}$$

the kinematic relation (1.32) must be supplemented by an additional condition. If the stress $\vec{\tau}$ were due to molecular viscosity, this would be provided at a rigid boundary by vanishing of the tangential velocity component (no slip). In practice, however, $\vec{\tau}$ is a turbulent stress arising from scales of motion smaller than those being considered explicitly ; the additional lower boundary condition is then given by an empirical law for $\vec{\tau}$:

$$\vec{\tau}(z = z_0) = C\rho_0|\vec{V}_0|\vec{V}_0 \tag{1.34}$$

where $\vec{V}_0$ signifies the horizontal component of $\vec{v}$ (at a height about 10 meters above the ground), and C is a number thought to be around 0.002.

Boundary conditions at the top of the atmosphere are much vaguer because we are ignoring the real processes occuring at heights above 100 km. They will be discussed later.

The incorporation of vertical fluxes of "heat" H due to small-scale turbulence into the heating rate q is represented formally by

$$q_{turb.} = -\frac{1}{\rho}\frac{\partial H}{\partial z} \tag{1.35}$$

Simple empirical relations between H and $\partial s/\partial z$ are used. The lower boundary conditions are then supplemented by an expression similar to (1.34) :

$$H_{z=z_0} = C'\rho_0 c_p|\vec{V}_0|(T_g - T_0) \tag{1.36}$$

where T_g is the surface temperature of the underlying ground or water. Similar considerations apply to the small-scale turbulent flux of water vapor.

2. CONFORMAL MAPS

The approximate equations of motion (1.21) - (1.23) can be rewritten in vector form as

$$\frac{\partial \vec{v}}{\partial t} = -\frac{1}{\rho}\nabla p - \nabla\left(\frac{1}{2}\vec{v}^2\right) - \left(f\hat{k} + \operatorname{curl}\vec{v}\right)\times\vec{v} - g\hat{k} + \vec{F} \tag{2.1}$$

where

$$f = 2\Omega \sin\theta \tag{2.2}$$

is the Coriolis parameter and $\hat{k}$ is the vertical unit vector. We consider now the replacement of λ and θ by two other orthogonal curvilinear coordinates $x(\lambda, \theta)$ and $y(\lambda, \theta)$, with scale factors $h_x(\lambda, \theta)$ and $h_y(\lambda, \theta)$ which do not depend on z. (In the following description, we maintain consistently the approximation of replacing r by a. If variable r were retained, as in (1.8) - (1.12), the conformal scale factors defined below would all be multiplied by (r/a) and the Coriolis term $f\hat{k}$ in (2.1) would be replaced by $2\vec{\Omega}$.) The velocity $\vec{v}$ has components

$$\vec{v} = (u, v, w) = \left(h_x\frac{dx}{dt}, h_y\frac{dy}{dt}, \frac{dz}{dt}\right) \tag{2.3}$$

Using standard orthogonal curvilinear vector formulae for ∇ and curl, and the relation

$$\frac{d}{dt} = \frac{\partial}{\partial t} + \vec{v}\cdot\nabla = \frac{\partial}{\partial t} + \frac{u}{h_x}\frac{\partial}{\partial x} + \frac{v}{h_y}\frac{\partial}{\partial y} + w\frac{\partial}{\partial z} \tag{2.4}$$

the component form of (2.1) can be written as

$$\frac{du}{dt} = -\frac{1}{\rho h_x}\frac{\partial p}{\partial x} + v\left[f + \frac{1}{h_x h_y}\left(v\frac{\partial h_y}{\partial x} - u\frac{\partial h_x}{\partial y}\right)\right] + F_x \tag{2.5}$$

$$\frac{dv}{dt} = -\frac{1}{\rho h_y}\frac{\partial p}{\partial y} - u\left[f + \frac{1}{h_x h_y}\left(v\frac{\partial h_y}{\partial x} - u\frac{\partial h_x}{\partial y}\right)\right] + F_y \tag{2.6}$$

$$\frac{dw}{dt} = -\frac{1}{\rho}\frac{\partial p}{\partial z} - g + F_z \tag{2.7}$$

(Note that if $x = \lambda$, $y = \theta$, $h_x = a\cos\theta$, $h_y = a$, these reduce back to (1.21) - (1.23).)

Conformal maps are characterized by $h_x = h_y$. This introduces a greater symmetry into the equations.

Mercator coordinates are

$$x = a\lambda \tag{2.8}$$

$$y = a \ln\left(\frac{1+\sin\theta}{\cos\theta}\right) \tag{2.9}$$

$$h_x = h_y = h = \cos\theta = \operatorname{sech}\left(\frac{y}{a}\right) \tag{2.10}$$

$$\sin\theta = \tanh\left(\frac{y}{a}\right) \tag{2.11}$$

The component equations of motion are

$$(u, v, w) = \left(a\cos\theta\frac{d\lambda}{dt}, a\frac{d\theta}{dt}, \frac{dz}{dt}\right) \tag{2.12}$$

$$\frac{du}{dt} = -\frac{1}{\rho h}\frac{\partial p}{\partial x} + v\left(2\Omega + \frac{u}{a\cos\theta}\right)\sin\theta + F_x \tag{2.13}$$

$$\frac{dv}{dt} = -\frac{1}{\rho h}\frac{\partial p}{\partial y} - u\left(2\Omega + \frac{u}{a\cos\theta}\right)\sin\theta + F_y \tag{2.14}$$

$$\frac{dw}{dt} = -\frac{1}{\rho}\frac{\partial p}{\partial z} - g + F_z \tag{2.15}$$

This is only a relabelling of the latitude coordinate (cf. 1.21) - (1.23)) in which the polar singularities at $\theta = \pm\pi/2$ are mapped into $y = \pm\infty$

Stereographic coordinates are normally determined by projecting a point P on a sphere of radius a onto a plane which is tangent at the North or South Pole, by means of a straight line through P and the opposite pole. It is of some interest to generalize this by allowing the tangent plane to be located at the point $\lambda = 0, \theta = \theta_0$, and replacing the opposite pole in the projection line by the antipode $\lambda = \pi, \theta = -\theta_0$. The coordinates are

$$x = 2a\,\frac{\cos\theta\sin\lambda}{1 + \sin\theta_0\sin\theta + \cos\theta_0\cos\theta\cos\lambda} \tag{2.16}$$

$$y = 2a\,\frac{\cos\theta_0\sin\theta - \sin\theta_0\cos\theta\cos\lambda}{1 + \sin\theta_0\sin\theta + \cos\theta_0\cos\theta\sin\lambda} \tag{2.17}$$

$$h_x = h_y = h = \frac{1}{2}(1 + \sin\theta_0\sin\theta + \cos\theta_0\cos\theta\cos\lambda) \tag{2.18}$$

$$= \frac{4a^2}{4a^2 + x^2 + y^2}$$

$$\sin\theta = \frac{\sin\theta_0(4a^2 - x^2 - y^2) + 4ay\cos\theta_0}{4a^2 + x^2 + y^2} \tag{2.19}$$

The positive y-axis is directed northward from the origin at θ_0 along the meridian $\lambda = 0$. The positive x-axis extends to the east of the origin along the great circle which goes through the origin and is tangent to the latitude circle $\theta = \theta_0$. The horizontal unit vectors $\hat{x}$ and $\hat{y}$ in the x and y directions are related to the unit vectors $\hat{\lambda}$ and $\hat{\theta}$ in the eastward and northward directions by

$$\hat{x} = \hat{\lambda} \cos\alpha + \hat{\theta} \sin\alpha$$
$$\hat{y} = -\hat{\lambda} \sin\alpha + \hat{\theta} \cos\alpha \qquad (2.20)$$

where

$$\cos\alpha = \frac{\cos\theta_0 \cos\theta + (1 + \sin\theta_0 \sin\theta)\cos\lambda}{1 + \sin\theta_0 \sin\theta + \cos\theta_0 \cos\theta \cos\lambda}$$
$$\sin\alpha = \frac{-(\sin\theta_0 + \sin\theta)\sin\lambda}{1 + \sin\theta_0 \sin\theta + \cos\theta_0 \cos\theta \cos\lambda} \qquad (2.21)$$

The singularity is at the antipode $\lambda = \pi$, $\theta = -\theta_0$, which maps into the circle $x^2 + y^2 \to \infty$. The equations of motion are

$$u = h\frac{dx}{dt} = \left(a\cos\theta \frac{d\lambda}{dt}\right)\cos\alpha + \left(a\frac{d\theta}{dt}\right)\sin\alpha \qquad (2.22)$$

$$v = h\frac{dy}{dt} = -\left(a\cos\theta\frac{d\lambda}{dt}\right)\sin\alpha + \left(a\frac{d\theta}{dt}\right)\cos\alpha \qquad (2.23)$$

$$\frac{du}{dt} = -\frac{1}{\rho h}\frac{\partial p}{\partial x} + \left(f + \frac{uy - vx}{2a^2}\right)v + F_x \qquad (2.24)$$

$$\frac{dv}{dt} = -\frac{1}{\rho h}\frac{\partial p}{\partial y} - \left(f + \frac{uy - vx}{2a^2}\right)u + F_y \qquad (2.25)$$

$$\frac{dw}{dt} = -\frac{1}{\rho}\frac{\partial p}{\partial z} - g + F_z \qquad (2.26)$$

Note the absence of any singular behavior at the origin, $x = y = 0$

These stereographic equations are normally used only with $\theta_0 = \pm\pi/2$. However, they provide the proper interpretation for the so-called "tangent plane" equations - the system often referred to in text books in which Cartesian coordinates are used and the motion u, v anywhere in the tangent plane is considered as horizontal even though this plane quickly becomes non-horizontal a short distance from the point of tangency. The point here is to note first that because h^{-1} in (2.24) - (2.25) differs from 1 by $O(x^2/a^2)$ it has increased only by about one percent at a horizontal distance as great as 1000 km from the origin. The only important term of first order in (x/a) in fact is the variation of $f = 2\Omega\sin\theta$,

$$f = 2\Omega\sin\theta_0 + \frac{2\Omega\cos\theta_0}{a}y + O\left(\frac{x^2+y^2}{a^2}\right) \qquad (2.27)$$

since the $uy - vx$ terms are negligible in comparison :

$$\frac{1}{2a^2}(uy - vx) = O\left(\frac{u}{2a}\right)O\left(\frac{y}{a}\right) \ll \Omega\, O\left(\frac{y}{a}\right) \qquad (2.28)$$

In this sense the equations can be simplified to treat local motions ; we set $h = 1$, ignore the $uy - vx$ terms, and set f equal to

$$f = f_0 + \beta y$$
$$f_0 = 2\Omega \sin\theta_0 , \quad \beta = \frac{2\Omega\cos\theta_0}{a} \tag{2.29}$$

with the expectation that this will be a good approximation for distances as great as several thousand kilometers from the origin. This rather formal approximation gives the so-called "β-plane" system which Rossby (1939) intuitively recognized as being sufficient to represent the important dynamical effect of the variability of $2\Omega\sin\theta$. (This demonstration is in fact the only reason for introducing here the more complicated general stereographic system (2.16) - (2.19) ; when hemispheric or global numerical weather calculations are performed with a stereographic system, θ_0 is always $\pm\pi/2$.)

3. JUSTIFICATION OF HYDROSTATIC BALANCE

The pressure and density distribution can always be split into a mean field depending only on z plus a variable part :

$$p = \bar{p}(z) + \delta p(\lambda,\theta,z,t)$$
$$\rho = \bar{\rho}(z) + \delta\rho(\lambda,\theta,z,t) \tag{3.1}$$

Below 100 km, the variable part of p and ρ is small compared to the mean part. The mean fields are hydrostatic by definition,

$$\frac{d\bar{p}}{dz} = -\bar{\rho}g \tag{3.2}$$

but the variable part δp need not be. To replace the vertical equation of motion (1.23) by the hydrostatic relation for the total field,

$$\frac{\partial p}{\partial z} = -\rho g \tag{3.3}$$

we must show that the vertical acceleration dw/dt is less than the <u>variable</u> part of the pressure force :

$$\frac{dw}{dt} \ll -\frac{1}{\rho}\frac{\partial\delta p}{\partial z} \tag{3.4}$$

(We ignore friction and heating.) From (1.21) and (1.22) we first note that δp has a magnitude of

$$\delta p \sim \rho L \frac{U}{\tau} \qquad \text{if} \qquad f\tau < 1 \tag{3.5}$$

or

$$\delta p \sim \rho L f U \qquad \text{if} \qquad f\tau > 1 \tag{3.6}$$

Here

$$f = 2\Omega \sin\theta \sim 10^{-4} \sec^{-1} \tag{3.7}$$

and

$$\begin{aligned} L &= \text{horizontal scale length of the motion} \\ \tau &= \text{time scale of the motion} \\ U &= \text{magnitude of horizontal velocity} \end{aligned} \tag{3.8}$$

(We have assumed that $u/a\cos\theta = d\lambda/dt$ is negligible compared to 2Ω .)

In the absence of heating $(ds/dt = 0)$, changes in p are related to changes in density by $\frac{dp}{dt} = C_s^2 \frac{d\rho}{dt}$ where C_s is the speed of sound ($\sim$ 320 m sec^{-1}). The continuity equation (1.27) is then

$$\frac{1}{\rho C_s^2}\frac{dp}{dt} \sim \frac{1}{\rho C_s^2}\left(\frac{d\delta p}{dt} + w\frac{d\bar{p}}{dz}\right) = -\frac{\partial w}{\partial z} - \frac{1}{a\cos\theta}\left(\frac{\partial u}{\partial \lambda} + \frac{\partial v\cos\theta}{\partial\theta}\right) \tag{3.9}$$

Introducing H for the vertical scale of the motion and W for the size of w , this equation contains the terms

$$\frac{1}{\rho C_s^2}\frac{\delta p}{\tau} + \frac{g}{C_s^2}W + \frac{W}{H} + \frac{U}{L} + \frac{U}{L} \sim 0 \tag{3.10}$$

C_s^2/g is a length of about 10 km. We let H' denote the smaller of H or C_s^2/g .

Suppose first that the δp term in (3.10) is equal to or larger than U/L . Then $W \leqslant H'(\rho\tau C_s^2)^{-1}\delta p$, or

$$\frac{dw}{dt} \sim \frac{W}{\tau} \leqslant \frac{H'H}{C_s^2\tau^2}\left(\frac{1}{\rho}\frac{\delta p}{H}\right) \sim \frac{H'H}{C_s^2\tau^2}\left(\frac{1}{\rho}\frac{\partial\delta p}{\partial z}\right) \tag{3.11}$$

Under these conditions hydrostatic balance is a good approximation if $H'H/C_s^2\tau^2$ is small :

$$\tau \gg \frac{(H'H)^{1/2}}{C_s} \leqslant \frac{10^4}{300} \sim 30 \sec \tag{3.12}$$

We assume this to be satisfied, which amounts to ignoring the existence of vertically propagating acoustic waves.

Suppose on the other hand that U/L in (3.10) is greater than the δp term, so that

$$W \lessapprox H'(U/L) \tag{3.13}$$

(The inequality allows $\partial u/\partial\lambda$ and $\partial v\cos\theta/\partial\theta$ to partially cancel one another.) (3.5) and (3.6) lead now to

$$\frac{dw}{dt} \lessapprox \frac{H'H}{L^2}\frac{1}{\rho}\frac{\delta p}{H} \sim \frac{H'H}{L^2}\left(\frac{1}{\rho}\frac{\partial \delta p}{\partial z}\right) \quad \text{if } f\tau < 1 \tag{3.14}$$

or

$$\frac{dw}{dt} \lessapprox \frac{H'H}{L^2}\frac{1}{f\tau}\left(\frac{1}{\rho}\frac{\partial \delta p}{\partial z}\right) \qquad \text{if } f\tau > 1 \tag{3.15}$$

Thus in this case it is sufficient to have $H'H/L^2$ small compared to one for hydrostatic balance to be valid :

$$L^2 \gg H'H \tag{3.16}$$

This condition is also valid for the motions treated in large-scale dynamical meteorology. It is not satisfied by small-scale motions such as cumulus clouds and turbulence near the ground.

The original equations (1.21) - (1.23) and (1.27) - (1.30) contain five partial time derivatives for u , v , w , ρ and s , and the five independent initial data fields u^0, v^0, w^0, ρ^0 and s^0 are just sufficient to enable these time derivatives to be evaluated. For example

$$\begin{aligned} \left(\frac{\partial \rho}{\partial t}\right)^0 &= -\nabla\cdot(\rho^0 \vec{v}^0) \\ \left(\frac{\partial s}{\partial t}\right)^0 &= -\vec{v}^0\cdot\nabla s^0 + \left(\frac{q}{T}\right)^0 \end{aligned} \tag{3.17}$$

The loss of $\partial w/\partial t$ in the vertical equation of motion means that w^0 can no longer be specified independently ; w at any instant must always be such that the changes in ρ and s computed from (3.17) are compatible with the state relations (1.29) - (1.30) and the hydrostatic relation (3.3). L. Richardson (1922, chapter V) originally derived an equation to determine w in this way. Nowadays another method is used, based on pressure as a vertical coordinate instead of z (Eliassen, 1949). (Richardson's method is used by one numerical weather prediction group (Kasahara and Washington, 1967).)

4. PRESSURE COORDINATES

The positiveness of ρ in the hydrostatic relation

$$\frac{\partial \phi}{\partial p} = -\frac{1}{\rho} = -\frac{RT}{p} \tag{4.1}$$

allows us to replace z by p as a vertical coordinate if we now assume that (4.1) is exact. The geopotential ϕ will replace p in its role as a dependent variable in the equations. We introduce the notation

$$\varpi = \frac{dp}{dt} = \dot{p} \tag{4.2}$$

$$\vec{V} = \text{horizontal part of } \vec{v}$$

We also now change the meaning of ∇ : ∇ will no longer denote the 3-dimensional gradient operator but from here on ∇ will denote the horizontal gradient operator. To begin with, we will use subscript notation p and z to denote whether partial derivatives with respect to x, y, and t are carried out at constant p or constant z. These subscripts will later be ignored, with differentiation at constant p being generally understood.

The material derivative is now

$$\frac{d}{dt} = \left(\frac{\partial}{\partial t}\right)_p + \vec{V}\cdot\nabla_p + \varpi\frac{\partial}{\partial p} \tag{4.3}$$

It is readily shown first that (4.1) leads to

$$(\nabla p)_z = \rho(\nabla\phi)_p \tag{4.4}$$

$$\left(\frac{\partial p}{\partial t}\right)_z = \rho\left(\frac{\partial \phi}{\partial t}\right)_p$$

Application of (4.1) - (4.4) to $\phi = gz$ (with g constant) relates ϖ to w :

$$\frac{\varpi}{\rho} = \left(\frac{\partial \phi}{\partial t}\right)_p + \vec{V}\cdot\nabla_p\phi - gw \tag{4.5}$$

This is needed to introduce the lower boundary condition (1.32) into the pressure coordinate system.

The continuity equation in the new coordinates is expressed most readily by considering the motion of a small element of mass δM, contained in a volume initially located between two pressure surfaces p and $p+\delta p$:

$$\delta M = \rho h_x \delta x\, h_y \delta y \delta z = h_x \delta x\, h_y \delta y \left(\frac{\delta p}{g}\right) \tag{4.6}$$

Allowing the boundaries of this volume to move with the fluid, we have $d(\delta M)/dt = 0$, and this leads kinematically to the statement

$$\frac{\partial \varpi}{\partial p} = -\frac{1}{h_x h_y}\left(\frac{\partial h_y u}{\partial x} + \frac{\partial h_x v}{\partial y}\right) = -\nabla_p \cdot \vec{V} \tag{4.7}$$

(Recognizing that δx is defined at constant p in (4.6), we use relations such as $d(\delta x)_p/dt = \delta_p(dx/dt) = [\partial(dx/dt)/\partial x]_p \delta x = [\partial(u/h_x)/\partial x]_p \delta x$ in deriving (4.7).) The assumption that h_x and h_y do not depend on p (i.e., r) (because we are using (1.19) instead of (1.8)) and the assumed constancy of g are both necessary to derive such a simple result. Other than these assumptions, however, (4.7) is exact if the motion is hydrostatic.

The thermodynamic laws (1.28) - (1.30) can be combined with the hydrostatic relation to give

$$\left[\left(\frac{\partial}{\partial t}\right)_p + \vec{V}\cdot\nabla_p\right]\frac{\partial \phi}{\partial p} - \frac{\varpi}{\rho}\frac{\partial \ln\Theta}{\partial p} = -\frac{R}{p c_p}q \tag{4.8}$$

The following alternate expressions exist for the coefficient of ϖ in (4.8) :

$$-\frac{1}{\rho}\frac{\partial \ln\Theta}{\partial p} = \frac{1}{g\rho^2}\frac{\partial \ln\Theta}{\partial z} = \frac{R^2T^2}{gp^2}\frac{\partial \ln\Theta}{\partial z} = \frac{S}{p^2},$$

$$S = \frac{R^2T^2}{g^2}N^2 = \frac{R^2T}{g}\left(\frac{\partial T}{\partial z} + \frac{g}{c_p}\right) = \frac{R}{c_p}\frac{\partial \phi}{\partial Z} + \frac{\partial^2 \phi}{\partial Z^2} \tag{4.9}$$

In the last of these Z is proportional to $-\ln p$,

$$Z = -\ln\left(\frac{p}{p_{00}}\right), \quad dZ = -\frac{dp}{p} = \frac{g}{RT}dz, \tag{4.10}$$

a coordinate useful in theoretical studies of hydrostatic atmospheric oscillations (e.g., atmospheric tides). S has the dimension of (velocity)^2.

The horizontal equations of motion, (2.5) - (2.6) or (2.13) - (2.14) or (2.24) - (2.25) can be written in a common vectorial form, after using

(4.3) and (4.4) :

$$\left(\frac{\partial \vec{V}}{\partial t}\right)_p = -\nabla_p\left(\phi + \frac{1}{2}\vec{V}^2\right) - (f + \zeta_p)\hat{k}\times\vec{V} - \varpi\frac{\partial \vec{V}}{\partial p} + \vec{F} \qquad (4.11)$$

where

$$\begin{aligned}\zeta_p &= \frac{1}{h_x h_y}\left[\left(\frac{\partial h_y v}{\partial x}\right)_p - \left(\frac{\partial h_x u}{\partial y}\right)_p\right] \\ &= -\nabla_p\cdot(\hat{k}\times\vec{V})\end{aligned} \qquad (4.12)$$

is almost equal to the vertical component of the relative vorticity, $\hat{k}\cdot \mathrm{curl}\,\vec{v}$

The energy equations for this system are derived by dot multiplication of (4.11) with $\vec{V}$,

$$\left(\frac{\partial}{\partial t}\frac{1}{2}\vec{V}^2\right)_p = -\nabla_p\cdot\left(\phi + \frac{1}{2}\vec{V}^2\right)\vec{V} - \frac{\partial}{\partial p}\left[\varpi\left(\phi + \frac{1}{2}\vec{V}^2\right)\right] + \vec{V}\cdot\vec{F} + \varpi\frac{\partial\phi}{\partial p} \qquad (4.13)$$

and (4.8) with $-pC_p/R$:

$$\left(\frac{\partial}{\partial t}c_pT\right)_p = -\nabla_p\cdot(\vec{V}c_pT) - \frac{\partial}{\partial p}(\varpi c_pT) + q - \varpi\frac{\partial\phi}{\partial p}\,. \qquad (4.14)$$

In numerical solutions of these equations it is customary as an upper boundary condition (normally at $p = 0$) to set

$$\varpi = \frac{dp}{dt} = 0 \quad \text{at } p = 0 \qquad (4.15)$$

and to consider $\vec{V}$, ϕ and T (or Θ) finite as $p \to 0$. For example the thermodynamic equation (4.8) is often written for computational purposes as

$$\left(\frac{\partial\Theta}{\partial t}\right)_p = -\nabla_p\cdot\Theta\vec{V} - \frac{\partial}{\partial p}(\varpi\Theta) + \frac{q}{c_p}\left(\frac{p_{00}}{p}\right)^{R/c_p} \qquad (4.16)$$

When a vertical finite-difference grid $p = 0$, Δp , $2\Delta p$, - - is introduced for p , the computation of $(\partial\Theta/\partial t)_p$ at the next to the top level ($p = \Delta p$) will contain, for the term $\partial(\varpi\Theta)/\partial p$, the finite-difference expression

$$\left(\frac{\partial\varpi\Theta}{\partial p}\right)_{p=\Delta p} \sim \frac{(\varpi\Theta)_{p=2\Delta p} - (\varpi\Theta)_{p=0}}{2\Delta p} \qquad (4.17)$$

The product $(\varpi \Theta)_{p=0}$ is assumed to vanish. The small increment Δp at the top level from $p = \Delta p$ to $p = 0$ in such a model corresponds to an infinite increment in z, over which the hydrodynamic variables in the real atmosphere may undergo many oscillations. An expression like (4.17) is therefore likely to be considerably in error as an approximation to the true derivative $\partial/\partial p$ at $p = \Delta p$.

Fortunately, it appears that the climatological distribution of $\vec{V}$ in the stratosphere (strong westerlies in winter, moderate easterlies in summer) is such as to prevent the major large-scale motions which are present either above or below this layer from propagating appreciably through the stratosphere (Charney and Drazin, 1961 ; Charney and Pedlosky 1963), which to some extent then acts as a "lid" on the atmosphere below. These conclusions do not hold for all types of motion, e.g., gravity waves (Hines, 1963. See also Dickinson, 1968). Numerical prediction of such phenomena as polar stratospheric warmings and the equatorial biennial oscillation undoubtedly require finer resolution above the tropopause than has been customary in large-scale numerical studies (cf. Miyakoda, et al., 1970).

If we assume that

$$\lim_{p \to 0} \varpi \left(\tfrac{1}{2}\vec{V}^2 + c_p T + \phi\right) = 0 \tag{4.18}$$

equations (4.13) - (4.14) can be integrated over the entire atmosphere (vertically from $p = 0$ to $p = p_0(x, y, t)$) to give the following energy integral :

$$\frac{\partial}{\partial t}\int dA \int_0^{p_0} \left(\tfrac{1}{2}\vec{V}^2 + c_p T + g z_0\right) dp = \int dA \int_0^{p_0} \left(q + \vec{V}\cdot\vec{F}\right) dp \tag{4.19}$$

To summarize matters, we suppose that at $t = 0$, $\vec{V}$ and ϕ are known everywhere as functions of x, y and p, and indicate how the evolution of these fields can be evaluated. Six steps are involved.

a. Integrate the continuity equation

$$\frac{\partial \varpi}{\partial p} = -\nabla_p \cdot \vec{V} \tag{4.20}$$

downwards from $p = 0$, using the upper boundary condition

$$\bar{\omega}(p = 0) = 0 \qquad (4.21)$$

to get $\bar{\omega}$ everywhere.

b. The vertical velocity at the ground w_0 is

$$w_0 = \vec{V}_0 \cdot \nabla z_0 \qquad (4.23)$$

whence the relation

$$\frac{\bar{\omega}}{\rho} = \left(\frac{\partial \phi}{\partial t}\right)_p + \vec{V} \cdot \nabla_p \phi - g w \qquad (4.24)$$

enables $\partial\phi/\partial t$ to be determined at the bottom of the atmosphere $(p = p_0)$.

c. The thermodynamic equation

$$\frac{\partial}{\partial p}\left(\frac{\partial \phi}{\partial t}\right)_p = -\vec{V} \cdot \nabla_p \left(\frac{\partial \phi}{\partial p}\right) + \frac{\bar{\omega}}{\rho} \frac{\partial \ln \Theta}{\partial p} - \frac{R}{p c_p} q \qquad (4.25)$$

determines $\partial^2\phi/\partial t \partial p$. Using the value of $(\partial\phi/\partial t)$ at p_0 from step b, (4.25) enables $\partial\phi/\partial t$ to be determined everywhere by upward integration $(dp < 0)$, if the heating rate q is known.

d. The equation of motion

$$\left(\frac{\partial \vec{V}}{\partial t}\right)_p = -\nabla_p\left(\phi + \tfrac{1}{2}\vec{V}^2\right) - (f + \zeta_p)\hat{k} \times \vec{V} - \bar{\omega}\frac{\partial \vec{V}}{\partial p} + \vec{F} \qquad (4.26)$$

enables $\partial\vec{V}/\partial t$ to be determined if the friction $\vec{F}$ is known.

e. The motion of the lower boundary $p_0(x, y, t)$ in x, y, p - space is determined either by

$$\left(\frac{\partial p_0}{\partial t}\right)_z = \rho_0 \left(\frac{\partial \phi}{\partial t}\right)_{p = p_0} \qquad (4.27)$$

from (4.4), or directly from

$$\frac{\partial p_0}{\partial t} = \bar{\omega}(p_0) - \vec{V}_0 \cdot \nabla p_0 \qquad (4.28)$$

f. Values of $\vec{V}$ and ϕ at a short time interval Δt later may be computed by an extrapolation. For example

$$\phi(x, y, p, \Delta t) = \phi(x, y, p, 0) + \Delta t \frac{\partial \phi}{\partial t}(x, y, p, 0) \tag{4.29}$$

We have in this way reconstituted, at $t = \Delta t$, the fields we began with at $t = 0$. Iteration of the 6 steps will produce a finite forecast. There are difficulties in this tedious but basically simple procedure.

1. The observations defining the initial state at $t = 0$ are not accurate enough, and also do not cover the entire globe.
2. The formulae for q and $\vec{F}$ are only crude representations of the effect of small-scale motions.
3. The numerical procedure must be designed carefully- the form of expressions such as (4.29) and the size of Δx , Δy , Δp , Δt must be chosen properly ("truncation error" and "computational stability").

One awkward feature about the numerical treatment of pressure coordinates concerns the variation with time of the lower boundary $p_0(x, y, t)$. (Consider, for example, the problem of assigning a value to $\vec{V}$ or Θ at $(x, y, p_0, t + \Delta t)$ if $p_0(t+\Delta t) > p_0(t)$). This can be partially overcome by introducing a new coordinate

$$\sigma = \frac{p}{p_0} \tag{4.30}$$

in place of p. σ is always zero at $p = 0$ and always 1 at $p = p_0$, so that $\dot{\sigma} = d\sigma/dt$ is always zero at the top and bottom of the atmosphere. The material derivative is

$$\frac{d}{dt} = \left(\frac{\partial}{\partial t}\right)_\sigma + \vec{V}\cdot\nabla_\sigma + \dot{\sigma}\frac{\partial}{\partial \sigma} \tag{4.31}$$

and, with ξ denoting x , y , or t , we have

$$\left(\frac{\partial}{\partial \xi}\right)_p = \left(\frac{\partial}{\partial \xi}\right)_\sigma - \frac{\sigma}{p_0}\frac{\partial p_0}{\partial \xi}\frac{\partial}{\partial \sigma} \tag{4.32}$$

$$\frac{\partial}{\partial p} = \frac{1}{p_0}\frac{\partial}{\partial \sigma} \tag{4.33}$$

$$\varpi = \frac{dp}{dt} = p_0 \frac{d\sigma}{dt} + \sigma\left(\frac{\partial p_0}{\partial t} + \vec{V}\cdot\nabla p_0\right) \tag{4.34}$$

The equation of continuity (4.20) becomes

$$\frac{\partial p_0}{\partial t} + \nabla_\sigma \cdot p_0\vec{V} + p_0 \frac{\partial \dot{\sigma}}{\partial \sigma} = 0 \tag{4.35}$$

and integration of this from $\sigma = 0$ to $\sigma = 1$ gives

$$\frac{\partial p_0}{\partial t} = -\int_0^1 (\nabla_\sigma \cdot p_0\vec{V})\, d\sigma \tag{4.36}$$

These allow $\dot{\sigma}$ to be computed by integration, analogous to (4.20) :

$$\frac{\partial \dot{\sigma}}{\partial \sigma} = \frac{1}{p_0}\left[\int_0^1 (\nabla_\sigma \cdot p_0\vec{V})\, d\sigma - \nabla_\sigma \cdot p_0\vec{V}\right] \tag{4.37}$$

The remaining important changes are in the hydrostatic equation

$$\frac{\partial \phi}{\partial \sigma} = -\frac{RT}{\sigma} \tag{4.38}$$

and in the "pressure gradient" $\nabla_p \phi$ in the horizontal equation of motion

$$\nabla_p \phi = \nabla_\sigma \phi - \frac{\sigma}{p_0}\frac{\partial \phi}{\partial \sigma}\nabla p_0 \tag{4.39}$$

The major deviation of σ- surfaces from constant pressure surfaces is caused by mountains ; the Himalayas, for example, cause a variation of 500 mb. Meteorologists at the Geophysical Fluid Dynamics Laboratory in Princeton, New Jersey, have had the most experience with this system. They have found two modifications useful. Firstly, the material derivative (4.31) in the equations for u, v, and $\textcircled{H}$ should be combined with the continuity equation (4.35), e.g.,

$$p_0 \frac{du}{dt} + u \frac{dp_0}{dt} \longrightarrow \frac{\partial p_0 u}{\partial t} + \nabla_\sigma \cdot (p_0\vec{V}u) + \frac{\partial p_0 \dot{\sigma} u}{\partial \sigma} \tag{4.40}$$

This device (which can also be used in z or p coordinates) results in what is called the "flux" (or divergence) form of the prediction equations, and experience of many groups has tended to demonstrate that it is numerically more stable.

Secondly, the right side of (4.39) appears to introduce large truncation errors near the edges of the major mountains where ∇p_0 is large. The GFDL group has found it useful to directly evaluate $\nabla_p \phi$ at a point 0 instead of using (4.39) :

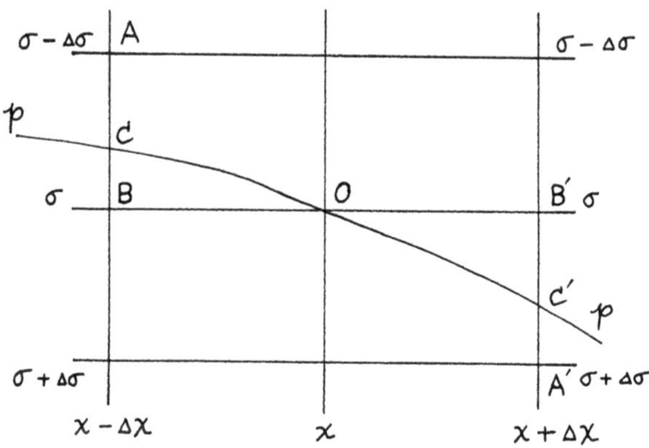

This can be done by first locating those points C and C' at the neighboring horizontal grid points which have by interpolation between A, B and A', B' the same value of p as point O , and then obtaining the corresponding values of ϕ at points C, C' by interpolation (Kurihara, 1968). An alternate method to reduce this truncation error might be to separate out from ϕ its major variation with p :

$$\phi'(x,y,p,t) \equiv \phi(x,y,p,t) - \bar{\phi}(p) \qquad (4.41)$$

where $\bar{\phi}(p)$ is a suitably defined simple function of p. (4.39) then becomes

$$\nabla_p \phi = \nabla_\sigma \phi' - \frac{\sigma}{p_0}\frac{\partial \phi'}{\partial \sigma}\nabla p_0 \qquad (4.42)$$

with a considerable reduction in the coefficient of ∇p_0 . A corresponding change would have to be made of course in the hydrostatic equation :

$$\begin{aligned} T &\equiv \bar{T} + T' \\ \bar{T} &= -\frac{p}{R}\frac{d\bar{\Phi}}{dp} = \bar{T}(\sigma p_0) \\ \frac{\partial \phi'}{\partial \sigma} &= -\frac{RT'}{\sigma} \end{aligned} \qquad (4.43)$$

This technique does not appear to have been tested yet.

5. DYNAMICAL BALANCE IN INITIAL DATA

The first numerical finite-difference computation of a time derivative $\partial/\partial t$ from real meteorological data was made more than 50 years ago by L.F. Richardson (1922). For one point in Central Europe he computed a surface pressure tendency ($\partial p_0/\partial t$) of +145 mb/6hr, whereas the observed change in the 6 hours preceding and 6 hours following the time of calculation was +0.2 mb and -1.0 mb. The source for this large error is to be found in the inaccuracy of the observations, and the way this enters can be examined most simply in a linear model of small amplitude motions on a resting stratified atmosphere. Friction and heating are ignored, and the ground is taken as level ($z_0 = 0$). The basic state is stratified $\overline{T} = \overline{T}(z) = \overline{T}(p)$, with

$$\overline{\phi} = \overline{\phi}(p) \quad , \quad p\frac{d\overline{\phi}}{dp} = -R\overline{T}$$
$$\overline{S}(p) = \frac{R^2\overline{T}}{g}\left(\frac{d\overline{T}}{dz} + \frac{g}{c_p}\right) \tag{5.1}$$

and $\overline{u} = \overline{v} = \overline{\varpi} = 0$. In Mercator coordinates (2.8) - (2.11), the linearized perturbation equations corresponding to (4.20) - (4. 28) are

$$\frac{\partial hu'}{\partial t} - fhv' = -\frac{\partial \phi'}{\partial x} \tag{5.2}$$

$$\frac{\partial hv'}{\partial t} + fhu' = -\frac{\partial \phi'}{\partial y} \tag{5.3}$$

$$h^2\frac{\partial \varpi'}{\partial p} + \frac{\partial hu'}{\partial x} + \frac{\partial hv'}{\partial y} = 0 \tag{5.4}$$

$$\frac{\partial}{\partial t}\left(\frac{\partial \phi'}{\partial p}\right) + \frac{\overline{S}}{p^2}\varpi' = 0 \tag{5.5}$$

At the ground $(p = p_0 , \overline{\rho} = \overline{\rho}_0 = p_0/R\overline{T}(p_0))$:

$$\varpi' = \overline{\rho}_0 \frac{\partial \phi'}{\partial t} \qquad at \ p = p_0 \tag{5.6}$$

For later orientation we first form an "energy" equation, multiplying (5.2) - (5.5) respectively by hu', hv', ϕ', and $(p^2\partial\phi'/\partial p)/\overline{S}$. We get, after some manipulation with derivatives,

$$\frac{1}{2} h^2 \frac{\partial}{\partial t}\left[u'^2 + v'^2 + \frac{1}{\bar{S}}\left(p \frac{\partial \phi'}{\partial p}\right)^2\right] = -h^2 \frac{\partial}{\partial p}(\varpi'\phi') - \frac{\partial}{\partial x}(\phi' h u') - \frac{\partial}{\partial y}(\phi' h v') \tag{5.7}$$

We integrate horizontally over the entire atmosphere and vertically from $p = 0$ to p_0. Introducing the upper boundary condition that $\varpi'\phi' = 0$ at $p = 0$*, and the lower boundary condition (5.6), we get

$$\frac{\partial}{\partial t}\iiint \frac{1}{2}\left[u'^2 + v'^2 + \frac{1}{\bar{S}}\left(p \frac{\partial \phi'}{\partial p}\right)^2 + \frac{1}{R\bar{T}_0}(\phi'^2)_{p=p_0}\right] h^2 \, dx\, dy\, dp$$

$$= -\iiint\left[\frac{\partial}{\partial x}(h\phi' u') + \frac{\partial}{\partial y}(h\phi' v')\right] dx\, dy\, dp \tag{5.8}$$

The right side vanishes (the integrand represents the divergence of the horizontal energy flux), and we see that the free motion described by (5.2) - (5.6) has a constant "energy". The energy density is

$$e = \frac{1}{2}\left[u'^2 + v'^2 + \frac{1}{\bar{S}}\left(p \frac{\partial \phi'}{\partial p}\right)^2\right] + \frac{1}{2R\bar{T}_0}(\phi'^2)_{p=p_0} \tag{5.9}$$

It is positive definite if $\bar{S} > 0$ and evidently consists of kinetic energy (only horizontal) plus two kinds of potential energy. One of the latter is proportional to $(p\, \partial\phi'/\partial p)^2$, i.e., the square of the temperature variance on an isobaric surface, and is recognizable as the simplified form of available potential energy defined by Lorenz (1955). The second form of potential energy we can think of as proportional to the variance of the surface pressure and exists even in cases where the perturbations have no temperature fluctuation $(\partial\phi'/\partial p) = 0$. Note that the stability of the system is ensured by positive $\bar{S}$; negative $\bar{S}$ would allow the two positive quantities $u'^2 + v'^2 + \frac{1}{R\bar{T}_0}(\phi'^2)_{p=p_0}$ and $(p\, \partial\phi'/\partial p)^2$ to increase simultaneously with time. For positive $\bar{S}$ we therefore expect only purely oscillatory oscillations, neither amplifying nor damping.

Free solutions of (5.2) - (5.6) are found by separation of variables. At this point we separate out only the vertical dependence.

---

* In forced problems, as in atmospheric tides, $\varpi'\phi'$ at $p = 0$ must be allowed to have a net negative value if it is not zero : the so-called "radiation condition".

$$
\begin{aligned}
h u' &= U(x,y,t)\,A(p) \\
h v' &= V(x,y,t)\,A(p) \\
\phi' &= H(x,y,t)\,A(p) \\
\bar{\omega}' &= W(x,y,t)\,B(p)
\end{aligned}
\tag{5.10}
$$

Substitution of these into (5.2) - (5.5) yields

$$\frac{\partial U}{\partial t} - fV = -\frac{\partial H}{\partial x} \tag{5.11}$$

$$\frac{\partial V}{\partial t} + fU = -\frac{\partial H}{\partial y} \tag{5.12}$$

$$h^2 W \frac{dB}{dp} + \left(\frac{\partial U}{\partial x} + \frac{\partial V}{\partial y}\right)A(p) = 0 \tag{5.13}$$

$$\frac{\partial H}{\partial t}\frac{dA}{dp} + W\frac{\bar{S}(p)}{p^2}B(p) = 0 \tag{5.14}$$

We can set

$$A(p) = \frac{dB}{dp} \tag{5.15}$$

so that (5.13) becomes

$$h^2 W + \frac{\partial U}{\partial x} + \frac{\partial V}{\partial y} = 0 \tag{5.16}$$

(5.14) then requires a <u>separation constant</u> c^2 :

$$\frac{\bar{S}}{p^2} B = -c^2 \frac{dA}{dp} \tag{5.17}$$

$$\frac{\partial H}{\partial t} = c^2 W \tag{5.18}$$

The problem now factors into one in x, y , t and one in p . The first is a set of three partial differential equations in x , y , t :

$$
\begin{aligned}
\frac{\partial U}{\partial t} - fV &= -\frac{\partial H}{\partial x} \\
\frac{\partial V}{\partial t} + fU &= -\frac{\partial H}{\partial y} \\
c^2\left(\frac{\partial U}{\partial x} + \frac{\partial V}{\partial y}\right) &= -h^2 \frac{\partial H}{\partial t}
\end{aligned}
\tag{5.19}
$$

u, V, H vary with x, y, t; f and h are known functions of y. The problem in p is an ordinary differential equation, most conveniently phrased in terms of $B(p)$, and obtained by combining (5.15) and (5.17) :

$$\frac{d^2B}{dp^2} + \frac{\bar{S}}{p^2c^2} B = 0 \tag{5.20}$$

The lower boundary condition (at $p = p_0$) for this is obtained from (5.6)

$$W B(p_0) = \rho_0 \frac{\partial H}{\partial t} A(p_0)$$

or, using (5.15) and (5.18),

$$B(p_0) = \bar{\rho}_0 c^2 \left(\frac{dB}{dp}\right)_{p_0} \tag{5.21}$$

The separation constant c^2 connects (5.19) with (5.20) - (5.21). We will find that it can take on many values. Note that its dimensions are those of $(\text{velocity})^2$. In the theory of atmospheric tides, c^2 is frequently set equal to g times a height $\tilde{h}$:

$$g\tilde{h} \equiv c^2 \tag{5.22}$$

$\tilde{h}$ is the so-called <u>equivalent depth</u>. This notation was introduced by <u>G. I. Taylor</u> (1936) who emphasized in this way that equations (5.19) are similar to those of a homogeneous ocean of depth $\tilde{h}$. We will not refer to equivalent depths.

To get solutions of (5.20) we must pick a suitable basic state, i.e., we must define $\bar{S}$ in (5.20) and $\bar{\rho}_0$. Before we do so, a change of variables turns out to be useful. The independent variable p is replaced by Z :

$$Z = -\ln\left(\frac{p}{p_0}\right); \quad p = p_0 e^{-Z}$$
$$\frac{dp}{dZ} = -p = -p_0 e^{-Z} \tag{5.23}$$

In a strictly isothermal atmosphere,

$$p = p_0 e^{-gz/RT}$$
$$\rho = \rho_0 e^{-gz/RT} \tag{5.24}$$

The height RT/g is frequently called the (density) scale height, even wher T is only approximately constant. RT/g is about 8 km for normal T values. Since T varies slowly with z, Z in (5.23) can therefore be thought of as being approximately a linear function of geometric height z, changing by +1 for each 8 km or so increase in z.

It is also convenient to replace B by D:

$$D = Be^{Z/2} = B\left(\frac{p}{p_0}\right)^{-1/2} \tag{5.25}$$

After some algebra, (5.20) transforms to

$$\frac{d^2D}{dZ^2} + \left(\frac{\bar{S}}{c^2} - \frac{1}{4}\right)D = 0 \tag{5.26}$$

and referring back to (4.9) we see that this can be written as

$$\frac{d^2D}{dZ^2} + \left[\frac{R}{c^2}\left(\frac{d\bar{T}}{dZ} + \kappa\bar{T}\right) - \frac{1}{4}\right]D = 0 \;;\; \kappa = \frac{R}{c_p} \tag{5.27}$$

The lower boundary condition (5.21) becomes

$$Z = 0: \quad \frac{dD}{dZ} = \left(\frac{1}{2} - \frac{R\bar{T}_0}{c^2}\right)D \tag{5.28}$$

Multiplication of (5.26) and (5.28) with D^* and integration with Z leads to

$$c^2 = \frac{R\bar{T}_0|D(0)|^2 + \int_0^Z \bar{S}|D|^2 dZ}{-D^*\frac{dD}{dZ} + \frac{1}{2}|D(0)|^2 + \int_0^Z \left(\left|\frac{dD}{dZ}\right|^2 + \frac{1}{4}|D|^2\right)dZ} \tag{5.29}$$

The internal energy density in (5.9) can be rewritten as

$$\begin{aligned}&\left[u'^2 + v'^2 + \frac{1}{\bar{S}}\left(p\frac{\partial\phi'}{\partial p}\right)^2\right]dp \\ &= \left[(U^2+V^2)\left(\frac{dD}{dZ} - \frac{1}{2}D\right)^2 + \frac{H^2}{c^4}\bar{S}D^2\right]\frac{dZ}{p_0^2}\end{aligned} \tag{5.30}$$

For simplicity we consider an isothermal atmosphere, so that

$$D \propto e^{\pm\mu Z} \;;\; \mu = \left(\frac{1}{4} - \frac{\kappa R T}{c^2}\right)^{1/2}, \; \text{Real}\,\mu \geqslant 0 \tag{5.31}$$

As an upper boundary condition we require finiteness of the curly bracket in (5.30). If c^2 were complex or real negative only the $\exp(-\mu Z)$ would be allowed in (5.31) ; whereupon $D^* dD/dz$ would vanish in (5.29) as $Z \to \infty$, proving instead that c^2 could only be real and positive. There are two cases.

Case I. c^2 greater than $4\kappa R\bar{T}$

$$D = (constant) \times \exp\left[-\sqrt{\frac{1}{4} - \frac{\kappa R\bar{T}}{c^2}}\, Z\right] \tag{5.32}$$

The lower boundary condition (5.28) requires that

$$c^2 = \frac{R\bar{T}}{1-\kappa} = \gamma R\bar{T} = c_1^2 \tag{5.33}$$

$$R = c_p - c_v \,, \quad \gamma = \frac{c_p}{c_v}$$

In this case the only allowable value for c is the speed of sound in this isothermal atmosphere. This is called a Lamb wave (Hydrodynamics, section 311a). D is proportional to $\exp\left[-\left(\frac{2-\gamma}{2\gamma}\right)Z\right]$ and we find

$$\begin{aligned} B &= p_0 e^{-Z/\gamma} = p_0 \left(\frac{p}{p_0}\right)^{1/\gamma} \\ A &= \frac{1}{\gamma} e^{\kappa Z} = \frac{1}{\gamma}\left(\frac{p}{p_0}\right)^{-\kappa} \end{aligned} \tag{5.34}$$

Case II. c^2 less than $4\kappa R\bar{T}$

This gives $D \propto \exp(\pm i m Z)$ where

$$m = \sqrt{\frac{\kappa R\bar{T}}{c^2} - \frac{1}{4}} \tag{5.35}$$

D is bounded for both roots. We set $D = K_1 \cos mZ + K_2 \sin mZ$ and the boundary condition (5.28) only determines the ratio of the constants K_1 and K_2. The result is

$$B = p_0 e^{-Z/2}\left[4\kappa m \cos mZ + (2\kappa - 4m^2 - 1)\sin mZ\right]$$
$$A = \left(\frac{1+4m^2}{2}\right) e^{Z/2}\left[2m\cos mZ + (2\kappa - 1)\sin mZ\right] \tag{5.36}$$
$$C_m^2 = \frac{\kappa RT}{m^2 + \frac{1}{4}} = \frac{10}{49(m^2+\frac{1}{4})}\, C_s^2$$

m is any real number from zero to ∞. This is a continuous spectrum, and C^2 varies from 0 when $m = \infty$ (short vertical wavelength) to a maximum of $40C_s^2/49$ for $m \to 0$ (large vertical wave length).

It is important to realize that the separation parameter C, although it has the dimension of a velocity, has not yet been shown here to have any relation to the horizontal phase velocity. This requires information about the solutions of (5.19). We will see that in one type of solution for (5.19) that this is the case - the horizontal phase velocity is very close to C. In another type of solution (Rossby waves), however, there will be only a very vague connection.

For orientation we may take in this isothermal atmosphere

$$C_s = 320 \ m\,sec^{-1}$$
$$0 \leq C_m \leq 288 \ m\,sec^{-1} \tag{5.37}$$

An isothermal atmosphere is a considerably simplified version of the real atmosphere. However, computations with realistic distributions of $\overline{T}$ (which model the temperature minima at the tropopause and at 80 km, and with $T \sim 270\,°K$ at 50 km and an increasing $\overline{T}$ above 80 km) also show only a single discrete value of $C \sim 320$ m sec^{-2}. Additional discrete values will appear if pronounced stable layers are introduced in the troposphere or if $\overline{T}$ at 50 km is taken to be unrealistically warm. If the hydrostatic approximation is not made, the vertical equation corresponding to (5.26) is more complicated, and many more discrete modes appear.

Having formed some idea about allowable values of the separation constant C^2, we return to the set (5.19) :

$$\frac{\partial U}{\partial t} - fV = -\frac{\partial H}{\partial x} \tag{5.19a}$$

$$\frac{\partial V}{\partial t} + fU = -\frac{\partial H}{\partial y} \tag{5.19b}$$

$$c^2\left(\frac{\partial U}{\partial x} + \frac{\partial V}{\partial y}\right) = -h^2\frac{\partial H}{\partial t} \tag{5.19c}$$

f and h are functions of y (i.e. latitude)

$$f = 2\Omega \sin\theta = 2\Omega \tanh\left(\frac{y}{a}\right)$$

$$h = \cos\theta = \operatorname{sech}\left(\frac{y}{a}\right) \tag{5.38}$$

$$\frac{d\theta}{dy} = \frac{h}{a} = \frac{\cos\theta}{a}$$

One approach, the most thorough, would be to find separable solutions for (5.19), of the form

$$L(y \text{ or } \theta) \times M(x \text{ or } \lambda) \times N(t) \tag{5.39}$$

$N(t)$ will have the form $\exp(-i\omega t)$, ω being a frequency (<u>it is not to be confused with</u> $\varpi = dp/dt$), and $M(\lambda)$ will be of the form $\exp(\pm i s\lambda)$ s = 0, 1, 2, ... A second-order differential equation for $L(\theta)$ will result. Boundary conditions at the two poles ($y = \pm\infty$) will allow only certain solutions for $L(\theta)$ to be valid. These will be discrete, and can be numbered by an index n , say

$$F(\omega, s, n, c_m^2) = 0 \tag{5.40}$$

For specific choices of the waves numbers s, n and m , a definite frequency $\omega(s, n, m)$ is determined.

This problem has been known for a long time, beginning with Laplace, Margules, and Hough, in the 19th century. (Laplace worked only with the forced tidal problem). Although the variable coefficients f and h^2

are well defined functions of y (i.e. θ), the solution of the differential equation for L is not trivial, and a thorough examination leading to (5.40) has only recently been finished (Longuet-Higgins, 1968).

Our approach instead will be an approximate one, based on the so-called "WKBJ" method. In this approach we assume that all three dependent variables $U(x,y,t)$, $V(x,y,t)$ and $H(x,y,t)$ in (5.19) can be written in the form

$$\begin{aligned} U &= Re\,\tilde{U}\,e^{i\psi} \\ V &= Re\,\tilde{V}\,e^{i\psi} \\ H &= Re\,\tilde{H}\,e^{i\psi} \end{aligned} \tag{5.41}$$

where

$\tilde{U}$, $\tilde{V}$, $\tilde{H}$ are slowly varying (complex) functions of x, y, t;

(5.41)'

the common phase ψ is real, and varies more rapidly with x, y, t than do the amplitudes $\tilde{U}$, $\tilde{V}$, $\tilde{H}$ or the coefficients f and h^2.

The picture of U at a fixed y, t might be a wiggly curve as follows :

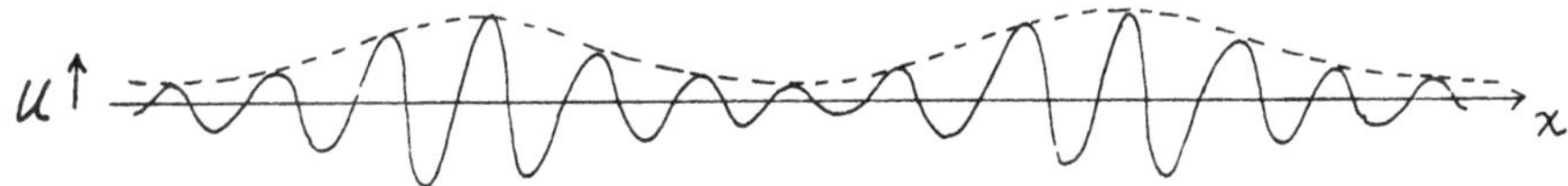

$|\tilde{U}|$ represents the envelope, or very nearly so. (This approach is equivalent to the arguments used in geometric optics, with its concepts of rays and group velocity. It is not exhaustive ; solutions of (5.19) exist which are not of the form (5.41)', just as geometric optics does not exhaust all the possibilities in "physical optics").

When (5.40) is introduced into (5.19), we obtain terms like

$$\frac{\partial U}{\partial t} = \left(\frac{\partial \tilde{U}}{\partial t} + i\,\tilde{U}\,\frac{\partial \psi}{\partial t}\right) e^{i\psi}$$

We interpret (5.41) as meaning

$$\frac{\partial \tilde{U}}{\partial t} \ll i\,\tilde{U}\,\frac{\partial \psi}{\partial t}$$

to the extent that <u>for a zero-order approximation</u>

$$\frac{\partial U}{\partial t} = i\,U\,\frac{\partial \psi}{\partial t} \tag{5.42}$$

Similar approximations hold, by assumption, for the other derivatives $\partial/\partial x$ and $\partial/\partial y$, and for V and H.
We now omit the special $(\tilde{\ })$ symbol on the "amplitudes" $\tilde{U}$, $\tilde{V}$, $\tilde{H}$ for convenience. It is also convenient to introduce the notation :

$$\begin{aligned} \omega &\equiv -\frac{\partial \psi}{\partial t} \\ \alpha &\equiv \frac{\partial \psi}{\partial x} \\ \beta &\equiv \frac{\partial \psi}{\partial y} \end{aligned} \tag{5.43}$$

ω, α and β take the role of a locally valid frequency and x- and y- wave numbers. They need not be constant ; but if they vary in x, y, t, they must obey the obvious relations

$$\frac{\partial \alpha}{\partial t} = -\frac{\partial \omega}{\partial x}, \quad \frac{\partial \beta}{\partial t} = -\frac{\partial \omega}{\partial y}, \quad \frac{\partial \alpha}{\partial y} = \frac{\partial \beta}{\partial x} \tag{5.44}$$

The following amplitude relations are obtained by inserting (5.41) into (5.19) and invoking the approximation (5.42)

$$\begin{aligned} \omega_0 U_0 - if V_0 &= \alpha_0 H_0 \\ if U_0 + \omega_0 V_0 &= \beta_0 H_0 \\ c^2 (\alpha_0 U_0 + \beta_0 V_0) &= h^2 \omega_0 H_0 \end{aligned} \tag{5.45}$$

The subscript zero is put in to emphasize the zero-order approximation. It will be dropped later. f and h still vary with y. U_0 and V_0 can be solved for from the first two equations,

$$\begin{aligned} (\omega_0^2 - f^2) U_0 &= (\omega_0 \alpha_0 + if\beta_0) H_0 \\ (\omega_0^2 - f^2) V_0 &= (\omega_0 \beta_0 - if\alpha_0) H_0 \end{aligned} \tag{5.46}$$

and these can be used to eliminate U_0 and V_0 in the third equation. The result is

$$\omega_0 \left[\omega_0^2 - f^2 - c^2 (\alpha_0^2 + \beta_0^2) h^{-2} \right] = 0 \tag{5.47}$$

This has three solutions for ω_0 . Two are given by

$$\omega_0^2 = f^2 + c^2 \frac{\alpha_0^2 + \beta_0^2}{h^2} \tag{5.48}$$

while the third one is simply

$$\omega_0 = 0 \tag{5.49}$$

Let a horizontal wave-number vector be defined by

$$\bar{k}_2 = \nabla \psi = \frac{1}{h} \left(\frac{\partial \psi}{\partial x}, \frac{\partial \psi}{\partial y} \right) = \frac{1}{h} (\alpha, \beta) \tag{5.50}$$

This may vary with position and time. The phase velocity of a plane wave is defined to be in the direction of $\nabla \psi$, with a magnitude

$$C_{ph} = \frac{\omega}{|\vec{k}_2|} = \frac{h\omega}{\sqrt{\alpha^2+\beta^2}} = \frac{-\frac{\partial\psi}{\partial t}}{|\nabla\psi|} \qquad (5.51)$$

An observer moving with this speed in the direction of $\vec{k}_2$ observes no change in the phase ψ :

$$\frac{\delta\psi}{\delta t} = \frac{\partial\psi}{\partial t} + C_{ph}\frac{\vec{k}_2}{|\vec{k}_2|}\cdot\nabla\psi = \frac{\partial\psi}{\partial t} - \frac{\partial\psi}{\partial t}\frac{|\vec{k}_2|^2}{|\vec{k}_2|^2} = 0$$

We now see that the separation parameter C^2 would correspond to the horizontal phase speed in (5.48) if f were zero. The solutions given by (5.46) and (5.48) are called inertia-gravity waves, hydrostatic in this case. Their frequency is always greater than or equal to the local Coriolis parameter.

Consider now the strange solution (5.49). It violates our original assumption (5.41)', since $\partial\psi_0/\partial t = 0$. Is $\partial/\partial t = 0$ an exact solution of (5.19) ? It is only if

$$V = \frac{1}{f}\frac{\partial H}{\partial x}$$

$$U = -\frac{1}{f}\frac{\partial H}{\partial y}$$

$$\frac{\partial U}{\partial x} + \frac{\partial V}{\partial y} = -\frac{1}{f^2}\frac{df}{dy}\frac{\partial H}{\partial x} = 0 \qquad (5.52)$$

(5.19) therefore has a steady solution only if

$$V = 0$$

$$H = H(y) \qquad (5.53)$$

$$fU = -\frac{dH}{dy}$$

i.e., purely zonal flow, with U in geostrophic balance with H (Lamb, 1932, p. 333). α is zero in this strict steady state. Suppose α were small. ω cannot be zero, but it is hard to see how $\omega = 0$ could suddenly be transformed into the $\omega^2 > f^2$ solution (5.48) by introducing a small α. We must evidently look closer at $\omega_0 = 0$.

Using U as an example, we will continue to treat horizontal derivatives as

$$\frac{\partial}{\partial x}(Ue^{i\psi}) = \left(\frac{\partial U}{\partial x} + i\alpha U\right)e^{i\psi}$$

$$= \left[(i\alpha_0 U_0) + \left(\frac{\partial U_0}{\partial x} + i\alpha_1 U_0 + i\alpha_0 U_1\right) + (\cdots) + \cdots\right]e^{i\psi} \tag{5.54}$$

where the first parentheses is the zero-order approximation, the second represents the first-order corrections, etc. For $\partial/\partial t$, however, we shall assume, as suggested by (5.49), that ω_0 is zero. To be specific, we write

$$\frac{\partial}{\partial t}(Ue^{i\psi}) = \left[(0) + (-i\omega_1 U_0) + \left(\frac{\partial U_0}{\partial t} - i\omega_1 U_1 - i\omega_2 U_0\right)\right]e^{i\psi} \tag{5.55}$$

where the parenthetical groupings denote again the zero-order, first-order, etc., terms. Note that although we have assumed ω is small, we still impose the fundamental approximation (5.42), that U, V, H vary more slowly with time than does the phase ψ.

The zero-order relations are

$$V_0 = i\frac{\alpha_0 H_0}{f} \tag{5.56a}$$

$$U_0 = -i\frac{\beta_0 H_0}{f} \tag{5.56b}$$

$$C^2(i\alpha_0 U_0 + i\beta_0 V_0) = 0 \tag{5.56c}$$

Observe, first, that these are mutually consistent, and second, that the first two are the geostrophic relations between U, V and the horizontal geopotential gradient.

The first-order relations are

$$i\omega_1 U_0 + fV_1 = \frac{\partial H_0}{\partial x} + i(\alpha_1 H_0 + \alpha_0 H_1) \tag{5.57a}$$

$$-i\omega_1 V_0 + fU_1 = -\frac{\partial H_0}{\partial y} - i(\beta_1 H_0 + \beta_0 H_1) \tag{5.57b}$$

$$\frac{\partial U_0}{\partial x} + \frac{\partial V_0}{\partial y} + i(\alpha_0 U_1 + \alpha_1 U_0 + \beta_0 V_1 + \beta_1 V_0) = \frac{h^2}{c^2} i\omega_1 H_0 \tag{5.57c}$$

The first two can be used to eliminate V_1 and U_1 in the third equation, and $\partial U_0/\partial x$ and $\partial V_0/\partial y$ can be expressed in terms of H_0 by differentiating (5.56a) and (5.56b). (In doing this, α_0 and β_0 are also differentiated, as is f.) The identities (5.44) can now be invoked, and a tremendous cancellation ensues (in particular, U_1, V_1, H_1, α_1, and β_1 all disappear).

The result is

$$\omega_1 = \frac{-\frac{2\Omega\cos\theta}{a}\frac{\alpha_0}{h}}{\frac{\alpha_0^2+\beta_0^2}{h^2} + \frac{f^2}{c^2}} \tag{5.58}$$

The numerator comes from the differentiation of f in (5.56a) :

$$\frac{df}{dy} = 2\Omega\frac{d\theta}{dy}\frac{d(\sin\theta)}{d\theta} = 2\Omega\frac{\cos^2\theta}{a} = \frac{2\Omega\cos\theta}{a}h \tag{5.59}$$

This variation of f with latitude appeared also in (5.52), where it led to the requirement that $\partial H/\partial x$ and V must both vanish for a steady state. (5.58) is consistent with this, since ω_1, V_0 and $(\partial H/\partial x)_0$ all vanish if $\alpha_0 = 0$

(5.58) is the frequency formula for Rossby waves in a resting atmosphere (no basic current).

[As an aside, we may note that this WKBJ approach could also have been performed directly with the original unseparated equations (5.2.) - (5.5), if the vertical boundary conditions were ignored and if the vertical wave length were suitably short. The result, in terms of geometrical vertical wave number γ,

$$\gamma \equiv \frac{\partial \psi}{\partial z} = \frac{\partial \psi}{\partial Z} \frac{\partial Z}{\partial z} = m \frac{g}{RT} \tag{5.60}$$

can be written down by using this definition together with

$$m^2 = \frac{S}{c^2} - \frac{1}{4} = \frac{N^2 R^2 T^2}{c^2 g^2} - \frac{1}{4} \tag{5.60'}$$

from (5.26) to replace c^2 in (5.58)

$$\omega_1 = \frac{-\frac{2\Omega\cos\theta}{a} \cdot \frac{\alpha}{h}}{\frac{\alpha^2+\beta^2}{h^2} + \frac{f^2}{N^2}\left(\gamma^2 + \frac{g^2}{4R^2T^2}\right)} \tag{5.61}$$

The presence of a uniform basic zonal current $\bar{u}$ would add a "Doppler" term $(\alpha/h)\,\bar{u}$ to the right side of this }

We can now formulate an artificial forecast problem. The following analysis is in the spirit of that by Hinkelmann (1951), except that we account for the variation of f. For simplicity we consider only one value of c^2. Secondly, we suppose that observations at $t = 0$ have determined a simple single wave-like field, in which

$$\begin{aligned} U(t=0) &= Re\, U^0 e^{i\psi^0} \\ V(t=0) &= Re\, V^0 e^{i\psi^0} \\ H(t=0) &= Re\, H^0 e^{i\psi^0} \end{aligned} \tag{5.62}$$

To make matters even more simple, we take U^0, V^0, H^0 as constants, and ψ^0, the phase at $t = 0$, as a linear function of x and y

$$\psi^0 = \alpha x + \beta y \qquad \begin{array}{l} \alpha = \text{constant} \\ \beta = \text{constant} \end{array} \tag{5.63}$$

The ensuing motion will consist of the three types of waves, gravity and Rossby. (We can now drop the zero and one subscript which reminded us of the WKBJ approximations). Each wave will be of the form ($j = I, II, III$)

$$\begin{aligned} H_j &= Re\, \bar{H}_j\, e^{i\psi_j} \\ U_j &= Re\, \bar{U}_j\, e^{i\psi_j} \\ V_j &= Re\, \bar{V}_j\, e^{i\psi_j} \\ \psi_j &= \alpha x + \beta y - \omega_j t \end{aligned} \tag{5.64}$$

For times not too large, $\bar{H}_j$, $\bar{U}_j$, $\bar{V}_j$, α, β, ω_j will be constants. [Dispersion effects, appearing in this example from the y-dependence of h^2 and f in (5.48) and (5.58), will slowly change these values. Dispersion enters also if α and β are not constant at $t = 0$ (Rossby, 1945 ; Eckart, 1960 ; Jacobs, 1967]. Assign j-values as follows.

j = I = Rossby wave :

$$\begin{aligned} \omega_I &= \frac{-\dfrac{2\Omega\alpha}{a}}{\dfrac{\alpha^2+\beta^2}{h^2} + \dfrac{f^2}{c^2}} \\ \bar{U}_I &= -\left(\frac{i\beta}{f}\right)\bar{H}_I \\ \bar{V}_I &= \left(\frac{i\alpha}{f}\right)\bar{H}_I \end{aligned} \tag{5.65}$$

j = II = east-ward moving gravity wave :

$$\omega_{II} = \sqrt{f^2 + \frac{c^2}{h^2}(\alpha^2+\beta^2)}$$

$$\bar{U}_{II} = \frac{\omega_{II}\alpha - if\beta}{\omega_{II}^2 - f^2}\bar{H}_{II} \qquad (5.66)$$

$$\bar{V}_{II} = \frac{\omega_{II}\beta - if\alpha}{\omega_{II}^2 - f^2}\bar{H}_{II}$$

j = III = west-ward moving gravity wave :

$$\omega_{III} = -\omega_{II}$$

$$\bar{U}_{III} = \frac{-\omega_{II}\alpha + if\beta}{\omega_{II}^2 - f^2}\bar{H}_{III} \qquad (5.67)$$

$$\bar{V}_{III} = \frac{-\omega_{II}\beta - if\alpha}{\omega_{II}^2 - f^2}\bar{H}_{III}$$

The ratios $\bar{U}_j/\bar{H}_j$ and $\bar{V}_j/\bar{H}_j$ come from (5.46) and (5.56).

At $t = 0$, the waves add up to the observed values :

$$\bar{U}_I + \bar{U}_{II} + \bar{U}_{III} = U^0$$

$$\bar{V}_I + \bar{V}_{II} + \bar{V}_{III} = V^0 \qquad (5.68)$$

$$\bar{H}_I + \bar{H}_{II} + \bar{H}_{III} = H^0$$

We replace each $\bar{U}_j$ and $\bar{V}_j$ here by its corresponding $\bar{H}_j$ and multiplicative factor from (5.65) - (5.67). Equation (5.68) then reduces to three simultaneous non-homogeneous equations for $\bar{H}_I$, $\bar{H}_{II}$ and $\bar{H}_{III}$. These can be solved for in terms of the observed variables U^0, V^0 H^0. The result, after some tedious algebra, is

$$\bar{H}_I = \frac{c^2}{\omega_{II}^2 k^2}\left[if\beta U^0 - if\alpha V^0 + \frac{f^2 k^2}{c^2} H^0\right] \qquad (5.69a)$$

$$\bar{H}_{II} = \frac{c^2}{2\omega_{II}^2 k^2}\left[(\alpha\omega_{II} - if\beta)U^0 + (\beta\omega_{II} + if\alpha)V^0 + (\alpha^2+\beta^2)H^0\right] \qquad (5.69b)$$

$$\bar{H}_{III} = \frac{c^2}{2\omega_{II}^2 k^2}\left[-(\alpha\omega_{II} + if\beta)U^0 + (-\beta\omega_{II} + if\alpha)V^0 + (\alpha^2+\beta^2)H^0\right] \tag{5.69c}$$

These formulas determine the $\bar{H}$ amplitude of each wave in terms of the observed quantities U^0, V^0, H^0. Each wave, with its U_j, V_j, and H_j fields then moves with frequency ω_j. The total H field is

$$H = Re\,(\bar{H}_I e^{i\psi_I} + \bar{H}_{II} e^{i\psi_{II}} + \bar{H}_{III} e^{i\psi_{III}})$$

A numerical computation, if it introduced no numerical error, would reproduce this development with time.

<u>In the extratropical atmosphere, the major part of the energy is in motion similar to the Rossby wave</u>, I. (It is true that our present solution (5.58) for ω_I gives a <u>west-ward</u> propagation for the Rossby waves, contrary to experience in middle latitudes ; but if we had inserted a basic uniform west-to-east zonal current, as is typical of middle latitudes, all waves would have received an added translation to the east). As an indication of the truth of the underlined statement, we can point out that U and V in the Rossby wave are <u>nearly</u> geostrophic.

$$V \approx \frac{1}{f}\frac{\partial H}{\partial x} \quad , \; U \approx -\frac{1}{f}\frac{\partial H}{\partial y} \tag{5.70}$$

just as winds on upper-level charts are <u>nearly</u> geostrophic. In the gravity waves, U and V are not geostrophic.

Another point is that the gravity waves II and III have high frequencies and large phase velocities :

$$|C_{ph}| \geqslant C \qquad \text{(gravity wave)} \tag{5.71}$$

To fix ideas, let us take wave lengths of 4000 km in middle latitudes :

$$f = 10^{-4}\,sec^{-1}, \; k = \cos\theta = 0.7$$

$$\frac{\alpha}{k} \sim \frac{\beta}{k} = \frac{2\pi}{4000\,km} = 1.6 \times 10^{-6}\,m^{-1}$$

$$c^2 \cong (300 \ m \ sec^{-1})^2 = 9 \times 10^4 \ m^2 sec^{-2}$$

These figures give :

$$\omega_I = - \frac{\frac{2\Omega\alpha}{a}}{\frac{\alpha^2+\beta^2}{h^2} + \frac{f^2}{c^2}} = - \frac{0.25 \times 10^{-16}}{5.1 \times 10^{-12} + 1.1 \times 10^{-13}}$$
$$= -0.5 \times 10^{-5} \ rad. \ sec^{-1} = 0.07 \ cycles \ day^{-1} \qquad (5.72)$$

and

$$\omega_{II} = - \omega_{III} = \sqrt{f^2 + \frac{c^2(\alpha^2+\beta^2)}{h^2}}$$
$$= \sqrt{10^{-8} + 46.1 \times 10^{-8}} \qquad (5.73)$$
$$= 9.3 \ cycles \ day^{-1}$$

Note how much larger ω_{II} is than ω_{III} under these typical circumstances.

If the "observations" U^0, V^0, H^0 had negligible error, they would presumably only represent a Rossby wave, and satisfy the geostrophic relations (5.56) (check that this gives $\bar{H}_{II}$ and $\bar{H}_{III}$ = 0 and $\bar{H}_I = H^0$ in (5.69).)

$$U^0 = - \frac{i\beta}{f} H^0 , \qquad V^0 = \frac{i\alpha}{f} H^0 \qquad (5.74)$$

What if U^0, V^0, H^0 have an error, though ? Consider the value of the"geopotential tendency", $\partial H/\partial t$, at $t = 0$:

$$\left(\frac{\partial H}{\partial t}\right)_{t=0} = Re\left[\frac{\partial}{\partial t}\left(\bar{H}_I e^{i\psi_I} + \bar{H}_{II} e^{i\psi_{II}} + \bar{H}_{III} e^{i\psi_{III}}\right)\right]_{t=0}$$
$$= - Re\left[i\left(\omega_I \bar{H}_I + \omega_{II} \bar{H}_{II} + \omega_{III} \bar{H}_{III}\right) e^{i(\alpha x + \beta y)}\right] \qquad (5.75)$$

Substituting from (5.69), with $\omega_{III} = -\omega_{II}$

$$\left(\frac{\partial H}{\partial t}\right)_{t=0} = -\frac{c^2}{h^2\omega_{II}^2} Re\left\{\left[(i\alpha\omega_{II}^2 - f\beta\omega_I)U^0 + (i\beta\omega_I^2 + f\alpha\omega_I)V^0 + \left(\frac{i f^2 h^2}{c^2}\omega_I\right)H^0\right]e^{i(\alpha x+\beta y)}\right\} \tag{5.76}$$

Suppose H^0 is observed correctly, but U^0 and V^0 differ from the "correct" values (5.74) by an error δU^0 and δV^0 :

$$U^0 = -\frac{i\beta}{f}H^0 + \delta U^0, \quad V^0 = \frac{i\alpha}{f}H^0 + \delta V^0 \tag{5.77}$$

We find by substitution in (5.76)

$$\left(\frac{\partial H}{\partial t}\right)_{t=0} = Re\left[-i\omega_I H^0 e^{i(\alpha x+\beta y)}\right] - \frac{c^2}{h^2\omega_{II}^2} Re\left\{\left[(i\alpha\omega_{II}^2 - f\beta\omega_I)\delta U^0 + (i\beta\omega_I^2 + f\alpha\omega_I)\delta V^0\right]e^{i(\alpha x+\beta y)}\right\} \tag{5.78}$$

The first term is the correct solution, the second is the error. We evaluate the latter with the same numbers used to get (5.72) and (5.73), and, for an example, we take the observational error as

$$|\delta U^0| = |\delta V^0| = 1 \text{ m sec}^{-1}$$

The ω_{II}^2 terms dominate in the error, so the magnitude of the error is

$$\delta\left(\frac{\partial H}{\partial t}\right)_{t=0} = \frac{c^2}{h^2}\sqrt{2}\,\alpha \times (1 \text{ m sec}^{-1}) \approx 0.3 \text{ m}^2\text{sec}^{-3}$$

In terms of an error in surface pressure tendency this corresponds to

$$\delta\left(\frac{\partial p_s}{\partial t}\right) \sim \rho_0\,\delta\left(\frac{\partial H}{\partial t}\right) \sim 0.3 \times 10^{-3} \text{ cb sec}^{-1}$$

$$\approx 11 \text{ mb per hour} \tag{5.79}$$

This error is much larger than typical rates of surface pressure change. What has happened is that the <u>slow</u> changes in H predicted by the "correct" geostrophic Rossby-wave solution have superimposed on them high frequency gravity wave fluctuations, the latter appearing in the "fore-cast" because of the seemingly innocuous errors of 1 m sec^{-1} in U^0 and V^0.

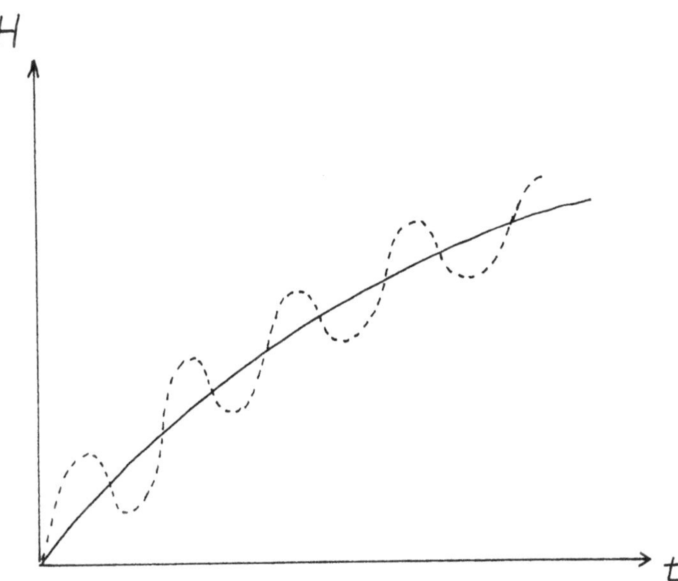

Superposition of high-frequency gravity waves (dashed) on the correct (solid) slowly changing value of H (or pressure) observed at a fixed point.

This is what evidently occurred in the initial tendency computation of surface pressure by L.F. Richardson referred to at the beginning of this section. The amplitude of the "erroneous" gravity waves is not very large. From (5.69b),

$$\delta \bar{H}_{II} \sim \frac{c^2 \alpha}{2 \omega_{II} h} \delta U$$

while the true H , $\bar{H}_I$, has a value equal to $(f/\beta) U$. Thus

$$\frac{\delta \bar{H}_{II}}{\bar{H}_I} = \frac{c^2}{2 f \omega_{II}} \frac{\alpha \beta}{h^2} \left(\frac{\delta U}{U}\right) \sim 1.5 \frac{\delta U}{U} \tag{5.80}$$

This is why the amplitude of the wiggles on the above figure is shown as small compared to the amplitude of the slowly varying correct solution.

The startling magnitude of the error (5.79) in $\partial H/\partial t$ arises from the high frequency associated with the gravity waves, rather than their amplitude.

The linearized system of equations (5.19) can be changed so that they do not include gravity waves as part of their solution. The procedure is similar to that used to get the Rossby-wave WKBJ solution. Let

$$f = f_0 + f_1(y) \qquad f_0 = \text{constant}$$
$$h = h_0 + h_1(y) \qquad h_0 = \text{constant} \tag{5.81}$$

represent the (slowly varying) latitude parameters. Then we set

$$U_0 = -\frac{1}{f_0}\frac{\partial H_0}{\partial y} , \quad V_0 = \frac{1}{f_0}\frac{\partial H_0}{\partial x} \tag{5.82}$$

as the zero-order solution of (5.19a) and (5.19b). (Recall that we kept $\partial/\partial t$ of first order, not zero-order). (5.19c) to zero-order is simply 0 = 0. The three first-order equations are

$$\frac{\partial H_0}{\partial t} - f_0 V_1 - f_1 V_0 = -\frac{\partial H_1}{\partial x} \tag{5.83a}$$

$$\frac{\partial V_0}{\partial t} + f_0 U_1 + f_1 U_0 = -\frac{\partial H_1}{\partial y} \tag{5.83b}$$

$$c^2\left(\frac{\partial U_1}{\partial x} + \frac{\partial V_1}{\partial y}\right) = -h_0^2\frac{\partial H_0}{\partial t} \tag{5.83c}$$

Solving for V_1 and U_1 from the first two of these for insertion into the third we get

$$\frac{f_0 h_0^2}{c^2}\frac{\partial H_0}{\partial t} = \frac{\partial}{\partial x}\left[\frac{\partial H_1}{\partial y} + \frac{\partial V_0}{\partial t} + f_1(y)U_0\right] + \frac{\partial}{\partial y}\left[-\frac{\partial H_1}{\partial x} + f_1(y)V_0 - \frac{\partial U_0}{\partial t}\right] \tag{5.84}$$

or, after substituting (5.82),

$$\left(\frac{\partial^2}{\partial x^2} + \frac{\partial^2}{\partial y^2} - \frac{f_0^2 h_0^2}{c^2}\right)\frac{\partial H_0}{\partial t} + \frac{df_1}{dy}\frac{\partial H_0}{\partial x} = 0 \tag{5.85}$$

For $H_0 \propto \exp[i(\alpha x + \beta y - \omega t)]$ we get

$$\omega = \frac{-\frac{df_1}{dy}\frac{\alpha}{h_0}}{\frac{\alpha^2+\beta^2}{h_0^2} + \frac{f_0^2}{c^2}} \tag{5.86}$$

as the only solution. Note the almost identical nature of this to the Rossby wave frequency.(5.58)

The result (5.85) is an equation giving the evolution of the geopotential field in terms of only itself - in principle measurements of the horizontal velocities are not required, and the source of the large initial error experienced by Richardson has been eliminated. Equations similar to it may be found in the papers by Rossby (1939), Charney (1947), and Obukhov (1949), each of whom arrived at it by somewhat different reasoning.

A generalization of (5.85) which includes non-linear and three-dimensional effects, and is known as the quasi-geostrophic theory, can be derived by means of a formal scale analysis applied to the original non-linear equations of motion (Monin, 1958 ; Charney, 1962 ; Phillips, 1963). This analysis assumes that

(a) the Rossby number is small :

$$R_o = \frac{1}{f_0 \tau} \sim \frac{U}{f_0 L} \ll 1 \tag{5.87}$$

U , τ , L being the size of the horizontal velocity and the time and horizontal space scales,

(b) the horizontal scale is small compared to the radius of the earth :

$$\frac{L}{a} \ll 1 \tag{5.88}$$

(c) the Richardson number $R_i = N^2(U/H)^2$ is large enough that

$$R_i R_o^2 = \frac{H^2 N^2}{f_0^2 L^2} \sim O(1) \tag{5.89}$$

In pressure coordinates this quasi-geostrophic system of equations is as follows

$$\frac{\partial \eta}{\partial t} + \vec{V}_0 \cdot \nabla \eta = \hat{k} \cdot \nabla \times \vec{F} - f_0 \frac{\partial}{\partial p}\left[\left(\frac{g}{\bar{N}\bar{T}}\right)^2 \frac{p}{RC_p} q\right] \quad (5.90)$$

$$\vec{V}_0 = \frac{1}{f_0} \hat{k} \times \nabla \phi \quad (5.91)$$

$$\eta = f + \zeta + f_0 \frac{\partial}{\partial p}\left[\left(\frac{g p}{R \bar{T} \bar{N}}\right)^2 \frac{\partial \phi}{\partial p}\right] \quad (5.92)$$

$$\zeta = \frac{1}{f_0} \nabla^2 \phi = \hat{k} \cdot \nabla \times \vec{V}_0 \quad (5.93)$$

with f_0 = constant, and $\bar{T}(p)$ and $\bar{N}(p)$ referring to a basic reference stratification. The divergence of $\vec{V}$ is given by

$$\nabla \cdot \vec{V} = -\frac{\partial \varpi}{\partial p} = -\frac{1}{f_0}\left[\frac{\partial \zeta}{\partial t} + \vec{V}_0 \cdot \nabla (f + \zeta) - \hat{k} \cdot \nabla \times \vec{F}\right] \quad (5.94)$$

It is small compared to the vorticity ζ , which in turn is small compared to f_0 (when (5.87) - (5.89) are satisfied) :

$$f_0 \gg \zeta \gg \nabla \cdot \vec{V} \quad (5.95)$$

Gravity waves have been eliminated, as in (5.85).

This quasi-geostrophic system formed the basis for the beginning of the modern practice of numerical weather prediction, and is still useful as a theoretical framework and even as a daily operational forecasting method. However, because it is subject to errors of the size of the supposedly small parameters appearing in (5.87) and (5.88), little accuracy can be expected from it in treating fronts and jet streams, motion of the very largest horizontal scales where $L \sim a$ (Burger, 1958), or low-latitude motions. Recent observational evidence (Yanai and Hayashi, 1969 ; Wallace and Kousky, 1968) suggests that much of the large-scale motion in equatorial latitudes is in the form of internal gravity waves, for which (5.95) is not valid. (With respect to the wave perturbation theory given above, the low-latitude wave-like motions in question appear to be part of the continuous vertical spectrum. They correspond to the longest possible latitudinal wave-lengths, however, and are thereby omitted in the short wave-length treatment of (5.40) - (5.58). (Matsuno, 1966 ; Longuet-Higgins, 1968 ; Dikii, 1970).

Current efforts in numerical weather prediction and simulation of the general circulation which use the hydrostatic system of equations without the geostrophic assumption therefore seem to be justified. The problem of the delicate balance between $\vec{V}$ and ϕ in the initial data for prediction is still present, however. Procedures exist for using the quasi-geostrophic theory to correct inconsistencies in initial data for motions that obey the requirements (5.87) - (5.89). A hopeful viewpoint is that the most energetic of the variable motions are the large-scale cyclones and anti-cyclones of extratropical latitudes, that these are described well by the quasi-geostrophic theory (Charney, 1947 ; Eady, 1949) as an instability process, and that the formation of fronts, jets, and even much of the large-scale motion in low latitudes is a secondary effect of the vigorous middle-latitude quasi-geostrophic motion. (Stone, 1966 ; Edelmann, 1963 ; Mak, 1969).

Some insight into the large-scale low-latitude motions (i.e., not hurricanes or tropical storms) can be obtained readily from a linearized analysis if we ignore the basic current $\bar{u}$ (Matsuno, 1966). Equations (5.19) can be simplified by an "equatorial beta-plane" approximation if $4\Omega^2a^2/c^2$ is large (Longuet-Higgins, 1968). We define

$$\begin{aligned} \epsilon &= \frac{4\Omega^2 a^2}{c^2} \quad ; \quad \sigma \equiv \frac{\omega}{2\Omega} > 0 \\ h &= \cos\theta \approx 1 \\ f &= 2\Omega \sin\theta \approx 2\Omega\theta \end{aligned} \tag{5.96}$$

Assuming that U, V and H vary as a function of θ times $\exp[i(s\lambda - \omega t)]$ with ω positive, s positive or negative, (5.19) becomes

$$\begin{aligned} \sigma U - i\theta V &= s\left(\frac{H}{2\Omega a}\right) \\ \theta U - i\sigma V &= -\frac{d}{d\theta}\left(\frac{H}{2\Omega a}\right) \end{aligned} \tag{5.97}$$

$$\epsilon\sigma\left(\frac{H}{2\Omega a}\right) = s\,U - i\,\frac{dV}{d\theta}$$

One solution has zero V. It is similar to a "Kelvin wave" (Lamb, 1945) and only moves eastward :

$$\omega = \frac{s\,c}{a} \quad ; \quad s > 0 \tag{5.98a}$$

$$(U, V, H) = (1, 0, c)\, e^{-\frac{\Omega a \theta^2}{c}} \tag{5.98b}$$

The other solutions satisfy the equation

$$\frac{d^2V}{d\theta^2} + \left(\epsilon\sigma^2 - s^2 - \frac{s}{\sigma} - \epsilon\theta^2\right)V = 0 \tag{5.98c}$$

which has solutions going to zero at large θ only if

$$\epsilon\sigma^2 - s^2 - \frac{s}{\sigma} = \epsilon^{1/2}(2n+1) \; ; \; n = 0, 1, 2, \cdots \tag{5.99a}$$

$$V = \mathcal{H}_n(\eta)\, e^{-\frac{1}{2}\eta^2} \tag{5.99b}$$

$$\eta = \epsilon^{1/4}\theta = \left(\frac{2\Omega a}{c}\right)^{1/2}\theta \tag{5.99c}$$

($\mathcal{H}$ is the Hermite polynomial).

A schematic picture of the frequency relations (5.98a) and (5.99a) is shown in the figure for a single value of C . To understand the three roots for σ given by (5.99a), that equation can be written as

$$\rho^3 - \rho - \chi = 0$$

$$\rho = \sigma\,\epsilon^{1/2}\left[s^2 + \epsilon^{1/2}(2n+1)\right]^{-1/2}$$

$$\chi = s\,\epsilon^{1/2}\left[s^2 + \epsilon^{1/2}(2n+1)\right]^{-3/2} \leqslant \left(\frac{4}{27}\right)^{1/2}\frac{2}{2n+1}$$

For $n \geqslant 1$, this cubic can be expanded as a convergent power series in χ,

$$\omega_1 = \frac{-2\Omega s}{s^2 + \frac{2\Omega a}{c}(2n+1)}(1 + \chi^2 + \ldots) \qquad (s<0)$$

$$\omega_{2,3} = c\left[\frac{s^2}{a^2} + \frac{2\Omega(2n+1)}{ac}\right]^{1/2}(1 \pm \frac{1}{2}\chi - \cdots)$$

The relation of these to the WKBJ solutions (5.65) - (5.67) is made clearer by noting that

$$\theta_t^2 = \frac{2n+1}{\epsilon^{1/2}} = \frac{c}{2\Omega a}(2n+1)$$

defines the "turning point latitude" at which the coefficient in (5.98c) becomes zero and poleward of which the solutions are no longer oscillatory in θ but decrease exponentially. Replacement of $(2n+1)$ by θ_t gives

$$\omega_1 \approx \frac{-\frac{2\Omega}{a}\left(\frac{s}{a}\right)}{\left(\frac{s}{a}\right)^2 + \frac{f_t^2}{c^2}} \qquad (s<0) \qquad (5.100a)$$

$$\omega_{2,3}^2 \approx \left(\frac{sc}{a}\right)^2 + f_t^2 \qquad (5.100b)$$

which correspond to (5.65) - (5.67) when $\beta = \partial f/\partial y = 0$ and $\cos\theta \approx 1$. For $n = 0$, the relevant solution of (5.99a) is

$$2\omega_0 = \frac{sc}{a} + \left[\left(\frac{sc}{a}\right)^2 + \frac{8\Omega c}{a}\right]^{1/2} \qquad (5.100c)$$

The solutions missed by the WKBJ method are therefore the Kelvin wave (5.98a) and the $n = 0$ solution of (5.100c), understandably so because these are the solutions of longest north-south wave length.

They seem to correspond roughly to the observed motions near the tropical tropopause. Thus, Wallace and Kousky (1968) (see also Holton and Lindzen, 1968) find good agreement between (5.98a) and their observed equatorial spectra of u at heights of around 20 km, which have peak amplitudes of ~ 10 m $\sec^{-1}$, periods of 10 - 15 days and some suggestion that s is small (~ 1), if c corresponds to a vertical wave length of

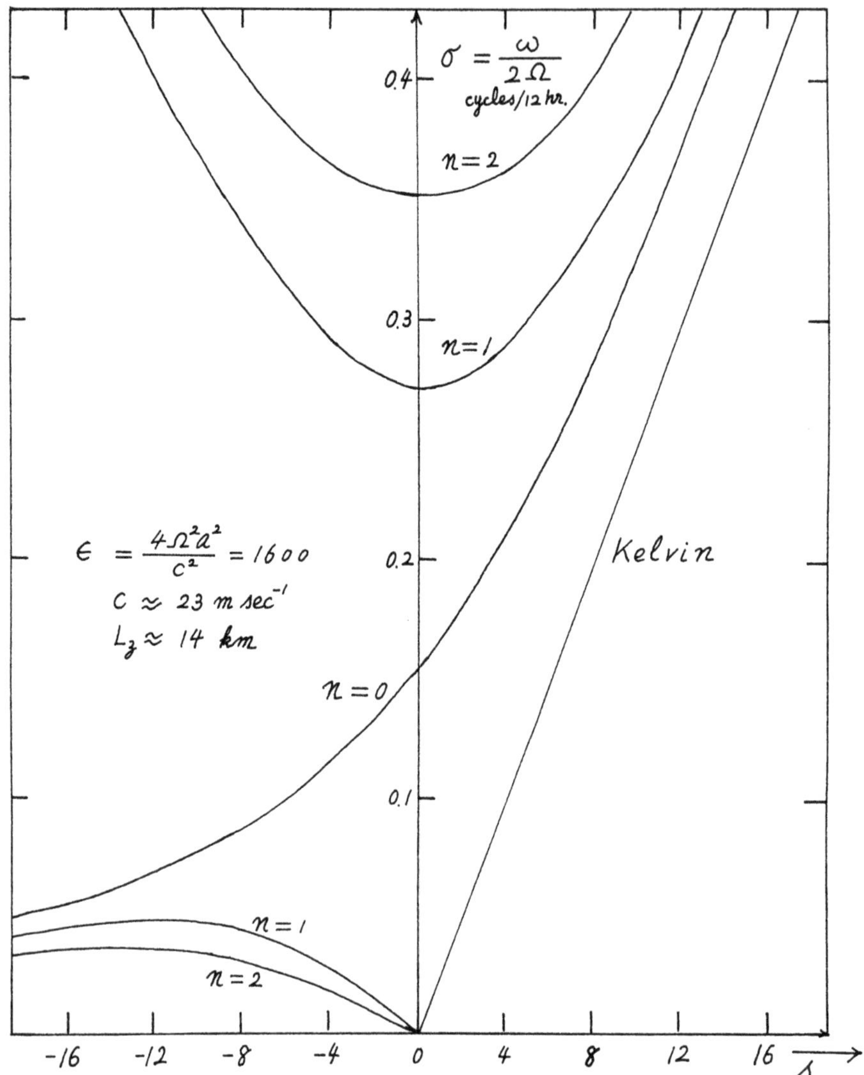

Frequency diagram showing σ as a function of zonal wave number s and north-south wave number n for a fixed vertical wave number.

$\sim$ 10 km (cf. (5.60).) In the equatorial spectra of v, Yanai and Hayashi (1969) report periods of 4 - 5 days with $s \sim -4$ (i.e., westward progression). This fits (5.100a) if $C \sim$ 23 m sec^{-1}, corresponding to a vertical wave length of about 12 km.

In the Kelvin wave, ζ is $-a^{-1}\partial u/\partial\theta$ and $\nabla\cdot\vec{V}$ is sU/a. Their ratio is :

$$\frac{|\nabla\cdot\vec{V}|_K}{|\zeta|_K} = \frac{\sigma}{\theta} \tag{5.101a}$$

For the n = 0 wave, V = $\exp(-\frac{1}{2}\eta^2)$.
From this it can be shown that

$$\frac{|\nabla\cdot\vec{V}|_{n=0}}{|\zeta|_{n=0}} = \frac{\sigma^3\epsilon\,\theta}{1-\epsilon\,\sigma^2\theta^2} \tag{5.101b}$$

Although the observed σ- values are small ($\sigma_K \sim 0.03$, $\sigma_0 \sim 0.1$) the above ratios are significantly less than one only at θ greater than about 25 degrees. These wave motions therefore do not satisfy the quasi-geostrophic condition (5.95) that $\zeta \gg \nabla\cdot\vec{V}$. The "balance equation" (Bolin, 1955 ; Houghton, 1968)

$$\nabla^2\phi = \nabla\cdot\left[f\,\nabla\Psi + \nabla^2\Psi\,\nabla\Psi - \frac{1}{2}\nabla(\nabla\Psi)^2\right], \quad \vec{V} = \hat{k}\times\nabla\Psi \tag{5.102}$$

is based on the assumption that $\nabla\cdot\vec{V} \ll \zeta = \nabla^2\Psi$
and has been used in low latitudes in place of the middle-latitude quasi-geostrophic relations to relate initial distributions of ϕ and $\vec{V}$ to one another (Krishnamurti, 1969). However (5.102) evidently does not apply to the dominant large-scale eddy motion in these latitudes near the tropopause.

6. NUMERICAL METHODS

The dynamical equations must be solved numerically as an initial value problem because of their complexity. Many different numerical methods can be employed, and the selection of "one" method in preference to another is as yet almost as much a matter of individual taste and experience as it is a matter of applying well-understood objective criteria. The following material is at most an incomplete summary of this "art" as it has so far developed in meteorological computations.

The most commonly used method (Richardson, 1922) uses a discrete grid mesh. For example, the discrete values of x, y, p, t,

$$
\begin{aligned}
x &= j\Delta x \\
y &= k\Delta y \\
p &= \ell\Delta p \\
t &= n\Delta t
\end{aligned}
\tag{6.1}
$$

at which j, k, ℓ and n are integers are the points at which the dependent variables (u , v, ϕ) are defined. The subscript notation

$$u_{jk\ell n} = u(j\Delta x, k\Delta y, \ell\Delta p, n\Delta t) \tag{6.2}$$

is commonly used to indicate this. It is sometimes convenient to displace one variable by half a grid interval in x , y, p and/or t . In such cases the subscripts j , k , ℓ and n may still be restricted to integer values for notational convenience if the corresponding change in the definition (6.2) is explicitly noted. For example,

$$u_{jk\ell n} = u\left[(j+\tfrac{1}{2})\Delta x, k\Delta y , \ell\Delta p , (n-\tfrac{1}{2})\Delta t\right] \tag{6.3}$$

An alternate method to the grid point representation (6.1) - (6.3) is based on using series of orthogonal functions for the spatial dependence. This is discussed later.

Given (6.2), the partial derivatives in the hydrodynamic equations are expressed as finite differences. The equations of numerical weather prediction are non-linear, but there is no general method to test the accuracy of solutions of non-linear finite-difference equations. In practice one examines the accuracy of linearized forms of the equations for a guide, and then supplements this by other considerations if possible, including numerical experimentation. As a simple example we will consider the linear advection equation

$$\frac{\partial \zeta}{\partial t} = -U \frac{\partial \zeta}{\partial x} \tag{6.4}$$

with U a known constant. Supposing that ζ is defined at integer values of j and n , we might consider the following <u>uncentered</u> expressions :

$$\frac{\partial \zeta}{\partial t} \longrightarrow \frac{\zeta_{j\,n+1} - \zeta_{j\,n}}{\Delta t}$$
$$\frac{\partial \zeta}{\partial x} \longrightarrow \frac{\zeta_{j\,n} - \zeta_{j-1\,n}}{\Delta x} \qquad (6.5.)$$

When substituted into (6.4) these give

$$\zeta_{j\,n+1} = \zeta_{j\,n} - \sigma(\zeta_{j\,n} - \zeta_{j-1\,n}) \qquad (6.6)$$

$$\sigma = \frac{U \Delta t}{\Delta x} \qquad (6.7)$$

which enables ζ to be computed at the new time step $n+1$ from known values at the previous time step n. An alternate formulation would be to use the centered expressions

$$\frac{\partial \zeta}{\partial t} \longrightarrow \frac{\zeta_{j\,n+1} - \zeta_{j\,n-1}}{2\Delta t}$$
$$\frac{\partial \zeta}{\partial x} \longrightarrow \frac{\zeta_{j+1\,n} - \zeta_{j-1\,n}}{2\Delta x} \qquad (6.8)$$

which produce the scheme

$$\zeta_{j\,n+1} = \zeta_{j\,n-1} - \sigma(\zeta_{j+1\,n} - \zeta_{j-1\,n}) \qquad (6.9)$$

Again, ζ can be computed at the new time step from known values at (two) previous time steps.

Many variations are possible, even for the simple equation (6.4). In all cases, however, we are interested in the accuracy of the numerical scheme - how close does the numerical finite-difference solution for a fixed time $T = N\Delta t$ approach the time solution for time T ? The errors will of course depend not only on the forms of the finite-differences, but also on the size of Δx and Δt . However, we will see that the error need not go to zero as Δt and Δx separately go to zero, even though the expressions for each derivative may become more exact as Δx and $\Delta t \rightarrow 0$. This possibility arises because as we let Δt go to zero, the number of time steps $N = T/\Delta t$ needed to compute from $t = 0$ until $t = T$ increases, and this increase in the number of time steps may cancel the increasing accuracy of the change computed for a single time step. With some exceptions, we will find that the errors for a fixed time T will decrease with Δx and Δt only if Δt approaches zero sufficiently rapidly with respect to Δx .

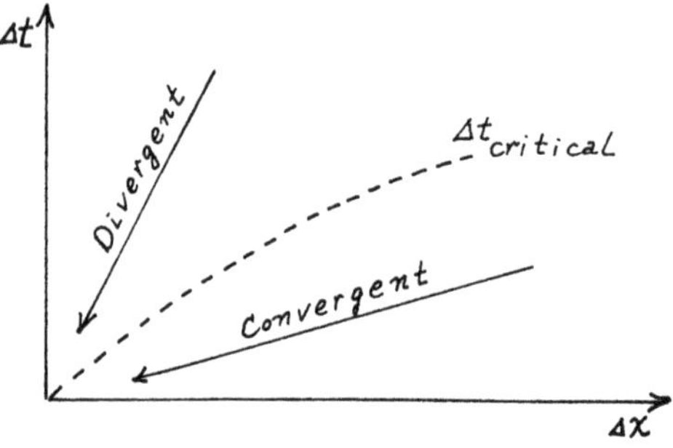

The existence of such a condition was first established by Courant, Friedericks and Lewy (1928).

To formulate matters more clearly, let us return to (6.4), whose true solution we know to be

$$\zeta(x,t) = \zeta(x - ut\ ,\ 0) \tag{6.10}$$

On an (x,t) diagram we have the following picture if we use the uncentered scheme (6.6) :

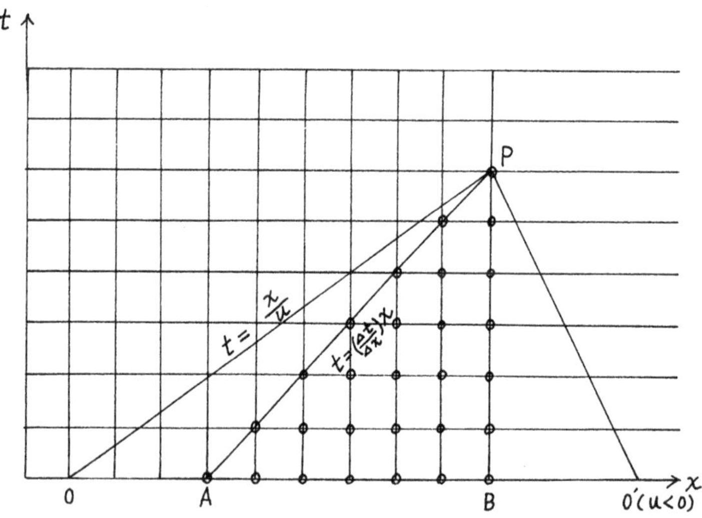

The numerical solution at P depends on the initial values between points A and B, whereas the true solution at P depends (only) on the value of ζ at point 0. In the example shown, $\Delta t/\Delta x$ is greater than $1/U$

$$\frac{U\Delta t}{\Delta x} = \sigma > 1$$

and no information about point 0 is made available to point P by the numerical process. Decreasing Δt and Δx will not let point P be affected by point 0 unless we arrange in this limiting process that

$$\frac{U\Delta t}{\Delta x} < 1 \tag{6.11}$$

We may note also that if U were negative, so that point 0 was situated to the right of B at 0', there is no choice of Δt in (6.6) which would numerically inform point P about the correct value of ζ it should have. We must therefore correct (6.11) to read

$$0 < \frac{U\Delta t}{\Delta x} = \sigma < 1 \tag{6.12}$$

as a necessary condition for convergence of the numerical solution of (6.6) to the true solution.

Let us consider another aspect of the uncentered numerical solution (6.6), the size of the numbers so generated. (We will see later that convergence and boundedness of the solutions are related). That formula says that the value of ζ_{j_0} at the time step $n+1$ is equal to the value of ζ obtained by interpolating along the straight line $\tilde{\zeta}_j$ defined by $\zeta_{j_0 n}$ and $\zeta_{j_0-1\,n}$:

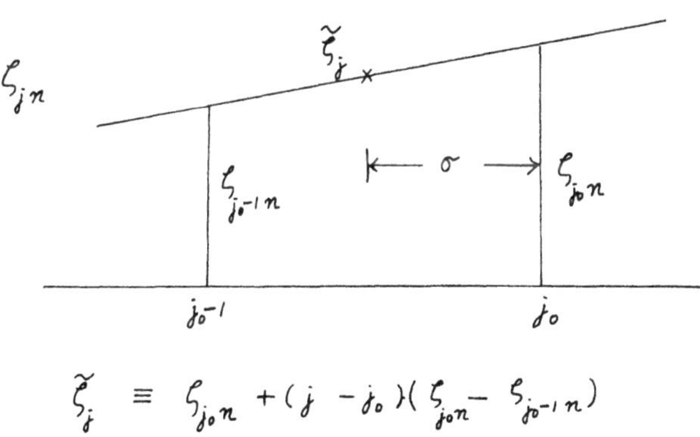

$$\tilde{\zeta}_j \equiv \zeta_{j_0 n} + (j - j_0)(\zeta_{j_0 n} - \zeta_{j_0-1\,n})$$

$$\zeta_{j_0 n+1} = \zeta_{j_0 n} - \sigma(\zeta_{j_0 n} - \zeta_{j_0 -1\, n}) = \tilde{\zeta}_j (j = j_0 - \sigma) \tag{6.13}$$

If σ is either negative or greater than 1, the value of $\zeta_{j_0 n+1}$ will be outside the range $(\zeta_{j_0-1\, n}, \zeta_{j_0 n})$ i.e., it will correspond to an extrapolation on $\tilde{\zeta}_j$. Considering now the entire field of ζ at all j-points, we see that violation of (6.12) will always produce somewhere a value of $\zeta_{j\, n+1}$ which is greater than the maximum value of ζ_{jn} and will also produce somewhere a value of $\zeta_{j\, n+1}$ which is algebraically less than the minimum value of ζ_{jn}. In other words, violation of the convergence criterion (6.12) for the scheme (6.6) will produce a finite-difference solution which increases in amplitude with n. (In case of negative U, (6.5) should clearly be changed to the upstream difference $\zeta_{j+1\, n} - \zeta_{jn}$.)

The boundedness of numerical solutions for finite-difference systems more complicated than (6.6) cannot be established as directly as this. An alternate method is therefore used. Suppose that ζ_{jn} can be represented by a Fourier series :

$$\zeta_{jn} = \sum_p Z_{pn}\, e^{ipj} \tag{6.14}$$

(6.6) becomes

$$\begin{aligned} Z_{p\, n+1} &= \lambda_p Z_{pn} \\ \lambda_p &= 1 - \sigma(1 - e^{-ip}) \end{aligned} \tag{6.15}$$

whence

$$Z_{pn} = (\lambda_p)^n Z_{p0}$$

We find that the eigenvalue λ_p has a magnitude

$$|\lambda_p|^2 = 1 - 4\sigma(1-\sigma)\sin^2\left(\frac{p}{2}\right) \tag{6.16}$$

Considering all values of p, we see that $|\lambda|^2 \leq 1$ requires $0 \leq \sigma \leq 1$ i.e., condition (6.12), as a condition for boundedness of (6.6).

The same Fourier procedure can be carried out for the centered formula (6.9). We obtain then

$$Z_{p\,n+1} = Z_{p\,n-1} - 2i\sigma \sin p \, Z_{p\,n} \tag{6.17}$$

By introducing an artificial new variable

$$W_{p\,n} \equiv Z_{p\,n-1} \tag{6.17'}$$

(6.17) may be written as the system

$$\begin{pmatrix} Z \\ W \end{pmatrix}_{p\,n+1} = \begin{pmatrix} -2i\sigma\sin p & 1 \\ 1 & 0 \end{pmatrix} \begin{pmatrix} Z \\ W \end{pmatrix}_{p\,n} \tag{6.18}$$

The two eigenvalues of the matrix are

$$\lambda_1 = e^{-i\theta}, \quad \lambda_2 = -e^{i\theta}, \quad \sin\theta = \sigma \sin p \tag{6.19}$$

θ must be real for $|\lambda| \leqslant 1$, which requires now that

$$|\sigma| = \frac{|U|\Delta t}{\Delta x} \leqslant 1 \tag{6.20}$$

if the solution is to be bounded for all p.

The centered scheme (6.9) is often called the "leap-frog" method. As presented here for the linear advection equation (6.4), it is analogous to the finite-difference method used by Charney, Fjørtoft and von Neumann (1950) in the first non-linear weather prediction computation with the barotropic vorticity equation. This system is based on the quasi-geostrophic approximation and an assumption that flow patterns do not change direction with height.

$$\begin{gathered} \frac{\partial \zeta}{\partial t} = -\vec{V} \cdot \nabla (f + \zeta) \\ \vec{V} = \hat{k} \times \nabla \Psi, \quad \zeta = \nabla^2 \Psi \end{gathered} \tag{6.21}$$

In order to examine linear versions of more complicated dynamical systems, we make use of a theorem by P. Lax (Richtmyer and Morton, 1967) :

Given a <u>properly-posed</u> linear initial-value problem, and a finite-difference approximation to it that satisfies the <u>consistency condition</u>, <u>stability of the difference equations</u> is the necessary and sufficient condition for <u>convergence</u>.

The four underlined concepts must be defined in order to apply the theorem. For our purposes the following remarks suffice.

<u>Properly posed</u> : The continuous equations and boundary conditions have a continuous solution.

<u>Consistency</u> : The finite-difference forms for the partial derivatives are consistent approximations to those derivatives.

<u>Stability</u> : For any allowable initial condition, the finite-difference solution for a fixed time $t = T$ is bounded as $\Delta t \to 0$

<u>Convergence</u> : The finite-difference solution for the fixed time T converges to the true solution as Δx, Δt (etc.) approach zero.

The theorem extends to all linear systems the relations between convergence and boundedness which have been established above for the simple system (6.4) - (6.7). It is useful because it allows us to establish convergence by examining separately the easier questions of consistency and stability (we shall always assume the problem is properly-posed). We are interested in convergence not because we will actually let the time and space increments go to zero, but because we want to be assured that if these increments are small, the error is small.

The theorem is however still difficult to apply for linear systems which have variable coefficients or awkward boundary conditions - stability of the difference equations is then difficult to establish. As a practical matter we are forced to simplify even further by assuming constant coefficients and boundary conditions simple enough (e.g. periodicity) so that the grid point values may be represented by Fourier series of the form (6.14). [In the hydrostatic system, it is necessary to consider a more exact treatment of the vertical coordinate, as discussed following (6.33).] Under such conditions, the consistency and stability tests become simpler.

<u>Consistency</u> : Subtract the continuous equation from the finite-difference equation, with each term in the latter evaluated by a Taylor series.

The finite-difference system is consistent if the remainder goes to zero with Δx and Δt .

Stability : Let the entire set of grid point values ζ_{jn} , say, be represented by a Fourier series with wave number $\underset{\sim}{p}$ (ζ here is not only a function of the "grid point vector" $\underset{\sim}{j}$, but insofar as it may represent all three variables (u , v , ϕ), ζ is itself a "vector" as is Z .)

$$\zeta_{\underset{\sim}{j}n} = \sum_{\underset{\sim}{p}} Z_{\underset{\sim}{p}n} e^{i\underset{\sim}{p}\cdot\underset{\sim}{j}}$$

When this is substituted into the finite-difference scheme, the system can be arranged in the form

$$Z_{\underset{\sim}{p}n+1} = G Z_{\underset{\sim}{p}n} \tag{6.22}$$

(A step similar to (6.17') may be needed to accomplish this.) G is a square matrix whose elements depend on $\underset{\sim}{p}$, Δt , and the space increments $\Delta x, \cdots$. Since the solution to (6.22) is

$$Z_{\underset{\sim}{p}n+1} = G^n Z_{\underset{\sim}{p}0}$$

there will be stability for any allowable initial condition if G^n is bounded for all $\underset{\sim}{p}$ as $\Delta t \to 0$. In general we can write

$$|\lambda|^n_{max} \leqslant \| G^n \| \leqslant \| G \|^n \tag{6.23}$$

where λ_{max} is the eigenvalue of G having the maximum absolute value and $\| \; \|$ denotes the bound of a matrix. (The three expressions are equal if G is a normal matrix).

The center expression in (6.23) is what we wish to bound as $\Delta t \to 0$ The von Neumann necessary condition for stability is that the left hand number, $|\lambda|^n_{max.}$ be bounded for all $\underset{\sim}{p}$ as n approaches infinity and Δt approaches zero for a fixed time $T = n\Delta t$. We find for $\Delta t \to 0$

$$|\lambda|^n_{max} = \text{finite} \leqslant e^c \; (\text{say})$$

$$|\lambda|_{max} \leqslant e^{\frac{c}{n}} = e^{\frac{c}{T}\Delta t} \sim 1 + \frac{c}{T}\Delta t$$

Thus,

$$|\lambda|_{max} \leqslant 1 + O(\Delta t) \tag{6.24}$$

is the von Neumann necessary condition for stability. The $O(\Delta t)$ term in (6.24) allows for a physical instability if there is one present in the continuous equations :

$$\lim_{\Delta t \to 0} (1 + K\Delta t)^{\frac{T}{\Delta t}} = e^{KT} \tag{6.25}$$

Investigation of the sufficient condition for stability, (the boundedness of $\|G\|^n$ in (6.23)) is seldom carried out. Some theorems are given in the book by Richtmyer and Morton ; in meteorological applications they seem to be satisfied by simply eliminating the equality in conditions such as (6.20).

Reflecting back on the uncentered solution (6.6) of the advection equation, the equation corresponding to (6.22) is (6.15). G is there a 1×1 matrix and our conclusion from (6.16) was in effect the von Neumann condition. Similarly, (6.18) is the equivalent of (6.22) for the centered solution (6.9). G there is a 2 x 2 matrix.

The consistency of (6.6) is readily established if we first restore it to its original form by dividing by Δt and then use a Taylor series expansion about $x = j\Delta x$, $t = n\Delta t$,

$$\frac{\zeta_{j\,n+1} - \zeta_{j\,n}}{\Delta t} = -\frac{u(\zeta_{j\,n} - \zeta_{j-1\,n})}{\Delta x}$$

$$\frac{1}{\Delta t}\left[\Delta t\frac{\partial \zeta}{\partial t} + \frac{1}{2}(\Delta t)^2\frac{\partial^2 \zeta}{\partial t^2} + \dots\right] = -\frac{u}{\Delta x}\left[\Delta x\frac{\partial \zeta}{\partial x} - \frac{1}{2}(\Delta x)^2\frac{\partial^2 \zeta}{\partial x^2} + \dots\right]$$

Subtracting $\partial\zeta/\partial t = -u\,\partial\zeta/\partial x$, we get

$$\frac{\Delta t}{2}\frac{\partial^2 \zeta}{\partial t^2} + O(\Delta t^2) = \frac{u}{2}\Delta x\frac{\partial^2 \zeta}{\partial x^2} + O(\Delta x^2)$$

which vanishes with Δt and Δx as required. A similar treatment of (6.9) leaves a smaller remainder :

$$\frac{1}{6}(\Delta t)^2 \frac{\partial^3 \zeta}{\partial t^3} + O(\Delta t^3) = -\frac{1}{6}(\Delta x)^2 U^2 \frac{\partial^3 \zeta}{\partial t^3} + O(\Delta x^3)$$

This higher order remainder of centered versus uncentered differences is basic and normally implies a higher accuracy for the centered form.

Linear finite-difference systems may have 1 of 3 types of computational stability : unstable for any choice of Δt (i.e., absolutely unstable), stable if Δt is limited (conditionally stable), or stable for any choice of Δt (absolutely stable). The systems (6.6) and (6.9) are conditionally stable. An absolutely unstable finite-difference version of the advection equation is easily obtained by combining the uncentered time derivative in (6.5) with the centered space derivative in (6.8), or vice versa. An absolutely stable version can be obtained by evaluating the space derivative at both time steps:

$$\zeta_{j\,n+1} = \zeta_{j\,n} - \frac{\sigma}{4}\left[(\zeta_{j+1} - \zeta_{j-1})_{n+1} + (\zeta_{j+1} - \zeta_{j-1})_n\right] \tag{6.26}$$

for which $\lambda = (2 - i\sigma \sin p)/(2 + i\sigma \sin p)$. This is an example of an <u>implicit</u> procedure and has $|\lambda| = 1$ for <u>all</u> Δt. Its convergence for any value of the ratio $\Delta t/\Delta x$ is understandable because each $\zeta_{j\,n+1}$ depends on the <u>entire</u> field of $\zeta_{j\,n}$. Although such methods seem attractive, the use of the larger values of Δt which they allow for stability seems, in meteorological problems, to lead rapidly to larger error as Δt is increased beyond the value allowed by a normal explicit method. Implicit methods also require considerably more computation for each time step. (Note the complications in (6.26) involved in determining $\zeta_{j\,n+1}$ as a function of $\zeta_{j\,n}$.)

Some idea of the accuracy of a scheme can be obtained by comparing the amplitude and frequency (or phase speed) of the linear finite-difference solution for a certain wave length with those of the continuous equation for the same wave length. Thus, if $\zeta(x,0)$ is

$$\zeta(x,0) = e^{i\alpha x} = e^{ipj}$$

$$p = \alpha \Delta x = 2\pi \left(\frac{\Delta x}{L}\right) \tag{6.27}$$

the solution of the continuous equation (6.4) is

$$\zeta(x,t) = e^{i\alpha(x - Ut)} \tag{6.28}$$

The centered leap-frog scheme (6.19) has a solution

$$\zeta_{jn} = A e^{i(pj - \theta n)} + (1 - A) e^{i(pj + \theta n + n\pi)}$$

$$\theta = \sin^{-1}(\sigma \sin p) \tag{6.29}$$

This satisfies the initial condition (6.27) and the leap-frog difference equation (6.9). The latter requires two time levels to be known before the iteration process can begin. One way commonly used to resolve this indeterminacy is to use an uncentered time step for the single step $n = 0$ to $n = 1$:

$$\zeta_{j1} = \zeta_{j0} - \frac{\sigma}{2}(\zeta_{j+1\,0} - \zeta_{j-1\,0}) \tag{6.30}$$

This gives $A = (1 + \cos\theta)/(2\cos\theta)$, whence

$$\zeta_{jn} = \frac{1}{2\cos\theta}\left[(1 + \cos\theta)\, e^{i(pj - \theta n)} - (1 - \cos\theta)\, e^{i(pj + \theta n + n\pi)}\right] \tag{6.31}$$

The exponent $i(pj - \theta n)$ can be written as

$$i\alpha(j\Delta x - U't), \qquad U' = \frac{\theta \Delta x}{p \Delta t} = \frac{U\theta}{p\sigma} \tag{6.32}$$

U' is the effective phase speed of the first term in (6.31). The following table compares this with the correct speed U for different values of p , when $\sigma = U\Delta t/\Delta x = 0.75$.

$\frac{L}{\Delta x} = \frac{2\pi}{p}$	$\frac{U'}{U}$	$\frac{1-\cos\theta}{1+\cos\theta}$
2	0	∞
3	0.45	7.35
4	0.72	0.20
6	0.90	0.14
8	0.95	0.08
12	0.98	0.04
24	0.995	0.01

The finite-difference solution moves the waves too slowly. About 8 grid intervals per wave length are necessary to get 95 % accuracy. The second part of the solution (6.31) is sometimes called the "computational wave". Except for the $e^{in\pi} = (-1)^n$ factor, it moves in the wrong direction. Its amplitude relative to the first term is given by the third column in the table.

The value of U'/U for the implicit scheme (6.26) is given by $\nu/\sigma p$ with $\tan\nu = 4\sigma\sin p\,(4-\sigma^2\sin^2 p)^{-1}$. The reduced accuracy in this absolutely stable scheme when $\sigma > 1$, which was referred to earlier, is shown by the following values for σ = 1, 2 and 4.

U'/U (implicit)

$\frac{L}{\Delta x} = \frac{2\pi}{p}$	$\sigma = 1$	$\sigma = 2$	$\sigma = 4$
2	0	0	0.25
3	0.39	0.34	0.25
4	0.59	0.50	0.35
6	0.78	0.68	0.50
8	0.86	0.78	0.61
12	0.94	0.89	0.75
24	0.98	0.97	0.91

The reduction in computational work achieved by taking $\Delta t > \Delta x/U$ in the implicit scheme (6.26) is not only counterbalanced by the additional labor in solving for $\zeta_{j\,n+1}$ in (6.26), but is accompanied by a less accurate value of U'. The centered leap-frog scheme has more accuracy than the simple uncentered scheme. If we consider the fixed time period $T = L/U$, the true solution is simply $exp(ipj)$. The corresponding finite-difference solutions can be expanded in

$p = 2\pi\Delta x/L$ for small p (i.e., small Δx). With $n = L/u\Delta t$ $= 2\pi/\sigma p$, one finds

Uncentered :

$$e^{ipj}\lambda^{n} \longrightarrow \left[1 - p\pi(1-\sigma) + O(p^2)\right] e^{ipj}$$

whereas the leap-frog solution (6.31) becomes

$$\left\{1 + p^2\left[\frac{\sigma^2}{4}(1 - \cos n\pi) + \frac{\pi i}{3}(1-\sigma^2)\right] + O(p^4)\right\} e^{ipj}$$

Although the leap-frog scheme is still used by many numerical prediction groups, it has some weaknesses. The basic structure of (6.9) is such that points on the j , n grid for which the sum $j+n$ is even are independent of those for which the sum $j+n$ is odd, with the exception of the link between them brought about by the special starting procedure (6.30). In other words, (6.9) almost amounts to two independent computations (Platzman, 1958). It is true that this independence will be modified somewhat by the special starting procedure (6.30) and by the non-linearities present in more realistic equations. However, experience has shown that non-linear numerical solutions produced with the leap-frog method tend to separate into even and odd time step solutions (Lilly, 1965). That is, the numbers at successive even time steps (n, n+2, n+4, ...) appear to progress smoothly with n and so do those at successive odd time steps, but the solutions at the successive time steps n and n+1 differ markedly. This of course means that the time rates of change must be incorrect. Most groups which use the leap-frog time extrapolation procedure therefore find it necessary to apply the special starting procedure (6.30) periodically, for example every 50th time step. This procedure evidently keeps the solutions similar at even and odd n .

The linear computational criteria when gravity waves are present can be seen by studying the linear system

$$\frac{\partial u}{\partial t} = -U\frac{\partial u}{\partial x} - V\frac{\partial u}{\partial y} + fv - \frac{\partial \phi}{\partial x}$$

$$\frac{\partial v}{\partial t} = -U\frac{\partial v}{\partial x} - V\frac{\partial v}{\partial y} - fu - \frac{\partial \phi}{\partial y} \qquad (6.33)$$

$$\frac{\partial \phi}{\partial t} = -U \frac{\partial \phi}{\partial x} - V \frac{\partial \phi}{\partial y} - C^2\left(\frac{\partial u}{\partial x} + \frac{\partial v}{\partial y}\right)$$

$$C^2, U, V, f = constants$$

which, apart from some change in notation, is a simplified form of (5.19) to which the effect of a uniform current in the basic state has been added. A more direct procedure would be to first formulate the hydrostatic equations in pressure coordinates (4.20) - (4.27) (or σ coordinates (4.35) - (4.39)) in terms of vertical finite-differences $p = \ell \Delta p$ (or $\sigma = \ell \Delta \sigma$) (the energetical consistency requirements concerning vertical finite-differences set forth by Lorenz (1960) would be significant here), linearize the resulting system of equations at each ℓ -level, and then separate out the ℓ -dependence from u , v and ϕ. This last step is similar to (5.10), and a second-order vertical difference equation analogous to (5.20) will emerge to define a discrete series of separation constants C^2 , equal in number to the number of discrete (u, v, ϕ). The largest value of c^2 in a multi-level model should be approximately equal to the "Lamb wave" value of about $(320 \text{ m sec}^{-1})^2$. Benwell and Bretherton (1968) compute values of 285, 111, 43.5, 26, 16 m $\sec^{-1}$ for the five largest values of in the 10-layer model used by Bushby and coworkers (Bushby, 1969). This approach is necessary rather than simply using for the vertical dependence the spatial Fourier spectrum assumption (6.22) which led to the von Neumann computational stability criterion ; in the hydrostatic system of equations a simple assumption of a trigonometrical vertical dependence would ignore the important effect of vertical boundary conditions in determining c^2.

Equations which are similar to (6.33), but in a non-linear form, are those which arise if the atmosphere is treated arbitrarily as a layer of homogeneous incompressible liquid, with a variable depth $D(\lambda, \theta, t)$

$$\frac{du}{dt} = -\frac{1}{a\cos\theta}\frac{\partial \phi}{\partial \lambda} + \left(2\Omega + \frac{u}{a\cos\theta}\right) v \sin\theta$$

$$\frac{dv}{dt} = -\frac{1}{a}\frac{\partial \phi}{\partial \theta} - \left(2\Omega + \frac{u}{a\cos\theta}\right) u \sin\theta \qquad (6.33a)$$

$$\frac{d\phi}{dt} = -\phi\,\nabla\cdot\vec{V} = -\frac{\phi}{a\cos\theta}\left(\frac{\partial u}{\partial\lambda} + \frac{\partial v\cos\theta}{\partial\theta}\right)$$

The correspondence is $\phi = gD,\ c^2 = g\bar{D}$. d/dt here is simply $\partial/\partial t + \vec{V}\cdot\nabla$. These are often used in exploratory tests of non-linear numerical integration schemes.

If u, v, ϕ are all defined at each grid point $x = j\Delta$, $y = k\Delta$ $t = n\Delta t$ and the centered "leap-frog" differences (6.8) are used, trigonometric finite-difference solutions have the form $\exp[i(pj + qk - \mu n)] \rightarrow \exp[i(\alpha x + \beta y - \omega t)]$ where

$$\sin\mu_1 = W,\ \sin\mu_2 = W + M,\ \sin\mu_3 = W - M$$

$$W = \frac{\Delta t}{\Delta}(U\sin p + V\sin q) \tag{6.34}$$

$$M^2 = (f\Delta t)^2 + \left(\frac{c\Delta t}{\Delta}\right)^2(\sin^2 p + \sin^2 q)$$

(Note that $\pi - \mu$ is also a solution, corresponding to the leap-frog "computational" mode present in (6.31).) $(f\Delta t)^2$ contributes little to M^2 and the von Neumann condition becomes

$$\frac{\sqrt{2}\,\Delta t}{\Delta}\left(c + \sqrt{U^2 + V^2}\right) < 1 \tag{6.35}$$

((U, V) has the maximum effect on $|\lambda|$ when $U/V = \sin p/\sin q$.) Δ is really the actual horizontal space increment $h\Delta x$, an important consideration in integrations on the entire globe with singular points in the coordinate system. The $\sqrt{2}$ factor appears almost always when both x and y coordinates are included. For Δ = 300 km, $|\vec{V}|$ = 80 m sec^{-1}, c = 320 m sec^{-1} Δt must be less than 9 minutes.

In the absence of the advective terms $(U\,\partial/\partial x + V\,\partial/\partial y)$ and the Coriolis terms in (6.33) it is easy to arrange the grid-point locations so that the computational mode is absent ; we simply use a version of the "staggered" type of arrangement suggested in (6.3) :

$$\phi_{jkn} \equiv \phi(j\Delta, k\Delta, n\Delta t)$$

$$u_{jkn} \equiv u((j+\tfrac{1}{2})\Delta, k\Delta, (n+\tfrac{1}{2})\Delta t) \tag{6.36}$$

$$v_{jkn} \equiv v(j\Delta, (k+\tfrac{1}{2})\Delta, (n+\tfrac{1}{2})\Delta t)$$

The obvious use of these in difference formulae centered in x, y, t (but apparently uncentered in j, k, n) leads to only three eigenvalues

$$\lambda = 1, e^{i\psi}, e^{-i\psi}; \sin\frac{\psi}{2} = \left(\frac{c\Delta t}{\Delta}\right)\left(\sin^2\frac{p}{2} + \sin^2\frac{q}{2}\right)^{1/2} \quad (6.37)$$

in place of the six values contained in (6.34). The continuous frequency ω is equal to $c(\alpha^2+\beta^2)^{1/2}$. The effective finite-difference frequency for (6.37) is $\omega' = \psi/\Delta t$, that for the leap-frog system (6.34) is $\omega'' = \mu/\Delta t$. Expanding ψ and μ for small p and q gives the following comparative measures of truncation error :

$$\frac{\omega''-\omega}{\omega} = \frac{1}{6}(p^2+q^2)\left[\left(\frac{c\Delta t}{\Delta}\right)^2 - (\cos^4\gamma+\sin^4\gamma)\right] + O(p^4, q^4)$$

$$\frac{\omega'-\omega}{\omega} = \frac{1}{24}(p^2+q^2)\left[\left(\frac{c\Delta t}{\Delta}\right)^2 - (\cos^4\gamma+\sin^4\gamma)\right] + O(p^4, q^4) \quad (6.38)$$

where $\tan\gamma = p/q = \alpha/\beta$. The greater efficiency of the staggered system eliminates the computational mode and is more accurate for the same amount of computation.

Unfortunately, when advection terms are present (and these are very important in the meteorological problem) it is not so easy to both eliminate the computational mode and retain the second-order truncation error typical of centered differences, by special location of the grid-point values. One method has been suggested by Marchuk(1964, 1969), who observed that equations such as (6.33) could be solved in successive steps by "splitting' or " partitioning" the operations on the right side of those equations into separate sequential operations. To illustrate with (6.33), let Z denote the column vector with components u, v, ϕ. Within each time step $0 \leq t \leq \Delta t$, one rewrites (6.33) as three successive operations :

$$\frac{\partial Z_1}{\partial t} = -U\frac{\partial Z_1}{\partial x}, \quad Z_1(x,y,0) = Z(x,y,0) \quad (6.39a)$$

$$\frac{\partial Z_2}{\partial t} = -V\frac{\partial Z_2}{\partial y} \quad , \; Z_2(x,y,0) = Z_1(x,y,\Delta t) \qquad (6.39b)$$

$$\frac{\partial Z_3}{\partial t} = -\begin{pmatrix} 0 & -f & \frac{\partial}{\partial x} \\ f & 0 & \frac{\partial}{\partial y} \\ c^2\frac{\partial}{\partial x} & c^2\frac{\partial}{\partial y} & 0 \end{pmatrix} Z_3 \, , \quad Z_3(x,y,0) = Z_2(x,y,\Delta t) \qquad (6.39c)$$

$$Z(x,y,\Delta t) = Z_3(x,y,\Delta t) \qquad (6.39d)$$

Δt should be small enough that the advective steps do not result in too great an imbalance between the $\vec{V}$ and ϕ fields. Marchuk shows that the combined system will be stable if each of the three steps is stable by itself (and does not contain computational modes). Marchuk suggests that second-order accuracy in the advective steps (6.39a) and (6.39b) be obtained by a two-step process. In the notation of (6.4) - (6.9) this would be an implicit "upstream differencing" step over $\Delta t/2$,

$$\zeta_{j\,n+\frac{1}{2}} = \zeta_{jn} - \frac{\sigma}{2}\left(\zeta_{j\,n+\frac{1}{2}} - \zeta_{j-1\,n+\frac{1}{2}}\right) \qquad (u>0) \qquad (6.40a)$$

followed by an explicit step

$$\zeta_{j\,n+1} = \zeta_{jn} - \frac{\sigma}{2}\left(\zeta_{j+1\,n+\frac{1}{2}} - \zeta_{j-1\,n+\frac{1}{2}}\right) \qquad (6.40b)$$

The eigenvalue for this combination is

$$\lambda = \frac{2+\sigma(1-\cos p) - i\sigma\sin p}{2+\sigma(1-\cos p) + i\sigma\sin p} \, , \qquad |\lambda| = 1 \qquad (6.40c)$$

The computational mode of the leap-frog procedure is absent since $\zeta_{j\,n+1}$ depends only on ζ_{jn} . Marchuk suggests that step (6.39c) also be an implicit step, but this will require the computationally expensive inversion of a three-dimensional elliptic operator. Crowley (1968) has tested the advection steps in a two-dimensional problem with good results (even better when higher-order differences are used).

An alternate method is due to Lax and Wendroff (1960 ; see also Phillips (1962), Leith (1965), Richtmyer (1962).) In its most convenient form it applies to the complete system of equations a two-step procedure like (6.40a) - (6.40b), except that the equivalent of (6.40a) is made explicit by having only ζ_n on the right side. If we consider, for example, the equation (4.26) for u in the symbolic form

$$\frac{du}{dt} = X(u, v, \phi) \tag{6.41}$$

the first step could be to obtain values at $t = (n + \frac{1}{2})\Delta t$ by an uncentered "upstream" derivative for du/dt

$$u_{jkln+\frac{1}{2}} = u'_{j'k'l'n} + \frac{\Delta t}{2} X_{jkln} \tag{6.42}$$

$$j' = j - \frac{u}{\Delta}\frac{\Delta t}{2}$$

$$k' = k - \frac{v}{\Delta}\frac{\Delta t}{2}$$

$$l' = l - \frac{\bar{\omega}}{\Delta p}\frac{\Delta t}{2}$$

in which $u'_{j'k'l'n}$ is obtained by interpolation among the u_{jkln} values. The second step is

$$u_{jkln+1} = u_{jkln} + \Delta t\left(-v\cdot\vec{\nabla}u - \frac{\partial\bar{\omega}u}{\partial p} + X\right)_{jkln+\frac{1}{2}} \tag{6.43}$$

This method damps all wavelengths, short waves more rapidly than long waves, and the gravity waves more rapidly than the geostrophic advective (or Rossby) wave. It can be adapted to a special staggered grid-system devised by Eliassen (1956), in which all variables are present at both full and half time steps, and as such has been tested on a non-linear hemispheric version of (6.33) by Phillips (1962) and is currently used in a 10-level baroclinic model by Bushby (1969).

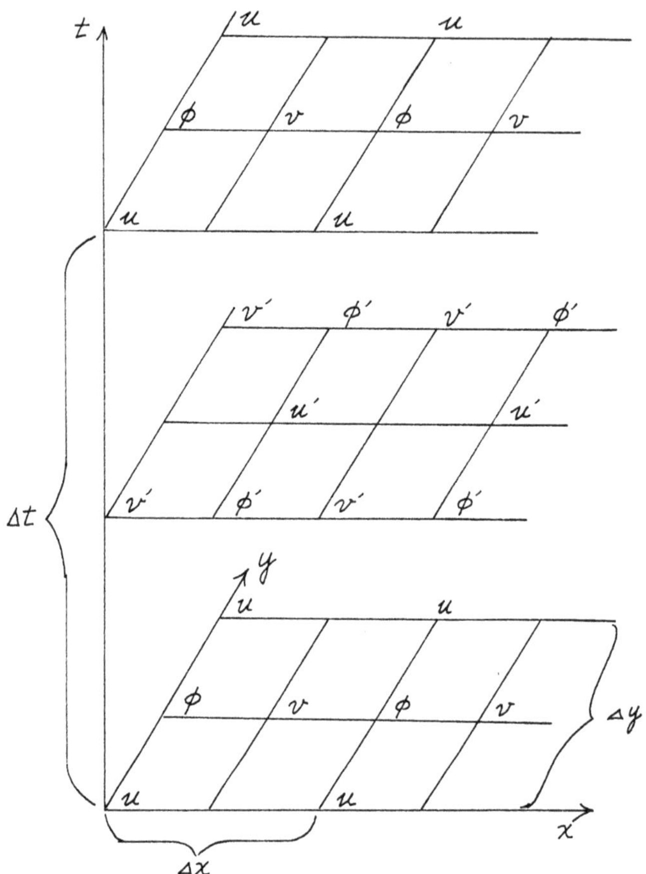

Eliassen staggered grid in x, y, t space

One of the adjacent figures illustrates this Eliassen grid in x, y, t space. The computational damping rate associated with a one-dimensional version of (6.33) in which $V = 0$ and $\partial/\partial y = 0$ is shown in the other figure. The rapid damping of very short waves is not without an advantage, since their phase speed is always in error.

Computational stability and truncation error analysis has so far been discussed only for linear equations. Experience has shown that satisfaction of the linear Lax stability theorem is not by itself sufficient to guarantee stability in integration of non-linear equations for reasonably long times (Phillips, 1959). To be more precise, one non-linear finite-difference formulation may give reasonably constant values of total energy in a closed system which should have constant total energy, while another finite-difference formulation may suddenly develop very large values of

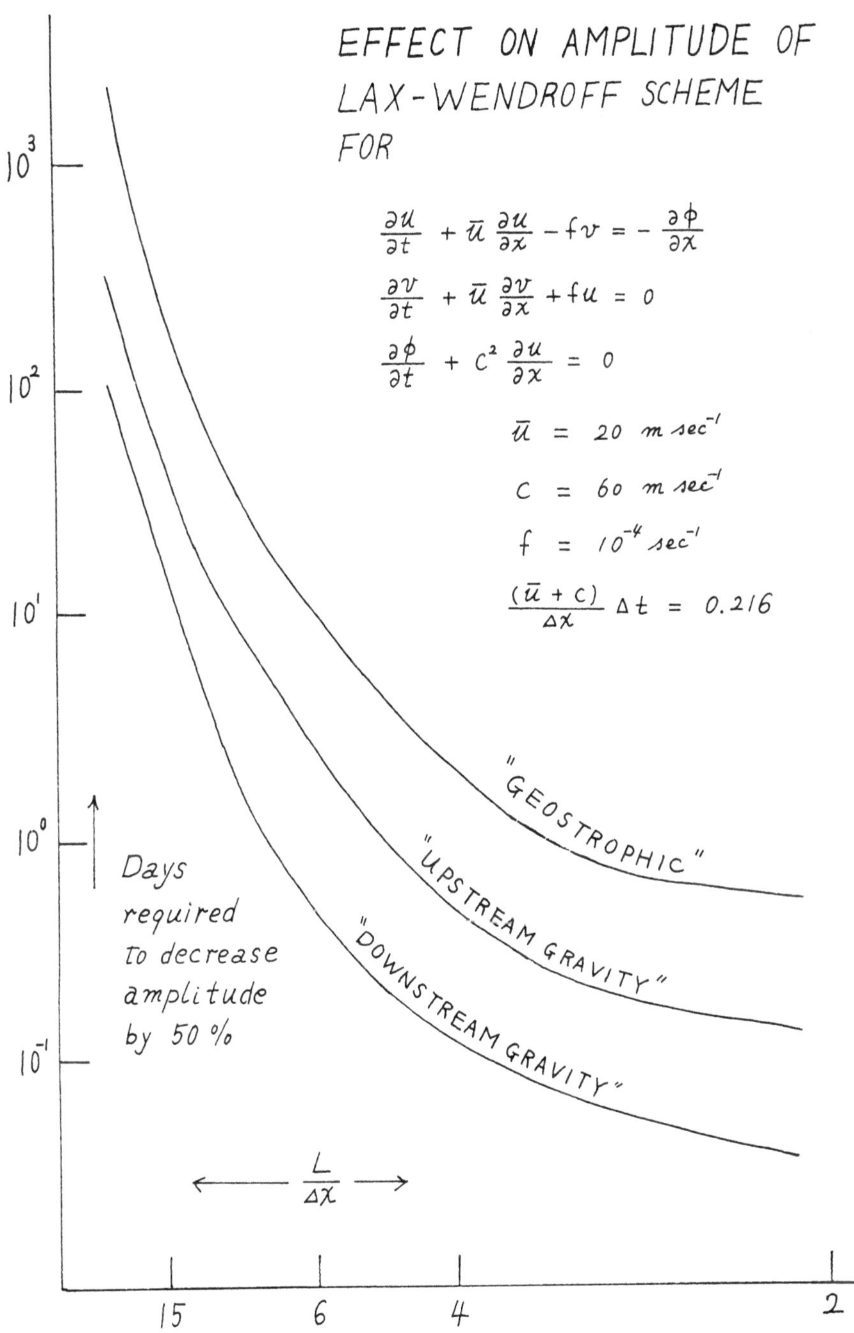
EFFECT ON AMPLITUDE OF
LAX-WENDROFF SCHEME
FOR
$\frac{\partial u}{\partial t} + \bar{u}\frac{\partial u}{\partial x} - fv = -\frac{\partial \phi}{\partial x}$
$\frac{\partial v}{\partial t} + \bar{u}\frac{\partial v}{\partial x} + fu = 0$
$\frac{\partial \phi}{\partial t} + c^2\frac{\partial u}{\partial x} = 0$
$\bar{u} = 20 \text{ m sec}^{-1}$
$c = 60 \text{ m sec}^{-1}$
$f = 10^{-4} \text{ sec}^{-1}$
$\frac{(\bar{u}+c)}{\Delta x}\Delta t = 0.216$
"GEOSTROPHIC"
"UPSTREAM GRAVITY"
"DOWNSTREAM GRAVITY"
10^3
10^2
10^1
10^0
10^{-1}
Days required to decrease amplitude by 50 %
$\frac{L}{\Delta x}$
15
6
4
2

the total energy, even though Δt in both formulations is chosen to satisfy the linear computational stability criterion. The linearly stable system which develops this symptom most rapidly seems to be the "leap-frog" method when the finite-differences are patterned after the advective form of the operator :

$$\frac{\partial s}{\partial t} = -\vec{V}\cdot\nabla s - \bar{\omega}\frac{\partial s}{\partial p} + \frac{ds}{dt} \tag{6.44}$$

If the finite differences are patterned instead after the divergence or flux forms,

$$\frac{\partial s}{\partial t} = -\nabla\cdot\vec{V}s - \frac{\partial \bar{\omega} s}{\partial p} + \frac{ds}{dt} \tag{6.45}$$

the leap-frog system appears to be less susceptible to "non-linear instability" (Arakawa, 1966 ; Lilly, 1965), at least if the separation of even and odd time step solutions is prevented. This stabilization appears to be due to the following circumstance (Bryan, 1966).

Consider a total region $R = \sum_j r_j$ consisting of J adjacent subregions r_j which are separated from one another by interfaces of area A_{jk}, $k = 1, \cdots, K_j$. We revert momentarily to 3-dimensional notation for the equations $d\alpha/dt = 0$, $\nabla\cdot\vec{v} = 0$. The continuity equation $\nabla\cdot\vec{v} = 0$ implies that for each sub-volume,

$$\sum_k c_{jk} A_{jk} = 0 \tag{6.46}$$

where C_{jk} is the outward normal component of $\vec{v}$ on the face A_{jk} of r_j. C_{jk} vanishes on the boundary of R.
We consider the prediction equation

$$\frac{\partial s}{\partial t} = -\vec{v}\cdot\nabla s = -\nabla\cdot\vec{v}s \tag{6.47}$$

in the following finite-difference <u>flux form</u> for the change of s in r_j:

$$r_j \frac{\partial s_j}{\partial t} = -\int s\vec{v}\cdot\hat{n}\,dA = -\sum_k C_{jk}\left(\frac{s_j + s_k}{2}\right)A_{jk} \tag{6.48}$$

$(s_j + s_k)/2$ the apparent value of s on A_{jk}, has been defined as the mean of s_j and s_k in the adjacent volume lying across A_{jk} from r_j. If we sum (6.48) over j, all terms in the last expression occur in cancelling pairs except those on the boundary of R and these vanish individually. Thus we conserve the mean value of s in

$$\frac{\partial}{\partial t}\sum_j r_j s_j = 0 \tag{6.49}$$

If (6.48) is multiplied by s_j before summing we get

$$\frac{1}{2}\frac{\partial}{\partial t}\sum_j r_j s_j^2 = -\frac{1}{2}\sum_j s_j^2 \sum_k c_{jk} A_{jk} - \frac{1}{2}\sum_j \sum_k s_j s_k c_{jk} A_{jk} = 0 \tag{6.50}$$

since the first sum vanishes for each j by (6.46), while the second sum vanishes again by mutual cancellation between adjacent subregions. If s is now thought of as being a horizontal velocity component, we see that this flux formulation of the advection terms will not increase the total kinetic energy. (Arakawa (1966) has extended this type of finite differences for the two-dimensional non-divergent system (6.21) by a special formulation of the $\vec{\nabla}\cdot\nabla$ operation so that the integrals of both $(\nabla\psi)^2$ and $(\nabla^2\psi)^2$ are conserved.) This conclusion is weakened by the leap-frog time procedure because

$$s_{jn}\left(\frac{\partial s}{\partial t}\right)_{jn} = \frac{s_{jn}(s_{jn+1} - s_{jn-1})}{2\Delta t} \neq \frac{s_{jn+1}^2 - s_{jn-1}^2}{2\Delta t} \tag{6.51}$$

Presumably this weakening effect is cancelled sufficiently by the practice referred to earlier, of periodically restarting the leap-frog scheme with an uncentered time step so as to keep the solutions at alternate time steps from separating.

Fischer and Ehlers (1968) tested (6.33a) to compare several finite-difference schemes. The flux form for du/dt and dv/dt was not used, but the variables were "staggered" in space. They reported an unstable increase in energy after one week with the unmodified leap-frog scheme, whereas a Lax-Wendroff scheme was still "stable" after 18.4 weeks.

An analysis by Phillips (1959) suggested that the numerical instability in the advective form of the leap-frog system was associated with "aliasing" ; when two wave numbers p_1 and p_2 interact non-linearly to form

$p_3 = p_1 + p_2$,

$$p_3 = p_1 + p_2 = 2\pi\Delta\left(\frac{1}{L_1} + \frac{1}{L_2}\right) = \frac{2\pi\Delta}{L_3} \tag{6.52}$$

the grid system interprets any value of p_3 greater than π (i.e., $L_3 < 2\Delta$) as the wave number $p_3 - \pi$. The periodic elimination of the wave numbers $p > \pi/2$ $(L < 4\Delta)$ by Fourier analysis restricts $p_1 + p_2$ to values less than π and did eliminate the instability in a test computation. Smagorinsky and collaborators (1965) have used instead a non-linear eddy viscosity to introduce a horizontal diffusion of momentum representing the effect of sub-grid-scale motions. In Cartesian coordinates this would have the form

$$F_x = \frac{\partial}{\partial x}(KD_1) + \frac{\partial}{\partial y}(KD_2); \quad F_y = \frac{\partial}{\partial x}(KD_1) - \frac{\partial}{\partial y}(KD_2) \tag{6.53a}$$

$$D_1 = \frac{\partial u}{\partial x} - \frac{\partial v}{\partial y}, \quad D_2 = \frac{\partial v}{\partial x} + \frac{\partial u}{\partial y} \tag{6.53b}$$

$$K \equiv \frac{k_0^2}{2}\left(D_1^2 + D_2^2\right)^{1/2}\Delta^2 \tag{6.53c}$$

where Δ is the horizontal mesh spacing. A value of the constant k_0 equal to about 0.4 was found to give smooth fields and long-term computational stability in a leap-frog time integrational based on the flux form of $d\vec{V}/dt$ (Expressions said to be suitable for F_x, F_y in spherical coordinates may be found in papers by Saint-Guilly (1956), Bryan (1969) and MacCracken (1969).) Williams (1969), on the other hand, found in a numerical simulation of flow in a heated rotating annulus that the ordinary constant molecular viscosity ν was sufficient for long-term numerical stability. In his example, however Δ was small enough that the "grid point Reynolds number"

$$R_\Delta = \frac{u\Delta}{\nu} \tag{6.54}$$

was about unity, indicating a dominant effect of viscosity on grid-scale motions. The non-linear expression (6.53c) for K would also give a value near unity for $u\Delta/K$ if $D_1 \sim D_2 \sim u/\Delta$, i.e., if the velocity field were not smooth on the scale of Δ , it would be rapidly smoothed.

(Further discussion of the physical significance of two-dimensional eddy viscosity coefficients is given in a paper by Leith (1969).)

For completeness it should be pointed out that frictional terms in the equations can produce numerical instability (in the sense of the linear Lax theorem) if they are evaluated at the mid time step of a leap-frog procedure, i.e.,

$$u_{n+1} = u_{n-1} + 2\Delta t (F_x)_n$$

is absolutely unstable. An uncentered treatment of this term

$$u_{n+1} = u_{n-1} + 2\Delta t (F_x)_{n-1}$$

is stable if Δt is small enough. Similar considerations will apply to eddy diffusion of entropy and to radiation.

Relatively little experience has been obtained yet in numerical integrations over the entire globe. The usual curvilinear coordinate systems, such as spherical coordinates or the several conformal projections, all have singularities where a scale factor goes to zero. Resolution of this by simple omission of the singularity as a grid point is undoubtedly not only numerically bad, but because uniform coordinate intervals Δx, Δy correspond to very small space increments $h_x \Delta x$, $h_y \Delta y$ as the singularity is approached, the limitation on Δt set by the computational stability criterion (6.35) becomes very severe even if the singular point itself is not used. Several schemes which have been proposed and partially tested are as follows.

1. Phillips (1962) proposed the use of a Mercator projection in low latitudes with two polar stereographic projection in high latitudes, each of which partially overlapped the Mercator projection. When combined with the Eliassen staggered grid, the Lax-Wendroff system (6.42) - (6.43), and careful interpolation at the overlapping boundaries, this gave good results on a simple initial flow pattern with the idealized non-linear equations (6.33a). Lateral viscosity or other smoothing devices (other than that inherent in the Lax-Wendroff procedure) were not used.

2. Spherical coordinates with regular grid increments $\Delta\theta$ and $\Delta\lambda$ are used by several groups. When written in flux form, the system (6.33a) is close enough to the multi-level baroclinic equations to indi-

cate the procedures

$$\frac{\partial}{\partial t}(\phi u) = -\nabla\cdot(\phi\vec{V}u) - \frac{\phi}{a\cos\theta}\frac{\partial\phi}{\partial\lambda} + v\phi\sin\theta\left(2\Omega + \frac{u}{a\cos\theta}\right) \tag{6.55a}$$

$$\frac{\partial}{\partial t}(\phi v) = -\nabla\cdot(\phi\vec{V}v) - \frac{\phi}{a}\frac{\partial\phi}{\partial\theta} - u\phi\sin\theta\left(2\Omega + \frac{u}{a\cos\theta}\right) \tag{6.55b}$$

$$\frac{\partial\phi}{\partial t} = -\nabla\cdot\vec{V}\phi = -\frac{1}{a\cos\theta}\left(\frac{\partial u\phi}{\partial\lambda} + \frac{\partial v\phi\cos\theta}{\partial\theta}\right) \tag{6.55c}$$

The NCAR group (Washington and Kasahara, 1970) define u, v and ϕ at each grid point. They overcome the smallness of $\Delta s_\lambda = a\cos\theta\,\Delta\lambda$ by gradual omission of intermediate grid points in λ at high latitudes. The $\cos\theta$ within the θ-derivative in $\nabla\cdot\phi\vec{V}u$ and $\nabla\cdot\phi\vec{V}v$ enables knowledge of u and v at the pole to be dispensed with in computing $\partial\phi u/\partial t$ and $\partial\phi v/\partial t$ at the neighboring points. But ϕ at the pole must be known to evaluate $\partial\phi/\partial\theta$ in (6.55b), and its direct computation there from (6.55c) is not possible. The NCAR group simply replaces ϕ at the pole by an average of the ϕ-values near the pole, thereby evading the computation of any $\partial/\partial t$ at the pole. Washington and Kasahara report that small irregularities arise near the pole from this scheme, but believe this treatment to be satisfactory. Arakawa and Mintz (Langlois and Kwok, 1969) use a slightly different procedure in which u, v are defined at grid points corresponding to the white squares of a checkerboard whereas ϕ is defined on the black squares (this is the arrangement used by Richardson). The pole is a ϕ-point, whose value is given by averaging the subpolar values. However, instead of eliminating grid points in λ near the pole to reduce Δs_λ, at each time step a special longitudinal averaging procedure is applied to λ-derivatives at high latitudes to minimize or remove the computational instability which would otherwise result from violating (6.35).

3. Kurihara (1965) devised a grid with a uniform value of $\Delta\theta$, but on which each latitude circle θ_j has a different number of grid points in the interval $0 \leqslant \lambda \leqslant 2\pi$ beginning with four at $\theta = \pm\left(\frac{\pi}{2} - \Delta\theta\right)$ and increasing by four for each successively equatorward θ_j. (Δs_λ) then varies in size from $(a\Delta\theta)$ to $\frac{\pi}{2}(a\Delta\theta)$. (The finite-difference formulae for ∇ and $\nabla\cdot$ are obviously complicated on this non-regular grid ; reference should be

made to the original paper). Kurihara originally treated the polar point as if it were on a polar stereographic grid with the four adjacent points at $\theta = \frac{\pi}{2} - \Delta\theta$ located on the x- and y- axes, and the proper interpretation made of the different components of $\vec{V}$. A later publication however (Kurihara, 1968) suggests that the polar point be eliminated completely. Evidently the matter is still not clear.

A more basic mapping of the sphere has been designed by Sadourny et al (1968) and by Williamson (1969). This net is obtained by using arcs of great circles ("geodesics") to divide the surface of the sphere into spherical triangles. Williamson made tests with the equations (6.33a) on a simple wave-like initial pattern ; the overall shape of the flow field was maintained correctly (except for an expected phase-lag) but small-scale irregularities developed by 6 days. Williamson ascribes this to a small first-order truncation error associated with a slight but unavoidable variation in the size of the triangle sides and suggests that this will be reduced by taking triangles of size less than 2 1/2 degrees.

The numerical methods discussed so far have used a grid-point representation in space and time. An alternate method is to use a finite series of differentiable and complete orthogonal functions for the dependence on space, retaining finite-differences, however, for the dependence on time. (Orszag (1970b) refers to the origination of this technique by Galerkin (1915) for some non-meteorological problems. It seems to have been introduced into meteorological practice by Silberman (1954) and Lorenz (1960). It is often referred to as the <u>spectral</u> method.) Thompson (1969) gives a simple illustrative example based on the two-dimensional non-divergent vorticity equation

$$\frac{\partial}{\partial t}\nabla^2\psi = -\hat{k} \times \nabla\psi \cdot \nabla(\nabla^2\psi) = -\hat{k}\cdot\nabla\psi \times \nabla(\nabla^2\psi) \tag{6.56}$$

Suppose this is to be solved in a region S on whose boundary B the stream function ψ is zero. In this case it is convenient to use functions $\phi_i(x,y)$ for which

$$\begin{aligned} &\nabla^2\phi_i = -\alpha_i^2\phi_i \\ &\phi_i(B) = 0 \\ &\int \phi_i \phi_j \, dS = \delta_{ij} \begin{cases} = 0 & i \neq j \\ = 1 & i = j \end{cases} \end{aligned} \tag{6.57}$$

We then set

$$\psi(x,y,t) = \sum_{i=1}^{N} A_i(t)\,\phi_i(x,y) \tag{6.58}$$

Substitution of this into the vorticity equation (6.56) gives

$$\begin{aligned}\sum_i \dot{A}_i \alpha_i^2 \phi_i &= -\hat{k} \times \sum_{p=1}^{N} A_p \nabla\phi_p \cdot \sum_{q=1}^{N} A_q \alpha_q^2 \nabla\phi_q \\ &= -\sum_{p=1}^{N}\sum_{q=1}^{N} \alpha_q^2 A_p A_q (\hat{k} \times \nabla\phi_p \cdot \nabla\phi_q)\end{aligned} \tag{6.59}$$

Multiplication of this by ϕ_i and integration over S gives

$$\alpha_i^2 \dot{A}_i = \sum_{p=1}^{N}\sum_{q=1}^{N} \alpha_q^2 A_p A_q I_{pqi} \tag{6.60a}$$

$$I_{pqi} = -\int \phi_i (\hat{k} \times \nabla\phi_p \cdot \nabla\phi_q)\, dS \tag{6.60b}$$

I_{pqi} is called an "interaction coefficient".

The important point here is that although as a function of x, y the entire expression on the right side of (6.59) will in general require more than N of the functions ϕ_i for its expression as a series of the form (6.58), the procedure leading to (6.60) ignores any contribution from the right side of (6.59) which is orthogonal to the finite set of functions already decided upon in (6.58). This has the following consequence, seen most readily by examining the special case $N = 3$ where (6.60b) has the property that $I_{123} = I_{231} = I_{312} = -I_{132} = -I_{321} = -I_{213} = C$ say,

$$\begin{aligned}\alpha_1^2 \dot{A}_1 &= C(\alpha_2^2 - \alpha_3^2) A_2 A_3 \\ \alpha_2^2 \dot{A}_2 &= C(\alpha_3^2 - \alpha_1^2) A_1 A_3 \\ \alpha_3^2 \dot{A}_3 &= C(\alpha_1^2 - \alpha_2^2) A_1 A_2\end{aligned} \tag{6.61}$$

By appropriate multiplication and addition of these we find that

$$\alpha_1^2 A_1 \dot{A}_1 + \alpha_2^2 A_2 \dot{A}_2 + \alpha_3^2 A_3 \dot{A}_3 = 0 \tag{6.62a}$$

$$\alpha_1^4 A_1 \dot{A}_1 + \alpha_2^4 A_2 \dot{A}_2 + \alpha_3^4 A_3 \dot{A}_3 = 0 \tag{6.62b}$$

The original equation (6.56) with the boundary condition $\psi(B) = 0$ has the property that the total kinetic energy is constant,

$$\frac{1}{2} \frac{\partial}{\partial t} \int (\nabla\psi)^2 dS = 0 \tag{6.63}$$

and that the integral of any function of the vorticity , $f(\nabla^2\psi)$, is constant

$$\frac{\partial}{\partial t} \int f(\nabla^2\psi)\, dS = 0 \tag{6.64}$$

The kinetic energy represented by the truncated sum (6.58) is

$$\frac{1}{2} \sum_{p=1}^{N} \sum_{q=1}^{N} A_p A_q \int \nabla\phi_p \cdot \nabla\phi_q dS = -\frac{1}{4} \sum_{p=1}^{N} \sum_{q=1}^{N} A_p A_q \int (\phi_p \nabla^2\phi_q + \phi_q \nabla^2\phi_p) dS$$
$$= \frac{1}{4} \sum_{p=1}^{N} \sum_{q=1}^{N} A_p A_q (\alpha_q^2 + \alpha_p^2) \delta_{pq} = \frac{1}{2} \sum_{i=1}^{N} A_i^2 \alpha_i^2$$

and (6.62a) shows (for $N = 3$) that this is constant, as a parallel to (6.63). The truncated system (6.58) - (6.60) does not have the complete generality of (6.64) for any $f(\nabla^2\psi)$, but it does match (6.64) for the special case $f = (\nabla^2\psi)^2$: the integral of $(\nabla^2\psi)^2$ for the truncated sum (6.58) is

$$-\sum_{p=1}^{N} \sum_{q=1}^{N} A_p A_q \alpha_q^2 \alpha_p^2 \int \phi_p \phi_n dS = -\sum_{i=1}^{N} \alpha_i^4 A_i^2$$

For $N = 3$ (6.62b) shows that this is constant. These properties are also true for $N > 3$ and are highly desirable, as insuring "stability" of the numerical system. They depend upon the orthogonality of ϕ_i and their satisfaction of the boundary conditions.

The system (6.60a) is a system of N non-linear first-order ordinary differential equations in time, for whose numerical integration many quite accurate iterative methods have been developed (e.g., Runge-Kutta). However, the number of terms on the right side of each of the N equations (6.60a) can be large when ϕ is not a simple trigonometric function (i.e., the product of two ϕ's may require many ϕ's for its expression as a sum of ϕ's so that a single product $A_p A_q$ will appear in many of the N equations in (6.60a). (The papers by Silberman (1954), Kubota (1960), Baer and Platzman (1961), Ellsaesser (1966), and Orszag (1970b) discuss the computation of I when ϕ is a spherical harmonic. Ellsaesser, for example, finds 146, 721 non-zero values of I for N = 629 and cross-equator symmetry). Simpler time extrapolation formulae have therefore been used (Lilly, 1965).

The simple system (6.61) may be linearized meaningfully about $A_1 = \bar{A}_1$ (constant), $A_2 = A_3 = 0$, if, as is implicit in using the linearized Lax theorem, we assume that ϕ_1 is the largest scale ($\alpha_1^2 < \alpha_2^2$ or α_3^2) and varies relatively slowly with time.

$$\dot{A}_2 = c\bar{A}_1\left(\frac{\alpha_3^2 - \alpha_1^2}{\alpha_2^2}\right)A_3 = aA_3$$
$$\dot{A}_3 = -c\bar{A}_1\left(\frac{\alpha_2^2 - \alpha_1^2}{\alpha_3^2}\right)A_2 = -bA_2 \tag{6.65}$$

This has the time dependences

$$e^{\pm i\omega t}, \quad \omega = (ab)^{1/2} = \text{real} \tag{6.66}$$

The leap-frog time integration applied to (6.65) gives solutions of the form $\exp(\pm in\theta)$, $\exp[\pm in(\theta + \pi)]$ with $t = n\Delta t$, $\cos\theta = 1 - 2ab(\Delta t)^2$. θ is real if

$$\Delta t < \frac{1}{\omega} \tag{6.67}$$

i.e., the time interval must be less than about 1/6 of the true period in order to at least reproduce the non-exponential character of the true solution. (Obvious possibilities exist for using the new values of $A_{i\,n+1}$ in the products on the right side of (6.60a) as soon as they have been computed.) The finite-difference time integration of (6.60a) is therefore still subject to a limitation on Δt. In practice this limitation has been determined best by experimentation (Baer, 1964).

The principal use of orthogonal functions in large-scale meteorological problems has been with the quasi-geostrophic equations (5.90) - (5.94) or minor variants from it. Computations with smallish values of N have been used with "two-layer" formulations, most notably by Lorenz to model the

"Rossby regime" in laboratory experiments with a heated rotating annulus of water (1962 ; see also Merilees, 1968b) and to illustrate the determinism and predictability of atmospheric flow (1963). Ellsaesser (1966) considered comparison numerical solutions of the system

$$\frac{\partial}{\partial t}(\nabla^2\psi - K^2\psi) = -\hat{k}\times\nabla\psi\cdot\nabla(2\Omega\sin\theta + \nabla^2\psi) \tag{6.68}$$

(a non-linear version of (5.85)) on the sphere, with assumed symmetry between the two hemispheres. He compared solutions obtained by the spectral method with the results obtained with the operational finite-difference grid program then in use by the U.S. Weather Bureau. His results show that the spectral method using somewhat less than 300 spherical harmonics (corresponding to both zonal and latitudinal wave numbers from 0 - 12) gave essentially the same 36-hour forecast as the grid-point computation (with about 2000 grid points), but required significantly less computing time than the grid-point method. (The latter requires in one time step not only the simple computation of the right side of (6.68) by spatial finite-differences at each grid point, but the more time-consuming inversion of the operator $\nabla^2 - K^2$.)

It would indeed be surprising if spectral methods with N degrees of freedom were not more accurate than any of the second-order grid system methods with N grid points. The critical point involves rather the balance between this and the greater computation time associated with each added degree of freedom in the spectral method as N becomes large. Orszag (1970b), after making comparative tests on a solution of the Navier-Stokes equations for a three-dimensional viscous vortex concludes, for example, that comparable accuracies and computation times in that problem are obtained for an N = 16 x 16 x 16 Fourier representation and a 32 x 32 x 32 second-order staggered mesh grid point calculation. (This comparison is based on using the so-called "fast Fourier transform" method (Cooley and Tukey, 1965) which is restricted to $N = 2^n$, and in spherical geometry is available to favor the spectral method only for the longitudinal coordinate.)

Spectral methods have a special attraction in spherical geometry because unlike grid-systems, they are not complicated by the coordinate singularities. Reduction of the computation time is therefore important. Following Orszag (1970a), suppose that spherical harmonics $Y_n^m = P_n^m(\sin\theta)e^{im\lambda}$,

$-n \leqslant m \leqslant n \quad , 0 \leqslant n \leqslant N$ are used, so that

$$\begin{aligned} \nabla^2\psi &= \sum_{n=0}^{N} \sum_{m=-n}^{n} a_{nm} Y_n^m \\ \psi &= \sum_{n=1}^{N} \sum_{m=-n}^{n} b_{nm} Y_n^m \quad , \quad b_{nm} = -\frac{a_{nm}}{n(n+1)} \end{aligned} \tag{6.69}$$

(The radius of the earth "a" cancels out in (6.69) and (6.70). Note also that the total number of functions is now N^2 (or nearly so) in place of the N used earlier). The spherical non-divergent vorticity equation

$$\frac{\partial}{\partial t} \nabla^2\psi = -\left[2\Omega \frac{\partial\psi}{\partial\lambda} + \frac{\partial\nabla^2\psi}{\partial\lambda} \frac{\partial\psi}{\partial(\sin\theta)} - \frac{\partial\nabla^2\psi}{\partial(\sin\theta)} \frac{\partial\psi}{\partial\lambda} \right] \tag{6.70}$$

in spectral form analogous to (6.60) is

$$\dot{a}_{nm} = \frac{-2i\Omega m\, a_{nm}}{n(n+1)} + J_{nm} \tag{6.71}$$

where the non-linear term is

$$J_{nm} = \sum_{p=0}^{N} \sum_{q=0}^{N} \sum_{|r| \leqslant N} I(p, q, r, m-r \mid n, m)\, a_{pr}\, b_{q, m-r} \tag{6.72}$$

Since the number of a_{nm} is $O(N^2)$ and each J_{nm} contains $O(N^3)$ terms there are $O(N^5)$ computations involved at each time step, and there is a required storage of $O(N^5)$ interaction coefficients. (Many I's are zero, but most of these are already recognized as such in (6.72).) Orszag shows how the J_{nm} can be computed more efficiently by an ingenious method based on considering the exact spectral representation of the non-linear terms of (6.70) over the range $0 < n \leqslant 2N-1$ and the values it assumes at a series of regular grid points in λ, θ , so that a total of only $10\,O(N^3) + 40\, N^2 \log_2 N$ arithmetical operations are required instead of $O(N^5)$. (The $\log_2 N$ comes from a "fast Fourier" transform.) Storage is also reduced to $O(N^3)$. Orszag estimates that this revised spectral method is faster than the standard spectral method for equation (6.71) when $N > 10$.

Little application of spectral methods has as yet been made to solution of the non-geostrophic equations (4.20) - (4.27) (and none at all to the sigma system (4.29) - (4.39).) Kubota (1960) and Merilees (1968a) base their spectral method on a representation of the horizontal velocity by a streamfunction Ψ and a velocity potential χ :

$$\begin{aligned} \vec{V} &= \vec{V}_{\Psi} + \vec{V}_{\chi} = \hat{k}\times\nabla\Psi + \nabla\chi \\ \zeta &= \hat{k}\cdot \mathrm{curl}\,\vec{V} = -\nabla\cdot(\hat{k}\times\vec{V}) = \nabla^2\Psi \\ \nabla\cdot\vec{V} &= \nabla^2\chi = -\frac{\partial\varpi}{\partial p} \end{aligned} \tag{6.73}$$

Equations for the rate of change of Ψ and χ with time are obtained by applying the curl and divergence operators to (4.11) : the curl operation gives

$$\nabla^2\frac{\partial\Psi}{\partial t} = \nabla\cdot\hat{k}\times\left[\nabla(\phi+\tfrac{1}{2}\vec{V}^2)+(f+\zeta)\hat{k}\times\vec{V}+\varpi\frac{\partial\vec{V}}{\partial p}-\vec{F}\right] \tag{6.74a}$$

In expanded form this is

$$\begin{aligned}\nabla^2\frac{\partial\Psi}{\partial t} = &-\hat{k}\times\nabla\Psi\cdot\nabla(\zeta+f) - \nabla\cdot(\zeta+f)\nabla\chi - \nabla\cdot\left(\varpi\nabla\frac{\partial\Psi}{\partial p}\right) \\ &+ \nabla\cdot\left(\varpi\hat{k}\times\nabla\frac{\partial\chi}{\partial p}\right) - \nabla\cdot(\hat{k}\times\vec{F})\end{aligned} \tag{6.74b}$$

The divergence operation gives

$$\nabla^2\frac{\partial\chi}{\partial t} = -\nabla\cdot\left[\nabla(\phi+\tfrac{1}{2}\vec{V}^2)+(f+\zeta)\hat{k}\times\vec{V}+\varpi\frac{\partial\vec{V}}{\partial p}-\vec{F}\right] \tag{6.75a}$$

In expanded form this is

$$\begin{aligned}\nabla^2\frac{\partial\chi}{\partial t} = &-\nabla^2\left[\phi+\tfrac{1}{2}(\nabla\Psi)^2+\tfrac{1}{2}(\nabla\chi)^2+\nabla\Psi\cdot\nabla\chi\right] + \nabla\cdot(\zeta+f)\nabla\Psi \\ &-\nabla\cdot\left(\varpi\nabla\frac{\partial\chi}{\partial p}\right) - \hat{k}\times\nabla\chi\cdot\nabla(\zeta+f) + \hat{k}\cdot\nabla\varpi\times\nabla\frac{\partial\Psi}{\partial p} + \nabla\cdot\vec{F}\end{aligned} \tag{6.75b}$$

(If all terms depending on $\vec{F}$, χ , and ϖ are dropped in the expanded forms (6.74b) and (6.75b), the non-divergent vorticity equation (6.70) and the "balance equation" (5.102) result. These approximations are not being made now, however.) Together with the thermodynamic equation

$$\frac{\partial \Theta}{\partial t} = -\nabla \cdot \vec{V} \Theta - \frac{\partial}{\partial p}(\varpi \Theta) + \left(\frac{p_{00}}{p}\right)^{R/c_p} \frac{q}{c_p} \tag{6.76a}$$

the hydrostatic relation

$$\frac{\partial \phi}{\partial p} = -\frac{R}{p}\left(\frac{p}{p_{00}}\right)^{R/c_p} \Theta \tag{6.76b}$$

and the continuity equation

$$\varpi = -\int_0^p \nabla^2 \chi \, dp \tag{6.76c}$$

these form a complete dynamical system. Finite sums of spherical harmonics would be used to describe the four scalar variables Ψ, χ, Θ, ϕ with coefficients depending on p and t. Kubota and Merilees work with the expanded forms (6.74b) and (6.75b), and the time rate of change of the coefficients is computed by a generalization of the procedure discussed earlier for the non-divergent vorticity equation (6.70). Merilees (1968) has described the three basic types of interaction coefficients involved - one of which is analogous to the I of (6.72) - and verified that a fixed truncation in Y_n^m does conserve total energy, kinetic plus potential plus internal (in absence of q and $\vec{F}$). (His proof is based on using $\varpi = 0$ at a fixed constant pressure p_0 in place of the correct lower boundary conditions (4.27), and the related suppression of the gz_0 term in the energy integral (4.19). The extension to the correct lower boundary condition may not be trivial.)

Robert (1966) suggests instead that the functions

$$G_n^m = e^{im\lambda} (\cos\theta)^{|m|} (\sin\theta)^n \tag{6.77}$$

(where m is a positive or negative integer and n a non-negative integer) be used to represent the horizontal dependence of the four scalars Ψ, χ, Θ and ϕ. Although these functions are not orthogonal, they can be combined to form spherical harmonics, so that a scalar P (representing one of Ψ, χ, Θ or ϕ) may be represented as two equivalent finite series :

$$P = \sum_m \sum_n A_{nm} G_n^m = \sum_n \sum_m A'_{nm} Y_n^m \tag{6.78}$$

The two arrays of coefficients A_{nm} and A'_{nm} are linear functions of one another. The multiplication of two finite sums of G's gives another sum of G's, whose coefficients are simpler to compute than when spherical harmonic series are multiplied together. Robert also shows that the operations $\cos\theta \nabla P$ and $\nabla^2 P$ generate another sum of G's, so that $u \cos\theta$ and $v \cos\theta$ can also be represented in the form (6.78). These features make it possible to avoid the complicated expanded forms (6.74b) and (6.75b). Robert accomplishes this by first computing from the Ψ, χ, ϕ series the array of coefficients which represent the vector in curly brackets of (6.74a) and (6.75a). The equivalent of the $k \cdot \text{curl}$ and $\nabla\cdot$ operations can then be performed on these coefficient arrays, resulting in expressions for $\nabla^2 \partial\Psi/\partial t$ and $\nabla^2 \partial\chi/\partial t$, each as an untruncated array of coefficients of G's The truncation operation necessary to preserve the energy integral is carried out at this point by using the relations between A_{nm} and A'_{nm} implied by (6.78), once a fixed truncation in Y_n^m has been fixed. The advantage of this method over that described by Kubota and Merilees lies in being able to avoid the explicit expression of the right sides of (6.74b) and (6.75b) as complicated operations on Ψ and χ. The price paid is the final combined orthogonalization-truncation operation (6.78) needed in each time step. A successful use of this scheme in a 5-level global general circulation experiment has been described by Robert (1966), but a comparison with other methods with respect to computational time is not yet available.

REFERENCES

BOOKS AND CONFERENCE REPORTS

Bolin, B. and R. Garcia (eds.), 1967 : The global atmospheric research programme (GARP) ; report of the Stockholm study conference. ICSU/IUGG - WMO. World Meteorological Organization, Geneva.

Bykov, V. (ed.), 1969 : Lectures on numerical short-range weather prediction (WMO training seminar). Gidromet.Izdat, Leningrad (also in Russian).

Japan Meteorological Agency, 1969 : Proceedings WMO/IUGG symposium on numerical weather prediction in Tokyo November 26 - December 4, 1968. Japan Meteorological Agency, Tokyo.

Kibel, I., 1957 : Vvedenie v gidrodinamicheskie metody kratkosrochnovo prognoza pogody. Gosud. Izd. Tek.-Teor. Lit. (Transl. R. Baker 1963, MacMillan, New York.

Marchuk, G., 1969 : Chislennie metodi v prognoze pogodi. Gidromet. Izdat., Leningrad.

Meteorological Society of Japan, 1962 : Proceedings international symposium on numerical weather prediction. Meteorological Society of Japan, Tokyo.

Monin, A., 1969 : Prognoz pogodi kak zadacha fiziki. Izd. Nauka, Moscow.

Richardson, L., 1922 : Weather prediction by numerical process. Cambridge University Press, London. (Dover reprint, New York, 1965).

Thompson, P., 1961 : Numerical weather analysis and prediction. MacMillan New York.

REFERENCES IN TEXT

Arakawa, A., 1966 : Computational design for long-term numerical integration of the equations of fluid motion. Two-dimensional incompressible flow. Part I. J. Computational Physics, 1, 119-143.

Baer, F., 1964 : Integration with the spectral vorticity equation. J. Atmos. Sci., 21, 260-276.

Baer, F., and G. Platzman, 1961 : A procedure for numerical integration of the spectral vorticity equation. J. Meteor., 18, 393-401.

Benwell, G., and F. Bretherton, 1968 : A pressure oscillation in a 10-level atmospheric model. Quart. J. Royal Meteorol. Soc., 94, 123-131.

Bolin, B., 1955 : Numerical Forecasting with the barotropic model. Tellus, 7, 27-49.

Bryan, K., 1966 : A scheme for numerical integration of the equations of motion on an irregular grid free of nonlinear instability. Mon. Wea. Rev., 94, 39-40.

Bryan, K., 1969 : A numerical method for the study of the circulation of the world ocean. J. Computational Physics, 4, 347-376.

Burger, A., 1948 : Scale considerations of planetary motions of the atmosphere. Tellus, 10, 195-205.

Bushby, F., 1969 : Further developments of a model for forecasting rain and weather. WMO/IUGG Symp. on Numerical Weather Prediction, Tokyo, 1175-1184.

Charney, J., 1947 : The dynamics of long waves in a baroclinic westerly current. J. Meteor., 4, 135-163.

Charney, J., 1947 : On the scale of atmospheric motions. Geofys. Publ., Norske Vids.-Akad. Oslo, 17, n° 2.

Charney, J., R. Fjørtoft and J. von Neumann, 1950 : Numerical integration of the barotropic vorticity equation. Tellus, 2, 237-254.

Charney, J., and P. Drazin, 1961 : Propagation of planetary-scale disturbances from the lower into the upper atmosphere. J. Geophys. Res., 66, 83-109.

Charney, J., 1962 : Integration of the primitive and balance equations. Proc. Inter. Symp. Numerical Weather Prediction, pp. 131-152, Meteorological Society of Japan, Tokyo.

Charney, J., and J. Pedlosky, 1963 : On the trapping of unstable planetary waves in the atmosphere. J. Geophys. Res., 68, 6441-6442.

Cooley, J., and J. Tukey, 1965 : An algorithm for the machine calculation of complex Fourier series. Math. of Comp., 19, 297-301.

Courant, R., K. Friedericks and H. Lewy, 1928 : Über die partiellen differenzen gleichungen der mathematischen physik. Math. Annalen, 100, 32.

Crowley,W., 1968 : Numerical advection experiments. Mon. Wea. Rev., 96, 1-11.

Dickinson, R., 1968 : Planetary Rossby waves propagating vertically through weak westerly wind wave guides. J. Atmos. Sci., 25, 984-1002.

Dikii, L;, 1969 : Teoria kolebanii zemnoi atmosferi. Gidromet. Izdat., Leningrad, 196 pp.

Eady, E., 1949 : Long waves and cyclone waves. Tellus, 1(3), 33-52.

Eckart, C., 1960 : Hydrodynamics of oceans and atmospheres. Pergamon, New York, 290 pp.

Edelman, W., 1963 : On the behavior of disturbances in a baroclinic channel. Res. Report n° 2, Deutscher Wetterdienst, Offenbach.

Eliassen, A., 1949 : The quasi-static equations of motion with pressure as independent variable. Geofys. Publ., Norske. Vids.-Akad. Oslo, 17, n° 3.

Eliassen, A., 1945 : A procedure for numerical integration of the primitive equations of the two-parameter model of the atmosphere. Sci. Report, Dept. Meteorology, Univ. California at Los Angeles.

Ellsaesser, H., 1966 : Evaluation of spectral versus grid methods of hemispheric numerical weather prediction. J. Applied Meteor. 5, 246-262.

Fischer, G., and R. Ehlers, 1968 : Vergleich dreier Differenzen-Verfahren zur numerischen Integration der "primitiven Gleichungen" für ein barotropes divergentes Modell der Atmosphäre. Tellus, 20, 318-329.

Galerkin, B., 1915 : Rods and plates. Series occuring in various questions concerning the elastic equilibrium of rods and plates. Vestnik Inzhenerov, 19, 897-908.

Hines, C., 1963 : The upper atmosphere in motion. Quart. J. Royal Meteor. Soc., 89, 1-42.

Hinkelmann, K., 1941 : Der mechanismus der meteorologischen Larmes. Tellus, 3, 285-296.

Holton, J., and R. Lindzen, 1968 : A note on Kelvin waves in the atmosphere. Mon. Wea. Rev., 96, 385-386.

Houghton, D., 1968 : Derivation of the elliptic condition for the balance equation in spherical coordinates. J. Atmos. Sci., 25, 927-928.

Krishnamurti, T., 1969 : An experiment in numerical prediction in equatorial latitudes. Quart. J. Royal Meteo. Soc., 95, 594-620.

Kubota, S., 1960 : Surface spherical harmonic representation of systems of equations for analysis. Papers Meteor. Geophys. (Tokyo), 10, 145-166.

Kurihara, Y., 1965 : Numerical integration of the primitive equations on a spherical grid. Mon. Wea. Rev., 93, 399-415.

Kurihara, Y., 1968 : Note on finite difference expressions for the hydrostatic relation and pressure gradient force. Mon. Wea. Rev., 96, 654-656.

Lamb, H., 1945 : Hydrodynamics, 6th edition. Dover Publications, New York 738 pp.

Langlois, W. and H. Kwok, 1969 : Description of the Mintz-Arakawa numerical general circulation model. Tech. Rept. Dept. Meteorology, Univ. California at Los Angeles, 95 pp.

Lax, P., and B. Wendroff, 1960 : Systems of conservation laws. Comm. Pure and Applied Math. , 13, 217-237.

Leith, C., 1965 : Numerical simulation of the earth's atmosphere. In Methods in Computational Physics, vol. 4. Academic Press, New York.

Leith, C., 1969 : Two-dimensional eddy viscosity coefficients. WMO/IUGG Symp. on Numerical Weather Pre., Tokyo, 141-145.

Lilly, D., 1965 : On the computational stability of numerical solutions of time-dependent non-linear geophysical fluid dynamics problems. Mon. Wea. Rev., 93, 11-26.

Longuet-Higgins, M., 1968 : The eigenfunctions of Laplace's tidal equation over a sphere. Phil. Trans. Royal Soc., A 262, n° 1132.

Lorenz, E., 1955 : Available potential energy and the maintenance of the general circulation, Tellus, 7, 157-167.

Lorenz, E., 1960a : Maximum simplification of the dynamic equations. Tellus, 12, 243-254.

Lorenz, E., 1960b : Energy and numerical weather prediction. Tellus, 12, 364-373.

Lorenz, E., 1962 : Simplified dynamic equations applied to the rotating-basin experiments. J. Atmos. Sci., 19, 39-51.

Lorenz, E., 1963 : Deterministic nonperiodic flow. J. Atmos. Sci., 20, 130-141.

MacCracken, M., 1969 : The atmospheric conservation equations in flux divergence form. Report TID-4500, UC-53, Lawrence Radiation Lab., U. Calif. Livermore, 18 pp.

Mak, M.-K., 1969 : Laterally driven stochastic motions in the tropics. J. Atmos. Sci., 26, 41-64.

Marchuk, G., 1964 : Teoreticheskaya model prognoza pogodi. Akad. Nauk SSSR, Dokl., 155, 1062-1065.

Matsuno, T., 1966 : Quasi-geostrophic motions in the equatorial area. J. Meteorol. Soc. Japan, 44, 25-43.

Merilees, P., 1968a : The equations of motion in spectral form. J. Atmos. Sci., 25, 736-743.

Merilees, P., 1968b : On the transition from axisymmetric to non-axisymmetric flow in a rotating annulus. J. Atmos. Sci., 25, 1003-1014.

Miyakoda, K., et al., 1970 : Numerical simulation of the breakdown of a polar-night vortex in the stratosphere. J.Atmos. Sci., 27, 139-154.

Monin, A., 1958 : Izmenenia davlenia v baroklinnoi atmosfere. Izv. Akad. Nauk SSSR, Ser. Geofiz., 4, 497-514.

Obukhov, A., 1949 : K voprosu o geostroficheskom vetra. Izv. Akad. Nauk SSSR; Ser. Geograf. Geofiz., 13, n° 4.

Orszag, S., 1970a : Transform method for the calculation of vector-coupled sums : application to the spectral form of the vorticity equation. In press (also NCAR Manuscript 70-13).

Orszag, S., 1970b : Accuracy of numerical simulation of imcompressible flows. In press (also NCAR Manuscript 70-73).

Phillips, N., 1959 : An example of non-linear computational instability The Atmosphere and the Sea in Motion, B. Bolin, ed. Rockefeller Press, New York, 509 pp.

Phillips, N., 1962 : Numerical integration of the hydrostatic system of equations with a modified version of the Eliassen finite-difference grid. Proc. Inter. Symp. Numerical Weather Pred., Meteor. Soc. Japan, Tokyo.

Phillips, N., 1963 : Geostrophic motion. Rev. Geophysics, 1, 123-176.

Phillips, N., 1968 and Veronis G., 1968 : Comments (and Reply) on Phillips' proposed simplification of the equations of motion for a shallow rotating atmosphere. J. Atmos. Sci., 25, 1154-1157.

Platzman, G., 1958 : The lattice structure of the finite-difference and primitive equations. Mon. Wea. Rev., 86, 285-292.

Richtmyer, R., 1962 : A survey of difference methods for non-steady fluid dynamics. NCAR Tech. Note 63-2, Boulder.

Robert, A., 1966 : The integration of a low order spectral form of the primitive meteorological equation. J. Meteor. Soc. Japan (II) 44; 237-245.

Richtmyer, R., and K. Morton, 1967 : Difference methods for initial-value problems. Interscience, New York.

Rossby, C., and collab., 1939 : Relation between variations in the intensity of the zonal circulation of the atmosphere and the displacements of the semi-permanent centers of action. J. Marine Res., 2, 38-55.

Sadourny, R., A. Arakawa and Y. Mintz, 1968 : Integration of the nondivergent barotropic vorticity equation with an icosahedral-hexagonal grid for the sphere. Mon. Wea. Rev., 96, 351-356.

Saint-Guilly, B., 1956 : Sur la théorie des courants marins induits par le vent. Ann. Inst. Océanographique, 33, 1-64.

Silberman, I., 1954 : Planetary waves in the atmosphere. J. Meteor., 11, 27-34.

Smagorinsky, J., S. Manabe and J. Holloway, 1965 : Numerical results from a nine-level general circulation model of the atmosphere. Mon. Wea. Rev., 93, 727-768.

Stone, P., 1966 : Frontogenesis by horizontal wind deformation fields. J. Atmos. Sci., 23, 455-465.

Taylor, G., 1936 : The oscillations of the atmosphere. Proc. Roy. Soc. (A) 318-326.

Thompson, P., 1969 : Non-Eulerian methods of integration. In Lectures on numerical short-range weather prediction. (WMO). Gidromet. Izdat. Leningrad. 706 pp.

Wallace, J., and V. Kousky, 1968 : Observational evidence of Kelvin waves in the tropical stratosphere. J. Atmos. Sci., 25, 900-907.

Washington, W., and A. Kasahara, 1970 : A January simulation experiment with the two-layer version of the NCAR global circulation model. (In press NCAR MS 69-149).

Williams, G., 1969 : Numerical integration of the three-dimensional Navier-Stokes equations for incompressible flow. J. Fluid Mech., 37, 727-750.

Williamson, D., 1969 : A comparison of spherical grids for numerical integration of atmospheric models. Ph. D. thesis, Dept. of Meteorology, Mass. Inst. Technology, Cambridge.

Yanai, M., and Y. Hayashi, 1969 : Large-scale equatorial waves penetrating from the upper troposphere into the lower stratosphere. J. Meteor. Soc. Japan, 47, 167-182.

PLANETARY FLUID DYNAMICS

J.G. CHARNEY

Department of Meteorology
Massachusetts Institute of Technology
CAMBRIDGE, Massachusetts (U.S.A.)

CONTENTS

PLANETARY FLUID DYNAMICS

by

J.G. CHARNEY

Department of Meteorology
Massachusetts Institute of Technology
CAMBRIDGE - U.S.A.

I. INTRODUCTION

Definition of planetary motions

The term "planetary" will be applied to fluid motions on the earth whose space and time scales are so large that the earth's rotation may not be ignored. Such motions have properties which are not to be found in non-rotating systems. For example, the action of external forces such as gravity invariably bring into being Coriolis forces which in turn produce circulatory motions. If a stone is thrown into an infinite resting ocean, the gravitational oscillations engendered will radiate their energy to infinity and leave the ocean finally undisturbed ; if the stone is thrown into an infinite rotating ocean, some of the energy of the gravitational oscillations will be converted by the action of the Coriolis forces into rotational motions, and these will persist until they are dissipated by viscosity.

In a general way, one may define planetary motions as those whose characteristic phase and orbital periods are large compared to the half-pendulum day*. All large-scale atmospheric circulations down to the smallest extratropical and tropical cyclones fall into this category, as do all large-scale oceanic circulations down to the smallest meanders of the western boundary currents. Typical examples of non-planetary motions in the atmosphere and oceans are small-scale gravity waves, boundary-layer

* Half the period of a Foucault pendulum.

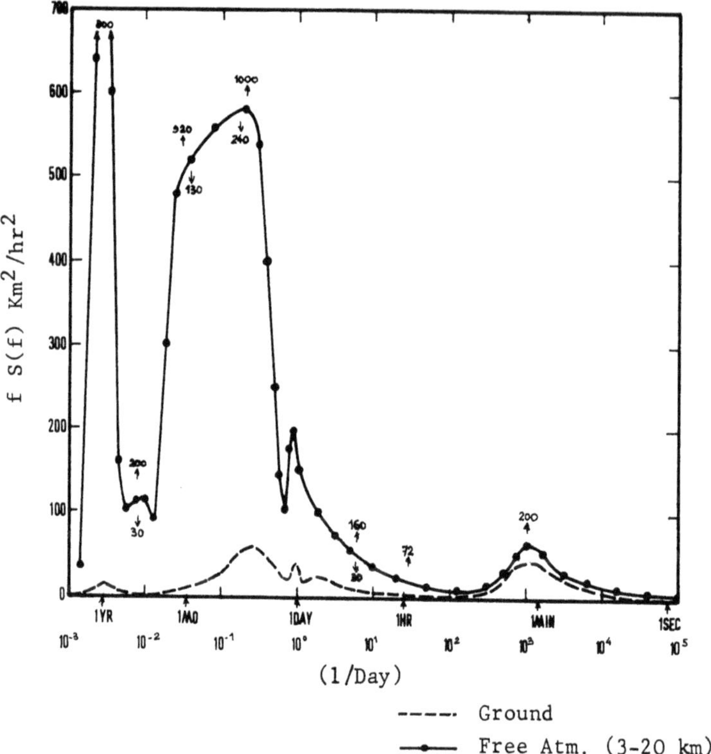

Figure 1.1

Average kinetic energy of E-W wind component in the free atmosphere (solid line) and near the ground (dashed line). Numbers indicate maximum values of kinetic energy at particular periods (after Vinnichenko, 1970 Tellus, 22, 158-166).

turbulence, small-scale convection, and flow around small obstacles.

General description of planetary motions in the atmosphere

Solar energy produces motions of planetary scale, and, although some of this energy cascades by various means into motions of smaller scale, the greater part remains at planetary scales. Indeed, it may be seen from Figure 1.1, which shows the kinetic energy spectrum of the east-west component of wind velocity at periods between 1 sec and 5 years, that most of the energy is at periods longer than one day.

Typical atmospheric circulations are represented in Figures 1.2 - 1.5, which show the isobaric pattern extrapolated to sea level and the height contours of the 500 mb, 100 mb, and 10 mb isobaric surfaces, all for the times 15 July 1958 and 15 January 1959. Because of the approximate balance between the horizontal components of the pressure and Coriolis forces, the isobaric and height contours may be regarded approximately as streamlines.* We note that the sea-level circulation is weaker in summer than in winter and that it consists of relatively strong cyclonic (counter-clockwise) vortices associated with low pressure centers and weaker anticyclonic (clockwise) vortices associated with high pressure centers. The main vortices can also be found at 500 mb (ca 5.5 km), displaced somewhat westward and imbedded in a mean circumpolar westerly (west to east) flow. At 100 mb (ca 16 km.) these vortices appear as wave perturbations of a large circumpolar cyclonic vortex, which, in summer, is surrounded by one or more anticyclonic vortices. Between 100 mb and 10 mb (ca 30 km) the circulation changes radically : in summer the 10 mb circulation is seen to consist of a circumpolar, axially-symmetric, anticyclonic vortex with relatively small superimposed wave perturbations ; in winter the circulation is asymmetric but far simpler than at 100 mb ; in the example shown it has but two cyclonic and two anticyclonic lobes.

Figures 1.6 and 1.7 show the mean circulations at sea level and 500 mb for July and January. The July sea-level circulation is dominated by strong anticyclones (highs) over the Pacific and Atlantic Oceans and cyclones (lows) over the Eurasian and North American continents. The winter

* This is the so-called "geostrophic approximation" which holds for slow, large-scale flows in the atmosphere and oceans.

International Geophysical Year world weather maps
Part I - Northern Hemisphere - 15 July, 1958.
1200 GMT. Sea Level

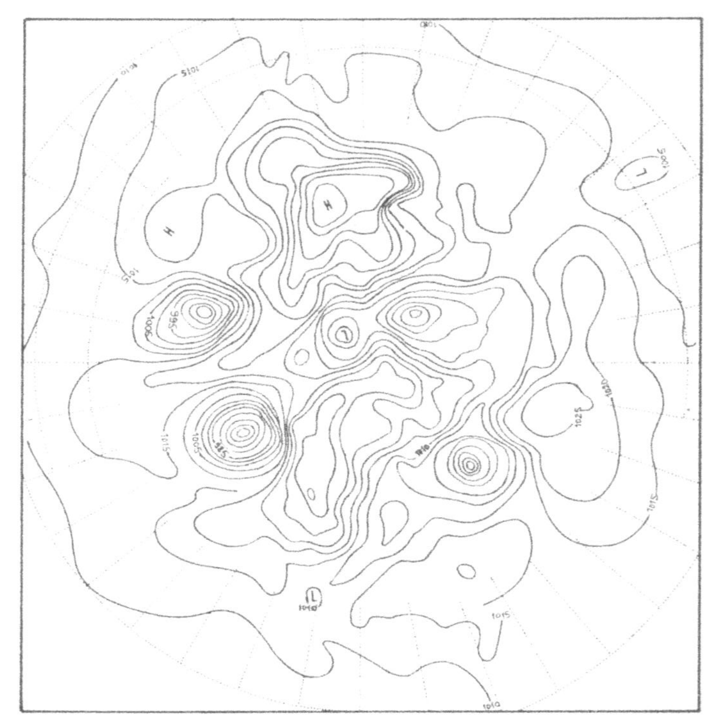

Synoptic weather map. Northern Hemisphere.
January 15, 1959
1230 GMT. Sea Level

Figure 1.2

International Geophysical Year World Weather Maps
Part I - Northern Hemisphere - JULY 15, 1958
1200 GMT 500 mb

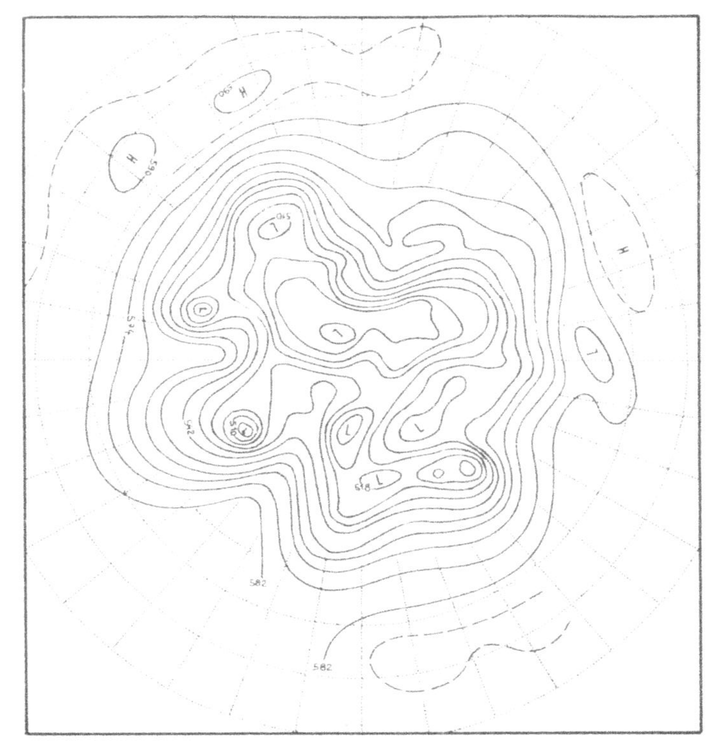

Synoptic Weather Map. Northern Hemisphere
JANUARY 15, 1959
1200 GMT 500 mb

Figure 1.3

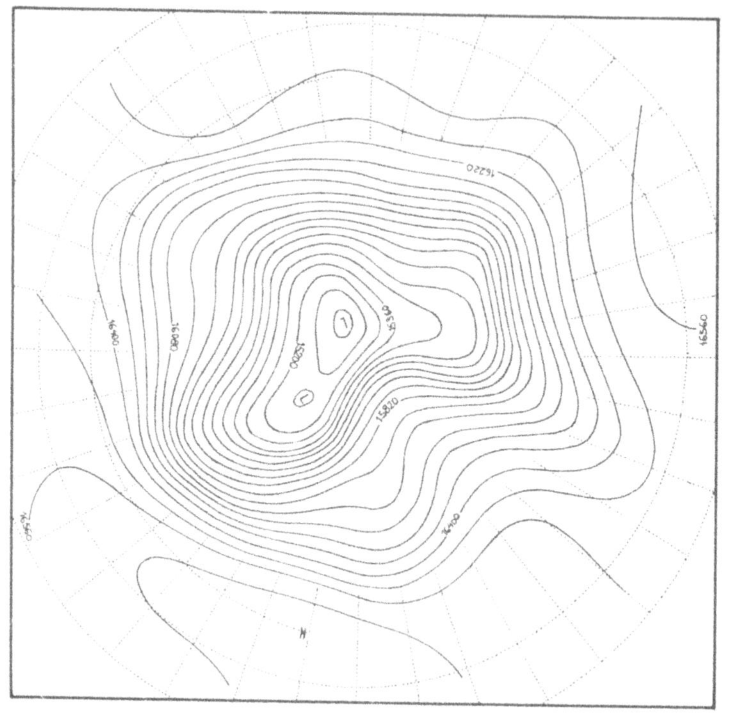

IGY Upper Air Chart Northern Hemisphere
100 mb 1200 GMT JULY 15, 1958

IGC Upper Air Chart Northern Hemisphere
100 mb 1200 GMT JANUARY 15, 1959

Figure 1.4

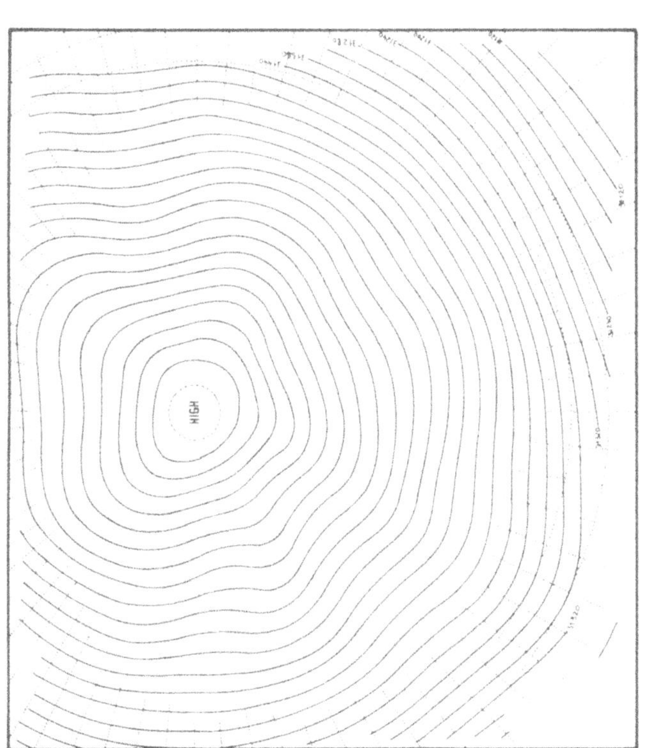

Constant pressure chart. 10 mb 1200 GMT
July 15, 1958

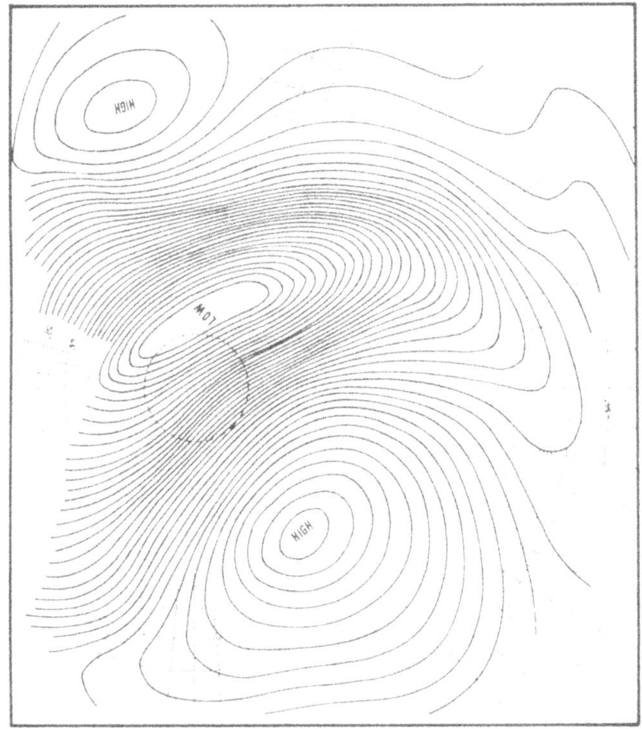

Constant pressure chart 10 mb 1200 GMT
January 15, 1959

Figure 1.5

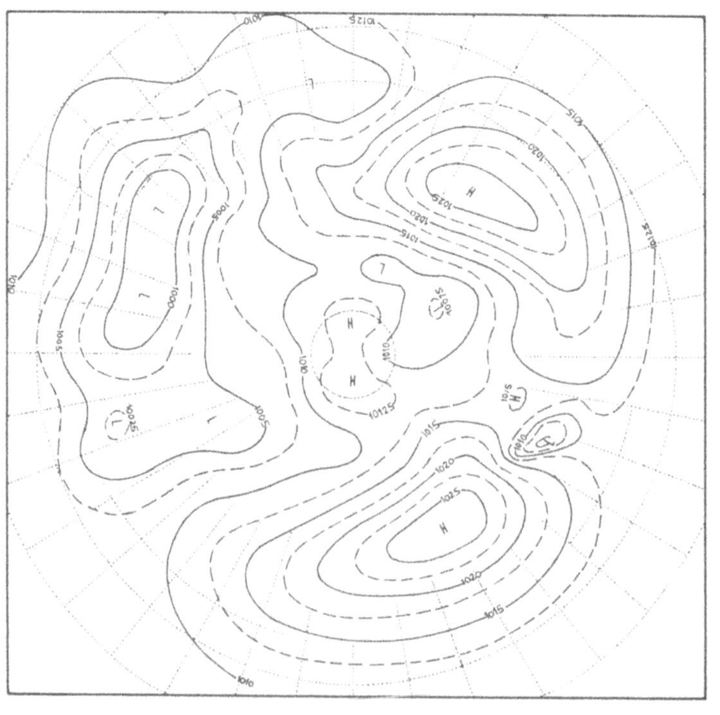

Normal Weather Chart Northern Hemisphere
Sea Level Pressure (mb) JULY

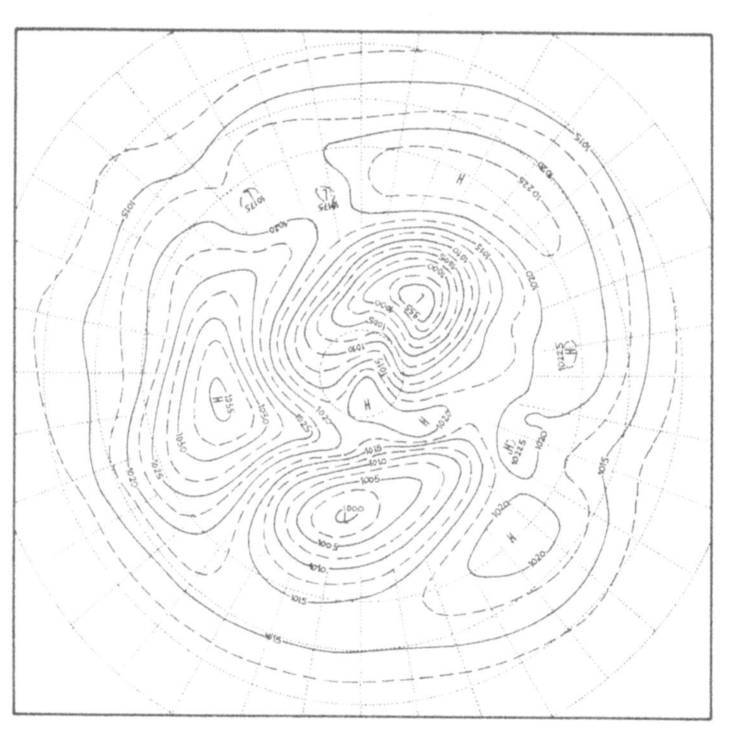

Normal Weather Chart Northern Hemisphere
Sea Level Pressure (mb) JANUARY

Figure 1.6

Normal Weather Chart Northern Hemisphere
500 mb Height (10's of ft) JULY

Normal Weather Chart Northern Hemisphere
500 mb Height (10's of ft) JANUARY

Figure 1.7

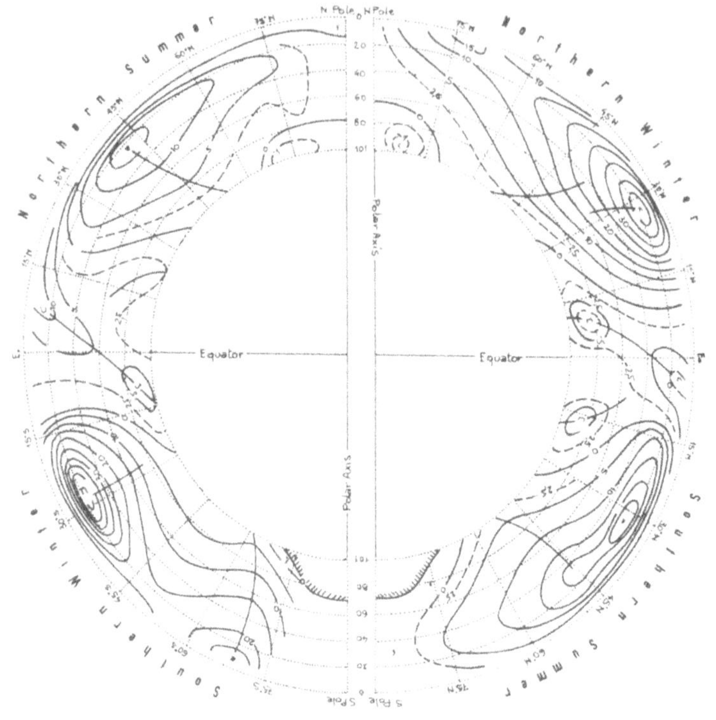

Figure 1.8

Mean zonal wind (m/sec) averaged over all longitudes, in summer and winter. Negative values denote mean easterly wind. (After Mintz, 1954 : Bull. Amer. Met. Soc., 35, 208-214).

sea-level circulation pattern is essentially reversed, with cyclones over the oceans and anticyclones over the continents. In both seasons these centers of action all but disappear at 500 mb, where the flow consists of a strong circumpolar cyclonic vortex with weaker superimposed standing wave perturbations ; the 100 mb mean charts (not shown) reveal similar patterns. Thus the relatively strong quasi-permanent centers at low levels are replaced at upper levels by standing wave-like perturbations of a circumpolar vortex ;here the transient wave and vortex perturbations which dominate the daily upper level charts are removed by the process of averaging.

A theory of the general circulation of the atmosphere must explain:

1) the presence of a mean circumpolar vortex at upper levels ;
2) quasi-permanent centers of action at low levels ;
3) the standing and transient waves and vortices superimposed on the mean circumpolar vortex at upper levels ;
4) the radical change in circulation pattern which occurs between 100 mb and 10 mb.

The axially-symmetric vortex

Figures 1.8 and 1.9 show meridional cross-sections of mean zonal wind and temperature up to 100 mb. They verify what may be inferred from the daily and mean constant level charts, namely, that the kinetic energy of the atmosphere increases upward from a minimum near sea-level to a maximum near the tropopause. The dominant flow may thus be described as an axially-symmetric circumpolar cyclonic vortex with superimposed wave and vortex perturbations. For this reason it is natural to begin the analysis of the atmospheric circulation with a discussion of the dynamics of the axially-symmetric vortex.

Historically, the first theoretical treatment of the general circulation of the atmosphere, the explanation of the trade winds by Hadley* (1735), assumed that the flow of the atmosphere was an axially-symmetric, circumpolar vortex. Moreover, his explanation of the trades was essentially correct : although the observed mid-latitude and polar circulations are

*Hadley, G., 1935 : Phil. Trans. Roy. Soc., 39, p. 58.

(a)

Winter Hemisphere Summer Hemisphere

JANUARY

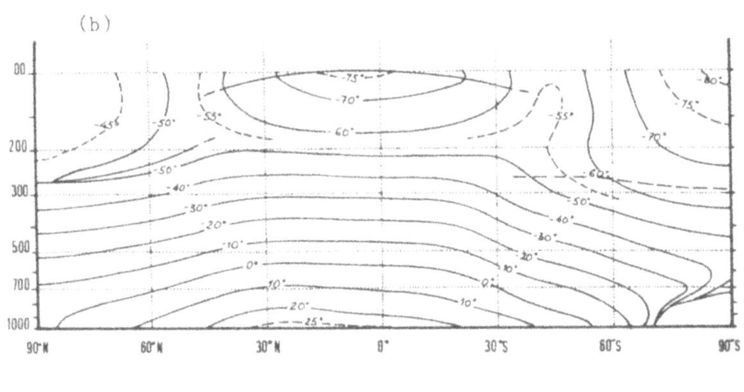

Summer Hemisphere Winter Hemisphere

JULY

Mean Temperature (°C) averaged around latitude circles for (a) January and (b) July. Heavier lines show approximate mean tropopauses. Palmen and Newton, 1969 : <u>Atmospheric Circulation Systems</u>. Int. Geophys. Ser. 13, Acad. Press, New York-London.

Figure 1.9

highly asymmetric, modern evidence shows that the low-latitude circulation is indeed maintained primarily by symmetric transport processes. Thus, it may be seen from Figures 1.10 - 1.13, which contrast the rates of change of temperature and zonal momentum due to the symmetric components of the circulation with the rates due to the horizontal eddy transports, that the low latitude convections ** of heat and momentum, level by level, are dominated by the symmetric meridional components of the flow, whereas the eddy transports become more important at higher latitudes. It will also be noted that the low-latitude changes in momentum occur primarily near the surface (ca.1000 mb) and the tropopause (ca.200 mb), where they are balanced, not so much by large-scale eddy transports, but by vertical turbulent diffusion processes. The boundary-layer character of the flow is also brought out in Figures 1.14 and 1.15, which show the mean meridional wind component in winter and summer.

The role of hydrodynamic instability in planetary motions

Later lectures will deal with perturbations of the axially-symmetric vortex and their interactions with this vortex. Perturbations which do not owe their origin to the direct action of external forces or of boundary constraints are usually caused by some form of hydrodynamic instability. It thus becomes important to study the varieties of hydrodynamic instability which occur in the atmosphere and oceans. The studies will make it possible to predict what motions will or will not occur in a given situation and will give insight into the nature of the acting forces. Moreover, an understanding of the small amplitude unstable perturbation will make it possible later to deal with the finite-amplitude motions that ultimately develop.

A considerable literature exists on the stability of non-rotating and rotating fluid systems. The hydrodynamic instabilities may be classified according to the forces which produces them. Among these are buoyancy forces associated with gravity, centrifugal and Coriolis forces associated with rotation, and inertial forces associated with shear. Viscous forces are generally stabilizing except in parallel flows of boundary

** The term "convection" is used here in its general sense as a transport with the particle motion, not in its specialized sense as turbulence caused by heating.

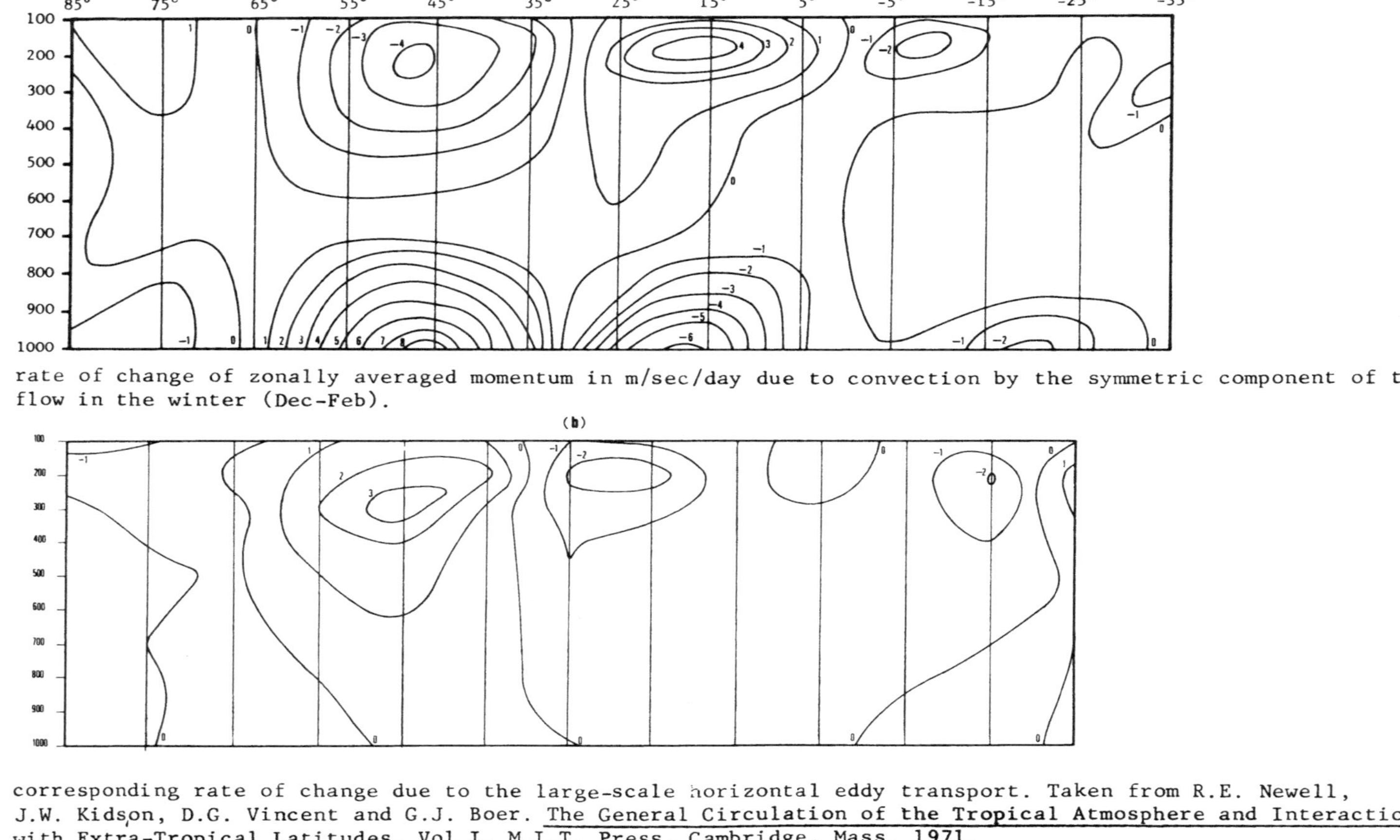

(a) rate of change of zonally averaged momentum in m/sec/day due to convection by the symmetric component of the flow in the winter (Dec-Feb).

(b) corresponding rate of change due to the large-scale horizontal eddy transport. Taken from R.E. Newell, J.W. Kidson, D.G. Vincent and G.J. Boer. The General Circulation of the Tropical Atmosphere and Interaction with Extra-Tropical Latitudes, Vol I, M.I.T. Press, Cambridge, Mass. 1971.

Figure 1.10

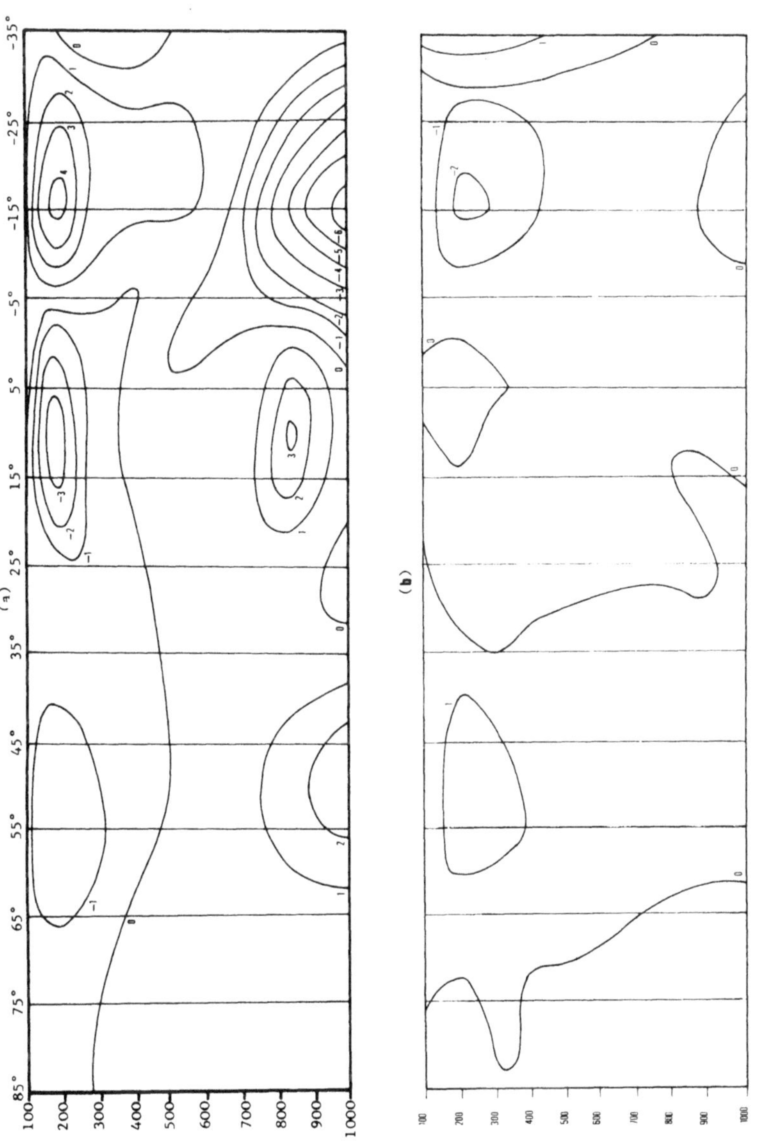

Figure 1.11

Same as for Figure 1.10 but for summer (June-August).

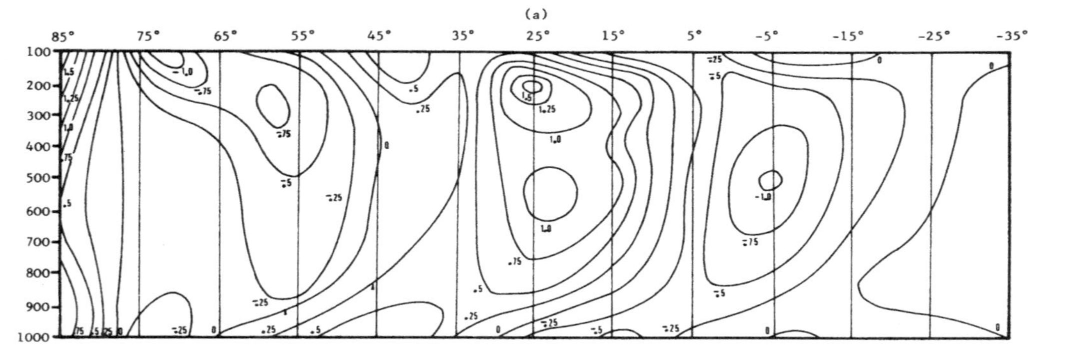

(a) rate of change of zonally averaged temperature in °K/day due to convection by the symmetric component of the flow in the winter (Dec-Feb).

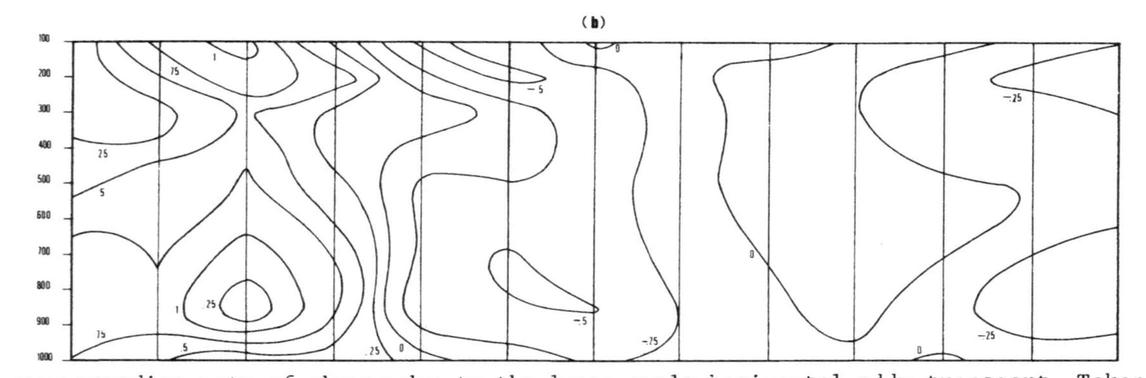

(b) corresponding rate of change due to the large-scale horizontal eddy transport. Taken from R.E. Newell, J.W. Kidson, D.G. Vincent and G.J. Boer. The General Circulation of The Tropical Atmosphere and Interaction with Extra-Tropical Latitudes, Vol. I, M.I.T. Press, Cambridge, Mass. 1971

Figure 1.12

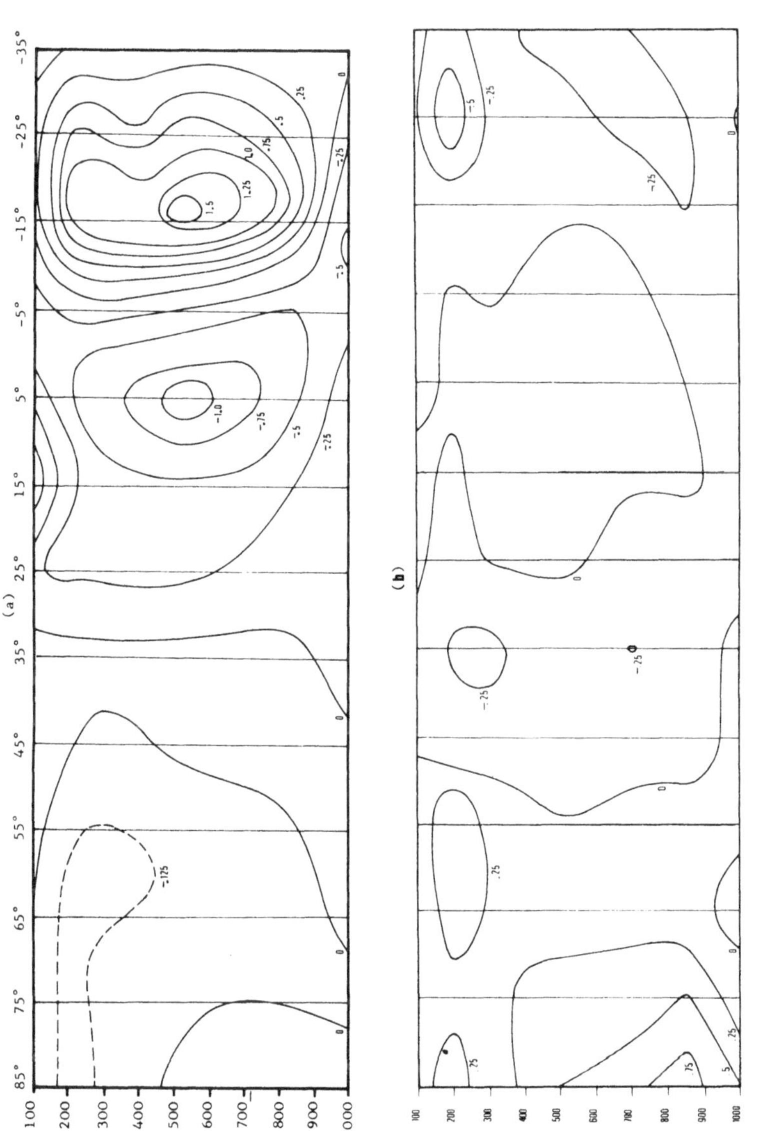

Figure 1.13 Same as for figure 1.12 but for Summer (June-August).

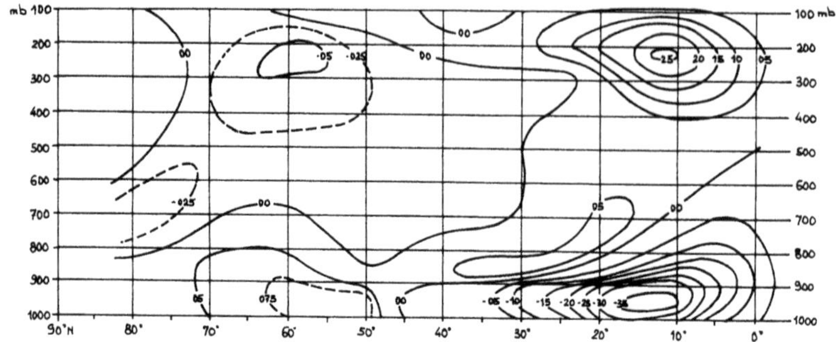

Mean meridional wind components (m/sec, positive for south wind) in the Northern Hemisphere during the winter season December-February. (After Palmén and Vuorela, 1963, Q.J.R.M.S., 89, 121-128).

Figure 1.14

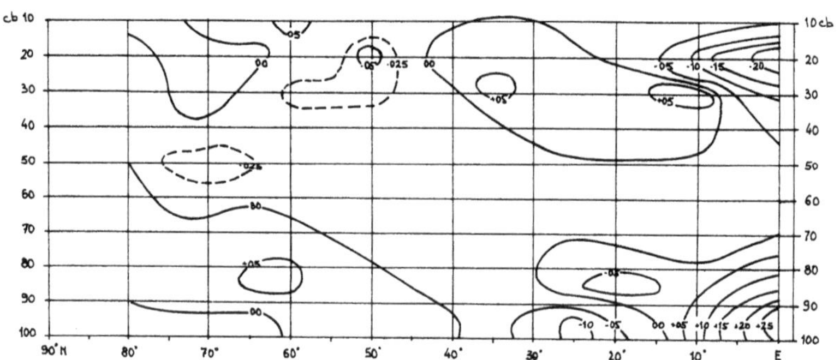

Same as Fig.1.14, for the summer season, June - August. (After Vuorela and Tuominen, 1964, Pure App. Geophys. 57, 167-180).

Figure 1.15

layer type, in tropical disturbances where frictionally induced secondary circulations contribute to the release of the heat of condensation (Chapter XIII), and in axially-symmetric motions where viscosity can cancel the stabilizing effect of the centrifugal forces (Chapter IV, p. 54).

The following is a classification of hydrodynamic instabilities and examples of naturally occuring motions for which they are responsible. The list, though incomplete, is fairly representative.

Principal forces	Examples in Nature	Principal Investigators
1. Gravitational	Free convection in atmosphere and oceans away from boundaries.	Halley (1686) Rumford (1797)
2. Gravitational (with heat of condensation)	Cumulus clouds	Von Bezold (1888) Bjerknes (1938) Höiland (1939) Haque (1952) Lilly (1960)
3. Gravitational-Viscous (with heat conduction)	Convection near rigid boundaries. Convection in earth's mantle and core ?	Bénard (1901) Rayleigh (1916a) Jeffries (1926)
4. Centrifugal		Rayleigh (1916b)
5. Centrifugal-Viscous		Taylor (1923)
6. Gravitational-Centrifugal	Limitation of anticyclonic vorticity in straight or circular currents.	Helmholtz (1888) Solberg (1936) Höiland (1941) Fjørtoft (1944, 1950) Ooyama (1966) McIntyre (1970a)

6a.	Gravitational-Centrifugal-Viscous		McIntyre (1970a)
7.	Gravitational-Coriolis-Viscous	Convection in stellar atmospheres ?	Chandrasekhar (1953)
8.	Shearing		Rayleigh (1880,1892) Taylor (1915) Tolimien (1935)
9.	Shearing-Coriolis		Kuo (1949) Fjørtoft (1950)
10.	Shearing-Viscous	Boundary turbulence in atmosphere and oceans with neutral gravitational stability	Orr (1906) Sommerfeld (1908) Heisenberg (1924) Lin (1945)
10a.	Shearing-Coriolis-Viscous	Roll vortices in planetary boundary layer	Faller & Kaylor (1966) Lilly (1966)
11.	Shearing-Gravitational (discontinuous density and velocity)	Billow clouds Ocean waves	Kelvin (1871) Helmholtz (1889) Miles (1957, 1959, 1960)
12.	Shearing-Gravitational (continuous density and velocity gradients)	Turbulence in the free atmosphere and oceans	Richardson (1920b) Taylor (1931) Goldstein (1931) Miles (1961) Howard (1961)

13.	Shearing-Gravitational-Coriolis (discontinuous density and velocity)	Frontal cyclones	Solberg (1928) Kochin (1932) Eliassen, E. (1960) Orlanski (1968)
14.	Shearing-Gravitational-Coriolis(uniform horizontal and vertical density gradients)	Large-scale waves and vortices in atmospheres(and oceans ?)	Charney (1947) Eady (1949) Fjørtoft (1950) Kuo (1952) Green (1960) Burger (1962, 1966) Arnason (1963) Miles (1964a, 1964b, 1964c, 1964d, 1965) Barcilon (1964) Pedlosky (1964c) Bretherton (1966a, 1966b) Stone (1966, 1970) Derome & Dolph(1970)
15.	Shearing-Gravitational-Coriolis (non-uniform horizontal and vertical density gradients)	Breakdown of jet like atmospheric and oceanic currents. Breakdown of Polar-Night Jet ? Waves and vortices in the Northwest African tropics ?	Charney-Stern (1962) Pedlosky (1964a, 1964b, 1965) McIntyre (1967,1970b) Brown (1968) Stone (1969)
16.	Gravitational-Coriolis (with heat of condensation)	Formation of tropical cyclones	Charney-Eliassen (1964)
17.	Shearing-Gravitational-Coriolis (with heat of condensation).	Formation of tropical cyclones.	Yamasaki (1969) Bates (1970) Chang (1971).

II. EQUATIONS OF MOTION

Let $\underline{V}$ be the velocity of a particle, p the pressure, ρ the density, $\alpha = 1/\rho$ the specific volume, Φ the gravitational potential, and $\underline{F}$ the force of friction per unit mass. The Navier-Stokes equations of motion may be written

$$D\underline{V}/Dt = -\alpha\nabla p - \nabla\Phi + \underline{F}\ , \tag{2.1}$$

and the equation of mass continuity

$$D\rho/Dt = -\rho\nabla\cdot\underline{V}\ ,$$

or (2.2)

$$D\alpha/Dt = \alpha\nabla\cdot\underline{V}\ ,$$

where ∇ is the gradient operator and D/Dt the time derivative following the motion of a particle :

$$D/Dt = \partial/\partial t + \underline{V}\cdot\nabla\ . \tag{2.3}$$

If Q is the rate of accession of heat from external sources, ε the rate of conversion of kinetic energy into heat by friction, and e the internal energy, all referred to unit mass, the first law of thermodynamics become

$$Q + \varepsilon = \frac{De}{Dt} + p\,\frac{D\alpha}{Dt}\ , \tag{2.4}$$

expressing the fact that the accession of heat is equal to the sum of the increase in internal energy and the work done in expansion. Ordinarily,

ε is a small quantity and may be ignored.

In Cartesian coordinates x_i $(i = 1,2,3)$, if F_i denotes the components of $\underline{F}$ and u_i the components of $\underline{V}$, then

$$F_i = -\alpha \frac{\partial F_{ij}}{\partial x_j}, \tag{2.5}$$

$$\varepsilon = \alpha F_{ij} \frac{\partial u_i}{\partial x_j}, \tag{2.6}$$

where

$$F_{ij} = \mu \left[\frac{2}{3} \delta_{ij} \frac{\partial u_k}{\partial x_k} - \left(\frac{\partial u_i}{\partial x_j} + \frac{\partial u_j}{\partial x_i} \right) \right] \tag{2.7}$$

is the frictional stress tensor, μ is the coefficient of viscosity, δ_{ij} is the Kronecker delta ($\delta_{ij} = 0$ if $i \neq j$; $\delta_{ij} = 1$ if $i = j$), and we employ the convention that quantities involving a repeated index are to be summed over that index.

When the motion contains turbulent eddies on smaller scales, one adds to F_{ij} the Reynolds stress tensor that results from averaging the equations of motion over a time interval which is small compared to the characteristic period of the large-scale motion but is large compared to the characteristic periods of the small-scale turbulence, i.e.,

$$F_{ij} = [F_{ij}] + [\rho] \, [u_i' u_j'] \tag{2.8}$$

the brackets denoting suitable ensemble averages and the primes turbulent fluctuations *. By the same token, eddy conduction of heat is included in the term Q in (2.4).

---

* It is assumed here that the density fluctuations may be ignored.

Perfect gases

We shall consider two types of fluids, perfect gases and incompressible fluids. For a perfect gas,

$$e = c_v T, \qquad p = \rho RT, \qquad R = c_p - c_v, \tag{2.9}$$

where c_p and c_v are the specific heats at constant pressure and volume respectively, T is the absolute temperature, and R is the gas constant referred to unit mass. In this case, neglect of frictional dissipation in the first law of thermodynamics (2.4) gives the following expressions for the rate of change of the specific entropy s :

$$\frac{Ds}{Dt} \equiv \frac{Q}{T} = \frac{c_v}{T}\frac{DT}{Dt} + \frac{p}{T}\frac{D\alpha}{Dt} = \frac{c_p}{T}\frac{DT}{Dt} - \frac{R}{p}\frac{Dp}{Dt} = \frac{c_p}{\alpha}\frac{D\alpha}{Dt} + \frac{c_v}{p}\frac{Dp}{Dt}, \tag{2.10}$$

or

$$\begin{aligned} s &= c_p \ln T - R \ln p + \text{constant}, \\ &= c_p \ln \alpha + c_v \ln p + \text{constant}, \\ &= -c_p \ln \rho + c_v \ln p + \text{constant}. \end{aligned} \tag{2.11}$$

The potential temperature θ of a parcel is defined as the temperature it would acquire if it were brought adiabatically to a standard pressure p_o. Thus

$$c_p \ln\theta - R \ln p_o = c_p \ln T - R \ln p, \tag{2.12}$$

or

$$s = c_p \ln\theta + \text{constant} \tag{2.13}$$

If θ does not vary throughout the fluid ρ, α, and T are functions of p alone. In this case the fluid is called autobarotropic or, for brevity, barotropic. Fluids for which the variables of state are not functions of p alone are said to the baroclinic.

For adiabatic motion equation (2.10) takes the form

$$\frac{c_p}{\alpha}\frac{D\alpha}{Dt} + \frac{c_v}{p}\frac{Dp}{Dt} = 0,$$

or (2.14)

$$-\frac{c_p}{\rho}\frac{D\rho}{Dt} + \frac{c_v}{p}\frac{Dp}{Dt} = 0,$$

giving

$$\left(\frac{Dp}{D\rho}\right)_s = \gamma RT = c^2, \tag{2.15}$$

where $\gamma = c_p/c_v$, and c is the velocity of sound.

Boussinesq fluids

In the case of an incompressible fluid, $D\rho/Dt = 0$, and the continuity equation (2.2) becomes :

$$\nabla \cdot \underline{V} = 0 \tag{2.16}$$

In a nearly incompressible fluid, however, small variations of density produced by heating may produce important buoyancy forces when coupled with gravity. In order to provide for such density changes, we may :

(a) ignore the density variations except when coupled with gravity in the equation of vertical motion i.e., replace (2.1) by

$$\rho_o \frac{D\underline{V}}{Dt} = -\nabla p - \rho\nabla\Phi + \rho_o \underline{F}$$

(b) retain equation (2.16) but make $D\rho/Dt$ equal to the thermally induced rate of change of density.

More generally, the density is given by some equation of state $\rho = \rho(p, T, n_1, n_2 \ldots)$ and depends upon pressure and temperature, and also upon other quantities n_i such as salinity, etc. The generalized expression for $D\rho/Dt$ then replaces the first law of thermodynamics (2.10).

This approximation, which consists in ignoring density variations, except when coupled with gravity in providing buoyancy forces which drive the motion in vertically stratified fluids, was introduced by BOUSSINESQ and so is called the "Boussinesq approximation". The fluid itself is often called a "Boussinesq fluid".

Let us consider, for example, the case of a nearly incompressible fluid like water, with a constant coefficient of thermal expansion γ, so that the equation of state is

$$\rho = \rho_{oo}\left[1 - \gamma(T - T_{oo})\right] \tag{2.17}$$

The accession of heat derives from conductive and convective processes only, as no expansion or compression of the fluid can take place. Thus

$$\frac{DT}{Dt} = \kappa \nabla^2 T \tag{2.18}$$

where κ is the coefficient of thermal conductivity of the fluid. In view of (2.17), this last equation is equivalent to :

$$\frac{D\rho}{Dt} = \kappa \nabla^2 \rho \tag{2.19}$$

and replaces the first law of thermodynamics.

Earth rotation

On a rotating sphere like the earth, which is sufficiently fluid so that over a long period of time surface tangential stresses are equalized, the mean surface becomes an equipotential surface for the combined gravitational and centrifugal forces. Thus, if G is the gravitational constant, M the earth's mass, r the radial distance from the center of the earth and R the perpendicular distance from the axis of rotation, we have approximately :

$$\Phi \simeq -\frac{GM}{r} + \frac{\Omega^2 R^2}{2}$$

and (2.20)

$$\nabla\Phi = g\underline{k}$$

where g is the acceleration of gravity and $\underline{k}$ is a unit vector normal to the mean surface of the earth.

In the next section, we shall deal with axially symmetric circulations on a rotating gravitating spheroidal annulus whose centrifugal forces, as on the earth, are so small that departures from true sphericity can be ignored. We may therefore (Figure 2.1) introduce a system of spherical coordinates in which the surfaces of the inner and outer spheres are represented by r = a and r = a + h, where a and h are constants.

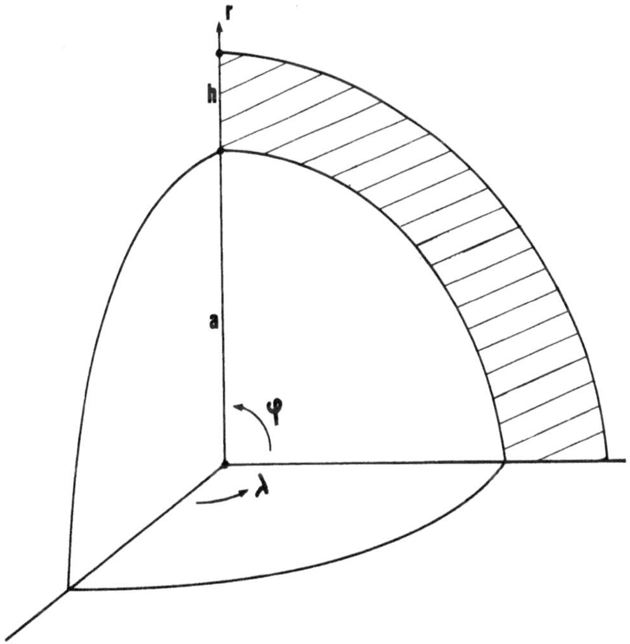

Figure 2.1

If λ is the longitude, ϕ the latitude and u, v and w, the velocity components directed toward increasing λ, ϕ and r respectively, the equations of motion and continuity for a Boussinesq fluid become :

$$u_t + \frac{vu_\phi}{r} + \frac{w(ru)_r}{r} - \frac{uv\tan\phi}{r} - 2\Omega(v\sin\phi - w\cos\phi) = \nu\left[\frac{(u_\phi\cos\phi)_\phi}{r^2\cos\phi} + \frac{(ru)_{rr}}{r} - \frac{u}{r^2\cos^2\phi}\right] \quad (2.21)$$

$$v_t + \frac{vv_\phi}{r} + \frac{w(rv)_r}{r} + \frac{u^2\tan\phi}{r} + 2\Omega u\sin\phi = -\frac{1}{\rho_{oo}r}p_\phi + \nu\left[\frac{(v_\phi\cos\phi)_\phi}{r^2\cos\phi} + \frac{(rv)_{rr}}{r} + \frac{2w_\phi}{r^2} - \frac{v}{r^2\cos^2\phi}\right] \quad (2.22)$$

$$w_t + \frac{vw_\phi}{r} + ww_r - \frac{u^2+v^2}{r} - 2\Omega u\cos\phi \quad (2.23)$$

$$= -\frac{g\rho}{\rho_{oo}} - \frac{p_r}{\rho_{oo}} + \nu\left[\frac{(rw)_{rr}}{r} + \frac{(w_\phi\cos\phi)_\phi}{r^2\cos\phi} - \frac{2v_\phi}{r^2} - \frac{2w}{r^2} + \frac{2v\tan\phi}{r^2}\right]$$

$$\frac{(v\cos\phi)_\phi}{r\cos\phi} + \frac{(r^2w)_r}{r^2} = 0 \quad (2.24)$$

Finally, the thermal equation (2.18) is :

$$T_t + \frac{vT_\phi}{r} + wT_r = \kappa\left[\frac{(T_\phi\cos\phi)_\phi}{r^2\cos\phi} + \frac{(rT)_{rr}}{r}\right] \quad (2.25)$$

The standard notation A_ϕ , A_r for partial derivatives, is used here and henceforth

$$A_\phi = \frac{\partial A}{\partial \phi}$$

$$A_r = \frac{\partial A}{\partial r}$$

$$A_{rr} = \frac{\partial^2 A}{\partial r^2} \qquad \text{etc} \ldots$$

Zonal symmetry is implied so that

$$u_\lambda = v_\lambda = T_\lambda = 0.$$

III. SYMMETRIC CIRCULATIONS IN IDEALIZED MODELS

Let us begin the treatment of the dynamics of planetary motions by analysing simple examples of thermally and mechanically driven flows on rotating spheres. The examples are not intended to correspond closely to observed motions in the atmosphere but rather to illustrate a number of dynamical and thermodynamical processes which do act in the atmosphere. In particular, we shall consider steady, axially symmetric circulations produced in a thin spherical annulus by differential heating or by differential rotation of the bounding surfaces of the annulus. The fluid is assumed to be incompressible, and the radiative eddy-diffusive processes transferring heat in the atmosphere are replaced by simple conduction.

The first example is that of a thermally induced circulation of a Boussinesq fluid in the spherical annulus shown in Figure 2.1. We suppose that the temperature is $T_o(\phi)$ at the lower boundary, $r = a$, and $T_o(\phi) + \delta T$ at the upper boundary, $r = a + h$.

Scale analysis

We anticipate that the relative motion produced by the horizontal thermal gradient will be so slow that the relative accelerations will be small compared to the Coriolis forces. In this case we may scale the horizontal velocity by a characteristic wind U derived from the imposed surface temperature gradient. The vertical coordinate is scaled by h and the vertical velocity by hU/a. The variable parts of the pressure and temperature are scaled geostrophically with $2\Omega aU/gh$ respectively. Neglecting terms multiplied by the small aspect ratio h/a, we then obtain the nondimensional equations :

$$Ro \left[\frac{v(u \cos \phi)_\phi}{\cos\phi} + wu_z\right] - v \sin \phi = E\, u_{zz}, \qquad (3.1)$$

$$Ro \left[vv_\phi + wv_z + u^2 \tan\phi\right] + u \sin\phi = Ev_{zz} - p_\phi , \qquad (3.2)$$

$$T = p_z , \qquad (3.3)$$

$$\frac{(v\cos\phi)_\phi}{\cos\phi} + w_z = 0, \qquad (3.4)$$

$$\sigma Ro\,(vT_\phi + wT_z) = ET_{zz}\,, \qquad (3.5)$$

where Ro is the Rossby number $U/2\Omega a$, E is the Ekman number $\nu/2\Omega h^2$, and σ is the Prandtl number ν/κ. It is assumed that $E < O(1)$, $Ro \leq O(E)$*, $\sigma = O(1)$, that the lower boundary is rigid with $T = T_o(\phi)$ (non-dimensional) and that the upper boundary is free with $T = T_o(\phi) + \delta T$ (non-dimensional), where δT, as well as T_o, is of order unity. The variation in height of the upper boundary may be ignored and w may be set equal to zero to order greater than E, providing $2\Omega aU/gh < O(E^{1/2})$. The boundary conditions then become

$$u(0) = v(0) = w(0) = u_z(1) = v_z(1) = w(1) = 0$$
$$T(0) = T_o(\phi)\,,\quad T(1) = T_o(\phi) + \delta T\,. \qquad (3.6)$$

Expanding all dependent variables in powers of $E^{1/2}$ we find

$$-v^{(o)}\sin\phi = 0\,, \qquad (3.7)$$

$$u^{(o)}\sin\phi = p_\phi^{(o)}, \qquad (3.8)$$

$$T^{(o)} = p_z^{(o)}, \qquad (3.9)$$

$$w_z^{(o)} = 0\,, \qquad (3.10)$$

* For the earth, the absolute velocity Ωa at the equator is 440 m sec^{-1} while the relative velocity U is at most 40 m sec^{-1}. Hence $Ro \sim 10^{-2}$ to 10^{-1}. The coefficient of turbulent viscosity is of order 10^5 to 10^6 $cm^2\,sec^{-1}$ and $h \sim 10$ km. Thus $E \sim 10^{-3}$ to 10^{-2} and $Ro = O(E^{1/2})$ rather than O(E). The approximation $Ro \leq O(E)$ is made for mathematical convenience. It will later be relaxed.

which state that in the interior of the fluid the meridional velocity vanishes, the vertical velocity is independent of height, and the zonal velocity is geostrophic to first order in $E^{1/2}$.

Eliminating $p^{(o)}$ between (3.8) and (3.9) we get the thermal wind equation

$$u_z^{(o)} \sin\phi = - T_\phi^{(o)} , \tag{3.11}$$

which states that $u_z^{(o)} \neq 0$ at $z = 1$. In view of (3.6) there must be a boundary layer at $z = 1$ in which u_z reduces to zero. It can be shown that the horizontal velocities in this layer, as well as its thickness, are of order $E^{1/2}$: so that the meridional and vertical mass transports are of order E. If $u^{(o)}(0)$ were different from zero, there would be an Ekman layer near the lower boundary whose thickness would be of order $E^{1/2}$ and whose horizontal velocities would be order unity. The mass transport in this layer would then be of order $E^{1/2}$, but since the transport elsewhere is of order E at most, we conclude that $u^{(o)}(0) = 0$.

The first order interior equations are of identically the same form as those of zero order and, like them, imply that to this order the meridional velocity is zero and the zonal velocity is geostrophic. Since the boundary layer pumping is at most of order E, the first order continuity equation $w_z^{(1)} = 0$ implies that $w^{(o)}$ and $w^{(1)}$, as well as $v^{(o)}$ and $v^{(1)}$, vanish identically in the interior. The first and second order thermal equations therefore become

$$T_{zz}^{(o)} = T_{zz}^{(1)} = 0, \tag{3.12}$$

which state that the temperature field is in conductive equilibrium to first order. Their solutions

$$T^{(o)} = T_o(\phi) + \delta z, \quad T^{(1)} \equiv 0, \tag{3.13}$$

may be substituted into (3.11) and the corresponding first order equation to give

$$u^{(o)}(z) = -zT_{o\phi}/\sin\phi$$
$$(3.14)$$
$$u^{(1)}(z) = u^{(1)}(0).$$

The solution for $u^{(o)}$ has a non-zero z-derivative at the upper boundary and therefore does not satisfy the condition of vanishing stress. Let boundary layer corrections be denoted by the tilde (~) and introduce the stretching substitution.

$$1 - z = \frac{E^{1/2}}{(\sin\phi)^{1/2}}\zeta.$$

The first order momentum equations for the upper boundary become

$$-\tilde{v}^{(1)} = \tilde{u}^{(1)}_{\zeta\zeta}, \qquad (3.15)$$

$$\tilde{u}^{(1)} = \tilde{v}^{(1)}_{\zeta\zeta},$$

with the conditions

$$u_z^{(o)} + E^{1/2}\tilde{u}_z^{(1)} = 0,$$

$$v_z^{(o)} + E^{1/2}\tilde{v}_z^{(1)} = 0,$$

or

$$\tilde{u}_\zeta^{(1)} = \frac{u_z^{(o)}}{(\sin\phi)^{1/2}}, \qquad (3.16)$$

$$\tilde{v}_\zeta^{(1)} = 0,$$

at the free surface, $\zeta = 0$, and

$$\tilde{u}^{(1)}(\infty) = \tilde{v}^{(1)}(\infty) = 0 \qquad (3.17)$$

at the base of the boundary layer. The solution of (3.15) subject to these boundary conditions is

$$\tilde{u}^{(1)}(\zeta) = \frac{u_z^{(o)}(1)}{(2\sin\phi)^{1/2}} e^{-\zeta/\sqrt{2}} \left(-\cos\frac{\zeta}{\sqrt{2}} + \sin\frac{\zeta}{\sqrt{2}}\right),$$

$$\tilde{v}^{(1)}(\zeta) = \frac{u_z^{(o)}(1)}{(2\sin\phi)^{1/2}} e^{-\zeta/\sqrt{2}} \left(\cos\frac{\zeta}{\sqrt{2}} + \sin\frac{\zeta}{\sqrt{2}}\right), \tag{3.18}$$

giving

$$\int_{1-\infty}^{1} \tilde{v}^{(1)} dz = \frac{E^{1/2} u_z^{(o)}(1)}{\sin\phi} = -\frac{E^{1/2} T_{o\phi}}{\sin^2\phi}, \tag{3.19}$$

and from (3.4)

$$w^{(2)}(1) = \frac{-1}{\cos\phi}\left[\frac{T_{o\phi}\cos\phi}{\sin^2\phi}\right]_\phi. \tag{3.20}$$

To find the first order interior flow we must consider the expansion terms of order E in (3.1) :

$$-v^{(2)}\sin\phi = u_{zz}^{(o)} = -T_{\phi z}^{(o)} = 0, \tag{3.21}$$

which gives $w_z^{(2)} = 0$ or $w^{(2)}(z) = w^{(2)}(1)$.

Now we consider the first order equations for the lower boundary layer. The stretching substitution

$$z = \frac{E^{1/2}}{(\sin\phi)^{1/2}}\zeta,$$

reduces them to the same form as (3.15) but with the boundary conditions,

$$\tilde{u}^{(1)} + u^{(1)} = 0, \tag{3.22}$$

$$\tilde{v}^{(1)} + v^{(1)} = \tilde{v}^{(1)} = 0 \,, \tag{3.22}$$

at $\zeta = 0$, and

$$\tilde{u}^{(1)}(\infty) = \tilde{v}^{(1)}(\infty) = 0 \,, \tag{3.23}$$

at the top of the boundary layer. Their solution,

$$\begin{aligned} \tilde{u}^{(1)} &= - u^{(1)}(0)\, e^{-\zeta/\sqrt{2}} \cos \frac{\zeta}{\sqrt{2}} \,, \\ \tilde{v}^{(1)} &= u^{(1)}(0)\, e^{-\zeta/\sqrt{2}} \sin \frac{\zeta}{\sqrt{2}} \,, \end{aligned} \tag{3.24}$$

then gives

$$\int_0^\infty \tilde{v}^{(1)} dz = \frac{E^{1/2}}{(2\sin\phi)^{1/2}} u^{(1)}(0) \,. \tag{3.25}$$

It follows from (3.21) that the meridional mass transport in the lower boundary layer is equal and opposite to that in the upper boundary layer. Hence from (3.19), (3.25) and (3.14) we obtain

$$u^{(1)}(0) = -\left(\frac{2}{\sin\phi}\right)^{1/2} u_z^{(o)}(1) = \left(\frac{2}{\sin^3\phi}\right)^{1/2} T_{o\phi} = u^{(1)}(z), \tag{3.26}$$

which completes the solution to order E.

The nature of the circulation is illustrated schematically in Figures 3.1, 3.2 and 3.3 which show the zonal velocity profile, the meridional velocity profile and the streamlines in the meridional plane respectively.

Taking the thermal driving

$$T_o = - T_m(\sin\phi)^n \qquad (n = 2, 3, 4, \ldots), \tag{3.27}$$

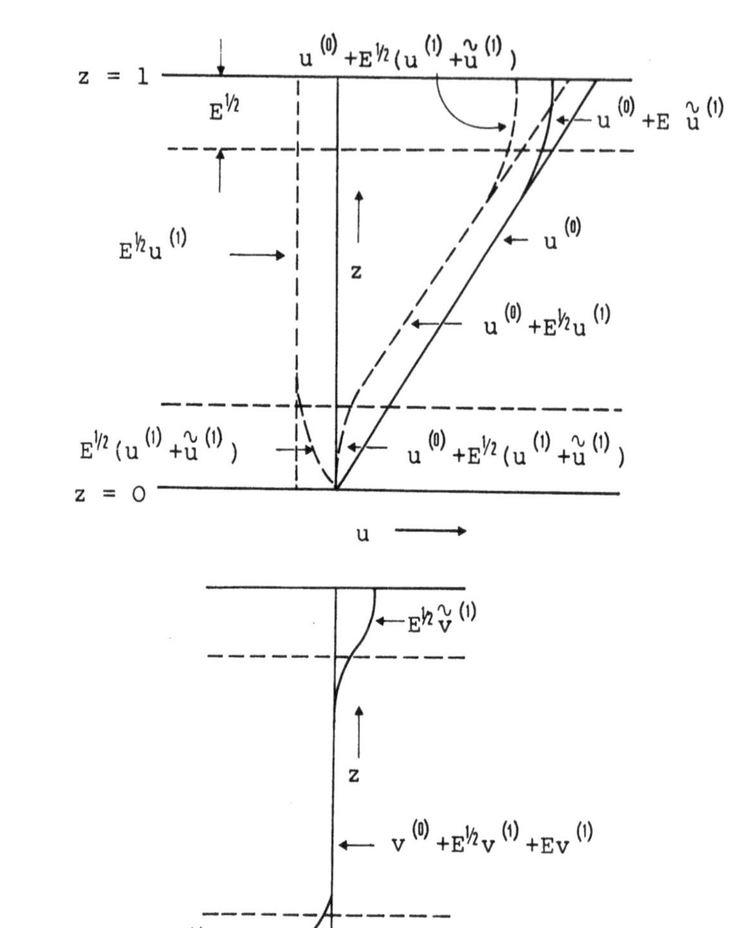

Figure 3.1

Figure 3.2

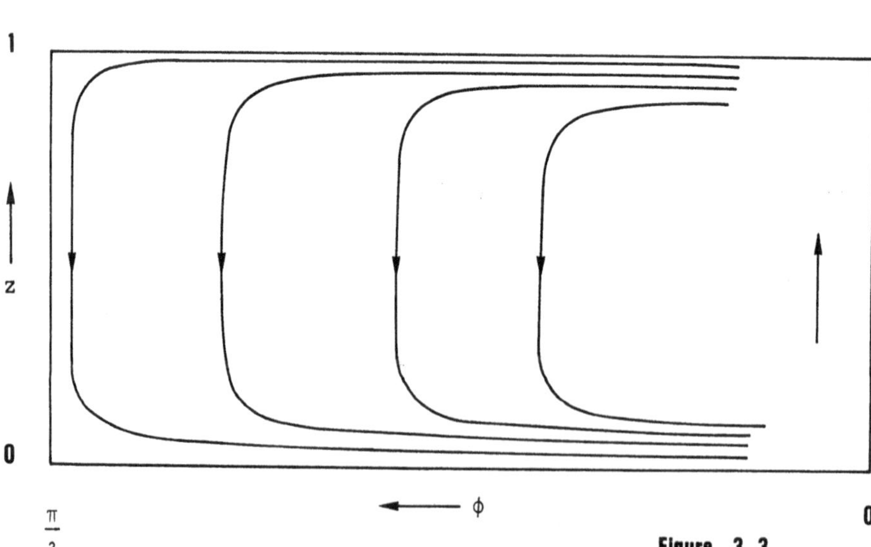

Figure 3.3

we find from (3.19) and (3.20) that

$$\int_{1-\infty}^{1} vdz = ET_m n(\sin\phi)^{n-3}\cos\phi$$

and

$$w = ET_m n\,[(n-3)(\sin\phi)^{n-4} - (n-1)(\sin\phi)^{n-2}].$$

If $n \geq 4$, the expansion is uniformly valid at all latitudes. If $n = 3$, the expansion is uniformly valid for $\sin\phi > O(E)$, but because the boundary layer transport remains finite at the equator, the terms multiplied by Ro in equation (3.2) become important for $\sin\phi \leq O(E)$. If $n = 2$, the Ekman pumping ceases to be $O(E)$ for $\sin\phi = O(E^{1/2})$. Thus it is only for $n \geq 4$ that our solution is uniformly valid. In all other cases, one must consider advective effects near the equator.

We wish to call attention in particular to the boundary layer character of the meridional mass transport, determined by the frictional-Coriolis force balance. This property of the flow is also observed in the atmosphere. (See Figures 1.14 and 1.15).

In the actual atmosphere, $Ro = O(E^{1/2})$. In this case we find that : (1) $v^{(o)}$ and $v^{(1)}$ are zero in the interior ; (2) $u^{(o)}$ but not $u^{(1)}$ satisfies the thermal wind equation (3.11) ; (3) $u^{(o)}(0) = 0$; and (4) $T^{(o)}$, but not $T^{(1)}$, is in conductive equilibrium. Again, the main meridional transport takes place in boundary layers. The details of the analysis for this case will not be presented here.

Problem 3.1 : Prove the statement on page 16 that $w(1) = 0$ to order E if $2\Omega aU/gh < O(E^{1/2})$.

Problem 3.2 : Find the circulation with the same thermal driving correct to order $E^{1/2}$ when the upper bounding sphere is also rigid.

Problem 3.3 : Consider the flow between two concentric rigid isothermal spherical shells with radii a and a+h (h<<a) rotating with the slightly differing angular velocities Ω and $\Omega+\omega$. A vertical density stratification is maintained by postulating that the temperature of the inner sphere

exceeds that of the outer sphere by the amount $(\Delta T)_V$. Define $U = \omega a$, $Ro = U/2\Omega a = \omega/2\Omega$ and assume that $Ro \leq O(E) < O(1)$. The horizontal temperature scale is given geostrophically as before by $(\Delta T)_H = 2\Omega aU/\gamma gh$, but now one has to consider the possibility that $\lambda = (\Delta T)_V/(\Delta T)_H$ is not of order unity. Two cases may be solved without elaborate mathematics :

$$\text{(i)} \quad \sigma Ro\lambda < O(E^{1/2}),$$

$$\text{(ii)} \quad \sigma Ro\lambda > O(1) \quad .$$

Show that in case (i) the interior flow is a solid rotation with angular velocity $\Omega + \frac{\omega}{2}$ and that in case (ii) it increases linearly with z from Ω to $\Omega + \omega$.

Solutions to problems

3.1 The forced w(h) must be smaller in magnitude than the boundary layer pumping whose magnitude is $O(\frac{Uh}{a} E)$. w(h) is obtained from the requirement that the flow be parallel to the free surface. If η is the height of the free surface, we have

$$w(h) = \frac{v(h)\eta_\phi}{a} \cong \frac{2\Omega}{g} uv \sin\phi$$

by geostrophy. Since $u = U\,O(1)$ and $v = U\,O(E^{1/2})$,

$$w(h) = \frac{2\Omega U^2}{g} O(E^{1/2}),$$

and the condition we seek is

$$\frac{2\Omega U^2}{g} O(E^{1/2}) < \frac{Uh}{a} O(E) \ ,$$

or

$$\frac{2\Omega aU}{gh} < O(E^{1/2}).$$

3.2 Equations (3.7) - (3.11) continue to apply, but now the velocity must be reduced to zero at both boundaries. The horizontal boundary layer velocities are therefore O(1), the transports $O(E^{1/2})$, and the boundary layer pumping $O(E^{1/2})$. Since the first order internal equations are identical to (3.7) - (3.11),

$$w_z^{(o)} = w_z^{(1)} = 0\ ,\quad w^{(o)}(z) = 0\ ,\quad w^{(1)}(z) = w_z^{(1)}.$$

It follows, as in the free-boundary case, that $T^{(o)}$ is in conductive equilibrium, so that $T^{(o)}(z) = T_o(\phi) + \delta z$.

The boundary layer corrections $\tilde{u}^{(o)}$, $\tilde{v}^{(o)}$ in the lower boundary layer satisfy equations (3.15), (3.22) - (3.25) with "1" replaced by "0". We obtain

$$\int_0^\infty \tilde{v}^{(o)} dz = \frac{E^{1/2} u^{(o)}(0)}{(2\sin\phi)^{1/2}} .$$

The upper boundary layer velocities satisfy similar equations, and we obtain

$$\int_{1-\infty}^{1} \tilde{v}^{(o)} dz = \frac{E^{1/2} u^{(o)}(1)}{(2\sin\phi)^{1/2}} .$$

Since $v^{(o)} = v^{(1)} \equiv 0$ in the interior, the condition that the net meridional transport vanish to order $E^{1/2}$ gives

$$u^{(o)}(0) = -\, u^{(o)}(1)\ ,$$

which, combined with the integral of (3.11),

$$\left[u^{(o)}(z) - u^{(o)}(0)\right]\sin\phi = -\, zT_{o\phi}$$

gives

$$u^{(o)}\sin\phi = (\tfrac{1}{2} - z)\, T_{o\phi}$$

To obtain $u^{(1)}$ and $T^{(1)}$, we have from (3.5)

$$T_{zz}^{(1)} = \sigma \frac{Ro}{E} w^{(1)} T_z^{(o)} = \frac{\sigma Ro\delta}{E} w^{(1)} ,$$

where $w^{(1)}$ is given by integration of the continuity equation :

$$E^{1/2} w^{(1)} = - \frac{1}{\cos\phi} \left[\cos\phi \int_0^\infty \tilde{v}^{(o)} dz\right]_\phi$$

$$= - \frac{E^{1/2}}{\sqrt{2}\cos\phi} \left[\frac{u^{(o)}(0)\cos\phi}{\sqrt{\sin\phi}}\right]_\phi$$

$$= - \frac{E^{1/2}}{2\sqrt{2}\cos\phi} \left(\frac{T_{o\phi}\cos\phi}{\sqrt{\sin\phi}}\right)_\phi$$

Since $w^{(1)}$ is independent of z, the solution for $T^{(1)}$ subject to the boundary conditions, $T^{(1)}(0) = T^{(1)}(1) = 0$, is

$$T^{(1)} = - \frac{z(1-z)}{2} \frac{\sigma Ro\delta w^{(1)}}{E} ,$$

from which we derive, by integration of the thermal wind equation,

$$u^{(1)}(z) - u^{(1)}(0) = \frac{1}{2} \left(\frac{z^2}{2} - \frac{z^3}{3}\right) \frac{\sigma Ro\delta}{E} \frac{w_\phi^{(1)}}{\sin\phi} ,$$

To obtain $u^{(1)}(0)$, we impose the condition that the net meridional mass transport, $\int_0^1 v\, dz$, must vanish to order E. Since $u^{(o)}$ is a linear function of z, $u_{zz}^{(o)} = 0$, and it follows from (3.1) that $v^{(2)}$ is zero in the interior. Hence $u^{(1)}(0)$ must be chosen so that the boundary-layer transports just compensate. As before we get

$$u^{(1)}(0) + u^{(1)}(1) = 0,$$

from which

$$- 2u^{(1)}(0) = \frac{1}{2} \left(\frac{1}{2} - \frac{1}{3} \right) \frac{\sigma Ro\delta}{E} \frac{w_\phi^{(1)}}{\sin\phi}$$

or

$$u^{(1)}(0) = \frac{1}{48\sqrt{2}} \frac{\sigma Ro\delta}{E\ \sin\phi} \left[\frac{1}{\cos\phi}\left(\frac{T_{o\phi}\cos\phi}{\sqrt{\sin\phi}}\right)_{\phi}\right]_{\phi}$$

3.3 The non-dimensional momentum and continuity equations remain the same as (3.1) - (3.4), but if we subtract from the temperature and pressure the conductive equilibrium temperature λz and the associated hydrostatic pressure $\frac{\lambda}{2}(z^2 - 1)$, in non-dimensional units, the thermal equation (3.5) becomes

$$\frac{\sigma Ro}{E}\left[vT + w(T_z + \lambda)\right] = T_{zz}.$$

Consider case (i). Expansion in powers of $E^{1/2}$ again gives equations (3.7) to (3.11), and since the boundary layer pumping is $O(E^{1/2})$ at most, the thermal equation implies that $T_{zz}^{(o)} = 0$, and the boundary conditions $T^{(o)}(0) = T^{(o)}(1) = 0$ require that $T^{(o)}$ = constant. It then follows from (3.11) that $u_z^{(o)} = 0$, and $u^{(o)}(0) = u^{(o)}(1)$ = constant. Since u vanishes at z = 0 and equals $\cos\phi$ at z = 1, there must be frictional boundary layers at z = 0 and 1. We find, as in example (3.2), that

$$\int_0^{\infty} \tilde{v}^{(o)} dz = \left[\frac{E}{2\sin\phi}\right]^{1/2} \left[u^{(o)}(1) - \cos\phi\right].$$

Since $u^{(o)}(0) = u^{(o)}(1)$, and the interior meridional transport vanishes to order E, the boundary layer transports must be equal and opposite. Hence

$$u^{(o)} = \frac{\cos\phi}{2},$$

which states that the interior zonal flow is a solid rotation whose angular velocity is the average of those of the bounding spherical shells.

With respect to the meridional flow, all of the north-south transport to order $E^{1/2}$ takes place in the boundary layers, but, this transport being divergent, integration of (3.5) gives

$$E^{1/2} w^{(1)} = - \frac{\left[\cos\phi \int_0^{\infty} \tilde{v}^{(o)} dz\right]_\phi}{\cos\phi} = \frac{E^{1/2}}{2\sqrt{2}\cos\phi} \left[\frac{\cos^2\phi}{\sin^{1/2}\phi}\right]_\phi$$

$$= - \frac{E^{1/2}}{4\sqrt{2}} \frac{1 + 4\sin^2\phi}{\sin^{3/2}\phi}$$

Thus there is a downward leakage from the upper to the lower boundary layer which increases in magnitude toward the equator where eventually the return flow takes place.

The equatorial dynamics is complicated by the presence of inertial effects and/or internal boundary layers and therefore will not be discussed here.

We note that the density stratification for case (i) is so weak that the circulation is the same as that for a homogeneous fluid. For this case and for vanishingly small Rossby number, the flow has been deduced analytically by I. Proudman * and K. Stewartson** . As in the present case, the interior flow is found to be in solid rotation with an angular velocity equal to the average of the angular velocity of the spherical shells. However, outside the cylinder of revolution tangent to the inner sphere at its equator, the fluid moves with the angular velocity of the outer spherical shell. This necessitates a shear layer across the cylinder. Such motions are examples of the so-called "Taylor-Proudman" theorem, according to which small disturbances of a homogeneous fluid in solid rotation have the property that the velocity components perpendicular to the axis of rotation are uniform along the axis. (The result that u is independent of z, rather than of $r \sin\phi$, is due to the neglect of the

* Proudman, I., 1956 : J. Fluid Mech., 1, 505-516

**Stewartson, K., 1957 : J. Fluid Mech. 3, 17-26 and
1966 : J. Fluid Mech. 26, 131-144.

term $2\Omega w \cos\phi$ in equation (2.20) as being small of order h/a. The error is negligible except in the immediate vicinity of the equator).

Case (ii) is one for which the density (or temperature) stratification is so great that vertical motions up to order E are prevented entirely, as may be seen from the thermal equation. This implies that there can be no boundary layers, and therefore that $u^{(o)}(0) = 0$ and, $u^{(o)}(1) = \cos\phi$. It follows from (3.4) that $v^{(2)} = 0$. Then equation (3.1) requires that $u_{zz}^{(o)} = 0$, which, with the conditions $u^{(o)}(0) = 0$, $u^{(o)}(1) = \cos\phi$, gives $u^{(o)} = z \cos\phi$, i.e., a linear shear of $u^{(o)}$ from the inner to the outer sphere. The interior temperature field is found from (3.11) to be $T^{(o)} = - 1/4 \sin^2\phi$, which completes the analysis for the interior flow. It may be noted that thermal boundary layers are required to reduce the horizontal temperature gradients to zero at the boundaries. For a fuller exposition see Pedlosky (J. Fluid Mech., 36 pt 2, 1969 - pp. 401-415).

IV STABILITY OF THE CIRCULAR VORTEX

Introduction

The next two chapters will deal with the general properties of circular vortices. Their stability properties will be discussed in Chapter IV, and the circulations generated by slowly or impulsively applied heat sources or frictional torques will be discussed in Chapter V.

Equations of motion. Condition for balance

It will be convenient to introduce a cylindrical coordinate system in which the z-axis coincides with the axis of rotation (Figure 4.1). Let λ be the azimuthal coordinate, R the distance from the axis of rotation, z the axial coordinate and u, v, and w the velocity components in the R, λ and z directions respectively. If Φ is the gravitational potential and $F^{(R)}$, $F^{(\lambda)}$, $F^{(z)}$ the components of the frictional force, the equations of motion become

$$\frac{Dv}{Dt} + \frac{vu}{R} = - \frac{\alpha}{R}\frac{\partial p}{\partial \lambda} - \frac{1}{R}\frac{\partial \Phi}{\partial \lambda} + F^{(\lambda)} , \qquad (4.1)$$

$$\frac{Du}{Dt} - \frac{v^2}{R} = - \alpha \frac{\partial p}{\partial R} - \frac{\partial \Phi}{\partial R} + F^{(R)} , \qquad (4.2)$$

$$\frac{Dw}{Dt} = - \alpha \frac{\partial p}{\partial z} - \frac{\partial \Phi}{\partial z} + F^{(z)} . \qquad (4.3)$$

It is assumed that the motion and the gravitational potential are independent of the azimuthal coordinate λ. In this case (4.1) may be written, after multiplication by R,

$$\frac{Dm}{Dt} = RF_{\lambda} , \quad (m = Rv) \qquad (4.4)$$

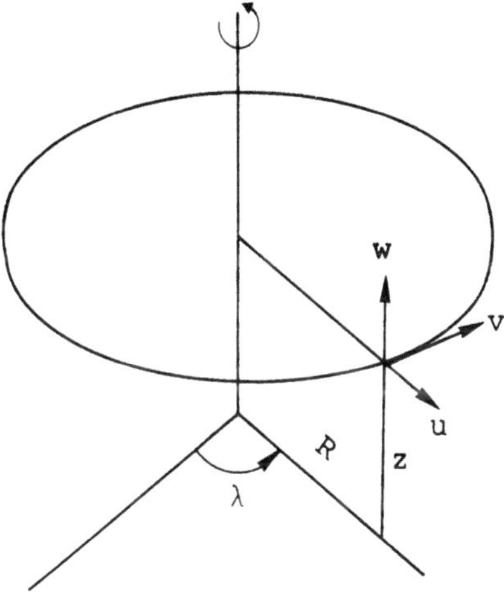

Figure 4.1

which states that the rate of change of the angular momentum of a ring of radius R and unit mass is equal to the applied torque ; thus m is conserved if the torque vanishes. The equations of motion in the R, z (meridional) plane may be combined in the form :

$$\frac{D\underline{V}_m}{Dt} = -m^2\nabla\chi - \nabla\Phi - \alpha\nabla p + \underline{F}_m \,, \qquad (4.5)$$

where ∇ is the gradient operator in the meridional plane, the subscript "m" denotes a vector in this plane, and $\chi = 1/(2\,R^2)$. If the vortex is in equilibrium (i.e., stationary), we have

$$-m^2\nabla\chi - \nabla\Phi - \alpha\nabla p = 0. \qquad (4.6)$$

This is the equation of balance in a circular vortex ; it states that the sum of the centrifugal, gravitational, and pressure forces is zero (Figure 4.2).

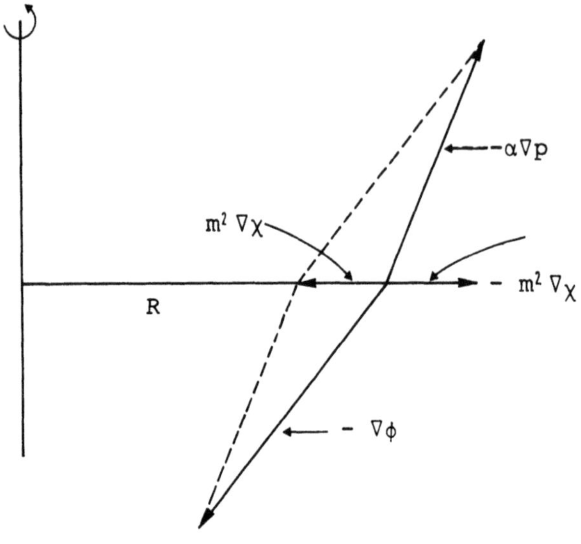

Figure 4.2

Taking the curl of (4.6), we get

$$\nabla m^2 \times (-\nabla\chi) + \nabla\alpha \times (-\nabla p) = 0 \ , \tag{4.7}$$

a form of the so-called thermal wind equation. It states that the torque couple $\nabla\alpha \times (-\nabla p)$ exerted on a volume element by the pressure force in the variable density field is balanced by the torque couple $\nabla m^2 \times (-\nabla\chi)$ exerted by the centrifugal force in the variable angular momentum field. In a barotropic field, the surfaces of constant specific volume and pressure coincide ($\alpha = \alpha(p)$, $\nabla\alpha \times \nabla p = 0$) : the pressure force exerts no torque, and consequently the centrifugal force exerts no torque. Hence ∇m^2 must be parallel to $\nabla\chi$, and v (or m) cannot vary in a direction parallel to the axis of rotation. This is at once a more special and a more general instance of the Taylor-Proudman theorem, which will be treated in Chapter VI. As will be seen, it is more special because the Taylor-Proudman theorem applies to asymmetric as well as symmetric motions, and more general because the Taylor-Proudman theorem applies only to. small deviations from solid rotation.

Condition for instability of a circular vortex with respect to axially-symmetric displacements in a compressible fluid *

Consider an axially-symmetric circular vortex in a state of equilibrium. The vortex is said to be unstable for axially-symmetric displacements if these displacements can amplify spontaneously. Let K be the total kinetic energy, P the total potential energy and I the total energy, i.e.,

$$K = \int_\tau \frac{1}{2} \underline{V}\,\underline{V}\,\rho d\tau$$

$$P = \int_\tau \Phi\rho d\tau \qquad (4.8)$$

$$I = \int_\tau e\rho d\tau \quad ,$$

where τ represents the total volume and $d\tau$ is an element of volume. Since the azimuthal and meridional components of the velocity are orthogonal, we may express the total kinetic energy K as the sum $K_m + K_z$ of its meridional and zonal components respectively :

$$K_z = \int \frac{v^2}{2} \rho dz = \int m^2\chi \, \rho d\tau \ ,$$

$$K_m = \int \frac{1}{2} \underline{V}_m \cdot \underline{V}_m \rho d\tau \ .$$

* Eliassen, A. and Kleinschmidt, E.: Handbuch der Physik, Vol. 48, 1-154.

Fjørtoft, R. (1950) : Stability Bibliography.

It is assumed that the gas is confined within a rigid container,† and friction and thermal sources and sinks are neglected. Under these circumstances the total energy of the system is conserved, i.e.,

$$K_m + K_z + P + I = \text{constant}. \tag{4.10}$$

Displacements of fluid particles from the equilibrium configuration are assumed to be infinitesimal and independent of the azimuthal coordinate. For such displacements the angular momentum m is conserved, so that the change of K_z is a function of the displacements alone. Moreover, according to (2.2) and (2.14) the changes of ρ and p, and therefore of the internal energy, also depend only on the displacements, so that the change of I as well as of K_z and P is a function solely of the displacements. We can therefore state that the stationary circular vortex is stable if the quantity

$$J \equiv K_z + P + I \tag{4.11}$$

is a minimum -just as a non-rotating system of particles is in stable equilibrium if the potential energy is a minimum. Thus, if the fluid is slightly displaced from an equilibrium configuration, the growth of K_m will be limited by the (infinitesimal) change of J from its equilibrium value : the smaller the initial increment of J, the smaller the subsequent growth of K_m.

To establish stability criteria, we seek the conditions under which J is a minimum. Its variations may be treated from either an Eulerian or a Lagrangian point of view. We adopt the latter and consider displacements of individual fluid (toroidal ring) parcels, for which the angular momentum m and the mass $\rho d\tau$ are conserved.

Let δ denote an individual change at a parcel and $\delta \underline{r}$ an arbitrary infinitesimal vector displacement field in the meridional plane with components δR and δz. Then, since $\delta(\rho d\tau) = 0$, the first variation in J

† A rigid boundary is assumed only for simplicity. The theorems to be derived will apply also to a fluid with free boundaries (see Fjørtoft (1944) in Stability Bibliography).

may be written

$$\delta J = \int_\tau (m^2\delta\chi + \delta\Phi + \delta e)\ \rho d\tau. \tag{4.12}$$

For fixed geometrical fields the change δ at a parcel is the same as the total differential d in the environment. Hence

$$\begin{aligned} \delta\chi &= d\chi = \nabla\chi\cdot\delta\underline{r}\ , \\ \delta\Phi &\equiv d\Phi = \nabla\Phi\cdot\delta\underline{r}\ . \end{aligned} \tag{4.13}$$

The changes $\delta\rho$, $\delta\alpha$, δe, and δp differ from the environmental changes $d\rho$, $d\alpha$, de, and dp, and are given by equations (2.2), (2.4) and (2.15) :

$$\begin{aligned} \delta\rho &= -\ \rho\nabla\cdot\delta\underline{r} \\ \delta\alpha &= \alpha\nabla\cdot\delta\underline{r} \\ \delta e &= -\ p\delta\alpha = -\ p\alpha\nabla\cdot\delta\underline{r} \\ \delta p &= c^2\delta\rho = -\ \rho c^2\nabla\cdot\delta\underline{r}\ . \end{aligned} \tag{4.14}$$

Hence, by Gauss's theorem and the assumption of a rigid boundary,

$$I = \int_\tau \delta e\rho d\tau = -\int_\tau p\nabla\cdot\delta\underline{r}d\tau = \int_\tau \alpha\nabla\cdot\delta r\rho d\tau$$

and therefore

$$\delta J = \int_\tau (m^2\nabla\chi + \nabla\Phi + \alpha\nabla p)\cdot\delta\underline{r}\ \rho d\tau \tag{4.15}$$

If J is to be a minimum, the first variation of J must vanish for arbitrary displacement vectors $\delta\underline{r}$. We may take $\delta\underline{r}$ to be zero except at an arbitrary small volume in the fluid and conclude that

$$m^2\nabla\chi + \nabla\Phi + \alpha\nabla p = 0, \tag{4.16}$$

i.e., that equilibrium of forces in a stationary vortex is both a necessary and a sufficient condition for J to have an extreme value.

In order to determine whether the extremum is a maximum, minimum, or minimax, we must compute the second term in the Taylor expansion of J. Let J_o be the value of J in the equilibrium configuration, then

$$J-J_o = \frac{dJ}{dt}dt + \frac{1}{2}\frac{d^2J}{dt^2}(dt)^2 + O(dt)^3 = \delta J + \frac{1}{2}\delta^2 J = \frac{1}{2}\delta^2 J, \qquad (4.17)$$

so that J is a minimum or maximum according as $\delta^2 J \gtrless 0$. Taking the second variation we get

$$\begin{aligned}\delta(\delta J) &= \int_\tau \left[m^2\delta(\nabla\chi\cdot\delta\underline{r}) + \delta(\nabla\Phi\cdot\delta\underline{r}) + \delta(\alpha\nabla p\cdot\delta\underline{r})\right]\rho d\tau \\ &= \int_\tau \left[m^2\nabla\chi + \nabla\Phi + \alpha\nabla p\right]\cdot\delta(\delta\underline{r})\rho d\tau \\ &\quad + \int_\tau \left[m^2\nabla(\delta\chi) + \nabla(\delta\Phi) + \delta\alpha\nabla p + \alpha\nabla(\delta p)\right]\cdot\delta\underline{r}\rho d\tau \\ &= \int_\tau \left[m^2\nabla(\delta\chi) + \nabla(\delta\Phi) + \delta\alpha\nabla p + \alpha\nabla(\delta p)\right]\cdot\delta\underline{r}\rho d\tau\,.\end{aligned} \qquad (4.18)$$

Substituting from (4.13) and (4.14), and denoting $\nabla\cdot\delta\underline{r}$ by D, we find

$$\begin{aligned}\delta^2 J &= \int_\tau \{\rho m^2\nabla(\nabla\chi\cdot\delta\underline{r}) + \rho\nabla(\nabla\Phi\cdot\delta\underline{r}) + D\nabla p + \nabla(-\rho c^2 D)\}\cdot\delta\underline{r}\, d\tau \\ &= \int_\tau \Big[\nabla\cdot\{[\rho m^2(\nabla\chi\cdot\delta\underline{r}) + \rho(\nabla\Phi\cdot\delta\underline{r}) - \rho c^2 D]\,\delta\underline{r}\} \\ &\quad - \left[(m^2\nabla\chi + \nabla\Phi + \alpha\nabla p)\cdot\delta r\right]\rho D \\ &\quad - \left[(\nabla(\rho m^2)\cdot\delta\underline{r})(\nabla\chi\cdot\delta\underline{r}) + (\nabla\rho\cdot\delta\underline{r})(\nabla\Phi\cdot\delta\underline{r}) + (\alpha\nabla p\cdot\delta\underline{r})(\frac{1}{c^2}\nabla p\cdot\delta\underline{r})\right] \\ &\quad + \rho c^2 D^2 + 2D\nabla p\cdot\delta\underline{r} + \frac{\alpha}{c^2}(\nabla p\cdot\delta\underline{r})^2\Big]d\tau.\end{aligned} \qquad (4.19)$$

The first term in the integrand vanishes by Gauss's theorem and the second by the balance condition. Substituting

$$\delta p = -\rho c^2 \nabla\cdot\delta\underline{r} = -\rho c^2 D \quad \text{and} \quad dp = \nabla p\cdot\delta\underline{r} \;, \tag{4.20}$$

we obtain finally

$$\delta^2 J = \int_\tau \left[\delta\underline{r} \cdot M \cdot \delta\underline{r} + \frac{\alpha^2}{c^2} (\delta p - dp)^2\right]\rho d\tau, \tag{4.21}$$

where M is the diadic defined by

$$M = -\frac{1}{\rho} \nabla(\rho m^2)\nabla\chi - \frac{\nabla\rho}{\rho}\nabla\Phi - \alpha\nabla p \frac{\alpha}{c^2} \nabla p \;,$$

That is, if $R = x_1$, $z = x_2$, $\delta\underline{r} = \{S_1, S_2\}$, then M is a Cartesian tensor whose components are

$$M_{ij} = -\frac{1}{\rho}\left[\frac{\partial(\rho m^2)}{\partial x_i}\frac{\partial\chi}{\partial x_j} + \frac{\partial\rho}{\partial x_i}\frac{\partial\Phi}{\partial x_j} + \alpha\frac{\partial p}{\partial x_i}\frac{1}{c^2}\frac{\partial p}{\partial x_j}\right](i,j = 1,2) \;,$$

and

$$\delta\underline{r}\cdot M\cdot\delta\underline{r} = M_{ij}S_iS_j .$$

M may also be written

$$\begin{aligned} M &= -\frac{1}{\rho}\nabla(\rho m^2)\nabla\chi - \frac{\nabla\rho}{\rho}\nabla\Phi - (\frac{\alpha}{c^2}\nabla p)(\alpha\nabla p) \\ &= -\nabla m^2\nabla\chi - \frac{\nabla\rho}{\rho}(m^2\nabla\chi + \nabla\Phi) - (\frac{\alpha}{c^2}\nabla p)(\alpha\nabla p) \\ &= (\nabla m^2)(-\nabla\chi) + (\frac{\nabla\rho}{\rho} - \frac{\alpha}{c^2}\nabla p)\,\alpha\nabla p \\ &= (\nabla m^2)(-\nabla\chi) + (\frac{\nabla\theta}{\theta})(-\alpha\nabla p) \;, \end{aligned} \tag{4.22}$$

by (2.11) and (2.13).

Summarizing : For a compressible perfect gas,

$$\delta J = \int_\tau \left[m^2\nabla\chi + \nabla\Phi + \alpha\nabla p\right]\cdot\delta\underline{r}\rho d\tau \tag{4.23}$$

$$\delta^2 J = \int_\tau \left[\delta\underline{r}\cdot M\cdot\delta\underline{r} + \alpha^2 c^{-2}(dp - \delta p)^2\right]\rho d\tau \tag{4.24}$$

$$M = (\nabla m^2)(-\nabla\chi) + (\nabla h)(-\alpha\nabla p)\ ;\ h = \ln\theta. \tag{4.25}$$

When $\delta J = 0$ and $\delta^2 J > 0$, J is a minimum and the vortex is stable. By inspecting the expression for $\delta^2 J$ we see that, since the term $\alpha^2 c^{-2}(dp - \delta p)^2$ is always positive, the vortex is stable if the quadratic form $\delta\underline{r}\cdot M\cdot\delta\underline{r}$ (or $M_{ij}S_i S_j$) is positive definite. If $\delta\underline{r}\cdot M\cdot\delta\underline{r}$ is negative definite, the vortex will be unstable providing displacements can be found for which $(dp - \delta p)^2$ is small, i.e., for which the individual change of pressure approximates its change in the environment. This may be done in an infinite variety of ways. Indeed, in an actual flow $dp - \delta p = -\frac{\partial p}{\partial t}\,dt$, and it may be shown that the quadratic term is negligible as long as the particle displacements are small compared to the density scale height of the fluid.* The stability criterion for a compressible fluid thus becomes

$$\delta\underline{r}\cdot M\cdot\delta\underline{r} \quad \begin{matrix} > 0 & \text{stable} \\ < 0 & \text{unstable} \end{matrix} \tag{4.26}$$

$$M = (\nabla m^2)(-\nabla\chi) + (\nabla h)(-\alpha\nabla p)\ ,$$

$$h = \ln\theta \quad \text{for a compressible perfect gas.}$$

If $\delta\underline{r}\cdot M\cdot\delta\underline{r}$ is indefinite, we shall again accept that it is always possible to find displacements for which the first integral in (4.24) is negative while the second is negligible. Thus, the stability of the vortex depends on the quadratic form, $\delta\underline{r}\cdot M\cdot\delta\underline{r}$.

---

* The assumption that displacements $\delta\underline{r}$ can be found for which $dp - \delta p = \nabla p\cdot\delta\underline{r} + \rho c^2\nabla\cdot\delta\underline{r}$ is arbitrarily small is analogous to the assumption that displacements $\delta\underline{r}$ can be found for which $\nabla\cdot\delta\underline{r}$ is zero in the incompressible case to be treated next.

Condition for instability of a circular vortex in an incompressible fluid

An incompressible fluid has no internal energy, and the displacements $\delta\underline{r}$ are constrained by the incompressibility condition,

$$\nabla\cdot\delta\underline{r} = 0 .$$

The condition for an extremum in $J = K_z + P$ subject to the above constraint must be found.

This problem is analogous to that of calculating the equilibrium of a system of particles subject to certain constraints, such as the particles being forced to move on given surfaces or to remain at given distances apart, etc. If U_k represents the potential energy of the k'th particle and $U = \sum U_k$ the total energy, then

$$\delta U = \sum \delta U_k = \sum \nabla U_k\cdot\delta\underline{r}_k = 0 \qquad (4.27)$$

in the state of equilibrium. But from this we cannot conclude that $\nabla U_k = 0$, because the $\delta\underline{r}_i$ are not independent. Let (4.27) be written in terms of the coordinates x_i [(x_1, x_2, x_3) referring to the first particle, (x_4, x_5, x_6) to the second, etc.], and let the corresponding components of $-\nabla U_k$ be denoted by F_i, then for a system of N particles

$$\sum_1^N \nabla U_k \cdot \delta\underline{r}_k = -\sum_1^{3N} F_i\delta x_i = 0. \qquad (4.28)$$

Suppose that the x_i's are connected by ℓ finite equations of constraint

$$f_k\,(x_1, x_2, \ldots\ldots, x_{3N}) = 0 \qquad (k = 1, 2, \ldots\ldots, \ell\,) \qquad (4.29)$$

Lagrange showed that the problem of minimizing U is equivalent to minimizing the function $U + \sum_1^{\ell} \lambda_k f_k$, where the λ_k are undetermined multipliers.

In the continuum analogy, instead of setting the first variation,

$$\delta J = \int_\tau (m^2\nabla\chi + \nabla\Phi)\cdot\delta\underline{r}\rho d\tau,$$

equal to zero, subject to the field of $\delta\underline{r}$ satisfying the condition $\nabla\cdot\delta\underline{r} = 0$, we get

$$\int_\tau \left[p(m^2\nabla\chi + \nabla\Phi) \cdot \delta\underline{r} + \lambda\nabla \cdot \delta\underline{r} \right] d\tau = 0 , \tag{4.30}$$

where $\delta\underline{r}$ is now a completely arbitrary displacement vector. Integration by parts and application of Gauss's theorem then gives

$$\int_\tau (m^2\nabla\chi + \nabla\Phi - \alpha\nabla\lambda) \cdot \delta\underline{r}\, \rho d\tau = 0 .$$

Since $\delta\underline{r}$ is now arbitrary,

$$m^2\nabla\chi + \nabla\Phi - \alpha\nabla\lambda = 0$$

is the necessary condition for an extremum. We see that λ may be identified with $-p$, so that the equilibrium condition is actually the same in both the compressible and incompressible cases. By analogy with the general dynamics of a system of particles where the Lagrange multipliers may be identified with the reactive forces due to rigid constraints, the pressure may be regarded as a reaction against the constraint of incompressibility* . It thus plays a different role from that in a compressible fluid where it is related to the other thermodynamical variables by an equation of state.

Proceeding as in the case of a compressible fluid, we get

$$\delta^2 J = \int_\tau \delta\underline{r}\cdot M\cdot\delta\underline{r}\rho d\tau ,$$

where

$$M = (\nabla m^2)(-\nabla\chi) + (\frac{\nabla\alpha}{\alpha})(-\alpha\nabla p) .$$

It is seen that the second variation of J becomes the same as the expression (4.24) providing $h = \ln\alpha$ and the term involving $(dp - \delta p)^2$ is

---

* See A. Sommerfeld, 1964 : Mechanics of Deformable Bodies, Academic Press, N.Y. - London, pp. 89-94.

absent. This result may be derived from (4.24) and (4.25) by letting c, the velocity of sound, tend toward infinity as it does when the fluid becomes more and more incompressible. Thus, we may treat both cases simultaneously, by defining $h = \ln\theta$ or $\ln\alpha$ according as we are dealing with a perfect gas or an incompressible fluid.

Heuristic derivation of the condition for stability for axially symmetric displacements in a stationary circular vortex

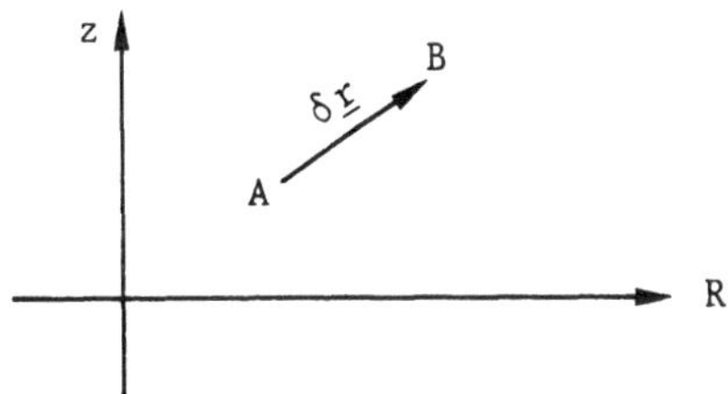

Let there be an axially symmetric displacement $\delta\underline{r}$, and assume that this does not disturb the pressure gradients. Due to this displacement, the forces will become unbalanced and the question to decide is whether the unbalanced forces do positive work (in which case there is instability) or negative work (in which case there is stability).

After the displacement, consider conditions at point B, and let Δ denote the local change at B :

The gravitational force, $-\nabla\Phi$, is unchanged

The centrifugal force becomes $-(m^2 + \Delta m^2)\nabla\chi$.

The pressure force becomes $-(\alpha + \Delta\alpha)\nabla p$.

Hence the unbalanced force $\underline{F}$ is

$$\underline{F} = -\Delta m^2\nabla\chi - \Delta\alpha\nabla p .$$

Since m^2 is conserved, $\Delta m^2 = -\nabla m^2 \cdot \delta \underline{r}$. Now consider $\Delta\alpha$:

In the incompressible case α is conserved, therefore

$$\frac{\Delta\alpha}{\alpha} = -\frac{\nabla\alpha}{\alpha} \cdot \delta\underline{r} = -\nabla h \cdot \delta\underline{r} \quad .$$

In the compressible case, since p does not change,

$$\frac{\Delta\alpha}{\alpha} = \frac{\Delta\theta}{\theta} = \Delta h = -\nabla h \cdot \delta\underline{r}$$

because $h = \ln\theta$ is conserved. Therefore

$$\underline{F} = (\nabla m^2 \cdot \delta\underline{r})(\nabla\chi) + (\nabla h \cdot \delta\underline{r})(\alpha\nabla p) \quad .$$

The criterion for stability is now that $\underline{F} \cdot \delta\underline{r} < 0$, i.e.,

$$(\nabla m^2 \cdot \delta\underline{r})(\nabla\chi \cdot \delta\underline{r}) + (\nabla h \cdot \delta\underline{r})(\alpha\nabla p \cdot \delta\underline{r}) < 0$$

or

$$\delta\underline{r} \cdot M \cdot \delta\underline{r} > 0 \quad ,$$

where M is defined as before.

The stability criterion for special cases

1. No rotation = m = 0. This is the case of equilibrium between the pressure and gravitational force. We have

$$-\nabla\Phi - \alpha\nabla p = 0 \; ,$$

$$M = (\nabla h)(-\alpha\nabla p) = (\nabla h)(\nabla\Phi) \; ,$$

$$\underline{F} \cdot \delta\underline{r} = (\nabla h \cdot \delta\underline{r})(\nabla\Phi \cdot \delta\underline{r}) \quad ,$$

and we may assume that the force of gravity, $-\nabla\Phi = -g$, acts in the negative z direction. Then

$$\underline{F}\cdot\delta\underline{r} = -gh_z(\delta z)^2 ,$$

and the restoring force per unit mass displacement is

$$\nu_g^2 \equiv g\frac{\partial h}{\partial z} ,$$

the square of the Brunt-Väisälä frequency ν_g. This is the frequency of a gravitational oscillation for which the displacements are primarily vertical. In the atmosphere $h = \ln\theta$ and the period, $T_g = 2\pi/\nu_g$, is about 10 minutes for average stability. In the oceans $h = \ln\alpha$ and a typical value of T_g is about 20 minutes for the thermocline region and about 50 minutes in the mixed surface layer and in the deep ocean *.

2. No gravity: This is the case of equilibrium between the pressure and centrifugal forces. We have

$$-m^2\nabla\chi - \alpha\nabla p = 0,$$

$$M = (\nabla m^2)(-\nabla\chi) + (\nabla h)(m^2\nabla\chi)$$

$$= \theta(\nabla\frac{m^2}{\theta})(-\nabla\chi) \text{ for a compressible fluid}$$

$$= \alpha(\nabla\frac{m^2}{\alpha})(-\nabla\chi) \text{ for an incompressible fluid,}$$

and

$$\delta\underline{r}\cdot M\cdot\delta\underline{r} = \begin{cases} \dfrac{\theta}{R^3}\,\delta(\dfrac{m^2}{\theta})\,\delta R & \text{for a compressible fluid} \\ \\ \dfrac{\alpha}{R^3}\,\delta(\dfrac{m^2}{\alpha})\,\delta R & \text{for an incompressible fluid.} \end{cases}$$

---

* See Eckart, C., 1960. Hydrodynamics of Oceans and Atmospheres, Pergamon Press, pp. 64-74.

The motion is stable or unstable according as $\frac{m^2}{\theta}$ (or $\frac{m^2}{\alpha}$) increases or decreases in the direction of increasing R.

The restoring force per unit displacement in the R-direction is given by

$$\nu_i^2 = \begin{cases} \dfrac{\alpha}{R^3}\dfrac{\partial}{\partial R}\left(\dfrac{m^2}{\alpha}\right) & \text{incompressible fluid} \\[2ex] \dfrac{\theta}{R^3}\dfrac{\partial}{\partial R}\left(\dfrac{m^2}{\theta}\right) & \text{perfect gas,} \end{cases}$$

where ν_i may be called the frequency of an inertial oscillation.

If the fluid is barotropic ($\alpha = \alpha(p)$, h = constant), there is no effect of gravity unless there are free surfaces or discontinuities and the stability criterion is the same as for pure rotation, i.e.,

$$\nu_i^2 = \frac{1}{R^3}\frac{\partial m^2}{\partial R} > 0 \quad .$$

If the vortex is a constant rotation with angular velocity ω, the restoring force per unit mass per unit displacement becomes

$$\nu_i^2 = 4\omega^2 \quad .$$

In particular, the frequency of oscillation for horizontal displacements in the centrifugal force field $\Omega^2 R\nabla R$ of the earth's rotation is

$$\nu_i = 2\Omega\sin\phi \equiv f \ ,$$

and the period of oscillation $2\pi/f$ is the so-called "half-pendulum day", half the period of revolution in the path of a Foucault pendulum.

Further illustrations and applications

In a homogeneous incompressible fluid of depth H acted upon by gravity, the restoring force on an entire vertical column of unit cross-section due to the displacement $\delta\eta$ of its free surface is $\rho g\delta\eta$. Hence the restoring force per unit mass per unit displacement of the free surface is given by

$$\nu_g^2 = \rho g\delta\eta/\rho H\ \delta\eta = \frac{g}{H} .$$

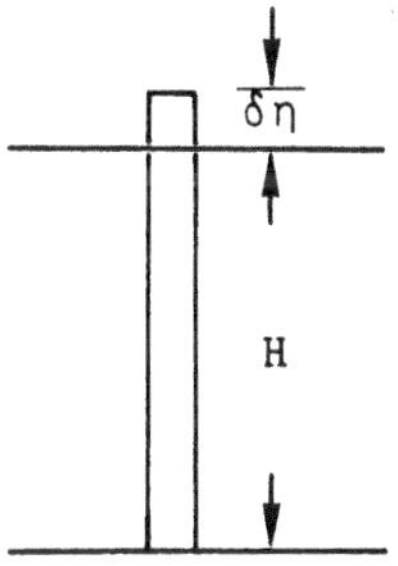

This also holds for a stratified fluid because the restoring force due to a displacement of the free surface is far greater than that due to internal displacements, i.e., $g/H \gg g\ \partial \ln\alpha/\partial z$ providing the vertical scale of variation of α or ρ greatly exceeds the depth H.

When the displacements are not all vertical the effect is to increase the mass without adding to the restoring force and therefore to decrease the frequency of oscillation. We may illustrate this effect by considering the oscillations in a U-tube of unit cross-section which is filled to the depth H by a homogeneous incompressible fluid. The force acting on the entire fluid due to a unit displacement of the free surface will be $2\rho g$. If 2L is the width of the horizontal arm, the total mass is $\rho(2H + 2L)$, and the frequency of small oscillations will be

$$\nu_g^2 = \frac{2\rho g}{\rho(2H + 2L)} = \frac{g}{H + L} ,$$

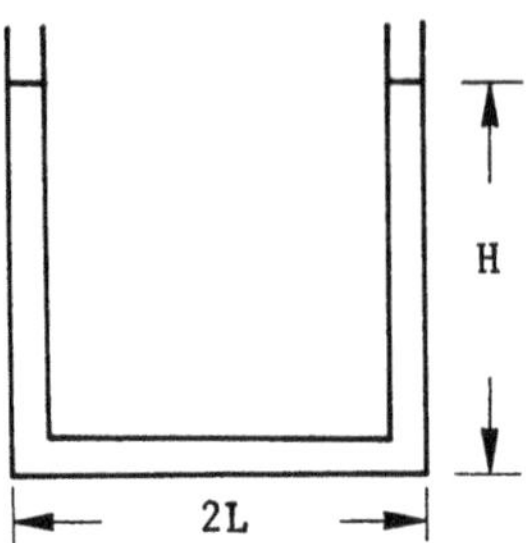

which is less than the Brunt Väisälä frequency because of the extra inert mass, $2\rho L$. This suggests the general rule that the Brunt-Väisälä frequency cannot be exceeded by gravity oscillations and is observed only for motions whose displacements are predominantly vertical.

When the rotational and buoyancy forces act simultaneously, their exact effects can be found only by the methods of continuum mechanics. However, an overall view of what is going on can sometimes be obtained by a particle approach. In illustration we will apply this approach to derive the scale relationship for motions suggested by some recent oceanographic observations in the Philippine Sea obtained by Stommel and Fedorov.* Continuous vertical soundings of pressure, temperature and salinity revealed lenticular bodies of homogeneous water imbedded in the thermocline region having horizontal dimensions L in the range 2 to 20 kilometers and vertical dimensions H in the range 2 to 40 meters. Whatever their cause, whether breaking of internal waves, salt convection† or some other mechanism, one may attempt to account for their relative dimensions by assuming that they are local vortices in equilibrium under the action

* H. Stommel and K.N. Fedorov, 1967 : Tellus XIX, 2, 306-325.

† A stable density stratification $\rho(T,s)$, in which the separate effect of the temperature T is stabilizing whereas that of the salinity is destabilizing, can be statically unstable because heat is diffused more rapidly than salt. (M.E. Stern, Tellus, XII, 1960.)

of rotational and buoyancy forces. If a system of particles is in equilibrium under the action of the forces $\underline{F}_i$ at the i'th particle, and $\delta\underline{r}_i$ are a set of virtual displacements, d'Alembert's principle states that $\sum \underline{F}_i \cdot \delta\underline{r}_i = 0$. This statement is non-trivial because the constraining forces (pressure here) do no work in the virtual displacement and may be disregarded. In the present case we assume that the typical buoyancy force is of the order $\nu_g^2 \delta z$, where δz is a typical vertical displacement, and the typical force in the earth's angular velocity field is $\nu_i^2 \delta R$, where δR is the typical horizontal displacement. Summing over all the particles, we have

$$HL\nu_g^2 \, (\delta z)^2 \sim HL\nu_i^2 \, (\delta R)^2 \ .$$

By continuity

$$\frac{\delta z}{\delta R} \sim \frac{H}{L} \ .$$

Hence

$$\frac{H^2}{L^2} \sim \frac{\nu_i^2}{\nu_g^2} \ .$$

As already remarked $2\pi/\nu_g \sim 20$ minutes, or $\nu_g \sim \frac{1}{2} \times 10^{-2}$ sec, and in the Philippine Sea at about 6° latitude $f = 10^{-5} \text{sec}^{-1}$. Hence

$$\frac{H}{L} \sim \frac{\nu_i}{\nu_g} \sim \frac{10^{-5}}{\frac{1}{2} \times 10^{-2}} = 2 \times 10^{-3} \ ,$$

giving, for example, L = 1 km for H = 2 m or L = 20 km for H = 40 m.

Stability of flow between rotating cylinders

An experimental and theoretical study of the stability of flow between coaxial rotating cylinders was carried out by G.I. Taylor (1923, Stability Bibliography). When the outer cylinder rotates in the same sense as the inner cylinder, but with so large an angular velocity that the square of the angular momentum increases outward, one expects the flow to be stable, viscosity acting as an additional stabilizing factor However, when the cylinders rotate in opposite senses, the square of the angular momentum first decreases and then increases. In this case the flow would be unstable near the inner cylinder in the absence of viscosity. It is found that viscosity causes the flow to be stable until the counter-rotation of the inner cylinder exceeds a certain critical value depending on the rotation of the outer cylinder. The stability curve is shown in Figure 4.3. Here Ω_1 and Ω_2 are the angular velocities of the inner and outer cylinders respectively, R_1 and R_2 are the corresponding radii, and ν is the kinematic coefficient of viscosity. The dashed line is approached by the stability curve for large Ω_1 and Ω_2. It corresponds to $\Omega_1 R_2^2 = \Omega_2 R_2^2$, i.e., to no gradient of angular momentum. The instability motions are cellular circulations in the meridional plane, as shown schematically in Figure 4.4.

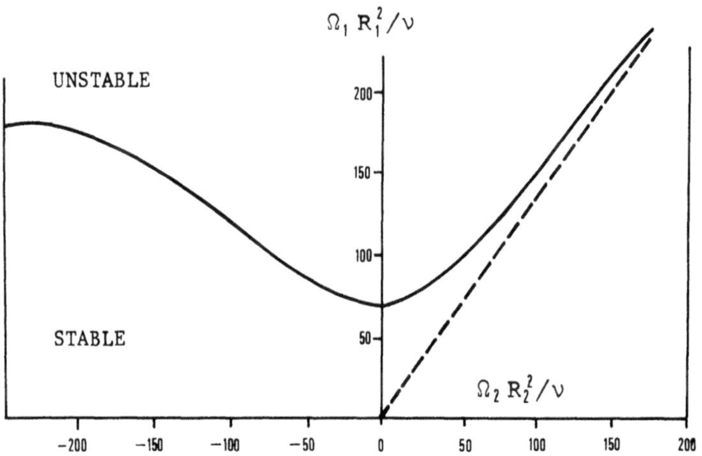

Figure 4.3

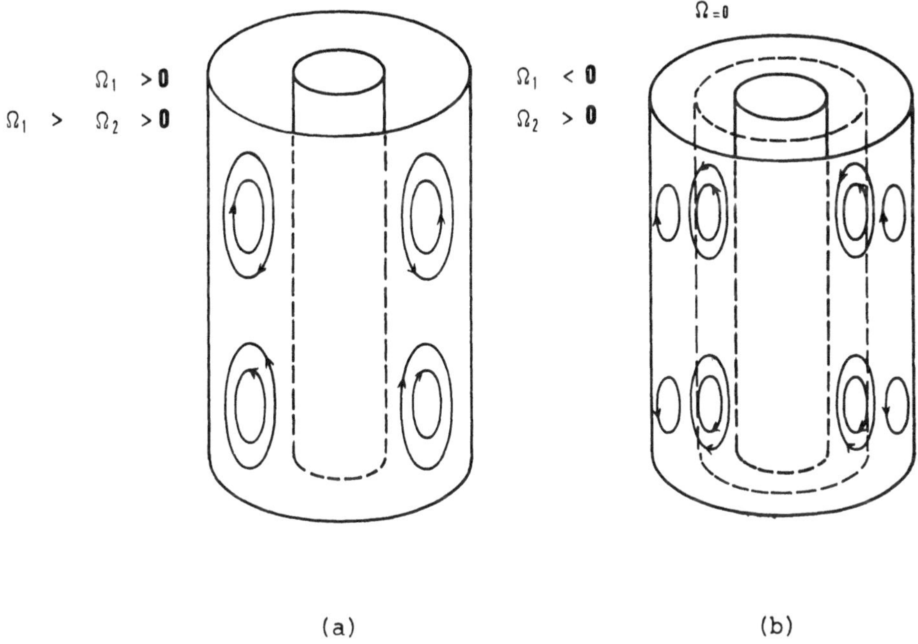

Figure 4.4

We note that in part (b) of Figure 4.4, corresponding to the case where Ω_1 and Ω_2 are of opposite sign, so that m^2 changes sign, the primary circulation in the unstable region ($\partial m^2/\partial R < 0$) extends slightly into the stable region ($\partial m^2/\partial R > 0$), where it induces a small secondary, frictionally-driven circulation. This is analogous to the phenomenon of penetrative convection where the instability motions in a gravitationally unstable region penetrate into a gravitationally stable region* . Indeed, the roll vortices produced by instability in a centrifugal force field in which the angular momentum decreases in the direction of the centrifugal

* See Veronis, 1963 : Penetrative convection. The Astrophysical Journal, 137, no. 2, pp. 641-663.

force are very closely analogous to the roll vortices produced by instability in a gravitational field in which the specific volume (entropy in a compressible fluid) decreases in the direction of gravity; and when the distance between the cylinders is small, and they are rotating in the same direction, the analogy to small amplitude convection between parallel plates is complete.*

The general criteria for stability in the circular vortex

The general stability criteria for both compressible and incompressible fluids have been shown to depend on the sign of the quadratic form

$$\underline{\delta r}\cdot M\cdot\underline{\delta r} = M_{ij}\, S_i S_j \tag{4.31}$$

$$M = (\nabla m^2)(-\nabla\chi) + (\nabla h)(-\alpha\nabla p),$$

where

$$h = \begin{cases} \ln\theta, & \text{perfect gas} \\ \ln\alpha, & \text{incompressible fluid.} \end{cases}$$

We first show that M is symmetric, i.e., that $M_{12} = M_{21}$. Setting

$$M = \underline{A}\,\underline{B} + \underline{C}\,\underline{D},$$

where

$$\underline{A} = \nabla m^2,\ \underline{B} = -\nabla\chi,\ \underline{C} = \nabla h,\ \underline{D} = -\alpha\nabla p,$$

we find

---

*Jeffries, H., 1928 : Some cases of instability in fluid motion. Proc. Roy. Soc. A 118, 185-208. The general problem of the onset of convection between parallel planes was first treated by Rayleigh (1916a, Stability Bibliography).

$$M_{12} - M_{21} = (A_1B_2 - A_2B_1) + (C_1D_2 - C_2D_1) = \underline{A} \times \underline{B} + \underline{C} \times \underline{D}$$

$$= (\nabla m^2) \times (-\nabla\chi) + (\nabla h) \times (-\alpha\nabla p)$$

$$= (\nabla m^2) \times (-\nabla\chi) + (\nabla\alpha) \times (-\nabla p) = 0 \text{ by (4.7).}$$

Hence

$$\delta\underline{r}\cdot M\cdot\delta\underline{r} = M_{11}S_1^2 + (M_{12} + M_{21})\, S_1S_2 + M_{22}S_2^2$$

is positive definite provided that both

$$\text{Det } M = M_{11}M_{22} - M_{12}M_{21}$$

and

$$\text{Trace } M = M_{11} + M_{22} > 0 .$$

The first condition guarantees that the form is definite (non-vanishing) and the second that it is positive.

For absolute instability we must have Det M > 0 and Trace M < 0. If Det M < 0, the vortex will be stable for some displacements and unstable for others. Thus, the general stability condition may be restated in the form

$$\begin{array}{lll} \text{Det } M > 0, \text{ Trace } M > 0 & & \text{stable} \\ \text{Det } M > 0, \text{ Trace } M < 0 & & \text{unstable} \\ \text{Det } M < 0, & & \text{conditionally unstable} \end{array} \tag{4.32}$$

Now

$$\text{Det } M = M_{11}M_{22} - M_{12}M_{21}$$

$$\text{Det } M = (A_1B_1 + C_1D_1)(A_2B_2 + C_2D_2) - (A_1B_2 + C_1D_2)(A_2B_1 + C_2D_1) ,$$

$$= (A_1C_2 - A_2C_1)(B_1D_2 - B_2D_1) = (\underline{A} \times \underline{C})\ (\underline{B} \times \underline{D})$$

$$= (\nabla m^2) \times (\nabla h) \bullet (-\nabla\chi) \times (-\alpha\nabla p) , \tag{4.33}$$

and

$$\text{Trace } M = M_{11} + M_{22} = A_1B_1 + C_1D_1 + A_2B_2 + C_2D_2$$

$$= \underline{A} \bullet \underline{B} + \underline{C} \bullet \underline{D} = (\nabla m^2) \bullet (-\nabla\chi) + (\nabla h) \bullet (-\alpha\nabla p) . \tag{4.34}$$

The significance of these equations is brought out by representing the vectors involved in a polar diagram. The $-\nabla\chi$ vector will be oriented in the radial direction and the $-\alpha\nabla p$ vector in a quasi-vertical direction. For definiteness let us confine ourselves to the case shown in Figures 4.5(a) and 4.5(b) where all the vectors lie in the first quadrant. For this case, Trace M $>$ 0, and the vortex is stable or conditionally unstable according as Det M $\gtrless$ 0. We see from the figure that (a) corresponds to absolute stability and (b) to conditional instability.

Let us consider the sign of each of the terms in

$$\delta\underline{r}\cdot M\cdot\delta\underline{r} = (\nabla m^2\cdot\delta\underline{r})(-\nabla\chi\cdot\delta\underline{r}) + (\nabla h\cdot\delta\underline{r})(-\alpha\nabla p\cdot\delta\underline{r}) . \tag{4.35}$$

The first term will be positive providing the displacement $\delta\underline{r}$ does not lie in the horizontally shaded areas in either diagram, for then $\nabla m^2\cdot\delta\underline{r}$ and $-\nabla\chi\cdot\delta\underline{r}$ will be both positive or both negative. Similarly, the second term will be positive if the displacement lies outside the vertically shaded areas. Each term by itself represents a different stabilizing or destabilizing effect: the first the work done against the centrifugal force, the second the work done against the gravitational force. The motion is certainly stable if the displacement lies outside both shaded areas. But what is **not** obvious is that in Figure 4.5(a), where Det M $>$ 0, it is stable in the other directions as well. If, however, Det M is negative, so that the two shaded areas overlap as in Figure 4.5(b), a displacement in the cross-hatched region where the areas overlap will give rise

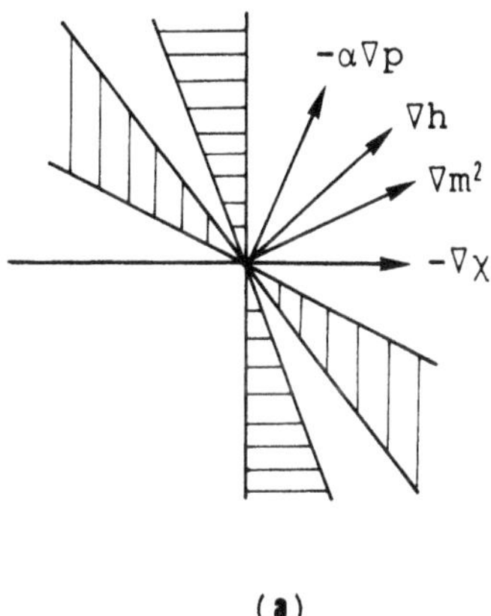

(a)

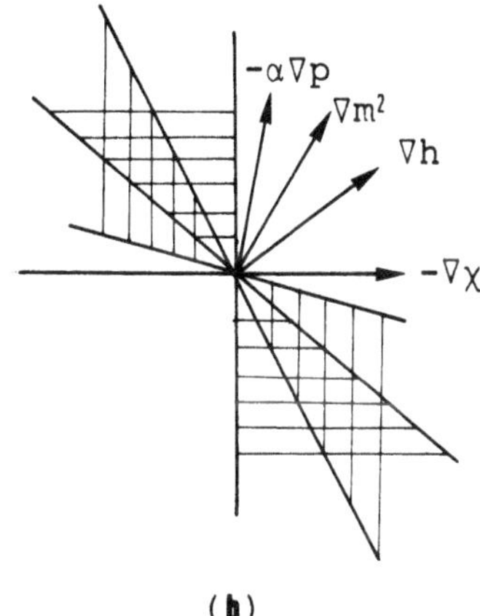

(b)

Figure 4.5

to both statically and dynamically destabilizing effects, and the motion is conditionally unstable.*

Suppose an overturning takes place as a consequence of instability. Fjørtoft (1950, Stability Bibliography) has shown that the new configuration will remain unstable. We may prove his result by deriving a new invariant of the motion: From the conservation of the Lagrangian volume element $\rho d\tau$ we obtain

$$\rho d\tau = \rho 2\pi R dR dz = \rho 2\pi R \frac{d(m^2)dh}{\frac{\partial(m^2,h)}{\partial(R,z)}} = \text{constant}, \qquad (4.36)$$

and therefore, since $d(m^2)$ and dh are also constants of the motion,

* In this case it is plausible that perturbations can be found satisfying the equations of motion for which the displacements are mainly in the destabilizing directions so that the instability can be realized ; but the displacement cannot be entirely so, and a rigorous proof of instability is still required for the case Det M $<$ 0. Such a proof has been given by Ooyama (1966, Stability Bibliography) for an incompressible fluid.

$$\frac{1}{\rho R}\frac{\partial(m^2,h)}{\partial(R,z)} = \frac{\nabla(m^2) \times \nabla h \cdot \underline{j}}{\rho R} = \text{constant*}, \qquad (4.37)$$

where $\underline{j}$ is a unit vector in the positive (counterclockwise) v direction. It follows from (4.33) and (4.34) that the overturning conserves $\frac{\nabla m^2 \times \nabla h}{\rho R}$ and therefore, in the situation represented by Figure 4.5(b), the sign of Det M.

If R and z are replaced as independent variables by R and h, equation (4.36) becomes

$$\frac{1}{\rho R}\frac{\partial(m^2,h)}{\partial(R,z)} = \frac{1}{\rho R}\frac{\partial(m^2,h)}{\partial(R,h)}\frac{\partial(R,h)}{\partial(R,z)} = \frac{1}{\rho R}\left(\frac{\partial m^2}{\partial R}\right)_h \left(\frac{\partial h}{\partial z}\right)_R = \text{constant}, \quad (4.38)$$

where the subscripts "h" and "R" refer to differentiation at constant h and R respectively. Instability in the atmospheric or oceanic case normally occurs when $\left(\frac{\partial h}{\partial z}\right)_R > 0$ and $\left(\frac{\partial m^2}{\partial R}\right)_h < 0$, i.e., when the centrifugal force is destabilizing for displacements which do not give rise to buoyancy forces. The transition from stability to instability is exemplified in Figures 4.5(a) and (b) which differ only in that ∇m^2 and ∇h are interchanged, i.e., in the sign of $\left(\frac{\partial m^2}{\partial R}\right)_h$.

Actually, it is rare to find a close approximation to a circular vortex in the atmosphere or oceans. However, any moderately straight or uniformly curved current in a rotating system may be regarded as part of a circular vortex, and one may calculate its stability with respect to transverse perturbations which do not vary, or vary only slowly, along the current. As will be shown in the next section, the quantity v + fx, where v is the current speed, x a transverse horizontal coordinate directed to the right of the current, and f twice the vertical component of the earth's angular velocity, is conserved in quasi-rectilinear motion with respect to the rotating earth. If one considers displacements in the Coriolis force field, the instability criterion becomes

*This is a special case of the theorem of conservation of potential vorticity, $\nabla \times \underline{V} \cdot \nabla h/\rho$, proved in Chapter VI.

$$\left(\frac{d(v + fx)}{dx}\right)_h = \left(\frac{dv}{dx}\right)_h + f < 0 \ .$$

In strong currents, such as the atmospheric jet stream, hurricanes, or the oceanic Gulf Stream, this criterion is often approached or slightly exceeded on the anticyclonic shear side, and one might expect to find transverse roll-like vortices due to the instability. It cannot be said, however, that they have been found. One possible indication may be the parallel cloud bands, space some tens of km apart, which are sometimes observed in the anticyclonic shear zone of the atmospheric jet stream parallel to its axis.

Effect of viscosity and heat conduction

One ordinarily expects diffusion of momentum and heat to be stabilizing, but McIntyre (1970a, Stability Bibliography) has shown that their effects may be destabilizing. Let us consider, for example, a region where the centrifugal force acting on the field of angular momentum is by itself destabilizing but that gravity acting on the density field is stabilizing. Then if the Prandtl number ν/κ is very small, a displacement of a ring in a direction in which the centrifugal force is destabilizing may bo altogether destabilizing because the density difference between the ring and the environment will be quickly neutralized by heat conduction, whereas the angular momentum difference will not be neutralized by viscosity. Similarly, if the gravity force is stabilizing while the centrifugal force is destabilizing, a high Prandtl number may destabilize the flow.*

Normal mode analysis of stability

The problem of the stability of a circular vortex may also be treated by normal mode theory. Let us suppose for simplicity that the

* The situation has some analogy to the destabilizing effect on a stably stratified ocean of a positive vertical gradient of salt concentration, due to the greater molecular diffusivity of heat than of salt. See footnote, page 60.

fluid is Boussinesq and that the flow is so far from the axis of rotation in a system of cylindrical coordinates which rotates with the fluid, R and z may be regarded as Cartesian coordinates with gravity (gravitational plus centrifugal force) acting in the negative z direction.

If the flow is inviscid, and there are no heat sources, the equations of motion become

$$\frac{Du}{Dt} - 2\omega v = -\frac{1}{\rho_{oo}} p_R ,$$

$$\frac{Dv}{Dt} + 2\omega u = 0 ,$$

$$\frac{Dw}{Dt} = -\frac{1}{\rho_{oo}} p_z - \frac{g}{\rho_{oo}} \rho , \qquad (4.39)$$

$$u_R + w_z = 0 ,$$

$$\frac{D\rho}{Dt} = 0 ,$$

where the basic flow ($\bar{u} = 0$, $\bar{v}$, $\bar{w} = 0$, $\bar{p}$, $\bar{\rho}$) must satisfy the equilibrium conditions,

$$- 2\omega\bar{v} = -\frac{1}{\rho_{oo}} \bar{p}_R ,$$

$$(4.40)$$

$$0 = -\frac{1}{\rho_{oo}} \bar{p}_z - \frac{g}{\rho_{oo}} \bar{\rho} ,$$

or, by elimination of $\bar{p}$, the thermal wind equation,

$$2\omega\bar{v}_z = -\frac{g}{\rho_{oo}}\bar{\rho}_R \quad . \tag{4.41}$$

Consider small perturbations of this basic flow, u', v', w', p', ρ'. These satisfy the perturbation equations

$$\begin{aligned} u'_t - 2\omega v' &= -\frac{p'_R}{\rho_{oo}} \\ v'_t + u'\bar{v}_R + w'\bar{v}_z + 2\omega u' &= 0 \quad , \\ w'_t &= -\frac{p'_z}{\rho_{oo}} - \frac{\rho'}{\rho_{oo}} \quad , \\ u'_R + w'_z &= 0 \; , \\ \rho'_t + u'\bar{\rho}_R + w'\bar{\rho}_z &= 0 \; . \end{aligned} \tag{4.42}$$

Now suppose that the fluid is unbounded and that all gradients of mean flow quantities are constant.* In this case the normal modes are plane wave solutions of the form $\exp\left[i(k_1R + k_2z) + \sigma t\right]$, and we wish to investigate the conditions under which solutions with real, positive σ exist.

Problem 4.1: (a) Derive the stability criteria from the above method and show that they are analogous to the general criteria already found.

(b) In the case $\bar{v}_R = 0$, $\bar{\rho}_z < 0$, show that the condition for instability is Ri < 1, where

$$Ri = -\frac{g}{\rho_{oo}}\frac{\partial\bar{\rho}}{\partial z}\bigg/\left(\frac{\partial\bar{v}}{\partial z}\right)^2$$

is the Richardson number.

---

* We envisage that if instabilities occur, as in the Taylor counter-rotating cylinder, they will occur on the scale of the unstable region and therefore that the problem is essentially a local one, not dependent for its existence on boundaries. One can think of the gradient as varying on a scale which is much greater than the scale of the unstable motions.

Problem 4.2: Consider a stationary axisymmetric vortex in an inviscid, homogeneous, incompressible fluid with gravity acting parallel to the axis of rotation. The vortex is bounded below by a rigid horizontal plane and above by a free surface at the height h(R). Find the stability criteria for axially-symmetric displacements $\{\delta R, \delta z\}$ for which δR is independent of z, i.e., for which a vertical fluid column is displaced into a vertical fluid column. Prove :

(a) $$J = K + P = \rho \int \frac{h}{2}\left(gh + \frac{m^2}{R^2}\right) 2\pi R dR \;;$$

(b) $$\delta J = 0 \;;$$

(c) $$\delta^2 J = \rho \int \left\{ g \left(\frac{1}{R}\frac{d}{dR}(hR\delta R) \right)^2 + \frac{h}{R^3}(\delta R)^2 \frac{d(m^2)}{dR^2} \right\} 2\pi R dR \;,$$

where m = Rv. Hint : The displacements in question satisfy the continuity equation

$$\frac{dh}{dt} + h\nabla\cdot\underline{V} = 0 \quad \text{or} \quad \delta h + \frac{h}{R}\frac{d}{dR}(R\delta R) = 0 \;.$$

Solutions :

Problem 4.1(a): Substituting $(u', v', w', \frac{p'}{\rho_{oo}}, \frac{\rho'}{\rho_{oo}}) =$ $(U,V,W,P,R)\exp\left[i(kx+nz+\sigma t\right]$ in (4.42) we get

$$\sigma U - 2\omega V + ikP = 0$$

$$(2\omega + \bar{v}_R)U + \sigma V + \bar{v}_z W = 0$$

$$\sigma W + inP - R = 0$$

$$ikU + nW = 0$$

$$\frac{\bar{\rho}_R}{\rho_{oo}} U + \frac{\bar{\rho}_z}{\rho_{oo}} W + \sigma R = 0$$

As these equations are linear and homogeneous with constant coefficients, their determinant must vanish. We obtain the dispersion equation

$$(k^2 + n^2)\sigma^2 - g\frac{\bar{\rho}_z}{\rho_{oo}}k^2 - (-g\frac{\bar{\rho}_z}{\rho_{oo}} + 2\omega\bar{v}_z)kn + 2\omega(2\omega + \bar{v}_R)n^2 = 0 ,$$

which gives the stability condition

$$-g\frac{\bar{\rho}_z}{\rho_{oo}}k^2 - (-g\frac{\bar{\rho}_z}{\rho_{oo}} + 2\omega\bar{v}_z)kn + 2\omega(2\omega + \bar{v}_R)n^2 \begin{matrix} > 0 & \text{stable} \\ < 0 & \text{unstable.} \end{matrix}$$

By (4.41) the determinant of this quadratic form gives the criterion,

$$\text{Det} = \frac{2\omega g}{\rho_{oo}}\left[-\bar{\rho}_z(2\omega + \bar{v}_R) + \bar{\rho}_R\bar{v}_z\right] = -\frac{2\omega g}{\rho_{oo}}\nabla\bar{m} \times \nabla\bar{\rho}\cdot\underline{j} \quad \begin{matrix} > 0 & \text{stable} \\ < 0 & \text{unstable ,} \end{matrix}$$

where $\bar{m} = 2\omega R + \bar{v}$. This criterion is seen to be closely analogous to (4.32), (4.33).

Problem 4.1(b) : If $\bar{v}_R = 0$ and $\bar{\rho}_z < 0$, (4.41) gives

$$\text{Det} = 4\omega^2\left[-\frac{g\bar{\rho}_z}{\rho_{oo}} + (v_z)^2\right] = 4\omega^2(v_z)^2(Ri - 1) ,$$

so that Ri < 1 corresponds to instability.

Problem 4.2(a) : It follows from the torque balance relationship (4.7) that the angular momentum $m = Rv$ is independent of z at equilibrium. The kinetic energy per unit horizontal area is therefore $\frac{\rho h m^2}{2R^2}$, and the potential energy is $\int_o^h \rho g z dz = \frac{\rho g h^2}{2}$. Hence

$$J = K + P = \rho\int_A \frac{h}{2}(gh + \frac{m^2}{R^2})dA ,$$

where $dA = 2\pi R dR$.

(b) The force balance relationship (4.6) states that

$$\rho \frac{m^2}{R^3} - \frac{\partial p}{\partial R} = 0 ,$$

$$\rho g - \frac{\partial p}{\partial z} = 0 .$$

Integration of the second from 0 to h gives $p = \rho gh$ and substitution in the first gives

$$\frac{m^2}{R^3} - g \frac{dh}{dR} = 0 .$$

Since hdA and m^2 are conserved for a variation in which δR is independent of height, we obtain

$$\delta J = \rho \int_R (g \frac{\delta h}{2} - \frac{m^2}{R^3} \delta R) h 2\pi R dR .$$

Substitution of the mass-continuity relationship

$$\delta h = - \frac{h}{R} \frac{d}{dR} (R \delta R) ,$$

and integration by parts gives

$$\delta J = \rho \int_R (g \frac{dh}{dR} - \frac{m^2}{R^3}) \delta R h 2\pi R dR$$

which is seen to vanish at equilibrium.

(c) The second variation gives

$$\delta^2 J = \rho \int_A (g \frac{dh}{dR} - \frac{m^2}{R^3}) \delta(\delta R) h dA + \rho \int_A \left[g \frac{d}{dR} (\delta h) - m^2 \delta (\frac{1}{R^3}) \right] \delta R h dA .$$

The first integral vanishes by the equilibrium condition. In the second we substitute

$$\delta h = - \frac{h}{R} \frac{d}{dR} (R \delta R) = - \frac{1}{R} \frac{d}{dR} (h R \delta R) + \frac{dh}{dR} \delta R$$

and set $\frac{1}{R}\frac{d}{dR}(hR\delta R) \equiv D$, $\frac{1}{2R^2} \equiv \chi$. Then, successively,

$$\delta^2 J = \rho\int_R \left[-g\frac{dD}{dR} + g\frac{dh}{dR}\delta R + m^2\frac{d}{dR}\left(\frac{d\chi}{dR}\delta R\right)\right] h\delta R 2\pi R dR$$

$$= \rho\int_R \frac{d}{dR}\left[\left(-gD + g\frac{dh}{dR}\delta R + m^2\frac{d\chi}{dR}\delta R\right)hR\delta R\right]2\pi dR$$

$$+ \rho\int_R \left[gD^2 + \left(-g\frac{dh}{dR} - m^2\frac{d\chi}{dR}\right)D\delta R - h\frac{d(m^2)}{dR}\frac{d\chi}{dR}(\delta R)^2\right]2\pi R dR$$

$$= \rho\int_R \left[gD^2 + \frac{h}{R^3}(\delta R)^2\frac{d(m^2)}{dR}\right]2\pi R dR ,$$

V. SLOW THERMALLY AND FRICTIONALLY DRIVEN CIRCULATIONS IN A CIRCULAR VORTEX* AND THE MECHANISM OF ADJUSTMENT

Equilibrium theory

It was recognized by Buys Ballot** that motions are quasi-geostrophic in the atmosphere and by Ferrel*** that this is also true of the oceans. However, classical works on the fluid dynamics of planetary-scale motions ignored this constraint on the velocity. The planetary fluid dynamics of the atmosphere and oceans arose, in a sense, when the geostrophic constraint on the motion of the atmosphere was incorporated into the theory †. An excellent review of the subject has been given by Phillips ††.

The equations of motion should be written and scaled according to the relevant non-dimensional parameters as in other branches of fluid mechanics. Such an analysis yields the conclusion that the large-scale motions are indeed quasi-geostrophic and quasi-hydrostatic. The systematic use of the geostrophic and hydrostatic approximations is then found to greatly simplify the treatment of large-scale motions.

The simplest system that we may treat to gain physical insight is that of axially-symmetric displacements in a circular vortex. In a crude sense, the earth's own atmosphere is a circumpolar vortex driven by solar heating and retarded by surface friction. What is characteristic of such motions is that the time scale of the driving forces is large in comparison with the natural periods of oscillation of the system, and that the vortex is itself stable so that self-excited oscillations do not arise. As a consequence, the accelerations are small and the motion passes

---

* Eliassen, A. : 1951 : Astrophysica Norvegica, Vol. V, no. 2, p. 19.

** Buys Ballot, C. H. D., 1857 : Jaarboek, Met. Inst. Nederlands.

*** Ferrel, W., 1856 : Nashville J. Med. Surg. II(4 and 5), and 1876 : Amer. J. Sci., 8. Both of Ferrel's papers are reprinted in the Abbe Collection of Translations, Smithsonian Miscellaneous Collections (2 vols.).

† Charney, J. G. 1948 : Geofys. Publikasjoner, 17, 2 ; Eliassen, A., 1949 : Geofys. Publ., 17, 3 ; Obukhov, A., 1949 : Izv. Akad. Nauk USSR, Ser. Geograf. Geofiz., 13, 4.

†† Phillips, N. A., 1963 : Rev. Geophys., 1, 2,

through a succession of states of balance among the pressure, gravitational, centrifugal, and external forces. The situation may be illustrated by the linear oscillator (spring) driven by a periodic force. If x is the displacement, ω^2 the restoring force per unit mass per unit displacement, and $e^{i\nu t}$ the driving force,

$$x + \omega^2 x = e^{i\nu t} . \tag{5.1}$$

The solution of this equation is

$$x = c_1 e^{i\omega t} + c_2 e^{-i\omega t} + \frac{e^{i\nu t}}{\omega^2 - \nu^2} , \tag{5.2}$$

where c_1 and c_2 are constants. Suppose that transient terms are damped out by a small amount of friction so that we are left with only the forced motion. Suppose also that the frequency of the driving force is much smaller than the natural period, i.e., $\nu^2 << \omega^2$; then

$$x \simeq \frac{e^{i\nu t}}{\omega^2} , \tag{5.3}$$

which corresponds to the solution of (5.1) with the acceleration term omitted, i.e., to a balance of the driving force and the restoring force.

A geophysical illustration is given by the equilibrium theory of the tides *. In this case, as in the case of the linear oscillator, the acting forces produce instantaneous displacements (or velocities), not accelerations. This behavior is unlike that of the original Newtonian systems in which the initial positions and velocities of the particles can be arbitrarily specified ; here they cannot.

Equations of slow motion of a circular vortex **

We assume that the transverse circulations produced in a circular vortex by the action of axially-symmetric heat sources or frictional torques are so slow that the meridional accelerations and the meridional frictional sources may be ignored. The force equilibrium equation (4.6) will continue to apply in an approximate sense. How do we deal with the dynamics of such a system ?

---

* Lamb, H. : Hydrodynamics, 6th edition, Appendix to Chapter VIII.

** The theory to be presented is due to A. Eliassen, 1951, loc. cit.

For simplicity, suppose that the fluid is quasi-incompressible, quasi-homogeneous and Boussinesq.† If we assume that the disturbing forces are symmetrically distributed about the axis, the axial symmetry of the vortex will not be destroyed. The frictional forces will in general exert a torque causing time changes in the angular momentum of the fluid particles. We shall employ (4.4) in the form

$$\frac{Dm^2}{Dt} = 2mRF_\lambda \tag{5.4}$$

to describe the time changes in the angular momentum. The heat sources and sinks will bring about small variations in density which, when coupled with gravity, will become important for driving the circulation. The density changes will be governed by equation (2.17) :

$$\frac{D\rho}{Dt} = S \ , \tag{5.5}$$

where S is the thermally induced rate of change of density. As a result of frictional torques and heating the balance of the vortex will be disturbed, and meridional motions will take place to restore the balance.

Gravity is assumed, for simplicity, to act paral l to the axis of rotation. With the introduction of the cylindrical coordinates R and z, the force balance condition (4.6) may be resolved into its components. In the radial direction,

$$\frac{m^2}{R^3} - \alpha \frac{\partial p}{\partial r} = 0 \ . \tag{5.6}$$

Since the density is not coupled with gravity in this equation, we may use the Boussinesq approximation and write

$$\frac{\rho_{oo}m}{R^3} - \frac{\partial p}{\partial R} = 0 \ , \tag{5.7}$$

where ρ_{oo} is a constant mean density.* The vertical equation of motion is given by

$$- \rho g - \frac{\partial p}{\partial z} = 0 \ , \tag{5.8}$$

† The compressible vortex is treated by Eliassen (see footnote on page 76) using pressure as a vertical coordinate.

* This assumption implies that the percentage variation of m^2 is much greater than that of ρ. If the flow is a small deviation from solid rotation, the constant density surfaces in solid rotation (which are parallel to the constant pressure surfaces) must have a much smaller slope than those in the actual flow.

where ρ is here treated as a variable density since it is coupled with gravity.

Elimination of the pressure between (5.7) and (5.8) gives

$$g \frac{\partial \rho}{\partial R} = \frac{\rho_{oo}}{R^3} \frac{\partial m^2}{\partial z} , \qquad (5.9)$$

which is the so-called thermal wind equation for our system.

Equations (5.4) and (5.5) may be written

$$\frac{\partial m^2}{\partial t} + u \frac{\partial m^2}{\partial R} + w \frac{\partial m^2}{\partial z} = 2mRF_{\lambda} , \qquad (5.10)$$

and

$$\frac{\partial \rho}{\partial t} + u \frac{\partial \rho}{\partial R} + w \frac{\partial \rho}{\partial z} = S . \qquad (5.11)$$

Combining (5.7) and (5.10), we get

$$\frac{\partial}{\partial R} \left(\frac{\partial p}{\partial t} \right) + \frac{\rho_{oo} u}{R^3} \frac{\partial m}{\partial R} + \frac{\rho_{oo} w}{R^3} \frac{\partial m^2}{\partial z} = \frac{2\rho_{oo} m F_{\lambda}}{R^2} . \qquad (5.12)$$

In the same manner (5.8) and (5.11) may be written

$$- \frac{\partial}{\partial z} \left(\frac{\partial p}{\partial t} \right) + gu \frac{\partial \rho}{\partial R} + gw \frac{\partial \rho}{\partial z} = gS . \qquad (5.13)$$

The equation of continuity of mass in cylindrical coordinates is given by

$$\frac{\partial}{\partial R} (Ru) + \frac{\partial}{\partial z} (Rw) = 0 . \qquad (5.14)$$

Hence we may define a stream function ψ for the meridional velocity such that

$$Ru = \frac{\partial \psi}{\partial z} , \quad Rw = - \frac{\partial \psi}{\partial R} . \qquad (5.15)$$

The curves ψ = constant are the streamlines of the meridional motion in the R - z plane.

Substitution of (5.15) into equations (5.13) and (5.12) gives

$$-\frac{\partial}{\partial z}\left(\frac{\partial p}{\partial t}\right) + A\frac{\partial\psi}{\partial R} + B\frac{\partial\psi}{\partial z} = E \qquad (5.16)$$

and

$$\frac{\partial}{\partial R}\left(\frac{\partial p}{\partial t}\right) + B\frac{\partial\psi}{\partial R} + C\frac{\partial\psi}{\partial z} = F \qquad (5.17)$$

where $A \equiv -\frac{g}{R}\frac{\partial\rho}{\partial z}$, $B \equiv \frac{g}{R}\frac{\partial\rho}{\partial R} = -\frac{\rho_{oo}}{R^4}\frac{\partial m^2}{\partial z}$ (by the thermal wind equation),

$C \equiv \frac{\rho_{oo}}{R^4}\frac{\partial m^2}{\partial R}$, $E \equiv gS$, and $F \equiv \frac{2\rho_{oo}mF_\lambda}{R^2}$.

Cross-differentiating (5.16) and (5.17) we get

$$L(\psi) \equiv \frac{\partial}{\partial R}\left(A\frac{\partial\psi}{\partial R} + B\frac{\partial\psi}{\partial z}\right) + \frac{\partial}{\partial z}\left(B\frac{\partial\psi}{\partial R} + C\frac{\partial\psi}{\partial z}\right) = \frac{\partial E}{\partial R} + \frac{\partial F}{\partial z} \equiv \mu(R,z)\ ; \qquad (5.18)$$

ψ = 0 at a rigid boundary.

This differential equation for the stream function is seen to be of the elliptic, parabolic, or hyperbolic type according as the sign of $AC - B^2$ is greater than, equal to, or less than zero, respectively. We have

$$AC - B^2 = \frac{\rho_{oo}g}{R^5}\left(\frac{\partial m^2}{\partial z}\frac{\partial\rho}{\partial R} - \frac{\partial m^2}{\partial R}\frac{\partial\rho}{\partial z}\right), \qquad (5.19)$$

but from (4.33),

$$\text{Det } M = (\nabla m^2) \times (\nabla h) \cdot (-\nabla\chi) \times (-\alpha\nabla p)\ , \qquad (5.20)$$

which, by (4.6), may be written

$$\text{Det } M = (\nabla m^2) \times (\nabla h) \cdot (-\nabla\chi) \times (\nabla\Phi)$$
$$= -\frac{g}{R^3}\left\{\frac{\partial m^2}{\partial z}\frac{\partial h}{\partial R} - \frac{\partial m^2}{\partial R}\frac{\partial h}{\partial z}\right\} \quad . \tag{5.21}$$

Since the Boussinesq approximation gives

$$\frac{\partial h}{\partial R} \simeq \frac{1}{\rho_{oo}}\frac{\partial \rho}{\partial R} \quad \text{and} \quad \frac{\partial h}{\partial z} \simeq -\frac{1}{\rho_{oo}}\frac{\partial \rho}{\partial z} \quad , \tag{5.22}$$

we get

$$\text{Det } M = \frac{g}{\rho_{oo}R^3}\left\{\frac{\partial m^2}{\partial z}\frac{\partial \rho}{\partial R} - \frac{\partial m^2}{\partial R}\frac{\partial \rho}{\partial z}\right\} = \frac{R^2}{\rho^2_{oo}}\ (AC - B^2\ ), \tag{5.23}$$

and we obtain the result that <u>the differential equation of the stream function (5.18) is of the elliptic type for all stable vortices</u>. The converse of this statement is also true if, in addition, Trace M < 0.

If an unstable vortex were acted upon by sources of heat and frictional torques the ensuing motion would not in general be in a state of balance, and the above analysis would not apply. Since (5.18) has been derived on the assumption that a state of balance with respect to meridional motion does exist, it may be applied <u>only</u> to stable vortices.

We note that whereas the meridional velocities in the general circular vortex can be specified arbitrarily, they are here completely determined by the instantaneous sources of heat and angular momentum and by the distribution of m^2 and h (or, from (5.8) and (5.9), by p only).

If we had eliminated ψ between (5.16) and (5.17), we would have obtained an elliptic equation in $\partial p/\partial t$ with coefficients expressible in terms of p and its space derivatives only. Thus the acting forces also determine the instantaneous time derivative of the pressure.

The Green's function for the system (5.18)

Equation (5.18) is solved by a Green's function :

$$\psi(R,z) = \int_\tau G(R_o z_o\ ;\ R,z)\mu(R_o,z_o)dR_o dz_o \ ,$$

or, if the point (R,z) is denoted by the vector $\underline{r}$,

$$\psi(\underline{r}) = \int_\tau G(\underline{r}_o \cdot \underline{r})\, \mu(\underline{r}_o)\, dR_o dz_o . \qquad (5.24)$$

Eliassen (loc.cit.) shows that the function G has a logarithmic singularity at the point $\underline{r} = \underline{r}_o$ whose principal part is

$$G_1 = \frac{1}{2\pi\delta_o} \ln\left[C_o(R - R_o)^2 - 2B_o(R - R_o)(z - z_o) + A_o(z - z_o)^2\right]^{1/2}, \qquad (5.25)$$

$$(\delta_o = A_o C_o - B_o^2) .$$

Thus G may be expressed as the sum of G_1 and a function which is regular at $\underline{r}_o$, so that in the vicinity of $\underline{r}_o$ the behavior of $G(\underline{r};\underline{r}_o)$ is given by that of $G_1(\underline{r};\underline{r}_o)$.

Examples of circulation in the vicinity of point sources

We now present some examples of motions produced by point sources of heat and momentum. The function G_1 represents the solution (apart from a constant factor) of (5.18) when $\mu(R,z)$ vanishes everywhere except at the point (R_o,z_o), and there becomes infinite.

Consider a point source $\mu(R_o,z_o)$ at (R_o,z_o). We wish to determine the effects of such a discontinuity (jump) in the frictional torques or heat sources. To find the streamlines of the meridional motion in the vicinity of (R_o,z_o), we set G_1 = constant in the vicinity of the point.

For the case of vanishing radial density gradient we have

$$B_o = \left(\frac{g}{R}\frac{\partial\rho}{\partial R}\right)_o = -\frac{\rho_o}{R^4}\left(\frac{\partial m^2}{\partial z}\right)_o = 0 ,$$

and

$$C_o(R - R_o)^2 + A_o(z - z_o)^2 = \text{constant} ,$$

Hence the streamlines in the meridional plane are conformal, concentric ellipses. If an elliptical streamline intersects the line $R = R_o$ at z_1 and the line $z = z_o$ at R_1, we have

$$\frac{R_1 - R_o}{z_1 - z_o} = \frac{A_o}{C_o} = \frac{\sqrt{-\frac{g}{\rho_{oo}}\left(\frac{\partial\rho}{\partial z}\right)_o}}{\sqrt{\frac{1}{R_o^3}\left(\frac{\partial m^2}{\partial R}\right)_o}} = \frac{\nu_g}{\nu_i}$$

where ν_g is the frequency of a gravitational oscillation and ν_i is the frequency of an inertial oscillation. Thus, if the gravitational stability is weak compared to the inertial stability ($\nu_g \ll \nu_i$), the streamlines are long, narrow ellipses with their major axes nearly parallel to the axis of the vortex, whereas if the inertial stability is weak compared to the gravitational stability, the streamlines will be long, narrow ellipses with their major axes nearly perpendicular to the axis of the vortex. For the rotating earth

$$\frac{\nu_g}{\nu_i} \sim \begin{cases} 100 & \text{in the lower atmosphere} \\ 50 & \text{in the ocean thermocline} \\ 20 & \text{in the surface and deep waters of the ocean.} \end{cases}$$

Hence the streamlines in the vicinity of a discontinuity in E or F would be of the latter type.

As a second example, consider a case in which the vortex is <u>almost</u> unstable. We are treating a limiting case in which $A_oC_o \simeq B_o^2$, or

$$C_o(R - R_o)^2 - 2B_o(R - R_o)(z - z_o) + A_o(z - z_o)^2$$

$$\simeq \frac{1}{A_o}\left[B_o(R - R_o) - A_o(z - z_o)\right]^2 = \text{constant.}$$

Thus the streamlines are highly elongated ellipses which may be approximated by the straight lines

$$\frac{R - R_o}{z - z_o} = \frac{A_o}{B_o} = - \frac{\left(\frac{\partial\rho}{\partial z}\right)_o}{\left(\frac{\partial\rho}{\partial R}\right)_o} \simeq \frac{B_o}{C_o} = - \frac{\left(\frac{\partial m^2}{\partial z}\right)_o}{\left(\frac{\partial m^2}{\partial R}\right)_o} ,$$

i.e., by lines of constant density or angular momentum, which in this case nearly coincide. Since δ may be regarded as a measure of the stability of the vortex, it follows from (5.25) that for a given jump in E or F, the meridional circulation will be strongest where the stability is weakest, a result which is intuitively apparent.

As a third example, consider a unit point source of heat, in the sense that $\int_\tau E \, dRdz = 1$. Writing,

$$\psi(R,z) = \int_\tau G \frac{\partial E}{\partial R_o} dR_o dz_o ,$$

and integrating by parts, we obtain

$$\psi(R,z) = - \int_\tau \frac{\partial G}{\partial R_o} E \, dR_o dz_o ,$$

since G = 0 at the boundary of τ . In the vicinity of the point (R_o,z_o) ,

$$\psi(R,z) \simeq - \frac{\partial G}{\partial R_o} = \frac{1}{2\pi\delta_o} \frac{C_o(R - R_o) - B_o(z - z_o)}{C_o(R - R_o)^2 - 2B_o(R - R_o)(z - z_o) + A_o(z - z_o)^2}$$

whence the streamlines ψ = constant are given by the one-parameter family of curves ,

$$C_o(R-R_o)^2 - 2B_o(R-R_o)(z-z_o) + A_o(z-z_o)^2 + k[C_o(R-R_o) - B_o(z-z_o)] = 0,$$

which represents a family of ellipes passing through the point (R_o,z_o) and tangent to the line, whose slope, obtained by differentiating with respect to R and evaluating at (R_o,z_o), is

$$\left(\frac{dz}{dR}\right)_o = \frac{C_o}{B_o} = -\frac{(\partial m^2/\partial R)_o}{(\partial m^2/\partial z)_o} \quad ,$$

i.e., to the line of constant angular momentum through R_o, z_o (Figure 5.1).

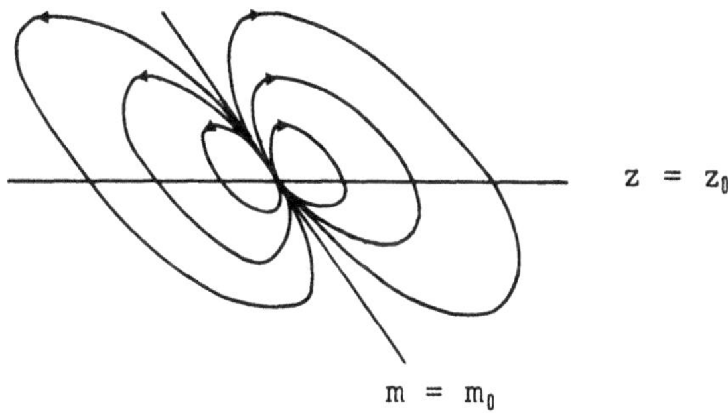

Figure 5.1

The streamlines passing through (R_o, z_o) are tangent to the line of constant angular momentum through this point because the heating impulse changes the density but not the angular momentum. One may show that the dual result is obtained for a point source of angular momentum: the streamlines are tangent to the line of constant density (entropy for a compressible fluid).

Specialization to quasi-rectilinear flow

As in the discussion of the stability of a circular vortex, we may suppose the flow to take place far from the axis of rotation and introduce Cartesian coordinates. The balance equations then take the form

$$\frac{Dv}{Dt} + 2\omega u = F \tag{5.26}$$

$$-2\omega v = -\frac{p_x}{\rho_{oo}} \tag{5.27}$$

$$0 = -\frac{p_z}{\rho_{oo}} - \frac{g\rho}{\rho_{oo}} \tag{5.28}$$

$$u_x + w_z = 0 \tag{5.29}$$

$$\frac{D\rho}{Dt} = S \tag{5.30}$$

where R has been replaced by x, and $F^{(\lambda)}$ by F*. Equation (5.26) may also be derived from the angular momentum equation

$$\frac{D}{Dt}\left[(R_o + x)(\omega(R_o + x) + v)\right] = (R_o + x)F \ ,$$

which reduces to

$$\frac{D}{Dt}(v + 2\omega x) = F \tag{5.26'}$$

by neglect of x in comparison to R_o and of the small quadratic terms in v and F.

Homogeneous ocean

A particularly simple case is that of a homogeneous incompressible ocean. In this case (5.28) may be integrated to give

$$p = \rho g(h - z) \ , \tag{5.31}$$

where h is the height of the free surface of the fluid. In terms of h (5.27) becomes

$$v = \frac{g}{f}\frac{\partial h}{\partial x} \ , \tag{5.32}$$

which shows that v is independent of depth. If, as we shall assume, F is independent of depth as well, it follows that u must also be. The

---

* These equations govern flow on a cone tangent to the earth at latitude ϕ_o (f-plane), if $\omega = \Omega \sin\phi_o$ (=f/2) and the small Coriolis force term $2\Omega v \cos\phi_o$ is ignored in the hydrostatic equation.

continuity equation may therefore be integrated with respect to z to give

$$h\, u_x + w \Big|_0^h = 0$$

or

$$\frac{Dh}{Dt} + h\, u_x = 0\ . \tag{5.33}$$

However, in order to allow the possibility that mass may be added at the free surface, let δx be a moving strip element and assume that $h\ \delta x$ increases at the rate $S\ \delta x$. Then

$$\frac{D}{Dt}(h\ \delta x) = S\ \delta x\ ,$$

or, since $\frac{1}{\delta x}\frac{D(\delta x)}{Dt} = u_x$,

$$\frac{Dh}{Dt} + hu_x = S\ . \tag{5.34}$$

This equation, together with (5.26) and (5.32), now constitute the system governing the motion.

In analogy to the more general case dealt with in the previous section, we may combine the equation of conservation of momentum and mass to obtain a single equation for the meridional velocity, the counterpart equation to (5.18) for the stream function. Setting $f = 2\omega$ and eliminating h_t and v, we obtain

$$\frac{\partial^2(hu)}{\partial x^2} - \frac{1}{gh}\left(f^2 + g\frac{\partial^2 h}{\partial x^2}\right)(hu) = -\frac{f}{g}F + \frac{\partial S}{\partial x}\ . \tag{5.35}$$

Now consider the effect of disturbing the balance by changing the momentum at the rate F or adding mass at the rate S to a resting ocean. In the former case v changes and the Coriolis force associated with this change causes the column to be deflected to the right of the motion. A readjustment of mass takes place simultaneously so that the proper height distribution for a state of geostrophic balance is established (Fig. 5.2.).

A rapid addition of momentum or mass in a limited region would generate gravity waves which would radiate outwards and cause a loss of energy to the region. In the present formalism, we implicitly suppose that the sources of momentum and heat are so slowly acting that no such loss of energy occurs. If the disturbance are small, we may set

$$h = H + h' ,$$
$$u = u' , \tag{5.36}$$
$$v = v' ,$$

where H is the constant depth of the undisturbed ocean, and the perturbation form of (5.35) becomes

$$\frac{\partial^2 u'}{\partial x^2} - \frac{u'}{\lambda^2} = -\frac{f}{gH} F + \frac{1}{H}\frac{\partial S}{\partial x} , \tag{5.37}$$

where

$$\lambda \equiv \sqrt{\frac{gH}{f^2}}$$

is called the "radius of deformation" after C-G. Rossby,* for reasons soon to appear.

Consider the case of a constant force $F = B$ applied to the strip $|x| \lessgtr a$. Then

$$F = \begin{cases} B & |x| \leqslant a , \\ 0, & |x| > a , \end{cases} \qquad S = 0, \tag{5.39}$$

---

* Rossby, C.-G., 1938 : J. Mar. Res., 1 - 2, pp. 239-263.

and (5.37) becomes

$$\frac{\partial^2 u'}{\partial x^2} - \frac{u'}{\lambda^2} = \begin{cases} -\dfrac{B}{\lambda^2 f} , & |x| \leq a \\ 0, & |x| > a \ . \end{cases} \tag{5.40}$$

The condition of zero mass flow at infinity implies that

$$u' = 0 \text{ at } x = \pm\infty , \tag{5.41}$$

and continuity of mass at $x = \pm a$ that

$$[u'] = 0 \text{ at } x = \pm a \ . \tag{5.42}$$

We also require the pressure to be continuous at $\pm a$. This is equivalent to saying that the height of the free surface is continuous, or by the perturbation form of (5.34)

$$[u'_x] = 0 \text{ at } x = \pm a \ . \tag{5.43}$$

The solution to (5.40) subject to these conditions is

$$u' = \frac{B}{f}\begin{cases} e^{-x/\lambda} \sinh \dfrac{a}{\lambda} , & x > a , \\ 1 - e^{-a/\lambda} \cosh \dfrac{x}{\lambda} , & |x| \leq a , \\ e^{x/\lambda} \sinh \dfrac{a}{\lambda} , & x < -a , \end{cases} \tag{5.44}$$

and

$$h' = -\int_0^t H \frac{\partial u}{\partial x} dt = \frac{BHt}{f\lambda} \begin{cases} e^{-x/\lambda} \sinh \dfrac{a}{\lambda} , & x > a , \\ e^{-a/\lambda} \sinh \dfrac{x}{\lambda} , & |x| \leq a , \\ -e^{x/\lambda} \sinh \dfrac{a}{\lambda} , & x < -a \end{cases} \tag{5.45}$$

The height distribution is shown in Fig. (5.2). The zonal velocities

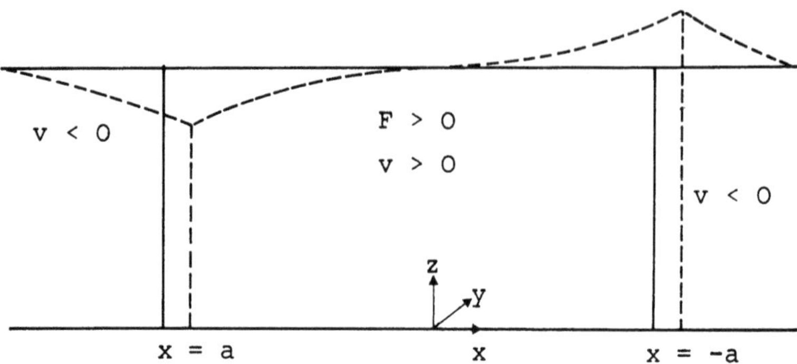

Figure 5.2

satisfying the geostrophic equation (5.32) are positive in the strip where the external force is applied, and negative on either side of the strip. The velocity transverse to the direction of the applied force provides the mass transport necessary to achieve the height distribution for which the motion induced by the external force is enabled to remain in geostrophic balance. The (infinitesimal) displacement of the central current to the right may be calculated from the transverse velocity u', viz., $x - x_o = \int_o^t u dt$.

The radius of deformation is seen to be the effective distance from the region of generation to which the disturbed flow extends. It may be written

$$\lambda^2 = \frac{gH}{f^2} = H^2 \frac{(g/H)}{f^2} = H^2 \frac{\nu_g^2}{\nu_i^2} \tag{5.46}$$

by virtue of the relation $\nu_g^2 = g/H$ (page 59), and is therefore the characteristic horizontal scale of motion corresponding to the vertical scale H.

Rossby's adjustment problem *

Rossby studied the mutual adjustment of mass and velocity distributions for current systems which are established impulsively. Consider again a homogeneous, incompressible ocean of depth H . A certain amount of momentum per unit mass, say V, is impulsively applied to an infinite parallel strip of width 2a.

External forces do not enter into this problem (except at the initial instant when the momentum V is imparted). The motion will therefore be governed by (5.26) with F = 0. Since $u = \frac{Dx}{Dt}$, this equation may be written

$$\frac{D}{Dt}(v + fx) = 0 \tag{5.47}$$

Let the subscript "o" denote a quantity in the initial configuration. Then by (5.47)

$$v + fx = v_o + fx_o \ , \tag{5.48}$$

where

$$v_o = \begin{cases} V, & |x| \leq a \ , \\ 0, & |x| > a \ , \end{cases} \tag{5.49}$$

In the final configuration, the velocity will be in a state of geostrophic balance :

$$v = \frac{g}{f}\frac{\partial h}{\partial x} \ . \tag{5.50}$$

Mass conservation is expressed in the Lagrangian form,

$$h_o \, dx_o = h dx \ , \tag{5.51}$$

* Rossby, C.-G., 1938 : On the mutual adjustment of pressure and velocity distributions in certain simple current systems (Part II). Journal of Marine Research, Vol. 1 - 2, pp. 239-263. Also, Bolin, B., Tellus, 1953.

or, since $h_o = H$,

$$h = H \frac{dx_o}{dx} . \tag{5.52}$$

Since time is not a variable in this analysis, $\partial h/\partial x = dh/dx$, and (5.50) may be combined with (5.52) to give

$$v = \frac{gH}{f} \frac{d^2x_o}{dx^2} . \tag{5.53}$$

If (5.53) and (5.49) are substituted in (5.48), we get

$$\frac{gH}{f} \frac{d^2x_o}{dx^2} + fx = v_o + fx_o ,$$

or

$$\frac{d^2(x - x_o)}{dx^2} - \frac{(x - x_o)}{\lambda^2} = \begin{cases} -\frac{V}{\lambda^2 f} , & |x| \leq a, \\ 0 , & |x| > a , \end{cases} \tag{5.54}$$

where again $\lambda^2 = \frac{gH}{f^2}$.

Mass continuity requires the displacement $x - x_o$ to vanish at large distances from the strip and to be continuous at its edges, i.e.,

$$x - x_o = 0 \text{ at } x_o = \pm \infty. \tag{5.55}$$

$$[x - x_o] = 0 \text{ at } x_o = \pm a . \tag{5.56}$$

The pressure, and therefore h, must be continuous at $x_o = \pm a$.

$$[h] = [h_o \frac{dx_o}{dx}] = h_o [\frac{dx_o}{dx}] = 0$$

implies that

$$[\frac{d}{dx}(x - x_o)] = 0 \text{ at } x_o = \pm a. \qquad (5.57)$$

It is assumed that the displacements are so small that the boundary conditions may be applied at $x = \pm a$. We obtain

$$x - x_o = \frac{V}{f} \begin{cases} e^{-x/\lambda} \sinh \frac{a}{\lambda}, & x > a, \\ 1 - e^{-a/\lambda} \cosh \frac{x}{\lambda}, & |x| \leq a, \\ e^{x/\lambda} \sinh \frac{a}{\lambda}, & x < -a. \end{cases} \qquad (5.58)$$

Let us compare these displacements with those for the case of the slowly applied force. The latter may be calculated from the transverse velocity, i.e., from $x - x_o = \int_0^t u dt$. If (5.44) is integrated with respect to time, we get

$$x - x_o = \frac{Bt}{f} \begin{cases} e^{-x/\lambda} \sinh \frac{a}{\lambda}, & x > a, \\ 1 - e^{-a/\lambda} \cosh \frac{x}{\lambda}, & |x| = a, \\ e^{x/\lambda} \sinh \frac{a}{\lambda}, & x < -a, \end{cases}$$

for the slowly applied force. Thus the impulsive force will give the same displacement as a constant force B applied for a finite time t providing $V = Bt$.

This point is emphasized further when we calculate the perturbation height h' for the impulsive case. From (5.63) we have

$$H + h' = H \frac{dx_o}{dx},$$

or

$$h' = -H \frac{d}{dx} (x - x_o)$$

$$= \frac{VH}{f\lambda} \begin{cases} e^{-x/\lambda} \sinh \frac{a}{\lambda}, & x > a, \\ e^{-a/\lambda} \sinh \frac{x}{\lambda}, & |x| = a, \\ -e^{x/\lambda} \sinh \frac{a}{\lambda}, & x < -a, \end{cases} \qquad (5.59)$$

which is identical to (5.45) if $\mathbf{V} = Bt$.

It was suggested in the previous section that the impulsive force would generate gravity waves and thereby cause a loss of energy when the waves radiate outward ; a greater amount of energy is required for the impulsive force to produce the same configuration as the slowly applied force. This energy difference may be computed.

Consider a unit length in the y-direction. In the initial configuration, a velocity V is impulsively applied to a strip of width 2a and depth H. The initial kinetic energy is given by

$$K_o = \frac{1}{2} (\rho\ 2a\ H)\ V^2 = \rho\ a\ H\ V^2 . \qquad (5.60)$$

The kinetic energy in the final configuration is

$$K_f = \int_{-\infty}^{\infty} \frac{1}{2} \rho v^2 H dx . \qquad (5.61)$$

Substituting the velocity v obtained geostrophically from the perturbation depth h',

$$v = V \begin{cases} -e^{-x/\lambda} \sinh \frac{a}{\lambda}, & x > a, \\ e^{-a/\lambda} \cosh \frac{x}{\lambda}, & |x| \leqslant a, \\ -e^{x/\lambda} \sinh \frac{a}{\lambda}, & x < -a, \end{cases} \qquad (5.62)$$

and performing the integrations, we find

$$K_f = K_o \left[\frac{\lambda}{4a}(1 - e^{-2a/\lambda}) + \frac{e^{-2a/\lambda}}{2}\right]. \qquad (5.63)$$

The final potential energy is

$$P_f = \int_{-\infty}^{\infty}\int_{H}^{H+h'} \rho gz\,dz\,dx = \int_{-\infty}^{\infty} \rho g \frac{h'^2}{2}\,dx$$

$$(5.64)$$

$$= K_o \left[\frac{\lambda}{4a}(1 - e^{-2a/\lambda}) - \frac{e^{-2a/\lambda}}{2}\right].$$

The energy of the gravitational oscillation is the difference between the initial energy and the final energy :

$$E_g = K_o - (K_f + P_f)$$

$$(5.65)$$

$$= K_o \left[1 - \frac{\lambda}{2a}(1 - e^{-2a/\lambda})\right].$$

Case I : If $\frac{\lambda}{a}$ is large, $\frac{2a}{\lambda}$ is small and $e^{-2a/\lambda}$ may be expanded in a Taylor's series,

$$e^{-2a/\lambda} = 1 - \frac{2a}{\lambda} + \frac{1}{2}\left(\frac{4a^2}{\lambda^2}\right) + 0\left(\frac{8a^3}{\lambda^3}\right), \qquad (5.66)$$

to give

$$E_g = K_o \frac{a}{\lambda}. \qquad (5.67)$$

Case II : If $\frac{\lambda}{a}$ is small, $\frac{2a}{\lambda}$ is large and $e^{-2a/\lambda} \simeq 0$. Hence

$$E_g = K_o \left(1 - \frac{\lambda}{2a}\right). \qquad (5.68)$$

Equations (5.67) and (5.68) may be given the following interpretation : Since λ is the scale of the horizontal displacement corresponding to the vertical scale H, a large value of λ/a means that there is ample room for lateral displacements to adjust the slope of the free surface to produce a pressure field corresponding geostrophically to the initial velocity distribution. A small value of λ/a signifies that appreciable changes of slope occur only in the vicinity of $x = \pm a$, i.e., no adjustment can take place in the main body of the initial current. In the former case there can be little change in the kinetic energy and little creation of potential energy. In the latter case the kinetic energy decreases to a small value, just equal to the potential energy. Most of the energy lost goes into gravity waves

A conservation equation may be derived from (5.48) by differentiation :

$$\frac{\partial v}{\partial x} + f = \left(\frac{\partial v}{\partial x_o} + f \right) \frac{dx_o}{dx} \qquad (5.69)$$

or, from (5.52),

$$\frac{\frac{\partial v}{\partial x} + f}{h} = \frac{\frac{\partial v_o}{\partial x_o} + f}{h_o} \quad . \qquad (5.70)$$

This is a special case of the theorem of conservation of "potential vorticity". (See Chapter VI.) Equation (5.70) determines the motion when combined with (5.50) and (5.51).

The Rossby adjustment problem may be generalized for a continuously stratified incompressible fluid as follows. (An analogous treatment can be given for a compressible fluid.) Conservation of angular momentum and density give

$$m = m_o \; , \qquad (5.71)$$

$$\rho = \rho_o \; , \qquad (5.72)$$

where $m = v + fx$ and $m_o = v_o + fx_o$. If the initial position of a fluid ring is denoted by the coordinates (x_o,z_o) and the final position by (x,z), the Lagrangian form of the continuity equation may be written

$$\frac{\partial(x_o,z_o)}{\partial(x,z)} = \frac{\rho}{\rho_o} = 1 \ , \tag{5.73}$$

and the generalization of (5.70) is obtained from (5.71) - (5.73) by the law for multiplication of Jacobian determinants :

$$\frac{\partial(m,\rho)}{\partial(x,z)} = \frac{\partial(m,\rho)}{\partial(m_o,\rho_o)} \times \frac{\partial(m_o,\rho_o)}{\partial(x_o,z_o)} \times \frac{\partial(x_o,z_o)}{\partial(x,z)} = \frac{\partial(m_o,\rho_o)}{\partial(x_o,y_o)} \ , \tag{5.74}$$

since

$$\frac{\partial(m,\rho)}{\partial(m_o,\rho_o)} = \frac{\partial(x_o,z_o)}{\partial(x,z)} = 1.$$

Equation (5.74) becomes equivalent to (4.37) for a Boussinesq fluid and with Cartesian geometry. It determines the adjustment when combined with (5.73) and the equations for balance in the adjusted state :

$$v = \frac{1}{\rho_{oo} f} \frac{\partial p}{\partial x} \ , \tag{5.75}$$

$$\rho = -\frac{1}{g} \frac{\partial p}{\partial z} \ . \tag{5.76}$$

One obtains

$$\left(\frac{\partial v}{\partial x} + f\right)\frac{\partial \rho}{\partial z} = \left(\frac{\partial v_o}{\partial x_o} + f\right)\frac{\partial \rho_o}{\partial z} \equiv q_o \ ,$$

or, by substitution from (5.75) and (5.76) and linearization,

$$\frac{\partial^2 p'}{\partial x^2} + \frac{f^2}{N^2}\frac{\partial^2 p'}{\partial z^2} = -\frac{gfq'_o}{N^2} = \rho_{oo} f \frac{\partial v'_o}{\partial x} - \frac{gf^2}{N^2}\frac{\partial \rho'_o}{\partial z} \ , \tag{5.77}$$

where $N^2 = \nu_g^2 = -(g/\rho_{oo})(\partial\rho_{oo}/\partial z)$. This is to be compared with the equation for slow, thermally or frictionally induced motions in a circular vortex with the same geometry. For small amplitudes the system of equations (5.26) to (5.30) combines to give

$$\frac{\partial^2}{\partial x^2}\left(\frac{\partial p'}{\partial t}\right) + f^2 \frac{\partial}{\partial z}\left(\frac{1}{N^2}\frac{\partial}{\partial z}\right)\frac{\partial p'}{\partial t} = \rho_{oo}\, f \frac{\partial F}{\partial x} - gf^2 \frac{\partial}{\partial z}\left(\frac{S}{N^2}\right) , \quad (5.78)$$

and if we set $u = \psi_z$, $w = -\psi_x$,

$$\frac{\partial^2 \psi}{\partial x^2} + f^2 \frac{\partial}{\partial z}\left(\frac{1}{N^2}\frac{\partial \psi}{\partial z}\right) = f \frac{\partial F}{\partial z} - \frac{g}{\rho_{oo}}\frac{\partial S}{\partial z} . \quad (5.79)$$

If F and S are constant in time, and $Ft = v_o'$, $St = \rho_o'$, time integration of (5.78) gives

$$\frac{\partial p'}{\partial x^2} + f^2 \frac{\partial}{\partial z}\left(\frac{1}{N^2}\frac{\partial p'}{\partial z}\right) = \rho_{oo} f \frac{\partial v_o'}{\partial x_o} - gf^2 \frac{\partial}{\partial z}\left(\frac{\rho_o'}{N^2}\right) \quad (5.80)$$

which is seen to be identical to (5.77) if N^2 is independent of z, but not otherwise.

Problem 5.1. : (a) Mass is added slowly at a constant rate S to the strip $|x| \leq a$ on an infinite, rotating, homogeneous, incompressible ocean of depth H. Find the resultant motion and slope of the free surface at an arbitrary time t.

(b) If the mass St is added impulsively find the resultant motion.

(c) Show that in this case little energy goes into gravity waves if $\lambda/a \ll 1$, but that most of it goes into gravity waves if $\lambda/a \gg 1$. Explain why.

Problem 5.2. : As an illustration of the adjustment process in a stratified fluid, let us return to the example on page 43 and, instead of considering the equilibrium of a lenticular blob of homogeneous water under the action of gravitational and rotational forces, suppose that the blob is not initially in equilibrium but that it comes to equilibrium after an adjustment process. To simplify matters suppose that the blob is in the form of a cylinder with rectangular cross-section, that the stratification of the ocean can be represented by two deep layers with densities ρ' and ρ'' (see figure), that the blob has density $\rho = (1/2)(\rho' + \rho'')$, width $2a$, and depth 2Δ which is small in comparison with the depths H' and H'' of the deep layers. In this case the horizontal mass convergences in these layers, and therefore their pressure perturbations, p' and p'' as well as the free surface and interface height perturbations, will be negligible. Find the final shape of the blob and show that the ratio of depth to width is of the same order as before.

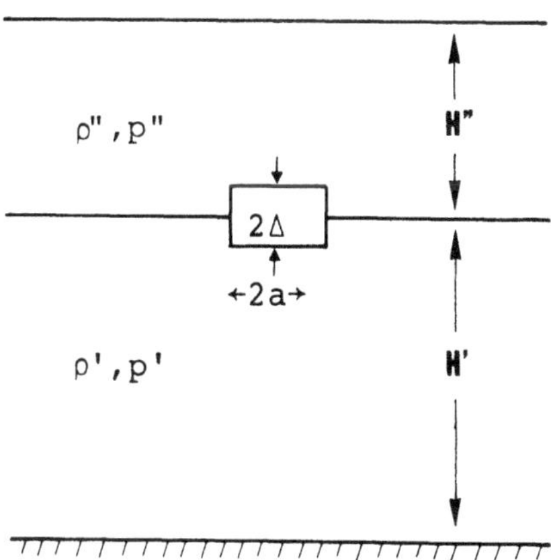

Solution to Problem 5.2.

<u>Solution to Problem 5.2</u> :

Denoting the post adjustment pressure and depth of the blob by p and $2h$, we must have, hydrostatically and from symmetry,

$$p = \rho'' g(H'' - h) + \rho g(H' + h - z)$$

$$= g'h + \text{linear function of } z \qquad \left(g' = \frac{\rho - \rho''}{\rho} g = \frac{\rho' - \rho}{\rho} g\right)$$

It follows from

$$\frac{dv}{dx} + f = \left(\frac{dv_o}{dx_o} + f\right) \frac{dx_o}{dx} ,$$

and

$$v = \frac{1}{\rho f} \frac{dp}{dx} = \frac{g'}{f} \frac{dh}{dx}$$

that

$$\frac{g'}{f} \frac{d^2h}{dx^2} + f = f \frac{h}{\Delta} ,$$

or

$$\frac{d^2h}{dx^2} - \frac{h}{\lambda^2} = - \frac{\Delta}{\lambda^2} \qquad (\lambda^2 = g'\Delta/f^2).$$

The blob will acquire a horizontal dimension 2b in the adjusted state and will not exist at $|x| > b$. The symmetric solution with this property is

$$h = \begin{cases} \Delta \left(1 - \dfrac{\cosh \frac{x}{\lambda}}{\cosh \frac{b}{\lambda}}\right), & |x| \leq b \\ 0, \ |x| > b \end{cases}$$

Continuity of mass requires

$$\int_{-b}^{b} h dx = \int_{-a}^{a} \Delta dx_o ,$$

which gives

$$\frac{b}{\lambda} - \tanh \frac{b}{\lambda} = \frac{a}{\lambda}$$

Assume, for example, that $a \sim 1000$ m, $\Delta \sim 100$ m initially and set $\Phi \sim 6°$,

$f \sim 10^{-5}\,\mathrm{sec}^{-1}$, $\frac{\rho'-\rho''}{\rho} \sim 2\times10^{-3}$, so that $\lambda \sim 1.6\times10^{4}$ m and $a/\lambda \sim 1/16$. We obtain $\frac{b}{\lambda} \gtrsim \frac{1}{2}$, or $b \sim 8$ km and $h \sim 11$ m at $x = 0$. The blob spreads out horizontally and shrinks vertically until the ratio $\frac{h}{b} \sim \frac{\nu_i}{\nu_g} \sim \frac{10^{-5}}{10^{-2}} \sim 10^{-3}$, as in the example on page 61.

VI. TWO-DIMENSIONAL ASYMMETRIC MOTION

Circulation and vorticity theorems

Consider a material chain of particles in a fluid. As the fluid moves the chain may be distorted and stretched but always consists of the same particles. Let c be a closed curve made up of such a chain of particles. The circulation C of the velocity around this curve is defined as

$$C \equiv \int_c \underline{V}\cdot d\underline{r} \equiv \oint V\cdot d\underline{r} \quad , \tag{6.1}$$

where $\underline{V}$ is the velocity of the fluid and $d\underline{r}$ is a vector line element of c (Fig. 6.1). From

$$\begin{aligned} \frac{D\underline{V}}{Dt}\cdot d\underline{r} &\equiv \frac{D}{Dt}(\underline{V}\cdot d\underline{r}) - \underline{V}\cdot\frac{D}{Dt}(d\underline{r}) \\ &\equiv \frac{D}{Dt}(\underline{V}\cdot d\underline{r}) - \underline{V}\cdot d\underline{V} \\ &\equiv \frac{D}{Dt}(V\cdot d\underline{r}) - d(\frac{V^2}{2}) \ , \end{aligned} \tag{6.2}$$

we get

$$\oint \frac{D\underline{V}}{Dt}\cdot d\underline{r} = \frac{D}{Dt}\oint \underline{V}\cdot d\underline{r} - \oint d(\frac{V^2}{2}) = \frac{DC}{Dt} \ . \tag{6.3}$$

Multiplying the equations of non-viscous motion (2.1) by $d\underline{r}$ and integrating around c, we obtain

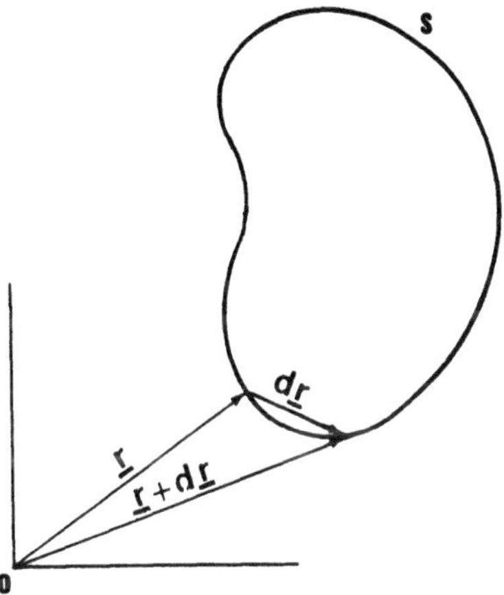

Figure 6.1

$$\oint \frac{D\underline{V}}{Dt} \cdot d\underline{r} = -\oint \nabla\Phi \cdot d\underline{r} - \oint \alpha\nabla p \cdot d\underline{r}$$

$$= -\oint d\phi - \oint \alpha dp = -\oint \alpha dp \; . \qquad (6.4)$$

Hence

$$\frac{DC}{DT} = -\oint \alpha dp \; . \qquad (6.5)$$

This is V. Bjerknes' Circulation Theorem.

If the fluid is either homogeneous-incompressible or barotropic [$\alpha = \alpha(p)$], we obtain Kelvin's Circulation Theorem :

$$\frac{DC}{Dt} = 0 \; . \qquad (6.6)$$

A line whose tangent everywhere coincides in direction with the vorticity $\nabla \times \underline{V}$ is called a vortex line. Helmholtz's theorem states that a vortex line moves as a material line if the fluid is homogeneous-incompressible or barotropic. To prove this theorem, consider a material surface which is everywhere tangent to the vorticity at a given instant. A surface with this property is called a vortex surface. Let c be an arbitrary closed material curve contained in this surface at the initial time. If σ is that part of the surface enclosed by the curve, we have, by Stokes' theorem,

$$C \equiv \oint \underline{V} \cdot d\underline{r} = \int_{\sigma} \nabla \times \underline{V} \cdot \underline{n} \, d\sigma \quad , \qquad (6.7)$$

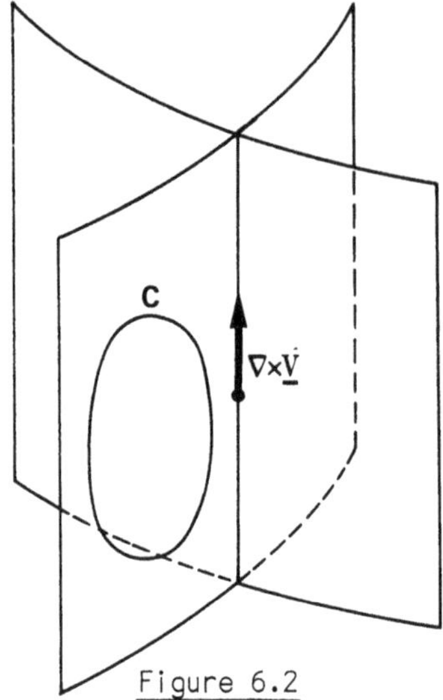

Figure 6.2

where $\underline{n}$ is the oriented unit vector normal to the surface and $d\sigma$ is the surface element of area. Since $\nabla\times\underline{V}\cdot\underline{n} = 0$ on the vortex surface, $C = 0$ initially, and since $DC/DT = 0$, C must remain zero. Since $C = 0$ for arbitrary curves c, it follows from Stokes' theorem that the material surface must remain a vortex surface, and a vortex line (determined by the intersection of two vortex surfaces) must remain a vortex line. This proves Helmholtz's theorem. (See Fig. 6.2.)

A vortex surface in the form of a tube is called a vortex tube. It follows from Stokes' theorem that the circulation around all cross-sectional curves of the tube are constant, so if the tube has an infinitesimal cross-section $\underline{n}d\sigma$, the vortex strength, $\nabla\times\underline{V}\ \underline{n}d\sigma$, is constant along the tube. (See Fig. 6.3.)

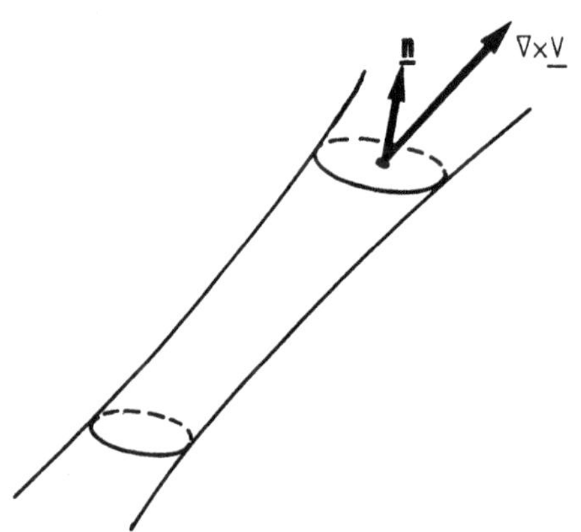

Figure 6.3

The analogue of angular momentum in symmetric motion is circulation or vorticity in asymmetric motion. Although circulation is not in general conserved for the inviscid, adiabatic motion of a stratified fluid, there is another quantity, potential vorticity, which is conserved.

A general proof of the theorem of conservation of potential vorticity which was discussed at the end of Section 5 will now be given. We assume that the specific entropy $s = s(p,\alpha)$. Consider a material infinitesimal curve c lying entirely in a surface of constant specific entropy and the material right circular cylinder, contained between the isentropic surfaces s = constant and s + ds = constant, intersecting the s surface in the curve c. Since s does not vary on the isentropic surface, we have $\alpha = \alpha(s,p)$ = function of p, and by V. Bjerknes' circulation theorem,

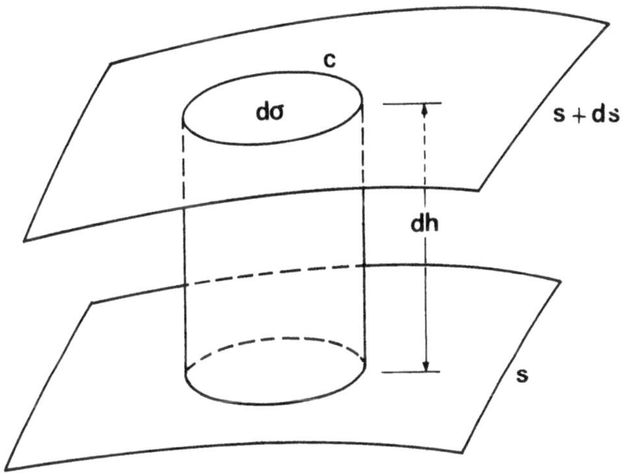

Figure 6.4

$$\frac{DC}{Dt} = \oint \alpha dp = 0 \quad , \tag{6.8}$$

where C is the circulation about the curve c. Further, from conservation of mass we have

$$\rho d\sigma\, dh = \text{constant}, \tag{6.9}$$

where $d\sigma$ is the area enclosed by the curve c and dh is the distance between the s and s + ds surfaces. The change in specific entropy from the top of the cylinder to the bottom is also constant :

$$ds \equiv |\nabla s| \, dh = \text{constant} \; . \tag{6.10}$$

Also, if Stokes' theorem is applied to the circulation about the (infinitesimal) curve c, we get from (6.8)

$$\nabla \times \underline{V} \cdot \underline{n} \, d\sigma = \text{constant} \; , \tag{6.11}$$

where $\underline{n}$ is a unit vector normal to the isentropic surface. Since $\underline{n}$ is in the direction of ∇s,

$$\underline{n} = \frac{\nabla s}{|\nabla s|} \; . \tag{6.12}$$

Combining equations (6.9) through (6.12), we obtain

$$\frac{\nabla \times \underline{V} \cdot \nabla s}{\rho} = \text{constant} \tag{6.13}$$

or

$$\frac{D}{Dt}\left(\frac{\nabla \times \underline{V} \cdot \nabla s}{\rho}\right) = 0 \; . \tag{6.14}$$

The quantity $\rho^{-1}(\nabla \times \underline{V} \cdot \nabla s)$ is called the potential vorticity.*

If the fluid is a perfect gas, we have

$$\frac{D}{Dt}\left(\frac{\nabla \times \underline{V} \cdot \nabla \theta}{\rho}\right) = 0 \; , \tag{6.15}$$

θ being the potential temperature, or, in general, any function of the entropy. If the fluid is incompressible, density, and therefore $\nabla \times \underline{V} \cdot \nabla \rho$, is conserved.

*Rossby, C.-G, 1940 : Quart J. Roy. Met. Soc., 66, Suppl. 68-87.
Ertel, H., 1942 : Met. Z., 59, 277-281.

Specialization to two-dimensional flows

Consider two-dimensional flow. Let $\underline{V}$ be the velocity of the fluid parallel to the plane of motion, $\nabla \times \underline{V}$ the vorticity, $\underline{k}$ a unit vector normal to the plane of the motion, and $\zeta \equiv \nabla \times \underline{V} \cdot \underline{k}$ the magnitude of the vorticity of the fluid.

1. Incompressible flow in two-dimensions. If c is a small closed curve in the plane of the motion, the constancy of its circulation C leads by Stokes' theorem to

$$\nabla \times \underline{V} \cdot \underline{k}\, d\sigma = \text{constant} , \tag{6.16}$$

where $d\sigma$ is the element of area enclosed by c (cf. (6.7). Since $d\sigma$ = constant in incompressible flow, we have ζ = constant, or

$$\frac{D\zeta}{Dt} = \left(\frac{\partial}{\partial t} + \underline{V}\cdot\nabla\right)\zeta = 0 , \tag{6.17}$$

where, because of the incompressibility,

$$\underline{V} = \underline{k} \times \nabla\psi \quad \text{and} \quad \zeta = \underline{k} \cdot \nabla \times \underline{V} = \nabla^2\psi .$$

2. Two-dimensional compressible flow. In this case $\rho d\sigma$ = constant, and we have

$$\frac{D}{Dt} \left(\frac{\zeta}{\rho}\right) = 0 . \tag{6.18}$$

3. Two-dimensional homogeneous-incompressible flow with a free surface. Consider a fluid acted upon by gravity and bounded by a free surface. The reference plane is horizontal. If ζ is again the vertical component of the absolute vorticity, we have

$$\zeta\, d\sigma = \text{constant} \tag{6.19}$$

for a vertical column of cross-sectional area $d\sigma$. If h is the height of the column, $h\,d\sigma$ = constant so that ζ/h = constant, or

$$\frac{D}{Dt}\left(\frac{\zeta}{h}\right) = 0 \ . \qquad (6.20)$$

The fluid behaves in all respects as if it were two-dimensional and compressible with h taking the place of ρ, i.e., compressibility is represented by the variation of the height of the free surface.

4. Two-dimensional, incompressible relative flow of a fluid in a rotating system, (Fig. 6.5). If $\underline{\Omega}$ is the angular velocity of rotation, $\underline{R}$ the polar vector, and $\underline{r}$ the radius vector of a particle fixed in the rotation system, then the velocity $\underline{V}_e$ of a point fixed in the system is given by

$$\underline{V}_e = \underline{\Omega} \times \underline{R} = \underline{\Omega} \times \underline{r} \ . \qquad (6.21)$$

The absolute velocity of a particle $\underline{V}_a$ is the sum of the relative velocity $\underline{V}_r$ and the velocity of a point fixed in the system $\underline{V}_e$, i.e.,

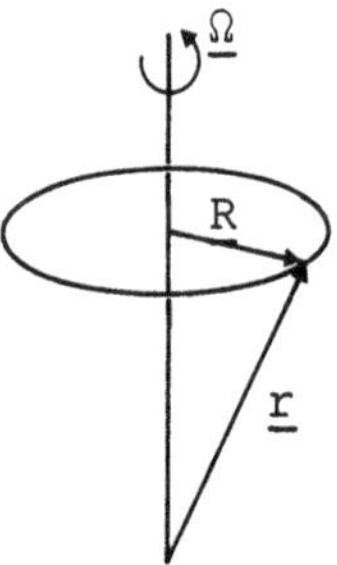

Figure 6.5

$$\underline{V}_a = \underline{V}_r + \underline{\Omega} \times \underline{r} \qquad (6.22)$$

Applying Kelvin's circulation theorem to an arbitrary curve c, we have

$$\frac{DC_a}{Dt} = \frac{DC_r}{Dt} + \frac{D}{Dt}\oint \underline{\Omega} \times \underline{r} \cdot d\underline{r} = 0 \quad . \tag{6.23}$$

But, by Stokes' theorem

$$\oint \underline{\Omega} \times \underline{r} \cdot d\underline{r} = \int_\sigma \nabla \times (\underline{\Omega} \times \underline{r}) \cdot \underline{n} d\sigma = \int_\sigma 2\underline{\Omega} \cdot \underline{n} \; d\sigma = 2\Omega\Sigma \ , \tag{6.24}$$

where Ω is the magnitude of the angular velocity vector and Σ is the projection of c onto a plane normal to the axis of rotation. Thus $C_a = C_r + 2\Omega\Sigma$, and

$$\frac{DC_a}{Dt} = \frac{D}{Dt}(C_r + 2\Omega\Sigma) = 0 \quad . \tag{6.25}$$

We wish to apply this result to two-dimensional flow relative to a rotating sphere. For a small curve in the plane of motion, (6.33) becomes, by Stokes' theorem,

$$\zeta_r d\sigma + 2\Omega d\sigma \sin\phi = \text{constant}, \tag{6.26}$$

where ζ_r is the relative vorticity component normal to the surface of the sphere, $d\sigma$ is the element of area bounded by the curve, and ϕ is the latitude. If the fluid is held gravitationally to the rotating sphere, and h is the height of the free surface, $h d\sigma$ = constant, and

$$\frac{D}{Dt}\left(\frac{\zeta_r + 2\Omega\sin\phi}{h}\right) = 0 \ . \tag{6.27}$$

If the upper surface is rigid, the flow is horizontally incompressible, and we have

$$\frac{D}{Dt}(\zeta_r + 2\Omega \sin\phi) = 0 \ . \tag{6.28}$$

Since the vorticity of a fluid moving with solid rotation is twice the angular velocity ($\nabla \times (\underline{\Omega} \times \underline{r}) = 2\,\underline{\Omega}$), (6.28) states that for a two-dimensional, incompressible flow the component of absolute vorticity normal to the plane of motion (the surface of the sphere) is constant following the motion of a fluid element.

The Taylor-Proudman theorem and Taylor's experiments

Proudman * pointed out that small steady disturbances of a homogeneous, incompressible fluid must be two-dimensional. To test this prediction, Taylor performed a series of remarkable experiments.** In one he rotated a vertical cylindrical tank filled with water until the water came to relative rest and then disturbed it slightly while ink was being poured into it. The ink gradually became drawn out into long, thin vertical sheets which remained accurately parallel to the axis of rotation. In another experiment, Taylor drew a sphere slowly along the floor of the cylinder; at right angles to the axis of rotation he observed that a spot of injected dye would move aside for the moving sphere, not over it. It was as if the water cylinder over the sphere were moving bodily with the sphere.

The proof that $\underline{V}_h$ is independent of height may be obtained from the equations of motion for a rotating system. Consider a system rotating with the constant angular velocity $\underline{\Omega}$. The relation (6.22) holds for any vector in the rotating system. In particular, it holds for the absolute velocity $\underline{V}_a$. Thus, dropping the subscript r, we have

$$\left(\frac{D\underline{V}_a}{Dt}\right) = \frac{D\underline{V}_a}{Dt} + \underline{\Omega} \times \underline{V}_a$$

$$= \frac{D\underline{V}}{Dt} + \underline{\Omega} \times \underline{V} + \underline{\Omega}\times(\underline{V} + \underline{\Omega} \times \underline{r}) ,$$

$$= \frac{D\underline{V}}{Dt} + 2\underline{\Omega} \times \underline{V} - \underline{\Omega}^2 R , \tag{6.29}$$

where Ω is the magnitude of the angular velocity vector. The Eulerian equations of motion (2.1) for a homogeneous-incompressible, inviscid fluid moving under the action of gravity thus become

* Proudman, J., 1916 : Proc. Roy. Soc. A, 92, pp. 408-424.

**Taylor, G. I., 1912 : Proc. Roy. Soc. A, 93, pp. 99-113.
1921 : Proc. Roy. Soc. A. 100, pp. 114-121.

$$\frac{D\underline{V}}{Dt} + 2\Omega\underline{k} \times \underline{V} = -\nabla P \ , \quad (P = p/\rho + gz - \frac{\Omega^2 R^2}{2}) \qquad (6.30)$$

where $\underline{k}$ is a unit vector pointing upwards against gravity and directed along the axis of rotation, and g is the acceleration of gravity. The component equations of (6.30) are

$$\frac{D\underline{V}_h}{Dt} + 2\Omega\underline{k} \times \underline{V}_h = -\nabla_h P \qquad (6.31)$$

$$\frac{Dw}{Dt} = -\frac{\partial P}{\partial z} \ . \qquad (6.32)$$

Proudman dealt with small disturbances for which $\partial/\partial t = 0$, and because the acceleration terms are the product of small quantities, the equations of motion reduce to*

$$2\Omega\underline{k} \times \underline{V}_h = -\nabla_h P \qquad (6.33)$$

$$0 = \frac{\partial P}{\partial z} \ . \qquad (6.34)$$

Taking the vertical derivative of (6.33), we get

$$2\Omega\underline{k} \times \frac{\partial \underline{V}_h}{\partial z} = -\nabla_h \frac{\partial P}{\partial z} = 0 \ , \qquad (6.35)$$

which states that the horizontal motion is independent of height. Taking the curl of (6.33), we obtain

$$\nabla_h \cdot \underline{V}_h = 0 \ , \qquad (6.36)$$

and it follows from

$$\frac{\partial u}{\partial x} + \frac{\partial u}{\partial y} + \frac{\partial w}{\partial z} = 0$$

* We note that it is not necessary to assume that $\partial/\partial t = 0$ but only that the time scale is the advective time scale, i.e., that $\partial/\partial t$ has the order of $\underline{V}\cdot\nabla$.

that $\frac{\partial w}{\partial z} = 0$. Hence, if $w = 0$ at the lower boundary, $w \equiv 0$ everywhere, and the motion is strictly two-dimensional.

If the lower boundary is sloped, steady motion must follow bottom contours. If the level lines intersect side boundaries, non-geostrophic boundary layers must occur. If the slopes of the boundaries are small, non-steady motion can be horizontally non-divergent only to the first order of small quantities. The motion is only quasi-geostrophic and quasi-nondivergent.

Taylor's actual proof of the two-dimensionality proceeded as follows : Consider a material curve c in the rotating tank. If the relative velocities are small, DC_r/Dt in (6.25) will be a quantity of second order in these velocities, whereas $D\Sigma/Dt$ will be of first order. Hence, to this order, we must have

$$\frac{D\Sigma}{Dt} \simeq 0$$

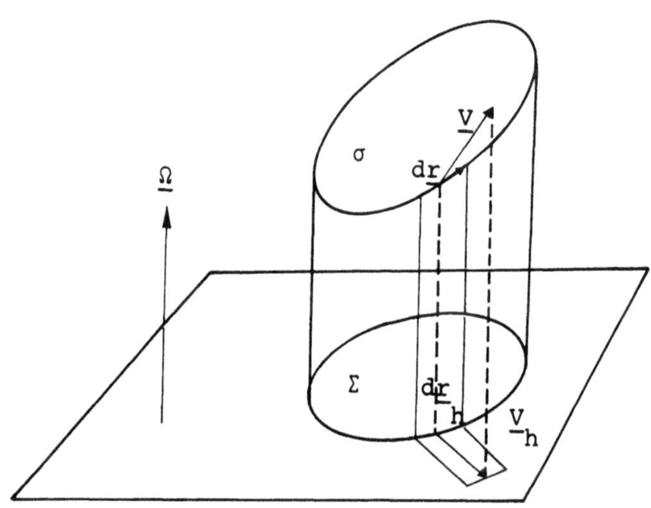

Figure 6.6

Let $\underline{V}$ be the relative velocity of a particle and $d\underline{r}$ a vector line element of the curve c. If $\underline{V}_h$ and $d\underline{r}_h$ are the components of these vectors in the horizontal plane (Fig. 6.6), and $\underline{k}$ is a unit vector parallel to $\underline{\Omega}$,

$$\frac{D\Sigma}{Dt} = \oint \underline{k}\cdot\underline{V}_h \times d\underline{r}_h = \oint \underline{k}\cdot\underline{V}_h \times d\underline{r} = \oint \underline{k} \times \underline{V}_h\cdot d\underline{r} \ ,$$

or, by Stokes' theorem,

$$\oint \underline{k} \times \underline{V}_h\cdot d\underline{r} = \int_\sigma \nabla \times (\underline{k} \times \underline{V}_h) \cdot \underline{n}\, d\sigma = 0,$$

where σ is the surface embraced by the curve c and $\underline{n}$ is a unit vector normal to the surface. By a vector identity,

$$\int_\sigma \left[\underline{k}\,(\nabla\cdot\underline{V}_h) - (\underline{k}\cdot\nabla)\,\underline{V}_h\right] \cdot \underline{n}\, d\sigma = 0 \ .$$

Since the surface σ is arbitrarily small and arbitrarily oriented, this relation requires that

$$\underline{k}(\nabla\cdot\underline{V}_h) - (\underline{k}\cdot\nabla)\,\underline{V}_h \equiv 0 \ .$$

The vectors $\underline{k}$ and $\underline{V}_h$ are orthogonal; hence each term may separately be set equal to zero. The vertical component gives

$$\frac{\partial \underline{V}_h}{\partial z} = 0 \ ,$$

and the horizontal component gives

$$\nabla\cdot\underline{V}_h = 0 \ ,$$

as before.

Quasi-geostrophic flow of a homogeneous fluid

The slow, quasi-geostrophic motions observed in the Taylor experiment can be distinguished from other possible types of motion, epitomized by gravity-inertial oscillations, by their modes of propagation and excitation. For example, the tidal oscillations of a uniform ocean covering a sphere are of two kinds (Lamb, *Hydrodynamics*, 1932, p. 350). Those of the "First Class" are essentially gravity oscillations, and for these the frequency change produced by the rotation* decreases with the speed of rotation, whereas the frequencies of the "Second Class" diminish to zero as the speed of rotation becomes small, and the oscillation becomes a steady rotational motion.

We may illustrate the two classes with the following examples. Consider first a uniform homogeneous ocean at rest with respect to an infinite rotating plane with gravity acting normal to the plane. Assume that the depth of the ocean is so small in relation to the horizontal scale of motion that the motion is quasi-hydrostatic and the horizontal velocity is independent of depth. If H is the undisturbed depth of the ocean, η the height perturbation of the free surface, and f/2 the angular speed of rotation of the plane, the linearized equations for small oscillations become

$$\frac{\partial u}{\partial t} - fv = -g\frac{\partial \eta}{\partial x} ,$$

$$\frac{\partial v}{\partial t} + fu = -g\frac{\partial \eta}{\partial y} , \qquad (6.37)$$

$$\frac{\partial \eta}{\partial t} + H\left(\frac{\partial u}{\partial x} + \frac{\partial v}{\partial y}\right) = 0 .$$

Setting

$$(u,v,\eta) = (U,V,N)e^{i\sigma t} ,$$

We obtain

$$i\sigma U - fV = -g\frac{\partial N}{\partial x} , \qquad (6.38)$$

---

*If the exact perturbation equations are used, the frequencies of the gravity oscillations are split by the rotation in an analogous manner to the Zeeman splitting of atomic lines in a magnetic field, or the splitting of the normal frequencies of oscillation of a rotating elastic earth. See for example G. Backus and F. Gilbert, 1961 : Proc. Roy. Soc., 47, 362-371.

$$i\sigma V + fU = -g\frac{\partial N}{\partial y}, \qquad (6.38)$$
$$i\sigma N + H\left(\frac{\partial U}{\partial x} + \frac{\partial V}{\partial y}\right) = 0 .$$

or, by elimination of U and V,

$$i\sigma\left(\frac{\partial^2 N}{\partial x^2} + \frac{\partial^2 N}{\partial y^2} + \frac{\sigma^2 - f^2}{gH}N\right) = 0 . \qquad (6.39)$$

Suppose first that $\sigma \neq 0$: then (6.39) becomes a Helmholtz equation for the eigenfunction N and the eigenvalues σ. For an enclosed basin, we see that the eigenfrequencies σ_i are related to the frequencies s_i of oscillation of the corresponding nonrotating system by the equation

$$\sigma_i^2 - f^2 = s_i^2 \quad (i = 1,2, \ldots) , \qquad (6.40)$$

i.e., σ_i^2 is larger than s_i^2 by the square of the inertial frequency f^2. This is because we now have, in addition to the gravitational restoring force per unit vertical displacement $\frac{g}{H}$, the horizontal restoring force per unit horizontal displacement f^2 .

For the infinite plane, (6.39) admits of the plane wave solution $N \sim e^{i(kx + \ell y + \sigma t)}$, $\sigma^2 = f^2 + gH(k^2 + \ell^2)$. (See next example.)

The second alternative, $\sigma = 0$, is not trivial. It implies that $\partial/\partial t = 0$, so that equations (6.37) become

$$-fv = -g\frac{\partial \eta}{\partial x} ,$$
$$fu = g\frac{\partial \eta}{\partial y} , \qquad (6.41)$$
$$\frac{\partial u}{\partial x} + \frac{\partial v}{\partial y} = 0 ,$$

i.e., the flow is geostrophic and horizontally non-divergent. This is a perfectly possible solution. Indeed, it is the solution implied by the statement made in the first paragraph of these lectures, that if a stone is thrown into an infinite rotating ocean, a rotational flow will result which will persist until dissipated by viscosity.

We may illustrate the second type of motion further by considering motion on a "β-plane", in which the geometry is Cartesian but the dynamical effect of the earth's curvature is retained through a variable Coriolis parameter. If ϕ_o is a mean latitude and a the radius of the earth, we set $\beta = 2\Omega\cos\phi_o/a$ and $f = f_o + \beta y$, where $f_o = 2\Omega\sin\phi_o$. The equations for the vorticity $\zeta = v_x - u_y$, x-momentum and continuity become

$$\zeta_t + f(u_x + v_y) + \beta v = 0 \quad ,$$

$$u_t - f\,v + g\eta_x = 0 \quad , \qquad (6.42)$$

$$\eta_t + H(u_x + v_y) = 0 \quad .$$

If we now ignore the variation of f in the first and second equations in the W-K-B sense and look for plane wave solutions of the form

$$u,v,\eta = (U,V,N)\exp\left[i(kx + \ell y) + i\sigma t\right] \quad ,$$

the condition for the existence of a solution is

$$\begin{vmatrix} ikf_o + \sigma\ell & if_o\ell - k\sigma + \beta & 0 \\ i\sigma & -f_o & ikg \\ ikH & i\ell H & +i\sigma \end{vmatrix} = 0 \quad ;$$

which yields a cubic equation for the velocities :

$$(-\sigma + \frac{\beta}{k})(\sigma^2 - gHk^2) + (f_o{}^2 + gH\ell^2)\sigma = 0 \quad . \qquad (6.43)$$

For $\beta/k \ll k\sqrt{gH}$, its roots may be approximated by setting $|\sigma| \gg \beta/k$ to recover the first class motions,

$$\sigma^2 = f_o{}^2 + gH(\ell^2 + k^2) \quad , \qquad (6.44)$$

as in the previous example, and by setting $|\sigma| \ll k\sqrt{gH}$ we obtain the second class motions,

$$\sigma = + \frac{\beta k}{k^2 + \ell^2 + \frac{f_o^2}{gH}} \ . \tag{6.45}$$

We note that in the latter case $\sigma \to 0$ for $\Omega \to 0$.

The two types of motion discussed above are examples of the two classes of tidal motions to which we have already referred. They may be characterized by their time scales : for the first, $\partial/\partial t > f$; for the second, $\partial/\partial t = 0 \ll f$. If we had assumed in the beginning that $\partial/\partial t \ll f$, we would have obtained equations (6.41) directly.

In the following we shall deal with quasi-geostrophic motions of a homogeneous-incompressible ocean on a rotating earth. The governing inviscid equations may be written

$$(\frac{\partial}{\partial t} + \underline{V}\cdot\nabla + w\frac{\partial}{\partial z})\underline{V} + 2\Omega\sin\phi'\underline{k} \times \underline{V} + 2\Omega\cos\phi\ \underline{wi} = -\nabla(p/\rho+gz) \ , \tag{6.46}$$

$$(\frac{\partial}{\partial t} + \underline{V}\cdot\nabla + w\frac{\partial}{\partial z})w - 2\Omega\cos\phi u = -\frac{\partial}{\partial z}(p/\rho + gz) \ , \tag{6.47}$$

$$\nabla\cdot\underline{V} + \frac{\partial w}{\partial z} = 0 \ , \tag{6.48}$$

where ∇ now denotes the horizontal del-operator and $\underline{V}$ the horizontal velocity component. We anticipate that the flow will be quasi-geostrophic, hence independent of height, and characterized by time scales which are large compared to the period of earth's rotation. However, among these motions there remain two possible characteristic time scales. When the horizontal gradient of the vertical component of the relative vorticity is at least of the same order of magnitude as that of the earth's rotation, the time scale will be advective; that is, if T, L, and U are the characteristic time, horizontal distance, and velocity scales, respectively, $T \sim L/U = T_a$. When the gradient of relative vorticity is small in comparison with that of the earth's rotation, T is the time for a particle moving meridionally at the speed U to change the earth's vorticity $2\Omega\sin\phi$ by the magnitude of the relative vorticity, U/L, i.e. $\frac{2\Omega UT}{a} \sim U/L$, or $T \sim a/2L\Omega = T_\beta$, where a is the radius of the earth.

In effect, the flow will be governed by (6.27), which, is we drop the subscript "r", may be written

$$(\frac{\partial}{\partial t} + \underline{V}\cdot\nabla)(\zeta + f) = \frac{\zeta+f}{h}(\frac{\partial}{\partial t} + \underline{V}\cdot\nabla)h \qquad (6.49)$$

where $f = 2\Omega\sin\phi$. The two time scales are defined by

$$\frac{\partial\zeta}{\partial t} \sim \underline{V}\cdot\nabla\zeta ,$$

or

$$\frac{\partial\zeta}{\partial t} \sim \underline{V}\cdot\nabla f .$$

A third time scale is obtained if the gravitational stability of the free surface is so weak that changes of potential vorticity are dominated by changes in h associated with geostrophic currents. In this case either

$$(\frac{\partial}{\partial t} + \underline{V}\cdot\nabla)h \sim 0 ,$$

in which case, since $\underline{V}\cdot\nabla h = 0$ by geostrophy, the flow is stationary, or

$$\frac{\partial h}{\partial t} \sim \frac{h}{f}\underline{V}\cdot\nabla f ,$$

which gives

$$T = \frac{2\Omega La}{gH} = Fr\, T_\beta \equiv T_h , \qquad (6.50)$$

where Fr is the rotational Froude number,

$$Fr = \frac{4\Omega^2 L^2}{gH} ,$$

and H is the mean depth of the ocean.

We shall now derive the characteristic simplifications of the equations for slow, relative, asymmetric motions. The first category of motion includes large-scale shear instabilities in the atmosphere and oceans and inertial boundary layers in the oceans. The second includes the so-called Rossby waves of large scale in the oceans generated by transient with stresses, the latter being analogous to the tides of the second class. The third includes motions of very large horizontal scale relative to the vertical scale.

Notation : Let us denote non-dimensional variables by prime superscripts. We shall assume a priori that all non-dimensional variables are of order unity. This assumption will be valid as long as the scaling is consistent. For a homogeneous fluid it is convenient to introduce the variable

$$\pi = p/\rho + gz \, . \tag{6.51}$$

We will have occasion to use the following non-dimensional numbers :

$$Fr = \frac{4\Omega^2 L^2}{gH} \quad = \text{ rotational Froude number}$$

$$Ro = \frac{U}{2\Omega L} \quad = \text{ Rossby number}$$

$$\delta = \frac{D}{L} \quad = \text{ aspect ratio}$$

$$s = \frac{L}{a}$$

$$I = \frac{T_a}{T_\beta} \quad = \frac{L}{a} \; / \; Ro \; .$$

Here, L and H are the horizontal and vertical length scales, respectively, U is the horizontal velocity scale, Ω is the earth's angular speed of rotation, and a denotes the radius of the earth.

Introducing the above characteristic magnitudes and, in addition the magnitudes W for the vertical velocity and P for π, we define the non-dimensional quantities :

$$\nabla = \frac{1}{L}\nabla' \; , \quad z = Dz' \; , \quad t = Tt' \; , \quad \underline{V} = U\underline{V}' \; , \; \pi = P\pi' \; ,$$

$$w = Ww' \; .$$

The following explicit assumptions are made :

$$\left.\begin{array}{ll} \text{i. Ro} & < O(1) \\ \text{ii. } \phi & = O(1) \end{array}\right\} \quad \text{(quasi-geostrophic)}$$

$$\begin{array}{lll} \text{iii. } \delta & \leqslant O(Ro^2) & \text{(quasi-hydrostatic)} \\ \text{iv. Fr} & \leqslant O(1) & \\ \text{v. } L/a & \leqslant O(Ro) & (\beta\text{-plane}) \; . \end{array} \tag{6.52}$$

The magnitude of π follows from the quasi-geostrophic character of the flow (assumptions i and ii) :

$$\frac{P}{L} = 2\Omega U \quad \text{or} \quad P = 2\Omega LU \; . \tag{6.53}$$

Assumption iv will imply that the time scales T are either T_a or T_β . If Fr $> O(1)$, it may be shown that $T = T_h$ as defined by (6.50). The magnitude of w follows from the hydrostatic relation (assumptions iii and iv) at $z = 0$:

$$p = \rho g h \quad , \tag{6.54}$$

where h is the depth of the ocean. From the geostrophic relation (6.53) we get

$$w = \frac{Dh}{Dt} = \frac{1}{g}\frac{D\pi}{Dt} \sim \frac{P}{gT} \; , \text{ or } \; W = \frac{P}{gT} = \frac{2\Omega LU}{gT} \; . \tag{6.55}$$

Assumption (v) permits the curvature of the earth's surface to be ignored but the dynamical effect of the variation of the Coriolis parameter to be retained. Expanding the sine and the cosine of ϕ about ϕ_o, we have

$$\begin{aligned} \sin\phi &= \sin\phi_o + \frac{L}{a}\cos\phi_o \quad (y' - y_o') \; + O(\frac{L}{a})^2 \\ \cos\phi &= \cos\phi_o - \frac{L}{a}\sin\phi \quad (y' - y_o') \; + O(\frac{L}{a})^2 \end{aligned} \tag{6.56}$$

We now distinguish two classes of motion according as $T_a/T_\beta \leqslant O(1)$ or $> O(1)$.

Case I : $\frac{I}{\ }$ = T_a/T_β = $(L/a)/Ro \lesssim O(1)$; $T = T_a$ = L/U (advective time scale $Fr = \frac{4\Omega^2 L^2}{gH} \lesssim O(1)$. Inserting the non-dimensional variables into the equations of motion with $T = T_a$, and dropping primes, we obtain

$$Ro(\frac{\partial}{\partial t} + \underline{V}\bullet\nabla + RoFrw\frac{\partial}{\partial z})\underline{V} + \frac{D}{L}\,FrRow\,(\cos\phi_o + \ldots)\underline{i} +$$

$$\underline{k}\times\underline{V}\,(\sin\phi_o + \frac{L}{a}\cos\phi_o(y - y_o) + \ldots) = -\nabla\pi\,. \tag{6.57}$$

$$\frac{D^2}{L^2}Ro^2Fr(\frac{\partial}{\partial t} + \underline{V}\bullet\nabla + \ldots)w - \frac{D}{L}(\cos\phi_o + \ldots)u = -\frac{\partial\pi}{\partial z}$$

$$\nabla\bullet\underline{V} + FrRo\frac{\partial w}{\partial z} = 0\,.$$

Next, expand $\underline{V}$, w, and π in powers of Ro, i.e.,

$$\begin{aligned}\underline{V} &= \underline{V}^{(o)} + \underline{V}^{(1)}Ro + \underline{V}^{(2)}Ro^2 + \ldots\\ \pi &= \pi^{(o)} + \pi^{(1)}Ro + \pi^{(2)}Ro^2 + \ldots\\ w &= w^{(o)} + w^{(1)}Ro + w^{(2)}Ro^2 + \ldots\end{aligned} \tag{6.58}$$

Equation (6.57) must be satisfied identically by the coefficients of each power of Ro. The zeroth order equations give

$$\underline{k}\times\underline{V}^{(o)}\sin\phi_o = -\nabla\,\pi^{(o)}\,, \tag{6.59}$$

$$0 = -\,\partial\pi^{(o)}/\partial z\,, \tag{6.60}$$

$$\nabla\bullet\underline{V}^{(o)} = 0\,. \tag{6.61}$$

It follows from (6.59) and (6.60) that the motion is independent of z and from (6.61) that it is horizontally non-divergent, i.e., the fluid behaves as if it were a Taylor-Proudman column.

Equations (6.59) and (6.61) are seen to be consistent. This seemingly fortuitous circumstance is related to our scaling for w ; an inconsistency would have forced us to modify our scaling.

The first order balance is given by

$$(\frac{\partial}{\partial t} + \underline{V}^{(o)}\cdot\nabla)\underline{V}^{(o)} + \underline{k} \times \underline{V}^{(1)} \sin\phi_o + I \cos\phi_o (y - y_o)\underline{k} \times \underline{V}^{(o)} = - \nabla \pi^{(1)} , \tag{6.62}$$

$$0 = - \partial\pi^{(1)}/\partial z \quad , \tag{6.63}$$

$$\nabla\cdot\underline{V}^{(1)} + Fr \frac{\partial w^{(o)}}{\partial z} = 0 \ . \tag{6.64}$$

We note that the fluid is so strongly hydrostatic that π is independent of z both to zeroth and first order. Since $\underline{V}^{(o)}$ and $\pi^{(1)}$ are independent of z in (6.62), $\underline{V}^{(1)}$ must be also, and $w^{(o)}$ becomes a linear function of z.

Defining $h = H + \eta$, we find from (6.53) that the proper scaling for η is $\frac{P}{g} = 2\Omega LU/g$. Hence, integration of (6.64) with respect to height from z = 0 to z = h gives

$$(1 + \frac{FrRo\eta^{(o)}}{H})\nabla \cdot\underline{V}^{(1)} + Fr \frac{D\eta^{(o)}}{Dt} = 0 \quad ,$$

or

$$Fr \frac{\partial\eta^{(o)}}{\partial t} + \nabla\cdot\underline{V}^{(1)} = 0 \tag{6.65}$$

The variations in the height of the free surface are small kinematically but are important dynamically because of their coupling to gravity.

Operating on (6.62) with $\underline{k}\cdot\nabla\times$, we get

$$(\frac{\partial}{\partial t} + \underline{V}^{(o)}\cdot\nabla)\zeta^{(o)} + I \cos\phi_o v^{(o)} + \sin\phi_o \nabla \cdot\underline{V}^{(1)} = 0 \ ,$$

where $\zeta^{(o)} = \underline{k}\cdot\nabla \times \underline{V}^{(o)}$ is the geostrophic vorticity. Combination with

(6.65) then gives

$$(\frac{\partial}{\partial t} + \underline{V}^{(o)}\cdot\nabla)\left[\zeta^{(o)} + I\cos\phi_o(y-y_o) - Fr\sin\phi_o\eta^{(o)}\right] = 0. \qquad (6.66)$$

In consideration of (6.59) we may introduce the stream function $\psi = \pi^{(o)}/\sin\phi_o$ or $\eta^{(o)}/\sin\phi_o$ for $\underline{V}^{(o)}$, so that $\underline{V}^{(o)} = \underline{k}\times\nabla\psi$ and $\zeta^{(o)} = \nabla^2\psi^{(o)}$. Equation (6.66) then becomes

$$(\frac{\partial}{\partial t} + \underline{V}^{(o)}\cdot\nabla)\left[\nabla^2\psi + I(y-y_o)\cos\phi_o - Fr\,\psi\sin^2\phi_o\right] = 0 \qquad (6.67)$$

or in dimensional form,

$$(\frac{\partial}{\partial t} + \underline{V}_g\cdot\nabla)\left[\nabla^2\psi + \beta(y-y_o) - \frac{\psi}{\lambda^2}\right] = 0\ , \qquad (6.68)$$

where

$$\begin{aligned} f_o &= 2\Omega\sin\phi_o \\ \psi &= g\eta/f_o \\ \underline{V}_g &= \underline{k}\times\nabla\,\psi \qquad (6.69) \\ \beta &= 2\Omega\cos\phi_o/a \\ \lambda^2 &= gD/f_o^{\,2} \end{aligned}$$

Equation (6.68) could also have been derived directly from (6.49), which already embodies the approximation that the horizontal velocity is independent of height,* if we make the additional approximations :

$$\begin{aligned} \underline{V} &= \frac{g}{f_o}\,\underline{k}\times\nabla\,h\ ,\ \zeta = \frac{g}{f_o}\nabla^2 h\ , \\ f &= f_o + \beta(y-y_o) \qquad (6.70) \\ \eta &<< H\ , \end{aligned}$$

all of which have been justified by the foregoing scale analysis.

---

* Strictly, independent of distance parallel to the axis of rotation. The error so introduced is O(H/L) and is negligible except in the immediate vicinity of the equator.

Case II : $I = T_a/T_\beta \geq O(1)$; $T = T_\beta = a/2\Omega L$ (β-time scaling) ; $Fr = \frac{4\Omega^2 L^2}{gD} \leqslant O(1)$. In this case the non-dimensional equations of motion become

$$\frac{L}{a}\left(\frac{\partial}{\partial t} + \frac{1}{I}\underline{V}\cdot\nabla + RoFrw\frac{\partial}{\partial z}\right)\underline{V} + \frac{D}{a}Frw\,(\cos\phi_o + \ldots) +$$

$$+ \underline{k}\times\underline{V}\,(\sin\phi_o + \frac{L}{a}\cos\phi_o(y-y_o) + \ldots) = \nabla\pi \qquad (6.71)$$

$$\frac{D^2}{a^2}RoFr\left(\frac{\partial}{\partial t} + \frac{1}{I}\underline{V}\cdot\nabla + RoFrw\frac{\partial}{\partial z}\right)w - \frac{D}{L}(\cos\phi_o + \ldots)u = -\frac{\partial\pi}{\partial z}$$

$$\nabla\cdot\underline{V} + \frac{L}{a}Fr\frac{\partial w}{\partial z} = 0 \quad ,$$

and, if we expand in the small parameter L/a instead of Ro, the zeroth and first order equations are the same as those previously derived, except that the convective terms involving $\underline{V}\cdot\nabla$ are absent. In this case the dimensional potential vorticity equation is linear :

$$\frac{\partial}{\partial t}(\nabla^2\psi - \frac{\psi}{\lambda^2}) + \beta\frac{\partial\psi}{\partial x} = 0 \ . \qquad (6.72)$$

Motions in this category are independent of preexisting currents.

If $Fr \lesssim O(Ro)$ in case I or $\lesssim O(L/a)$ in Case II, we see from (6.64) or its analogue for Case II that the motion is horizontally non-divergent to first order in Ro or L/a. The free surface remains effectively undisturbed ; all effects from gravity vanish ; and the term ψ/λ^2 disappears from equations (6.68) and (6.72). The motion is quasi-geostrophic and non-divergent.

Finally, we drop the assumption that $Fr \lesssim O(1)$. If the motion is to continue geostrophic with $T = T_a$, and $L/a \lesssim O(Ro)$, it is easy to show that the right hand side of (6.49) dominates and, the flow is stationary, i.e., $\partial h/\partial t = 0$. Otherwise, if $L/a > O(Ro)$, the convective change of the earth's vorticity, βv, must be balanced by the planetary divergence, $-f\nabla\cdot\underline{V}$, and the time scale is determined by the balance between $\partial h/\partial t$ and $h\nabla\cdot\underline{V}$ in the continuity equation. This gives $T = T_h = Fr/2\Omega$. We must therefore consider

Case III : $L/a \simeq O(1)$, $T = T_h = Fr/2\Omega$, $Fr > O(Ro^{-1})$.* The non-dimensional equations of motion become

$$(Fr^{-1}\frac{\partial}{\partial t} + Ro\,\underline{V}\cdot\nabla + Ro\,w\frac{\partial}{\partial z})\underline{V} + \frac{D}{a}w\cos\phi\,\underline{i} + \underline{k}\times\underline{V}\sin\phi = -\nabla\pi, \tag{6.73}$$

$$\frac{D^2}{a^2}(Fr^{-1}\frac{\partial}{\partial t} + Ro\,\underline{V}\cdot\nabla + Ro\,w\frac{\partial}{\partial z})w + \frac{D}{a}u\cos\phi = -\frac{\partial\pi}{\partial z},$$

$$\nabla\cdot\underline{V} + \frac{\partial w}{\partial z} = 0 .$$

Expanding in powers of the larger of Ro or $(Fr)^{-1}$ in the usual way, we obtain

$$\underline{k}\times\underline{V}^{(o)}\sin\phi = -\nabla\pi^{(o)}$$

$$0 = -\frac{\partial\pi^{(o)}}{\partial z} \tag{6.74}$$

$$\nabla\cdot\underline{V}^{(o)} + \frac{\partial w^{(o)}}{\partial z} = 0 .$$

As before, $\underline{V}^{(o)}$ is independent of height, and integration of the last of (6.74) leads to

$$(\frac{\partial}{\partial t} + Fr\,Ro\,\underline{V}^{(o)}\cdot\nabla)\eta^{(o)} + (1 + FrRo\eta^{(o)})\nabla\cdot\underline{V}^{(o)} = 0 .$$

Dimensionally

$$\underline{V} = \frac{g\underline{k}\times\nabla h}{f}, \tag{6.75}$$

$$\frac{\partial h}{\partial t} + h\nabla\cdot\underline{V} = 0 . \tag{6.76}$$

---

* This class of motions was discussed by A.P. Burger, 1958 : Scale considerations of planetary motions of the atmosphere, Tellus, 10, 195-205.

Taking the divergence of (6.75) we get

$$\nabla \cdot \underline{V} = -\frac{g\beta}{f^2}\frac{\partial h}{\partial x} \quad \text{or} \quad \beta v + f\nabla \cdot \underline{V} = 0 ,$$

and substitution in (6.76) gives

$$\frac{\partial h}{\partial t} - \frac{g\beta h}{f^2}\frac{\partial h}{\partial x} = 0 ,$$

which, combined with (6.75), may be taken as determining the motion. Equation (6.77) states that a given elevalation h moves at the speed $c = -g\beta h/f^2$.

Since the motion cannot remain quasi-geostrophic near the equator, the planetary scale motions require a special treatment in its vicinity. *

If Ro is not small and the flow is not quasi-geostrophic, there remains one case which still permits a simple treatment. This is when $Fr < O(1)$ and H/L remains small. If we separate out the gravity oscillations by supposing that the time scale is either T_a or T_β and that the Froude number U^2/gH is small, we find that the flow is horizontally non-divergent and is governed by (6.28), the non-divergent vorticity equation on the sphere. A stream function ψ may be defined in terms of which this equation becomes

$$\left(\frac{\partial}{\partial t} + \underline{V}\cdot\nabla\right)(\nabla^2\psi + f) = 0 , \tag{6.78}$$

the equation originally dealt with by Rossby on the β-plane.

---

* See, for example, Matsuno (1968).

Rossby waves

We turn finally to the dispersive properties of small amplitude quasi-geostrophic waves in a homogeneous-incompressible ocean at rest with respect to a rotating sphere. We suppose that L/a is small so we can use the β-plane approximation. The motion is then governed by equation (6.68), which applies to finite amplitude motions in a non-resting ocean as well, if $T_\beta/T < O(L/a)$ or $Ro < O(L/a)^2$, and $Fr \lesssim O(1)$.

Such motions were first considered qualitatively by J. Bjerknes (1931)** in relation to the problem of cyclogenesis, and quantitatively by Rossby (1939)* in relation to the properties of long waves in the westerlies. They are commonly called "Rossby waves". An elegant treatment of their properties has been given by Longuet-Higgins (1964, 1965)**.

Assuming a plane wave solution of the form

$$\psi = e^{i(kx + \ell y + \sigma t)}$$

we obtain the dispersion relationship already given by (6.45), i.e.,

$$-c = \frac{\sigma}{k} = \frac{\beta k}{k^2 + \ell^2 + \lambda^{-2}} \qquad (6.79)$$

Rossby waves in the oceans are excited by the action of variable wind stresses. The travelling wave and vortex systems of the atmosphere can be shown to produce external waves of this type, despite the density stratification, because internal waves have much smaller time scales that the atmospheric systems and are therefore not excited. Atmospheric Rossby waves are dealt with in Chapter X.

** J. Bjerknes, 1937 : Meteor. Z., 54, 462-166.

* C.-G. Rossby, 1939 : J. Marine Res., 2, 38-55.

** M.S. Longuet-Higgins, 1964 : Proc. Roy. Soc. A., 279, 446-473.
1965 : Proc. Roy. Soc. A., 284, 40-54.

VII. THREE-DIMENSIONAL QUASI-GEOSTROPHIC MOTION

Introduction

Chapters V and VI were intended as introductions to the subject of balanced flow in a rotating, stratified fluid. The first dealt with balanced, axially-symmetric flows, whose changes are governed by the laws of conservation of angular momentum and entropy (density in an incompressible fluid) together with the condition of balance among the meridional forces. The second dealt with slow, large-scale, two-dimensional asymmetric flows, which are governed by the laws of conservation of potential vorticity and mass together with the condition that the pressure and Coriolis forces are continually in a state of balance. We wish now to generalize these balance principles to include large-scale three-dimensional motions of a rotating, stratified fluid. We will first extend the scale analysis in Chapter VI to quasi-geostrophic flow, and then to flows which may be non-geostrophic but which satisfy a condition of balance analogous to that for the axisymmetric flows of Chapter V whose convective accelerations, viz, centripetal accelerations, are not necessarily small. A generalization of this kind is needed for the treatment of fronts, jets, and in general all narrow zones of strong gradient in which the Rossby number is not necessarily small. Such motions remain distinguishable in time scale from gravity-inertia modes, and their analysis is greatly facilitated by the introduction of the balance equations.

The general quasi-geostrophic equations for the atmosphere*

Let L be a characteristic horizontal dimension, D a characteristic vertical dimension, U a characteristic particle velocity relative to the rotating earth, and Ω the angular velocity of the earth's rotation.

---

* Charney, J.G., 1962 : Integration of the primitive and balance equations. Int. Symp. on N.W.P., Tokyo, 1960. Charney, J.G., and M.E. Stern, 1961 : On the stability of internal baroclinic jets in a rotating atmosphere. J. Atmos. Sci., 19, 159-172. Also see Phillips, N.A., 1963 : Geostrophic motion. Rev. Geophysics, 1, 123-175.

We shall be concerned with atmospheric motions for which D/L and the Rossby number $Ro = \frac{U}{2\Omega L}$ are so small that departures from hydrostatic and geostrophic equilibrium may be neglected.

Let $\underline{V}$ be the relative horizontal particle velocity component, p the pressure, ρ the density, α the specific volume, θ the potential temperature, g the acceleration of gravity (assumed constant), φ the latitude, a the radius of the earth, ∇ the horizontal component of the gradient operator, z the upward vertical coordinate, and $\underline{k}$ a unit vertical vector. Horizontal space-time averages will be denoted by the subscript "s". It is assumed that p_s and α_s are hydrostatically related and that $\alpha_s = \frac{1}{\rho_s}$. In dimensional form, the equations of relative motion and continuity for an inviscid fluid moving under the influence of gravity are

$$(\frac{\partial}{\partial t} + \underline{V}\cdot\nabla + w\frac{\partial}{\partial z})\,\underline{V} + 2\Omega\sin\phi\;\underline{k}\times\underline{V} + 2\Omega\cos\phi\;w\underline{i} = -\alpha\nabla p\ , \qquad (7.1)$$

$$(\frac{\partial}{\partial t} + \underline{V}\cdot\nabla + w\frac{\partial}{\partial z})\,w - 2\Omega\cos\phi\;\underline{V}\cdot\underline{i} = -g - \alpha\frac{\partial p}{\partial z}\ , \qquad (7.2)$$

$$(\frac{\partial}{\partial t} + \underline{V}\cdot\nabla)\,\rho + \rho\nabla\cdot\underline{V} + \frac{\partial(\rho w)}{\partial z} = 0\ , \qquad (7.3)$$

where $\underline{i}$ is a unit vector directed from west to east.

On the assumption that the flow is adiabatic, the first law of thermodynamics may be written

$$(\frac{\partial}{\partial t} + V\cdot\nabla + w\frac{\partial}{\partial z})\,s = 0\ . \qquad (7.4)$$

For a perfect gas the specific entropy

$$s = c_p\ln\theta = c_p\ln\alpha + c_v\ln p + \text{constant}\ , \qquad (7.5)$$

where c_p and c_v are the specific heats at constant pressure and volume respectively. Thus the first law takes the form

$$(\frac{\partial}{\partial t} + \underline{V}\cdot\nabla)(\gamma^{-1}\ln p + \ln\alpha) + w\,\frac{\partial \ln\theta}{\partial z} = 0 \ , \tag{7.6}$$

where $\gamma = c_p/c_v$.

The following non-dimensional parameters are found to be relevant :

$$Ro \equiv \frac{U}{2\Omega L} \ , \qquad \delta \equiv \frac{D}{L} \ , \qquad \mu \equiv \frac{D}{H}$$

$$\kappa = D\,\frac{\partial \ln\theta_s}{\partial z} \ , \qquad \varepsilon = \frac{4\Omega^2 L^2}{gD\kappa} \ , \qquad \gamma = \frac{c_p}{c_v} \ , \tag{7.7}$$

where H is the scale height,

$$H \equiv \frac{p_s \alpha_s}{g} \ . \tag{7.8}$$

The non-dimensional parameters, ε and κ, may on occasion be replaced by the rotational Froude and Richardson numbers,

$$Fr \equiv \frac{4\Omega^2 L^2}{gD} = \varepsilon\kappa \ ,$$

$$Ri = \frac{gD\kappa}{U^2} = \varepsilon^{-1} Ro^{-2} \ . \tag{7.9}$$

The range of the non-dimensional parameters will be limited by explicit assumption so as to focus attention on the principal large-scale motions of the atmosphere. We are interested in motions which are

(a) quasi-hydrostatic : $\delta \ll 1$. ($\delta \leq O(Ro)^2$).

(b) quasi-geostrophic : $Ro \ll 1$, and the characteristic time scale T is not smaller in magnitude than the advective time scale L/U.

(c) stratified : $\varepsilon = Ri^{-1}Ro^{-2} \lesssim O(1)$.

(d) quasi-Boussinesq : $\kappa << 1$, $\kappa\varepsilon << 1$, $\frac{\mu}{\gamma} \lesssim O(1)$.

(e) of sufficiently moderate latitudinal scale so as to satisfy the beta-plane approximation : $\frac{L}{a} \lesssim O(Ro)$.

The characteristic magnitude P of the horizontal variation of the pressure is determined by assumption (b). We get

$$P = \frac{2\Omega UL}{\alpha_s(z)} = \mu\kappa\varepsilon Ro p_s(z) ,$$

and may therefore define a non-dimensional pressure p' such that

$$p = p_s(z)(1 + \mu\kappa\varepsilon Ro\, p') , \tag{7.10}$$

where p' is of order unity.

The order of magnitude of the density variations may be established in a similar manner. We get

$$\rho = \rho_s(z)\ (1 + \kappa\varepsilon Ro\rho') , \tag{7.11}$$

where ρ' is order unity. By the same procedure, we may show that

$$\alpha = \alpha_s(z)\ (1 + \kappa\varepsilon Ro\alpha') , \tag{7.12}$$

where α' is order unity and

$$\alpha' \simeq - \rho' .$$

Since the non-dimensional forms of the thermodynamic variables p and α (or p and ρ) have been prescribed, the non-dimensional form of lnθ is determined. We find

$$\begin{aligned} \ln\theta &= \ln\theta_s(z) + \kappa\varepsilon Ro\theta' \\ \theta' &\simeq \mu\gamma^{-1}p' + \alpha' . \end{aligned} \tag{7.13}$$

The characteristic scale W of the vertical velocity component may be determined from the thermal equation (7.6). Evaluating the terms in this equation with the aid of (7.10) and (7.11) we obtain

$$\frac{W}{(D/L)U} = \varepsilon Ro < O(1) \ . \qquad (7.14)$$

We are also concerned with the behavior of the trigonometric functions, $\sin\phi$ and $\cos\phi$. Setting $\Delta\phi = (L/a)(y' - y'_o) = (L/a)\Delta y'$, we get

$$\sin\phi = \sin\phi_o + \frac{L}{a}\cos\phi_o \Delta y' + O(\frac{L}{a})^2 \ ,$$

$$\cos\phi = \cos\phi_o - \frac{L}{a}\sin\phi_o \Delta y' + O(\frac{L}{a})^2 \ , \qquad (7.15)$$

Scaling for weak stratification

The case $\varepsilon > 1$ requires another scaling for w. In this case the local time and horizontal space derivatives in the thermal equations compensate, and the estimate (7.14) is no longer valid. A valid estimate is obtained from the potential vorticity equation (cf. Chapter VI). Imagine the continuously stratified atmosphere to be divided into a number of thin isentropic layers of depth Δz_i at the heights z_i. For such layers the horizontal velocity will be independent of depth and within each layer the potential vorticity $(\zeta_i + f)/\rho\Delta z_i$ will be conserved, so that

$$\frac{D\zeta_i}{Dt} \sim \frac{f}{\rho\Delta z_i}\frac{D(\rho\Delta z_i)}{Dt} = \frac{f}{\rho}\frac{\Delta(\rho w_i)}{(\Delta z)_i} \ ,$$

or, in orders of magnitude,

$$\frac{U^2}{L^2} \sim f\frac{W}{D} \ , \quad \frac{W}{U} \sim Ro\delta$$

The condition $\varepsilon > 1$ may be interpreted physically as stating that the characteristic slope of a particle trajectory, W/U, is less than that of an isentropic surface. The characteristic slope of the latter is obtained by substituting the thermal wind relationship, $\frac{\partial u}{\partial z} \simeq -\frac{g}{f}\frac{\partial \ln\theta}{\partial y}$, into the expression

$$\left(\frac{dz}{dy}\right)_\theta = -\frac{\dfrac{\partial \ln\theta}{\partial y}}{\dfrac{\partial \ln\theta}{\partial z}} .$$

We obtain

$$\left(\frac{dz}{dy}\right)_\theta \sim \frac{2\Omega U}{g\kappa} ,$$

and

$$\frac{\dfrac{w}{v}}{\left(\dfrac{dz}{dx}\right)_\theta} \sim \frac{\dfrac{UD}{2\Omega L^2}}{\dfrac{2\Omega U}{g\kappa}} = \frac{gD\kappa}{4\Omega^2 L^2} = \frac{1}{\varepsilon} .$$

Now if particles at A and B in Figure 7.1 are exchanged, the potential energy will decrease, whereas if particles at A' and B' are exchanged the potential energy will increase.

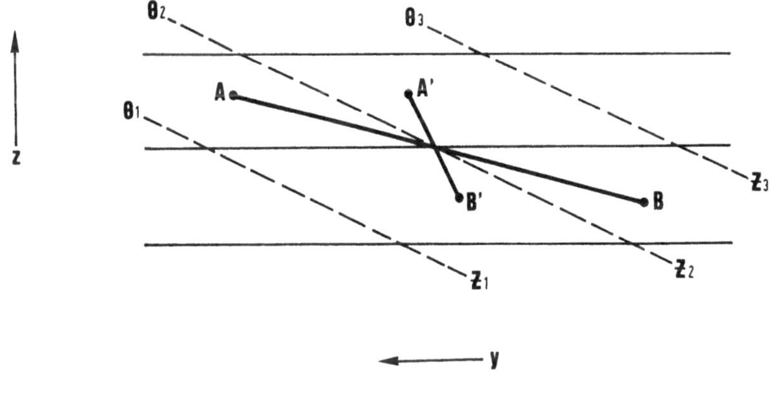

Figure 7.1

Thus the requirement $\varepsilon \geq O(1)$ describes the potential energy converting modes. If $\varepsilon < O(1)$, the fluid is so highly stratified that the flow has to be driven.

Non-dimensionalization of the equations of motion

In summary :

$$\nabla = \frac{\nabla'}{L}, \quad \underline{V} = U\underline{V}', \quad w = \delta U\varepsilon Ro w', \quad t = \frac{L}{U}t', \quad z = Dz',$$

$$\sin\phi = \sin\phi_o + \frac{L}{a}\cos\phi_o \Delta y' + O(\frac{L}{a})^2,$$

$$\cos\phi = \cos\phi_o - \frac{L}{a}\sin\phi_o \Delta y' + O(\frac{L}{a})^2,$$

$$p = p_s(z)\ (1 + \mu\kappa\varepsilon Ro p'), \tag{7.16}$$

$$\alpha = \alpha_s(z)\ (1 + \kappa\varepsilon Ro\ \alpha'),$$

$$\rho = \rho_s(z)\ (1 + \kappa\varepsilon Ro \rho'),$$

$$\ln\theta = \ln\theta_s(z) + \kappa\varepsilon Ro\theta' .$$

The non-dimensional forms of (7.1), (7.2), and (7.3) now become

$$Ro\ (\frac{\partial}{\partial t} + \underline{V}\cdot\nabla + \varepsilon Ro\ w\frac{\partial}{\partial z})\underline{V} + (\sin\phi_o + \frac{L}{a}\cos\phi_o\Delta y + \ldots)\underline{k}\times\underline{V} +$$

$$+ \delta\varepsilon Ro\ (\cos\phi_o - \frac{L}{a}\sin\phi_o\Delta y + \ldots)w\underline{i} = -(1 + \kappa\varepsilon Ro\ \alpha)\ \nabla p \tag{7.17}$$

$$\delta^2 Ro^2\ (\frac{\partial}{\partial t} + \underline{V}\cdot\nabla + \varepsilon Ro\ w\frac{\partial}{\partial z})w - \delta(\cos\phi_o - \frac{L}{a}\sin\phi_o\Delta y)\underline{V}\cdot\underline{i}$$

$$= -\alpha_s\ \{(1 + \kappa\varepsilon Ro\ \alpha)\frac{\partial}{\partial z}(\frac{p}{\alpha_s}) + \frac{\rho}{\alpha_s}\}, \tag{7.18}$$

$$\kappa\varepsilon Ro\rho_s\ (\frac{\partial}{\partial t} + \underline{V}\cdot\nabla)\rho + \rho_s(1 + \kappa\varepsilon Ro\rho)\nabla\cdot\underline{V} + \varepsilon Ro\frac{\partial}{\partial z}[\rho_s(1 + \kappa\varepsilon Ro\rho)w] = 0,$$

(7.19)

where for convenience, the primes are dropped. The first law of thermodynamics, (7.6), is transformed into

$$\frac{\mu\gamma^{-1}(\frac{\partial}{\partial t} + \underline{V}\cdot\nabla)p}{1 + \mu\kappa\varepsilon Ro\ p} + \frac{(\frac{\partial}{\partial t} + \underline{V}\cdot\nabla)\alpha}{1 + \kappa\varepsilon Ro\ \alpha} + w\left[1 + \varepsilon Ro \frac{\partial\theta}{\partial z}\right] = 0\ . \qquad (7.20)$$

The non-dimensional variables are now formally expanded in a power series in Ro. The resulting zero order equations of motion and continuity are found to be

$$\sin\phi_o\ \underline{k} \times \underline{V}^{(o)} = -\nabla p^{(o)} \qquad (7.21)$$

$$\frac{\alpha^{(o)}}{\alpha_s} = \frac{\partial}{\partial z}\left(\frac{p^{(o)}}{\alpha_s}\right) \cong -\frac{p^{(o)}}{\alpha_s} \qquad (7.22)$$

$$\nabla\cdot\underline{V}^{(o)} = 0\ , \qquad (7.23)$$

the latter being consistent with the non-divergent character of the zero-order motion implied by (7.21).

The first order horizontal equation of motion becomes

$$(\frac{\partial}{\partial t} + \underline{V}^{(o)}\cdot\nabla)\underline{V}^{(o)} + \sin\phi_o\ \underline{k} \times \underline{V}^{(1)} + \frac{L/a}{Ro}\cos\phi_o \Delta y \underline{k} \times \underline{V}^{(o)} = -\nabla p^{(1)}\ ; \qquad (7.24)$$

the vertical equation of motion remains the hydrostatic approximation,

$$\frac{\alpha^{(1)}}{\alpha_s} = \frac{\partial}{\partial z}\left(\frac{p^{(1)}}{\alpha_s}\right) \simeq -\frac{p^{(1)}}{\alpha_s}\ ; \qquad (7.25)$$

the first order continuity equation becomes

$$\rho_s\nabla\cdot\underline{V}^{(1)} + \varepsilon\frac{\partial}{\partial z}(\rho_s w^{(o)}) = 0\ ; \qquad (7.26)$$

and the first order thermodynamic equation may be written

$$(\frac{\partial}{\partial t} + \underline{V}^{(o)}\cdot\nabla)\theta^{(o)} + \frac{w^{(o)}}{\sin\phi_o} = 0 , \qquad (7.27)$$

where

$$\theta^{(o)} = \mu\gamma^{-1}p^{(o)} + \alpha^{(o)} . \qquad (7.28)$$

Equations (7.21) through (7.28) are the quasi-geostrophic equations for the atmosphere.

Law of conservation of potential vorticity

We note that the zero order equations are diagnostic rather than prognostic because of the absence of time derivatives. The time dependence is contained in the first order equations.

Equation (7.21) may be written

$$\underline{V}^{(o)} = \underline{k} \times \nabla\psi , \qquad (7.29)$$

where

$$\psi = \frac{p^{(o)}}{\sin\phi_o} . \qquad (7.30)$$

From (7.22) and (7.30) we obtain

$$\theta^{(o)} = \left[\frac{\mu}{\gamma}\psi + \alpha_s \frac{\partial}{\partial z}(\frac{\psi}{\alpha_s})\right]\sin\phi_o$$

$$= (\frac{\partial\psi}{\partial z} - \kappa\psi)\sin\phi_o \cong \frac{\partial\psi}{\partial z}\sin\phi_o . \qquad (7.31)$$

Similarly

$$T^{(o)} = (\frac{\partial\psi}{\partial z} - \frac{\partial \ln T_s}{\partial z}\psi)\sin\phi_o \cong \frac{\partial\psi}{\partial z}\sin\phi_o \cong \theta^{(o)} . \qquad (7.32)$$

In terms of ψ, Equation (7.27) becomes

$$(\frac{\partial}{\partial t} + \underline{V}^{(o)}\cdot\nabla)\,\frac{\partial\psi}{\partial z} + \frac{w^{(o)}}{\sin\phi_o} = 0\ . \qquad (7.33)$$

Taking the curl of (7.24), we get

$$(\frac{\partial}{\partial t} + \underline{V}^{(o)}\cdot\nabla)\left[\zeta^{(o)} + \beta'(y-y_o)\right] = -\sin\phi_o\nabla\cdot\underline{V}^{(1)} = \varepsilon\,\frac{\sin\phi_o}{\rho_s}\,\frac{\partial}{\partial z}\,(\rho_s w^{(o)}), \qquad (7.34)$$

where $\zeta^{(o)} = \nabla \times \underline{V}^{(o)}\cdot\underline{k} = \nabla^2\psi$, and $\beta' = (L/a)Ro^{-1}\cos\phi_o$, and substituting for $w^{(o)}$ from (7.33) we obtain finally

$$(\frac{\partial}{\partial t} + \underline{V}^{(o)}\cdot\nabla)q = 0\ , \qquad (7.35)$$

where

$$q = \nabla^2\psi + \beta'(y-y_o) + \frac{\sin^2\phi_o}{\rho_s}\,\frac{\partial}{\partial z}\,(\varepsilon\rho_s\,\frac{\partial\psi}{\partial z})\ . \qquad (7.36)$$

The dimensional forms of equations (7.35) and (7.33) are

$$(\frac{\partial}{\partial t} + \underline{V}\cdot\nabla)q = 0 \qquad (7.37)$$

and

$$(\frac{\partial}{\partial t} + \underline{V}\cdot\nabla)\psi_z + \frac{N^2}{f_o}\,w = 0\ , \qquad (7.38)$$

where

$$q = \nabla^2\psi + \frac{f_o^{\,2}}{\rho_s}\,\frac{\partial}{\partial z}\left(\frac{\rho_s}{N^2}\,\frac{\partial\psi}{\partial z}\right) + \beta(y - y_o)\ , \qquad (7.39)$$

and

$$\left.\begin{aligned} \psi &= \frac{p-p_s}{f_o\rho_s} \\ V &= \underline{k} \times \nabla\psi \\ \psi_z &\cong \frac{g}{f_o}(\ln\theta - \ln\theta_s) \cong \frac{g}{f_o}(\ln T - \ln T_s) \\ f_o &= 2\Omega\sin\phi_o \quad , \quad \beta = \frac{2\Omega\cos\phi_o}{a} \\ N^2 &= g\,\frac{\partial(\ln\theta_s)}{\partial z} \quad . \end{aligned}\right\} \tag{7.40}$$

Equations (7.37) and (7.38) together assert the conservation of potential vorticity, but q is not itself the potential vorticity. The potential vorticity is conserved at a particle, whereas q is conserved at the horizontal projection of a particle. We may call the latter "pseudo-potential vorticity" for want of a better name. However, it can be shown (Charney and Stern, loc. cit.) that if $\hat{q}$ is the potential vorticity,

$$\hat{q} = \frac{1}{\rho_s}\frac{d\theta_s}{dz}\left[\nabla^2\psi + \frac{f_o^{\,2}}{N^2}\frac{\partial^2\psi}{\partial z^2} + f_o + \beta(y-y_o)\right] , \tag{7.41}$$

and

$$\left(\frac{\partial}{\partial t} \text{ or } \nabla\right)_\theta \hat{q} = \frac{1}{\rho_s}\frac{d\theta_s}{dz}\left(\frac{\partial}{\partial t} \text{ or } \nabla\right)_z q , \tag{7.42}$$

where the subscripts θ or z indicate that the differentiation takes place in isentropic or level surfaces respectively. In particular, we note for later reference that the y-gradient of $\hat{q}$ in an isentropic surface is equal to the product of $\rho_s^{-1} d\theta_s/dz$ and the y-gradient of q in a level surface.

Heating and friction

If Q is the rate of accession of heat per unit mass, equations (7.37) and (7.38) become

$$\left(\frac{\partial}{\partial t} + \underline{V}\cdot\nabla\right)q = \frac{gf_o}{\rho_s}\frac{\partial}{\partial z}\left(\frac{\rho_s}{N^2}\frac{Q}{c_pT_s}\right) \tag{7.43}$$

and

$$\left(\frac{\partial}{\partial t} + \underline{V}\cdot\nabla\right)\psi_z + \frac{N^2}{f_o}w = \frac{g}{f_o}\frac{Q}{c_pT_s} \tag{7.44}$$

Friction may be introduced as a boundary layer phenomenon as in Chapter III. If ν is the kinematic coefficient of eddy viscosity, which, for simplicity, we assume to be constant, the same scaling arguments that were used to derive the general equations lead to the following balance of forces in the boundary layer

$$\underline{k}\times\underline{V} = -\frac{\nabla p}{f_o\rho_s} + \frac{\nu}{f_o}\underline{V}_{zz} \quad , \tag{7.45}$$

where the pressure gradient does not vary through the boundary layer. We may therefore write

$$-\frac{\nabla p}{\rho_s f_o} = \underline{k}\times\underline{V}_{go} \quad , \tag{7.46}$$

where $\underline{V}_{go}$ is the surface geostrophic wind. If we take the x-axis in the direction of $\underline{V}_g$, so that $\underline{V}_{go} = u_{go}\underline{i}$, the solution of (7.45) becomes

$$\begin{aligned} u &= u_{go}\left[1 - e^{-z/\sqrt{2}\,D_E}\cos(z/\sqrt{2}\,D_E)\right] \\ v &= u_{go}\,e^{-z/\sqrt{2}\,D_E}\sin(z/\sqrt{2}\,D_E) \quad , \quad \left(D_E^2 = \frac{2\nu}{f_o}\right) \end{aligned} \tag{7.47}$$

giving a transport

$$\rho_s(0)\int_0^\infty v\,dz = \rho_s(0)\,\frac{D_E u_{go}}{2} \tag{7.48}$$

to the left of the surface geostrophic wind.

Since equations (7.45) were first derived by Ekman (1907), it has become customary to call the surface boundary layer the "Ekman layer".

In general, if the geostrophic wind has an arbitrary direction, the mass transport may be written

$$\underline{M} = \frac{\rho_s D_E}{2}\,\underline{k} \times \underline{V}_{go} \,, \tag{7.49}$$

and integration of the continuity equation,

$$\nabla\cdot(\rho_s \underline{V}) + (\rho_s w)_z = 0 \,, \tag{7.50}$$

through the Ekman layer, gives

$$\nabla\cdot\underline{M} + \rho_s(0)\, w(0) = 0 \,, \tag{7.51}$$

where $w(0)$ denotes the vertical velocity at the <u>top</u> of the Ekman layer. Substituting from (7.49) we obtain

$$w(0) = \frac{D_E}{2}\,\zeta_g(0) \,, \tag{7.52}$$

where $\zeta_g(0) = \nabla^2\psi(0)$ is the vorticity of the surface geostrophic wind. Thus equation (7.52) replaces the inviscid condition $w = 0$ at a rigid horizontal boundary. As far as the fluid above the boundary layer is concerned, it "knows" only that there is a vertical mass transport of amount $D_E\zeta_g/2$ at $z = 0$.

The vertical transport of momentum, as distinguished from mass, does not appear as a boundary condition because in the geostrophic theory this quantity is small of order Ro. However, the vertical Ekman pumping also transports sensible and latent heat, and these transports must be taken into account in the thermal boundary condition.

In a later section we shall consider eddy conduction as a source of heat. In this case, the vertical flux of entropy per unit area is

$$\rho_s K \frac{\partial s}{\partial z} = \rho_s K c_p \frac{\partial \ln\theta}{\partial z} = \frac{c_p K}{g} f_o \rho_s \left(\frac{N^2}{f_o} + \psi_{zz}\right)$$

where K is the specific eddy-conductivity. We shall assume that K is constant ; hence

$$Q = -\frac{K c_p T_s f_o}{\rho_s g} \left[\rho_s\left(\frac{N^2}{f_o} + \psi_{zz}\right)\right]_z \quad .$$

Inserting this expression for Q in (7.43) and (7.44), and adding internal viscosity, we obtain

$$\left(\frac{\partial}{\partial t} + \underline{V}\cdot\nabla\right)q = -\frac{K f_o^{\,2}}{\rho_s}\left\{\frac{1}{N^2}\left[\rho_s\left(\frac{N^2}{f_o} + \psi_{zz}\right)\right]_z\right\}_z + \nu\nabla^2\psi_{zz} \ , \tag{7.53}$$

$$\left(\frac{\partial}{\partial t} + \underline{V}\cdot\nabla\right)\psi_z + \frac{N^2}{f_o} w = -\frac{K}{\rho_s}\left[\rho_s\left(\frac{N^2}{f_o} + \psi_{zz}\right)\right]_z \ . \tag{7.54}$$

The balance equations

We note that the scaling which led to the non-dimensional equations (7.17) - (7.20) remains valid if Ro is as large as O(1) providing the quantity $\varepsilon Ro = Ri^{-1} Ro^{-1}$, which multiplies w in the momentum and continuity equation, remains small, and, to a first approximation, the horizontal velocity remains non-divergent. If we express the horizontal velocity $\underline{V}$ as the sum of a solenoidal (non-divergent) and a potential (irrotational) vector,

$$\underline{V} = \underline{k} \times \nabla\psi + \nabla\sigma \equiv \underline{V}_\psi + \underline{V}_\sigma \ , \tag{7.55}$$

it follows from (7.14) and (7.19) that $\underline{V}_\psi$ and $\underline{V}_\sigma$ are scaled by U and εRoU, respectively, and that ψ and σ are scaled by LU and εRoLU, respectively. Substituting the scaled variables into (7.17) - (7.20) and taking the horizontal divergence and curl of the first we obtain

$$(\varepsilon Ro)\nabla\cdot(\underline{V}_\psi\cdot\nabla\underline{V}_\psi) - \varepsilon\nabla\cdot(\sin\phi\nabla\psi) + \varepsilon(\varepsilon Ro)\underline{k}\cdot\underline{V}_\sigma \times \nabla\sin\sigma +$$
$$+ (\varepsilon Ro)^2\nabla\cdot(\frac{\partial\underline{V}_\sigma}{\partial t} + \underline{V}_\psi\cdot\nabla\underline{V}_\sigma + \underline{V}_\sigma\cdot\nabla\underline{V}_\psi + w\frac{\partial\underline{V}_\psi}{\partial z}) +$$
$$+ (\varepsilon Ro)^3\nabla\cdot(\underline{V}_\sigma\cdot\nabla\underline{V}_\sigma + w\frac{\partial\underline{V}_\sigma}{\partial z}) = -\varepsilon\nabla^2 p \;, \qquad (7.56)$$

$$(\frac{\partial}{\partial t} + \underline{V}_\psi\cdot\nabla) + Ro^{-1}\underline{V}_\psi\cdot\nabla\sin\phi + \varepsilon\nabla\cdot(\underline{V}_\sigma\sin\phi) +$$
$$+ (\varepsilon Ro)\nabla\cdot\left[\zeta\underline{V}_\sigma + w\frac{\partial(\nabla\psi)}{\partial z}\right] + (\varepsilon Ro)^2\nabla\cdot\left[w\frac{\partial}{\partial z}(\underline{V}_\sigma\times\underline{k})\right] = 0 \;,$$
$$(7.57)$$

$$\rho_s\nabla\cdot\underline{V}_\sigma + \frac{\partial}{\partial z}(\rho_s w) = 0 \;, \qquad (7.58)$$

$$\left.\begin{aligned}&\left[\frac{\partial}{\partial t} + (\underline{V}_\psi + \varepsilon Ro\underline{V}_\sigma)\cdot\nabla\right]\theta + (1 + \varepsilon Ro\frac{\partial\theta}{\partial z})w = 0 \;,\\ &(\theta = \mu\gamma^{-1}p + \boldsymbol{\alpha}) \;,\end{aligned}\right\} \qquad (7.59)$$

where $\zeta = \underline{k}\cdot\nabla\times\underline{V} = \nabla^2\psi$, and the quasi-Boussinesq approximations, $\kappa << 1$, $\kappa\varepsilon << 1$, but not the β-plane approximation, has been made. The balance equations are obtained by omitting second and higher order terms in Ro. The resulting equations, in dimensional form, may be written

$$\left.\begin{aligned}&\nabla\cdot(f\nabla\psi) - \nabla\cdot(\underline{V}_\psi\cdot\nabla\underline{V}_\psi) - \underline{k}\cdot\underline{V}_\sigma \times \nabla f = \nabla^2\pi \;,\\ &\frac{\partial\zeta}{\partial t} + \nabla\cdot\left[(f + \zeta)(\underline{V}_\psi + \underline{V}_\sigma) + w\frac{\partial\nabla\psi}{\partial z}\right] = 0 \;,\\ &g(\ln\theta - \ln\theta_s) = \frac{\partial\pi}{\partial z} \;,\\ &\nabla\cdot(\rho_s\underline{V}_\sigma) + \frac{\partial}{\partial z}(\rho_s w) = 0 \;,\end{aligned}\right\} \qquad (7.60)$$

$$\left[\frac{\partial}{\partial t} + (\underline{V}_\psi + \underline{V}_\sigma)\cdot\nabla\right] \ln\theta + w\,\frac{\partial \ln\theta}{\partial z} = 0 \ , \qquad (7.60)$$

where

$$\underline{V}_\psi = \underline{k} \times \nabla\psi \ , \quad \underline{V}_\sigma = \nabla\sigma \ , \quad \zeta = \nabla^2\psi \ , \quad \pi = \frac{p - p_s}{\rho_s} \ .$$

It may be shown (Charney, 1962 loc. cit.) that these equations are energetically consistent. In the case of an axially symmetric vortex with sources of heat and angular momentum, the present balance equations reduce to Eliassen's formulation of the balance equations in Chapter V.

Although the balance equations are derived here as a kind of second order Rossby number expansion, the underlying justification for their validity probably goes deeper. Just as the flow in an axisymmetric vortex remains balanced as long as the sources and sinks of momentum and heat are slowly acting, and the vortex remains gravitationally and centrifugally stable (potential vorticity > 0), nonsymmetric flow remain balanced in the presence of slowly acting external forces as long as the potential vorticity is bounded away from zero. This is because the potential vorticity condition for instability in an axisymmetric vortex can be generalized to apply to non-symmetric flow as well : a part of the fluid for which the curvature of the streamlines does not change appreciably may be regarded as part of an axisymmetric vortex.

VIII. INSTABILITY OF A ZONAL VORTEX FOR ASYMMETRIC DISTURBANCES IN TWO DIMENSIONS

Stability of two-dimensional zonal flow

a. Fjørtoft's* version of Taylor's** proof of Rayleigh's† criterion. We consider a two-dimensional zonal flow on a sphere $\bar{u} = u(\phi)$, where ϕ is the latitude angle, and inquire into the stability of the flow for arbitrary two-dimensional perturbations.

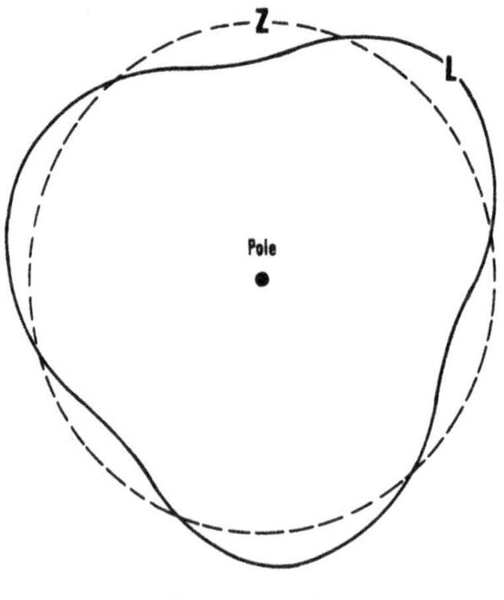

Figure 8.1

Let v be the meridional velocity component and define $u = \bar{u} + u'$, $v = \bar{v} + v'$ are perturbation velocities and $\bar{u}$, $\bar{v}$ are zonal averages. By continuity $\bar{v} = \frac{1}{2\pi}\int_0^{2\pi} v \, d\lambda = 0$. Let Z be a zonal circle and let L be the closed material curve at t = 0 which goes into Z at time t. By Stokes' theorem the circulation around Z is changed by exactly the amount of vorticity that flows into the area bounded by Z when L is deformed into Z. The net flux will consist of two parts : one due to the flux of mean

---

* R. Fjørtoft (1950, Stability Bibliography).

** G.I. Taylor (1915, Stability Bibliography).

† Lord Rayleigh (1880, Stability Bibliography).

vorticity and the other due to the flux of perturbation vorticity. If $a\delta\phi$ is the meridional displacement (a being the radius of the sphere), the first part is given by

$$-\int_0^{2\pi}\int_0^{\delta\phi}(\bar{\zeta}+\frac{\partial\bar{\zeta}}{\partial\phi}\eta)a\,d\eta\; a\cos\phi\, d\lambda = -\frac{1}{2}\int_0^{2\pi} a^2(\delta\phi)^2\frac{\partial\bar{\zeta}}{\partial\phi}\cos\phi\, d\lambda \quad ,$$

and the second by

$$\oint_L \underset{\sim}{V}'_o\cdot\delta\underline{r} - \int_0^{2\pi} u'a\cos\phi\, d\lambda = \oint_L \underset{\sim}{V}'_o\cdot\delta\underline{r} \quad ,$$

where $\underset{\sim}{V}'_o$ is the perturbation velocity at $t = 0$. Hence :

$$\delta C = C - C_o = -\frac{1}{2}\int_0^{2\pi} a^2(\delta\phi)^2\frac{\partial\bar{\zeta}}{\partial\phi}\cos\phi\, d\lambda + \oint_L \underset{\sim}{V}'_o\cdot\delta\underline{r} + O\left[(\delta\phi)^3\right]. \tag{8.1}$$

The total angular momentum is

$$\int_\sigma ua\cos\phi\, d\sigma = \int_{\frac{\pi}{2}}^{\frac{\pi}{2}} Ca^2\cos\, d\phi = \frac{1}{2\pi}\int_\sigma C\, d\sigma \quad , \tag{8.2}$$

where $d\sigma$ is the element of area and the integral is taken over the entire sphere. Since the total angular momentum is conserved, we must have

$$\int_\sigma \delta\, C\, d\sigma = 0 \ , \tag{8.3}$$

and therefore

$$-\pi\int_\sigma a^2(\delta\phi)^2\frac{\partial\bar{\zeta}}{\partial\phi}\cos\phi\, d\sigma + \int_\sigma\oint_L \underline{V}'_o\cdot\delta\underline{r}\, d\sigma = 0 \ . \tag{8.4}$$

Thus, if ζ is a monotonic function of ϕ, $\delta\phi$ must approach zero with $\underset{\sim}{V}'_o$ and we have the theorem : the vortex is stable if the vorticity is a monotonic function of latitude, and its converse, if the vortex is unstable the vorticity gradient must vanish at some latitude.

b. A second proof.* This time, to vary the monotony, we consider rectilinear flow between parallel walls at $y = 0$ and W. Since $\nabla \cdot \underline{V} = 0$, we may introduce a stream function ψ and let $\psi = \bar{\Psi}(y) + \psi'(x,y,t)$. The perturbation vorticity equation becomes

$$\mathfrak{D}\zeta' + v'\bar{\zeta}_y = 0 \quad , \tag{8.5}$$

where $\mathfrak{D} \equiv \partial/\partial t + \bar{u}\, \partial/\partial x$. Multiplying by v' and integrating over the whole domain, we get

$$\int_0^W \int_{-\infty}^{\infty} (\mathfrak{D}(v'\zeta') - \zeta'\mathfrak{D}v' + v'^2\bar{\zeta}_y)\, dxdy = 0 \quad ,$$

or (8.6)

$$\int_0^W \int_{-\infty}^{\infty} \zeta' a'\, dxdy = \int_0^W \int_{-\infty}^{\infty} v'^2\bar{\zeta}_y\, dxdy \quad ,$$

where a' is the perturbation acceleration $\mathfrak{D}v'$.

Now consider that an infinitisemal disturbance is imparted to the fluid such that a parcel is displaced from y_1 to y_2 in the time dt. Because of the conservation of ζ, the perturbation in ζ is given by

$$\zeta' = \bar{\zeta}(y_1) - \bar{\zeta}(y_2) = -\bar{\zeta}_y\, v'\, dt \quad .$$

If $\bar{\zeta}$ is a monotonic function of y, say, for definiteness, an increasing function, then ζ' will be positive or negative according as v' is negative or positive. Hence, by (8.6), the y-acceleration will be of the opposite sign to v'. The same conclusion is reached if $\bar{\zeta}$ is a monotonic decreasing function of y. Thus we derive the same result as before : the condition $d\bar{\zeta}/dy \neq 0$ is a sufficient condition for stability. If $d\bar{\zeta}/dy$ vanishes the restoring force also vanishes, and the possibility exists that a disturbance may grow. The condition $d\bar{\zeta}/dy = 0$ is thus a necessary condition for instability.

---

* The proof is a simplification of Lin's argument (1945, Stability Bibliography) which itself was based on a proof by von Karman.

The above proof could just as well have been given for a sphere. If the sphere is rotating, the vorticity $\bar{\zeta}$ must be replaced by $\bar{\zeta} + f$, where f is the Coriolis parameter, $2\Omega\sin\phi$. The gradient of f is itself a monotonic function of Φ and therefore acts as a stabilizing factor. Thus, if the relative vorticity is zero, the vortex is stable, and the restoring force due to the gradient of f gives rise to the neutral Rossby wave.

c. Rayleigh's* proof. Since $\zeta' = \nabla^2\psi'$, equation (8.5) becomes

$$(\frac{\partial}{\partial t} + \bar{u}\frac{\partial}{\partial x})\nabla^2\psi' + v'\bar{\zeta}_y = 0 \quad . \qquad (8.7)$$

Setting $\psi' = \Psi(y)\, e^{ik(x - ct)}$, we get

$$ik\,(\bar{u} - c)\,(\Psi_{yy} - k^2\Psi) + ik\bar{\zeta}_y\Psi = 0 \; . \qquad (8.8)$$

Multiplication by $-\Psi^*$, the complex conjuguate of Ψ, division by $ik(\bar{u} - c)$, and integration over y give

$$\int_0^W (|\Psi_y|^2 + k^2|\Psi|^2)dy = \int_0^W \frac{|\Psi|^2\bar{\zeta}_y}{\bar{u} - c}\,dy = \int_0^W \frac{|\Psi|^2\bar{\zeta}_y(\bar{u} - c^*)}{|\bar{u} - c|^2}\,dy \qquad (8.9)$$

or, equating real and imaginary parts,

$$\int_0^W (|\Psi_y|^2 + k^2|\Psi|^2)dy = \int_0^W \frac{|\Psi|^2\,\bar{\zeta}_y}{|\bar{u} - c|^2}\,(u - c_r)\,dy \; , \qquad (8.10)$$

$$0 = c_i \int_0^W \frac{|\Psi|^2\,\bar{\zeta}_y}{|\bar{u} - c|^2}\,dy \quad . \qquad (8.11)$$

We see from (8.11) that if $d\bar{\zeta}/dy \neq 0$ for all y, $c_i = 0$, and the wave is neutral. Hence as before, a necessary condition for instability is that $d\bar{\zeta}/dy = 0$ for some Y. This condition is also sufficient for instability

---

* Lord Rayleigh (1880, Stability Bibliography).

for certain types of symmetric velocity profiles.*

In the case of instability (8.10) becomes

$$\int_0^W (|\Psi_y|^2 + k^2|\Psi|^2)dy = \int_0^W \frac{|\Psi|^2 \bar{\zeta}_y}{|\bar{u} - c|^2} \bar{u}\, dy , \qquad (8.12)$$

so that another necessary condition for instability is that $\bar{u}$ and $\bar{\zeta}_y$ be essentially positively correlated.** Thus in Figure 8.2 the velocity profile (1) may be unstable whereas (2) is stable, although both have zeros in their associated vorticity profiles. The first profile generalizes the vortex sheet profile (3) which is known to be unstable. We now give a possible geophysical example.

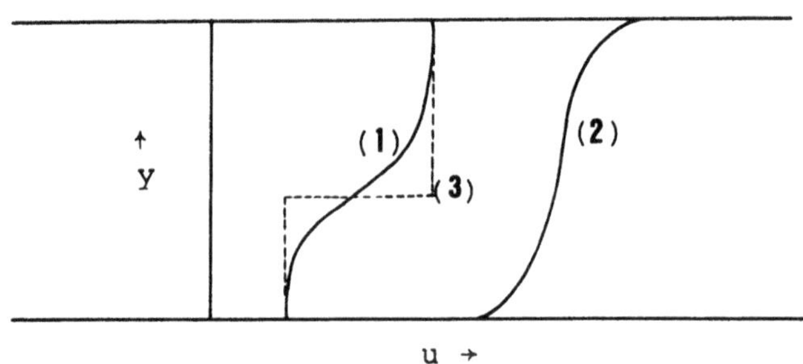

Figure 8.2

* Tollmien (1935, Stability Bibliography). See also C.C. Lin : Theory of Hydrodynamic Stability, Cambridge, 1955, pp. 122-123.

** Fjørtoft (1950, Stability Bibliography).

Formation of tropical depressions in the Intertropical Convergence Zone. This is a zone roughly paralleling the equator in which the Northeast Trades meet the Southeast Trades. If we consider that these wind regimes are roughly independent of longitude for long stretches, they will tend to transport angular momentum in such a way as to produce strong shear zones of the kind shown in Figure 8.3. We have

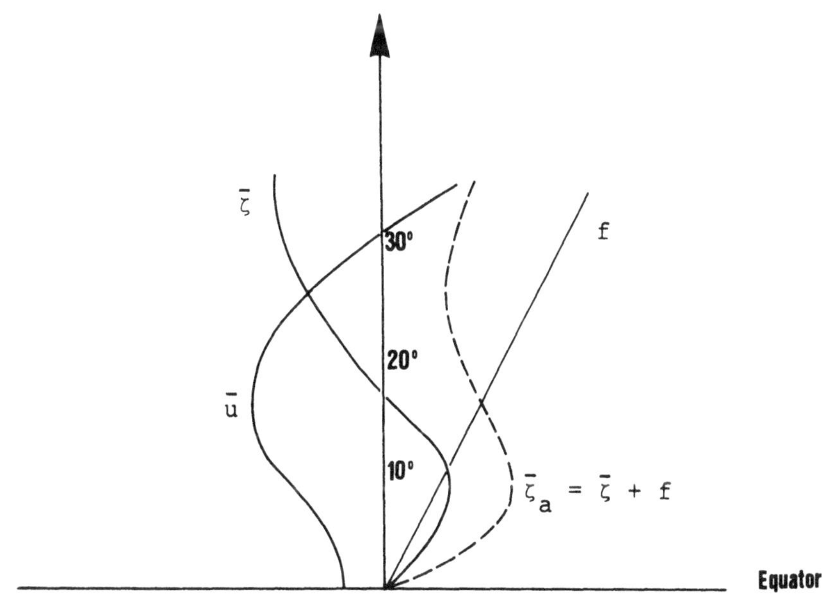

Figure 8.3

$$\zeta_a = -\frac{1}{a\cos\phi}\frac{d}{d\phi}(\bar{u}\cos\phi) + 2\Omega\sin\phi$$

$$\simeq -\frac{d\bar{u}}{dy} + 2\Omega\sin\phi$$

at low latitudes, and we see in the schematic profile that the maximum in $\bar{\zeta} \simeq -\bar{u}_y$ may also produce a maximum in $\bar{\zeta}_a = -\bar{u}_y + f$ and the possibility exists for instability, providing, of course, that the flow may be regarded as two-dimensional. The presence of weak horizontal temperature gradients and friction prevents this from being strictly the case, and the motion is also complicated by condensation. Nevertheless, the possibility remains an attractive one that the incipient tropical depression owes its origin essentially to shear instability, i.e., conversion of the kinetic energy of the mean flow into perturbation kinetic energy.*

d. Instability from the energy viewpoint. If the vorticity equation

$$\frac{\partial\zeta}{\partial t} + \underline{V}\cdot\nabla\zeta = 0 \qquad (8.13)$$

is multiplied by ψ, we find

$$\psi\nabla\cdot(\nabla\psi_t) + \psi\nabla\cdot(\zeta\underline{V}) = \nabla\cdot(\psi\nabla\psi_t) - \nabla\psi_t\cdot\nabla\psi + \nabla\cdot(\psi\zeta\underline{V}) ,$$

and integration over the fluid and use of Gauss's theorem gives

$$\int_\sigma \nabla\psi_t\cdot\nabla\psi \, d\sigma = \frac{d}{dt}\int_\sigma \frac{1}{2}\nabla\psi\cdot\nabla\psi \, d\sigma = 0 , \qquad (8.14)$$

which expresses the conservation of kinetic energy.

Looking at the flow in terms of mean and perturbation components, we have

$$\bar{\zeta}_t + \zeta'_t + (\underline{\bar{V}} + \underline{V}')\cdot\nabla(\bar{\zeta} + \zeta') = 0 \qquad (8.15)$$

where the bar denotes an x-average :

$$\overline{(\quad)} = \frac{1}{L}\int_0^L (\quad)dx .$$

(For definiteness we assume that the flow has the period L in the x-direction and is bounded by rigid walls at $y = y_1, y_2$). Taking the mean of (8.15), we get

* See Bates (1970, Stability Bibliography).

$$\bar{\zeta}_t + \underline{\bar{V}}\cdot\nabla\bar{\zeta} + \overline{\underline{V}'\cdot\nabla\zeta'} = 0 \quad , \tag{8.16}$$

and, proceeding as before, we obtain, after a number of integrations by parts with respect to x and y,

$$\frac{d}{dt}\int \frac{1}{2}|\nabla\bar{\psi}|^2\, dy \;=\; \int \bar{\psi}\,\overline{\underline{V}'\cdot\nabla\zeta'}\, dy = \int \bar{u}\,\tau_y\, dy \;. \tag{8.17}$$

where $\tau = -\overline{u'v'}$ is the Reynold's stress defined as the eddy flux per unit area of x-momentum in the negative y-direction.

Since the total kinetic energy is the sum of the mean and perturbation kinetic energies, i.e.,

$$\overline{|\underline{V}|^2} = \overline{|\underline{\bar{V}} + V'|^2} = \overline{|\underline{\bar{V}}|^2 + 2\underline{\bar{V}}\,\underline{V}' + |\underline{V}'|^2} = \overline{|\underline{V}|}^2 + \overline{|\underline{V}'|^2}$$

the rate of change of the perturbation kinetic energy is the negative of the rate of change of the mean kinetic energy, so that we also have

$$\frac{d}{dt}\int \frac{1}{2}\overline{|\nabla\psi'|^2}\, dy = -\int_y \bar{u}\,\tau_y\, dy = \int \tau\,\bar{u}_y\, dy \quad . \tag{8.18}$$

The integrand $\bar{u}\,\tau_y$ in (8.17) represents the rate of working of the Reynolds force and therefore the rate of increase in mean flow kinetic energy per unit volume. The integrand $\tau\bar{u}_y$ in (8.18) correspondingly represents the rate at which the eddy stresses increase the kinetic energy of the perturbation.* The integral of their sum over the whole volume must of course vanish since the Reynolds stresses do not change but only redistribute the total kinetic energy ; they do not work at the boundaries in the absence of frictions.

Equation (8.9) or (8.10) may now be given an energetic interpretation. The real part of a complex quantity being one-half the sum of the quantity and its conjugate, we may write

---

*If we were dealing with the stress $\tau = \mu\bar{u}_y$ due to molecular viscosity, $\tau\bar{u}_y$ would be the Stokes dissipation function ε defined on page 22.

$$\tau = -\overline{\text{Re } u' \text{ Re } v'}$$

$$= -\frac{1}{4}\overline{\left[\Psi_y e^{ik(x-ct)} + \Psi_y^* e^{-ik(x-c^*t)}\right]\left[ik\Psi e^{ik(x-ct)} - ik\Psi^* e^{-ik(x-c^*t)}\right]}$$

$$= -\frac{ik}{4}(\Psi\Psi_y^* - \Psi^*\Psi_y)e^{2kc_i t} \quad ,$$

and

$$K' = \int \frac{1}{2}\left[\overline{(\text{Re } u')^2} + \overline{(\text{Re } v')^2}\right] dy = \int \frac{1}{2}(|\Psi_y|^2 + k^2|\Psi|^2)e^{2kc_i t}\, dy \quad .$$

Multiplying (8.8) by Ψ^*, its conjugate by Ψ, and subtracting gives

$$-\frac{ik}{4}(\Psi\Psi_y^* - \Psi^*\Psi_y)_y = \frac{kc_i}{2}\frac{\bar{\zeta}_y|\Psi|^2}{|\bar{u} - c|^2} \quad ,$$

or

$$\tau_y = \frac{kc_i}{2}\frac{\bar{\zeta}_y|\Psi|^2}{|\bar{u} - c|^2}e^{2kc_i t} \quad . \tag{8.19}$$

Hence, if $c_i \neq 0$, we may interpret (8.9) multiplied by $kc_i e^{2kc_i t}$ as simply the perturbation kinetic energy equation (8.18), and we see from (8.19) that instability requires that the Reynolds force be different from zero and change sign in the fluid.

IX. BAROCLINIC INSTABILITY

Stability of three-dimensional quasi-geostrophic, zonal flow

Our starting point will be equations (7.29) - (7.36), which were derived for moderate scale, adiabatic motions in the atmosphere. It will be convenient to redefine Ro and ε by replacing 2Ω by $2\Omega\sin\phi_o$. If we drop the superscript "o", the zero order equations become

$$\left(\frac{\partial}{\partial t} + \underline{V}\cdot\nabla\right)\left[\nabla^2\psi + \beta'(y-y_o) + \frac{1}{\rho_s}\frac{\partial}{\partial z}\left(\varepsilon\rho_s\frac{\partial\psi}{\partial z}\right)\right] = 0\ , \qquad (9.1)$$

$$\left(\frac{\partial}{\partial t} + \underline{V}\cdot\nabla\right)\frac{\partial\psi}{\partial z} + w = 0\ , \qquad (9.2)$$

$$\underline{V} = \underline{k} \times \nabla\psi\ . \qquad (9.3)$$

General boundary conditions

(i) At the surface (z = 0), friction and topography will be neglected, as well as forced w-fields ; this latter is not necessary, but we are interested in free motions.

(ii) At the top of the atmosphere we may choose one of the following conditions :

a) the boundary is a rigid plane, or a very stable layer, so w = 0 there, or, at z = ∞,
b) the energy density and vertical energy flux is zero

(iii) The x-motion is cyclic ; F(x) = F(x + L)

(iv) Rigid walls exist at $y = y_1$, $y = y_2$

The stability problem

We are interested in investigating the stability of the dominant (zonal) component of the atmospheric flow, the goal being to discover criteria of instability and the modes of unstable breakdown. This in turn should give an insight into the energy conversion processes in the atmosphere.

Let us assume a steady zonal flow $\bar{u}(y,z)$ on the β-plane and impress on it a small perturbation. Using primes for perturbation quantities and bars for mean quantities, we set

$$\begin{aligned} u &= \bar{u} + u' = -\frac{\partial\bar{\psi}}{\partial y} - \frac{\partial\psi'}{\partial y} \,, \\ v &= v' = \frac{\partial\psi'}{\partial x} \,, \\ w &= w' \,, \end{aligned} \tag{9.4}$$

where

$$\begin{aligned} \psi &\equiv \bar{\psi}\,(y,z) + \psi'(x,y,z,t)\,, \\ \psi' &\ll \bar{\psi} \,. \end{aligned} \tag{9.5}$$

On substitution into (9.1), the linearized pseudo-potential vorticity equation becomes

$$\left(\frac{\partial}{\partial t} + \bar{u}\frac{\partial}{\partial x}\right)\left\{\nabla^2\psi' + \frac{1}{\rho_s}\frac{\partial}{\partial z}\left(\varepsilon\rho_s\frac{\partial\psi'}{\partial z}\right)\right\} + \frac{\partial\psi'}{\partial x}\frac{\partial\bar{q}}{\partial y} = 0, \tag{9.6}$$

where

$$\bar{q} \equiv \frac{\partial^2\bar{\psi}}{\partial y^2} + \beta'(y-y_o) + \frac{1}{\rho_s}\frac{\partial}{\partial z}\left(\varepsilon\rho_s\frac{\partial\bar{\psi}}{\partial z}\right) . \tag{9.7}$$

Boundary conditions

(i) At the rigid walls $y = y_1, y_2$, we have

$$\psi' = 0 ,$$
$$\frac{\partial \psi'}{\partial x} = 0 . \tag{9.8}$$

(ii) At a rigid horizontal boundary $w' = 0$, and the appropriate equation is derived from (9.2) in linearized form :

$$\left(\frac{\partial}{\partial t} + \bar{u}\frac{\partial}{\partial x}\right)\frac{\partial \psi'}{\partial z} + v'\frac{\partial}{\partial y}\left(\frac{\partial \bar{\psi}}{\partial z}\right) + w' = 0 , \tag{9.9}$$

or

$$\left(\frac{\partial}{\partial t} + \bar{u}\frac{\partial}{\partial x}\right)\theta' + v'\frac{\partial}{\partial y}\bar{\theta} + w' = 0 .$$

Since

$$\frac{\partial}{\partial y}\frac{\partial \bar{\psi}}{\partial z} \equiv \frac{\partial}{\partial z}\frac{\partial \bar{\psi}}{\partial y} = -\frac{\partial \bar{u}}{\partial z} ,$$

equation (9.9) becomes, with $w' = 0$,

$$\left(\frac{\partial}{\partial t} + \bar{u}\frac{\partial}{\partial x}\right)\frac{\partial \psi'}{\partial z} - \frac{\partial \psi'}{\partial x}\frac{\partial \bar{u}}{\partial z} = 0 , \tag{9.10}$$

or

$$\left(\frac{\partial}{\partial t} + \bar{u}\frac{\partial}{\partial x}\right)\theta' + \frac{\partial \psi'}{\partial x}\frac{\partial \bar{\theta}}{\partial y} = 0 . \tag{9.10'}$$

(iii) At $z = \infty$, we choose the boundary condition that the upward energy flux is zero. This implies that

$$\lim_{z\to\infty} \rho_s \overline{\psi' w'} = 0 \tag{9.11}$$

(iv) The motion is cyclic in x and t.

Because of (iv), and the fact that the coefficients of (9.6) are independent of x, we may assume that

$$\psi' = \Psi(y,z)e^{ik(x-ct)} \tag{9.12}$$

where c = (complex) wave speed $\equiv c_r + ic_i$.

Direct substitution into (9.6) leads to

$$(\bar{u} - c)\{\frac{\partial^2\Psi}{\partial y^2} - k^2\Psi + \frac{1}{\rho_s}\frac{\partial}{\partial z}(\varepsilon\rho_s\frac{\partial\Psi}{\partial z})\} + \Psi\frac{\partial\bar{q}}{\partial y} = 0\ , \tag{9.13}$$

and the boundary conditions become

(i) At $y = y_1, y_2$: Ψ and its complex conjugate $\Psi^* = 0$. (9.14)

(ii) At a rigid horizontal boundary :

$$(\bar{u} - c)\frac{\partial\Psi}{\partial z} - \Psi\frac{\partial\bar{u}}{\partial z} = 0 \qquad (\Psi_z = \theta)\ . \tag{9.15}$$

(iii) At the top of the atmosphere, we obtain from (9.11)

$$\lim_{z\to\infty}\rho_s\,\overline{Re(\psi')Re(w')} = \lim_{z\to\infty}\rho_s\,\Psi\Psi^*_z = \lim_{z\to\infty}\rho_s\,\Psi_z\,\Psi^* = 0 \tag{9.16}$$

Equations (9.13) to (9.16) constitute an eigenvalue problem in which we are given $\bar{u}(y,z)$ and ε and have to find the values of k, if any, for which $c_i > 0$. Equation (9.13) is a non-separable partial differential equation which resists direct attack. However, we may use integral techniques to give us general criteria.

Integral relationships

Divide (9.13) by $\bar{u}-c\neq 0$, multiply by $\rho_s\Psi^*$ and integrate over a meridional cross-section $y = y_1$ to y_2, $z = 0$ to ∞. Integrating by parts, in analogy with the two dimensional cases, we obtain

$$\int_0^\infty \int_{y_1}^{y_2} \rho_s\{|\Psi_y|^2 + \varepsilon|\Psi_z|^2 + k^2|\Psi|^2\}dydz =$$

$$- \int_{y_1}^{y_2} \varepsilon\rho_s \frac{|\Psi|^2}{\bar{u}-c} \frac{\partial\bar{u}}{\partial z}\Bigg|_{z=o} dy + \int_0^\infty \int_{y_1}^{y_2} \rho_s \frac{|\Psi|^2}{\bar{u}-c} \frac{\partial\bar{q}}{\partial y} dydz$$

$$= \int_0^\infty \int_{y_1}^{y_2} \rho_s \frac{|\Psi|^2}{|\bar{u}-c|^2} (\bar{u}-c_r+ic_i)\frac{\partial\bar{q}}{\partial y} dydz - \int_{y_1}^{y_2} \left[\rho_s\varepsilon \frac{|\Psi|^2}{|\bar{u}-c|^2} (\frac{\partial\bar{u}}{\partial z})(\bar{u}-c_r+ic_i)\right]_o dy \,, \tag{9.17}$$

where the subscript "o" now denotes quantities at z = 0. Equating the real and imaginary parts :

$$c_i \left[\int_{y_1}^{y_2}\left(\frac{\rho_s\varepsilon|\Psi|^2}{|\bar{u}_o-c|^2} \frac{\partial\bar{u}}{\partial z}\right)_o dy - \int_0^\infty \int_{y_1}^{y_2} \frac{\rho_s|\Psi|^2}{|(\bar{u}-c)|^2} \frac{\partial\bar{q}}{\partial y} dydz\right] = 0 \tag{9.18}$$

$$\iint\rho_s\{|\Psi_y|^2 + \varepsilon|\Psi_z|^2 + k^2|\Psi|^2\}dydz = \iint\rho_s \frac{|\Psi|^2 (\bar{u}-c_r)}{|\bar{u} - c|^2} \frac{\partial\bar{q}}{\partial y} dydz -$$

$$- \int \frac{\rho_s\varepsilon|\Psi|^2}{|\bar{u}_o-c|^2} (\frac{\partial\bar{u}}{\partial z})_o (\bar{u}_o - c_r) dy \tag{9.19}$$

These two equations will be discussed in turn.

I. Imaginary-part equation, (9.18)

This may be written

$$c_i\left[\int_{y_1}^{y_2} \varepsilon P_o(\frac{\partial\bar{u}}{\partial z})_o dy - \int_0^\infty \int_{y_1}^{y_2} P \frac{\partial\bar{q}}{\partial y} dydz\right] = 0 \,, \tag{9.20}$$

$$P \equiv \frac{\rho_s |\Psi|^2}{|\bar{u} - c|^2} \quad ,$$

or, since $\dfrac{\partial\bar{\theta}}{\partial y} = \dfrac{\partial}{\partial y}\dfrac{\partial\bar{\psi}}{\partial z} = \dfrac{\partial}{\partial z}\dfrac{\partial\bar{\psi}}{\partial y} = -\dfrac{\partial\bar{u}}{\partial z}$,

$$c_i\left[\int \varepsilon P_o\left(\frac{\partial\bar{\theta}}{\partial y}\right)_o dy + \iint P\,\frac{\partial\bar{q}}{\partial y}\,dydz\right] = 0 \,. \tag{9.21}$$

By definition

$$\frac{\partial\bar{q}}{\partial y} = \beta - \frac{\partial^2\bar{u}}{\partial y^2} - \frac{\varepsilon}{\rho_s}\frac{\partial\bar{u}}{\partial z}\frac{\partial\rho_s}{\partial z} - \frac{\partial}{\partial z}\left(\varepsilon\,\frac{\partial\bar{u}}{\partial z}\right) \,. \tag{9.22}$$

Since β and $-\varepsilon\rho_{sz}/\rho_s$ are always > 0, and $\bar{u}_z > 0$ usually, then, if $\bar{u}_{yy}$ and $\left(\varepsilon\bar{u}_z\right)_z$ are small,

$$\frac{\partial\bar{q}}{\partial y} \simeq \beta - \frac{\varepsilon}{\rho_s}\frac{\partial\rho_s}{\partial z}\frac{\partial\bar{u}}{\partial z} = \beta + \frac{\varepsilon}{\rho_s}\frac{\partial\rho_s}{\partial z}\frac{\partial\bar{\theta}}{\partial y} \tag{9.23}$$

Returning to (9.21), we can make the following classifications :

(a) If $(\theta_y)_o < 0$ (cooler toward the poles), and $\bar{q}_y > 0$, the integral in the square brackets of (9.21) can be zero, so that c_i can be different from zero. Instability in the flow is possible.

(b) If $(\bar{\theta}_y)_o = 0$ and $c_i \neq 0$, we must have $\bar{q}_y = 0$ somewhere in the fluid, i.e., if the surface temperature gradient vanishes, then a necessary condition for instability of the zonal flow is that the mean potential vorticity gradient in an isentropic surface vanish somewhere in the fluid ; or put in another way, $\bar{q}_y$ must change sign between y_1 and y_2.

(c) If $(\bar{\theta}_y)_o > 0$ (hotter poleward), but not so great that $\bar{q}_y < 0$, both integrals in (9.21) are > 0, and $c_i = 0$ (neutral stability). Instability could arise, however, if $\bar{q}_y$ were < 0, but this would be rare.

II. The real-part of equation (9.19)

If $c_i \neq 0$, its coefficient in (9.18) - the whole of the square brackets - is zero. But this coefficient is the same as that for c_r in (9.19). Therefore (9.19) simplifies to

$$\iint \rho_s\{|\Psi_y|^2 + \varepsilon|\Psi_z|^2 + k^2|\Psi|^2\}dydz = \iint \rho_s \frac{|\Psi|^2}{|\bar{u}-c|^2}\,\bar{u}\,\frac{\partial \bar{q}}{\partial y}\,dydz$$

$$- \int \left(\frac{\rho_s \varepsilon |\Psi|^2}{|\bar{u}-c|^2}\,\bar{u}\,\frac{\partial \bar{u}}{\partial z}\right)_o dy \qquad (9.24)$$

If $\bar{u}_o$ is sufficiently small or the ground is isentropic, the second integral on the right can be ignored, and it follows that the indicated weighted mean of $\bar{u}\,\bar{q}_y$ must be greater than zero. A similar theorem was proved for the two-dimensional case by Fjørtoft (1950, Stability Bibliography).

A proof of the necessary condition $\bar{q}_y = 0$ for the instability of an internal baroclinic jet $[(\bar{\theta}_y)_o = 0]$ may be given paralleling our variant of Lin's proof for the two-dimensional case.

Define

$$\mathfrak{D} = \frac{\partial}{\partial t} + \bar{u}\,\frac{\partial}{\partial x}$$

Then, by (9.6) and (9.10'),

$$\mathfrak{D}q' + v'\bar{q}_y = 0$$

$$\mathfrak{D}\,\theta' + v_o'\theta_{oy} = 0$$

where

$$q' = \nabla^2\psi' + \frac{1}{\rho_s}\,(\varepsilon\rho_s\psi_z')_z \quad ,$$

$$\theta' = \psi_z'$$

We have

$$- v'\mathcal{L}q' = q'\mathcal{D}v' - \mathcal{D}(v'q') = v'^2\bar{q}_y \quad ,$$

or

$$\int_\tau \rho_s q'\mathcal{D}v'd\tau = \int_\tau \rho_s v'^2\bar{q}_y d\tau + \int_\tau \rho_s\mathcal{D}(v'q')d\tau \ .$$

But

$$\rho_s v'q' = \rho_s\nabla\cdot(\psi_x'\nabla\psi') - \rho_s(\nabla\psi')_x\cdot\nabla\psi' + (\rho_s\varepsilon\psi_x'\psi_z') - \rho_s\varepsilon\psi_z'\psi_{xz}' \quad .$$

Hence

$$\int_\tau \rho_s q'\mathcal{L}v'd\tau = \int_\tau \rho_s v'^2\bar{q}_y \, d\tau - \int_\tau(\rho_s\varepsilon\psi_x'\psi_z')_o \, dy = \int_\tau \rho_s v'^2\bar{q}_y \, d\tau \ , \tag{9.25}$$

since by (9.10), $(\psi_z')_o = 0$. The proof now follows as in the two-dimensional case (see Chapter VIII).

<u>The physical causes of instability</u>

If $\bar{\theta}_{oy} \neq 0$, a formal analogy to the two-dimensional Rayleigh criteria may still be obtained by defining

$$\tilde{q} = q + \varepsilon\psi_z\delta(z) \equiv q + \varepsilon\theta_o\delta(z) \quad , \tag{9.26}$$

where $\delta(z)$ is the Dirac delta function. If the upper boundary, $z = 1$, is also rigid,

$$\tilde{q} = q + \varepsilon\theta_o \, \delta(z) - \varepsilon\theta_1 \, \delta(z - 1) \ . \tag{9.26'}$$

In terms of $\tilde{q}$ the general quasi-geostrophic equation may be written

$$(\frac{\partial}{\partial t} + \underline{V}\cdot\nabla)\tilde{q} = 0 \ , \tag{9.26''}$$

equations (9.21) and (9.26) become

$$c_i \iint P \, \bar{\tilde{q}}_y \, dydz = 0 \quad , \tag{9.27}$$

$$\iint \rho_s\{|\Psi_y|^2 + \varepsilon|\Psi_z|^2 + k^2|\Psi|^2\}dydz = \iint \rho_s \frac{|\Psi|^2}{|\bar{u}-c|^2}\, \bar{u}\, \tilde{\bar{q}}_y\, dydz\ , \quad (9.28)$$

and the generalization of (9.25) is

$$\int_\tau \rho_s\, \tilde{q}'\mathcal{D}v'\, d\tau = \int \rho_s\, v'\, \tilde{\bar{q}}_y\, d\tau\ . \quad (9.29)$$

Thus a necessary condition for instability is the vanishing of a weighted vertical average of $\tilde{q}_y$. Equation (9.29) may be given a similar interpretation to (8.6) for the two-dimensional problem.

The extended definition of $\tilde{q}$ in (9.26') is needed when the vertical scale is so large that boundary interactions are essential. It seems likely that a sufficiently strong jet-like zonal flow for which $\bar{q}_y$ vanishes in the interior would be unstable independently of the existence of surface temperature gradients. A possible example of such an instability may be the low latitude easterly wave that seems to originate in Northwest Equatorial Africa and propagate into the Atlantic. Burpee* has suggested that it forms as an instability of the low level easterly jet which exists in summer in this region.

Energy equations for the perturbation and the mean flow

To discuss the baroclinic stability problem further it will be useful to derive energy equations for the perturbations and the zonally averaged motions. For this purpose we utilize equations (7.33) and (7.34). Replacing 2Ω by $2\Omega\sin\phi_o$ in Ro and ε and dropping the superscript "o" we have

$$\nabla^2\psi_t = -\nabla\cdot\left[(\zeta + f)\underline{V}\right] + \frac{(\rho_s w)_z}{\rho_s}\ , \quad (9.29)$$

$$\psi_{tz} = -\nabla\cdot(\psi_z\underline{V}) - w\ . \quad (9.30)$$

---

* R. Burpee, J. Atmos. Sci. 29, 1, 1972, pp. 77-90.

Writing $\psi = \bar{\psi}(y,z,t) + \psi'(x,y,z,t)$, $w = \bar{w} + w'$, $\zeta = \bar{\zeta} + \zeta'$, and averaging with respect to x, we get

$$\nabla^2\bar{\psi}_t = -\nabla\cdot\left[(\bar{\zeta} + f)\bar{\underline{V}}\right] - \nabla\cdot(\overline{\zeta'V'}) + \frac{(\rho_s\bar{w})_z}{\rho_s}, \tag{9.31}$$

$$\bar{\psi}_{tz} = -\nabla\cdot(\bar{\psi}_z\bar{\underline{V}} + \overline{\psi_z'\underline{V}'}) - \bar{w}, \tag{9.32}$$

and by subtracting these from (9.29) and (9.30),

$$\nabla^2\psi_t' = -\nabla\cdot\left[(\bar{\zeta}+f)\underline{V}' + \zeta'\bar{\underline{V}} + \zeta'\underline{V}' - \overline{\zeta'\underline{V}'}\right] + \frac{(\rho_s w')_z}{\rho_s}, \tag{9.33}$$

$$\psi'_{tz} = -\nabla\cdot\left[\bar{\psi}_z\underline{V}' + \psi_z'\bar{\underline{V}} + \psi_z'\underline{V}' - \overline{\psi_z'\underline{V}'}\right] - w'. \tag{9.34}$$

Let us multiply the first by $-\rho_s\psi'$, the second by $\rho_s\varepsilon\psi_z'$, add and integrate over the fluid volume. After integration by parts and application of Gauss's theorem, we get

$$\frac{d}{dt}\iint \frac{1}{2}\left[\overline{(\nabla\phi')^2} + \varepsilon\overline{(\psi_z')^2}\right]\rho_s dydz =$$

$$= \iint \overline{\psi_x'\psi_y'}\,\bar{u}_y\rho_s\,dydz - \iint \overline{\varepsilon\psi_x'\psi_z'}\,\overline{\psi_{zy}}\,\rho_s\,dydz - \int\rho_s\varepsilon\overline{\psi'w'}dy\Bigg|_{z=z_1}^{z=z_2} \tag{9.35}$$

The left-hand integral represents the time rate of change of total perturbation energy, kinetic plus available potential. The first integral on the right hand side represents the rate of conversion of mean flow kinetic energy to perturbation kinetic energy by the Reynold's stress, $-\rho_s\,\overline{u'v'} = \rho_s\,\overline{\psi_x'\psi_y'}$, the second the rate of conversion of mean available potential energy* to perturbation available potential energy by the

*Available potential energy is that part of the potential plus internal energy which is available for conversion into kinetic energy by adiabatic overturning. For a general definition see E.N. Lorenz, Tellus, 7, 157-167, 1955.

eddy flow of heat, and the third is the rate at which the perturbation components of the pressure in the fluid between z_1 and z_2 do work on the surrounding fluid, i.e., the energy flow. We note that this latter quantity does not include mass flow of energy. If we take $z_1 = 0$ and $z_2 = \infty$, the last integral vanishes providing, as we assume, there is no escape of energy from the substantial atmosphere.

A similar procedure applied to equations (9.31) and (9.32) gives for the mean flow energy

$$\frac{d}{dt}\iint \frac{1}{2}[(\nabla\bar{\psi})^2 + \varepsilon\,\bar{\psi}_z^{\,2}]\rho_s dydz =$$

$$= \iint (\overline{\psi'_x\psi'_y})_y \bar{u}\rho_s dydz - \iint \varepsilon(\overline{\psi'_x\psi'_z})_y\bar{\psi}_z\rho_s dydz - \int \rho_s\varepsilon\overline{\psi}\overline{w}\, dy \Bigg|_{z_1}^{z_2} , \tag{9.36}$$

and we note that the first two right-hand integrals cancel those of (9.35) when (9.35) and (9.36) are added. This is because they represent interconversions of energy, not energy sources or sinks. Since the flow has been assumed adiabatic and non-viscous, sources and sinks are absent. They could easily have been included.

We note also that the perturbation energy equation could have been derived directly from the perturbation equations (9.6) and (9.9) since the triple product terms in the perturbation quantities do not contribute to the energy integrals.

The perturbation energy equation for the entire fluid is

$$\frac{d}{dt}\iint \frac{1}{2}[\overline{(\nabla\psi')^2} + \varepsilon\overline{(\psi'_z)^2}]\rho_s dydz =$$

$$= \iint \overline{\psi'_x\psi'_y}\;\bar{u}_y\rho_s dydz + \iint \varepsilon\;\overline{\psi'_x\psi'_z}\;\bar{u}_z\rho_s\; dydz\;, \tag{9.37}$$

and we see that the two main sources of energy for the perturbations are

(a) the horizontal shear of the mean flow ;

(b) the vertical shear of the mean flow, or, what is the same, the horizontal temperature gradient : $\frac{\partial\bar{\theta}}{\partial y} = -\frac{\partial\bar{u}}{\partial z}$, i.e., the baroclinicity of the mean flow. Horizontal shear in the mean flow represents a source of available kinetic energy just as vertical shear represents a source of available potential energy. Thus it may be shown by integral methods* that

$$c_i^2k^2 \leqq \frac{\{(\frac{\partial\bar{u}}{\partial y})^2 + \varepsilon(\frac{\partial\bar{u}}{\partial z})^2\}_{max}}{1 + \frac{\pi^2}{k^2(y_2-y_1)^2}} \tag{9.38}$$

That is, the growth rate is limited by the maximum amount of available horizontal shear and vertical shear of the mean flow. The longest waves ($k \to 0$) have the smallest possible growth rate ; shorter waves offer a greater opportunity for energy exchange between regions of high and low kinetic plus potential energy.

We note that the energy of vertical shear is not available as kinetic energy. This is because the vertical eddy stresses $-\rho_s \overline{u'w'}$, like the vertical mass flow of energy, are too small in quasi-geostrophic flow to convert this energy. They are of order Ro.

An energetic interpretation of the stability criterion (9.26) similar to that for two-dimensional flow may now be given. The Reynolds stress and eddy heat flow appearing in the energy equation are

$$\tau = -\rho_s \overline{u'v'} = \rho_s \overline{Re\psi_x' Re\psi_y'} = \rho_s \frac{ik}{4}(\Psi\Psi_y^* - \Psi^*\Psi_y)e^{2kc_it} \tag{9.39}$$

$$B = \rho_s\varepsilon \overline{v'\theta'} = -\rho_s\varepsilon \overline{Re\psi_x' Re\psi_y'} = \rho_s\varepsilon \frac{ik}{4}(\Psi\Psi_z^* - \Psi^*\Psi_z) . \tag{9.40}$$

---

* Pedlosky, J. Atmos Sci. 21, 2, pp. 201-219, 1964.

Multiplying (9.13) by $-\rho_s\Psi^*/(\bar{u} - c)$, its conjugate by $-\rho_s\Psi/(\bar{u} - c^*)$, and subtracting, we get

$$\tau_y + B_z = \rho_s \frac{kc_i}{2} \frac{|\Psi|^2}{|\bar{u} - c|^2} \bar{q}_y e^{2kc_i t} \tag{9.41}$$

Also, at a rigid horizontal boundary, we get from (9.15)

$$B_o = \rho_s \varepsilon \frac{kc_i}{2} \frac{|\Psi|^2}{|\bar{u} - c|^2} \bar{\theta}_{oy} e^{2kc_i t} \tag{9.42}$$

Combining, we obtain

$$\tau_y + B_z - B_o\delta(z) = \rho_s \frac{kc_i}{2} \frac{|\Psi|^2}{|\bar{u} - c|^2} \tilde{\bar{q}}_y e^{2kc_i t} . \tag{9.43}$$

Thus, if $c_i \neq 0$, equation (9.26) multiplied by $kc_i e^{2kc_i t}$ becomes the perturbation energy equation (9.37) written in the form

$$\frac{d}{dt} \iint \frac{1}{2} \left[\overline{(\nabla\psi')^2} + \varepsilon\overline{(\psi_z')^2}\right]\rho_s dydz = - \iint \left[\tau_y + B_z - B_o\delta(z)\right]\bar{u}dydz , \tag{9.44}$$

which is obtained by integration by parts. If the upper boundary is also rigid, $-B_o\delta(z)$ is replaced by $-B_o\delta(z) + B_1\delta(z-1)$ in (9.43) and (9.44).

<u>The Eady problem</u>

A simple but instructive application of the energy-momentum-heat flux relationship is to the stability of the zonal flow of a Boussinesq fluid (ρ_s = constant) confined between rigid horizontal planes with $\beta = 0$, ε = constant, $\bar{u}_z = \bar{\theta}_y = m$ = constant. This problem was solved by Eady (1949, Stability Bibliography). In this case $\bar{q}_y = 0$, and (9.6) becomes

$$\left(\frac{\partial}{\partial t} + \bar{u}\frac{\partial}{\partial x}\right) (\nabla^2\psi' + \varepsilon\psi'_{zz}) = 0 . \tag{9.45}$$

Since there are two rigid horizontal boundaries, we must define

$$\bar{\tilde{q}} = \bar{q} + \bar{\theta}_o \, \delta(z) - \bar{\theta}_1 \delta(z - 1)$$

in all equations containing $\tilde{q}$, and we see that it is the contribution from the surface delta-function terms that permits instability.

In view of the boundary conditions (9.46) and the fact that the coefficients of (9.45) and (9.47) are independent of y, we may separate the y-dependence by writing

$$\psi'(y,z) = \sin \ell(y - y_1) \, \Psi(z) e^{ik(x-ct)} \tag{9.46}$$

$$\ell = \frac{n\pi}{y_2 - y_1} \quad (n = 1, 2, 3 \ldots)$$

where Ψ satisfies the system

$$(mz - c)\left[\varepsilon\Psi_{zz} - (k^2 + \ell^2)\Psi\right] = 0 \, , \tag{9.47}$$

$$(mz - c)\Psi_z - m\Psi = 0 \quad (z = 0,1) \, . \tag{9.48}$$

The solution of (9.47) is

$$\Psi = A \sinh \alpha z + B \cosh \alpha z \quad , \tag{9.49}$$

$$\alpha^2 = \frac{k^2 + \ell^2}{\varepsilon} \quad .$$

Substituting in (9.48) we obtain two linear homogeneous equations in A and B. Setting the determinant of the coefficient equal to zero, we get

$$\alpha^2 c^2 - mc\alpha^2 + m^2\alpha \frac{\cosh\alpha}{\sinh\alpha} - m^2 = 0$$

or

$$c = \frac{m}{2} \pm \frac{m}{2}\sqrt{1 - \frac{4\cosh\alpha}{\alpha\sinh\alpha} + \frac{4}{\alpha^2}} \equiv \frac{m}{2} \pm \frac{m}{2}\sqrt{-R^2} \tag{9.50}$$

If the radicand is negative,

$$c = c_r + ic_i = \frac{m}{2}(1 \pm iR)$$

We see that the waves move at the mean current speed, that the critical value for α at which $c_i = 0$ is given by

$$\frac{\alpha_c^{\,2}}{4} - \alpha_c \cosh\alpha_c + 1 = 0 \quad , \tag{9.51}$$

and that instability will occur if

$$\frac{\alpha^2}{4} - \alpha \cosh\alpha + 1 < 0 \quad . \tag{9.52}$$

By using the identity

$$\tanh\alpha = \frac{2\tanh\frac{\alpha}{2}}{1 + \tanh^2\frac{\alpha}{2}} \quad ,$$

the quadratic (9.52) can be factored :

$$\left(\frac{\alpha}{2} - \tanh\frac{\alpha}{2}\right)\left(\frac{\alpha}{2} - \cosh\frac{\alpha}{2}\right) < 0 \; . \tag{9.53}$$

Since the first bracket is always > 0, we must have

$$\frac{\alpha}{2} < \cosh\frac{\alpha}{2}$$

or

$$\frac{\alpha}{2} < 1.1997 = \frac{\alpha_c}{2} \tag{9.54}$$

It turns out that when $\frac{\alpha}{2} = 0.8031 = \frac{\alpha_m}{2}$ the growth rate kc_i is a maximum for $\ell = 0$. We note that our $\frac{\alpha_c}{2}$ is Eady's α_c.

Recalling that $\alpha^2 = (k^2 + \ell^2)/\varepsilon$, the instability criterion may be written

$$(k^2 + \ell^2)_{dim} < \alpha_c^2 \frac{f_o^{\,2}}{N^2D^2} \simeq 5.76 \frac{f_o^{\,2}}{N^2D^2} \tag{9.55}$$

Taking $f_o = 10^{-4}\mathrm{sec}^{-1}$, corresponding to $\phi_o = 43°$, $N = 10^{-2}\mathrm{sec}^{-1}$, corresponding to a lapse-rate of about 6.5°/km, and D = 10 km, corresponding to the mean height of the tropopause, we get

$$L_m = \frac{2\pi}{\alpha_m}\frac{ND}{f_o} = 3900 \text{ km}$$

for the wave length of maximum instability at $\ell = 0$, or

$$L_m = \frac{2\pi\sqrt{2}}{\alpha_m}\frac{ND}{f_o} = 5500 \text{ km}$$

for the wave length of maximum instability at $\ell = k$.

Since there are walls at y_1 and y_2,

$$(k^2 + \ell^2)_{min} = \frac{\pi^2}{(y_2-y_1)^2} \quad , \tag{9.56}$$

and the flow becomes stable for

$$\frac{\pi^2}{(y_2-y_1)^2} > 5.76 \frac{f_o^{\,2}}{N^2D^2}, \text{ or } \frac{y_2 - y_1}{D} < 1.31 \frac{N}{f_o} .$$

This may be interpreted to mean that if the horizontal scale is too small relative to the vertical scale, the slope of the particle trajectories will be greater than that of the mean isentropic surfaces and hence too great to convert potential to kinetic energy. However, it may be asked why eigenfunctions with sufficiently small vertical scales are not permitted. The answer is not known for sure. It may perhaps be found in the fact that a non-zero $\bar{q}_y$ in (9.6) is required for vertical energy propagation and therefore for internal wave modes.

We have noted that in the Eady problem $\bar{q}_y \equiv 0$, so that the instability is closely associated with the existence of surface temperature gradients (cf. eqs (9.26'), (9.27)). The importance of surface temperature gradients in this instance is brought out by equations (9.41) and (9.42). Since there is a 90° phase lag between u' and v', $\tau = 0$, we have

$$\frac{\partial B}{\partial z} = \rho_s \frac{kc_i}{z} |\Psi|^2 \frac{\bar{q}_y}{|\bar{u} - c|^2} e^{2kc_it} \quad , \tag{9.57}$$

$$B = \varepsilon\rho_s kc_i \frac{|\Psi|^2\bar{\theta}_y}{|\bar{u} - c|^2} e^{2kc_it} \quad \text{at } z = 0,1 . \tag{9.58}$$

Thus if $\bar{q}_y = 0$ and $\bar{\theta}_y = 0$ at a rigid boundary, then B must vanish everywhere and, by (9.37), there can be no increase in perturbation energy.

An analogous theorem holds for the Reynolds stress in two-dimensional parallel flow. Here the Reynolds stress vanishes at a rigid boundary, so that if $\bar{\zeta}_y$ vanishes, the flow will be stable, i.e., two-dimensional linear shear between rigid boundaries is stable. However, an exact parallel to Eady's problem for small amplitudes turns out to be linear shear with free boundaries, where the condition of vanishing pressure gives

$$(\bar{u} - c)\Psi_y - \bar{u}_y\Psi = 0 \; . \tag{9.59}$$

In this case the Reynolds stress does not vanish at the boundaries, and we may state that linear shear between free boundaries is unstable for sufficiently large wave numbers.

Structure of Eady waves

The phase and amplitude structure of the unstable waves is shown in the following diagram.

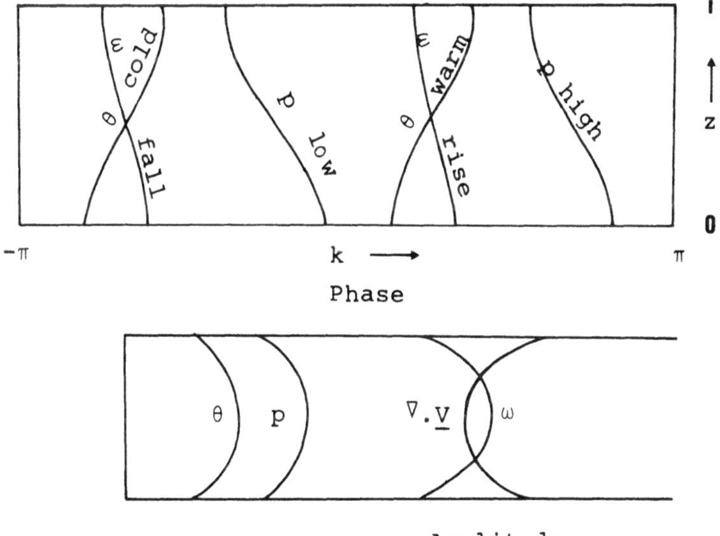

Figure 9.1

The phase difference between the θ and p fields results in the transport of warm air north and upward, and cold air south and down. This converts the available P.E. into K.E. of perturbations. Eady showed that if z = 1 corresponded to a tropopause, his waves of maximum growth rate were about the same size as observed long waves.

The Charney problem*

The assumptions $\beta = 0$ and ρ_s = constant are too restrictive for the atmosphere, and $\beta = 0$ is too restrictive for the oceans. If they are dropped, $\bar{q}_y \neq 0$, and we return to equation (9.13) with the boundary conditions (9.14) - (9.16). Because of its great stability, the free surface of the ocean may be taken as rigid. For the atmosphere we assume, as before, that $\bar{u}$ is a linear function of z and ε = constant. In a strict sense, the latter assumption requires that the atmosphere be isothermal. However, even if the temperature has a non-zero lapse-rate, the density distribution up to the stratosphere is well-approximated by that of an isothermal atmosphere with a suitable mean temperature. Accordingly we set

$$\rho_s = \rho_{so}\, e^{-z/H} , \qquad H = RT_m/g , \tag{9.60}$$

and rely on the exponential decrease of density to mitigate the unrealistic linear increase of $\bar{u}$ and constancy of ε through the tropopause. We anticipate that the unstable disturbance will be trapped in the lower atmosphere and therefore will be relatively insensitive to the structure of the atmosphere above the tropopause.

If the y-dependence is again separated as in (9.46), equation (9.13) becomes

$$\varepsilon(\Psi_{zz} - \frac{\Psi_z}{H}) - (k^2 + \ell^2)\Psi + \frac{\bar{q}_y}{\bar{u} - c}\Psi = 0 , \tag{9.61}$$

where Ψ satisfies the boundary conditions

$$\left(\frac{\Psi_z}{\Psi}\right)_0 = \left(\frac{\bar{u}_z}{\bar{u} - c}\right)_0 \tag{9.62}$$

at $z = 0$, and

$$\rho_s \Psi \Psi_z^* \longrightarrow \rho_s \Psi_z \Psi^* \longrightarrow \rho_s |\Psi|^2 \longrightarrow 0 \tag{9.63}$$

*Charney (1947, Stability Bibliography).

as $z \to \infty$. If H is taken as the vertical scale D, $\bar{q}_y$ becomes (cf. (9.7))

$$\bar{q}_y = \beta + \varepsilon m. \tag{9.64}$$

The substitutions

$$\alpha^2 = \frac{k^2 + \ell^2}{\varepsilon}$$

$$\delta^2 = \alpha^2 + \frac{1}{4}$$

$$\xi = 2\frac{\delta}{m}(\bar{u} - c) = 2\delta z + \frac{2\delta}{m}(\bar{u}_o - c)$$
$$= 2\delta z + \xi_o , \tag{9.65}$$

$$\chi = \Psi(z)\, e^{(\delta - \frac{1}{2})z}$$

$$\gamma = \frac{\beta\varepsilon}{m}$$

$$r = \frac{\gamma + 1}{2\delta}$$

reduce equations (9.61) - (9.63) to

$$\xi \frac{d^2\chi}{d\xi^2} - \xi \frac{d\chi}{d\xi} + r\chi = 0 \tag{9.66}$$

$$\left(\frac{\chi_\xi}{\chi}\right)_o = \frac{1}{2} - \frac{1}{4\delta} + \frac{1}{\xi_o} \tag{9.67}$$

$$\lim_{\xi\to\infty} |\chi|^2 e^{-\xi} = 0 \text{ at } \xi = \infty . \tag{9.68}$$

The problem becomes one of determining the eigenvalues ξ_o in terms of the non-dimensional wave number α and the non-dimensional parameter r (or γ) which depends on the ratio β/m. Because the solution is mathematically complicated and not particularly enlightening, it is placed as an appendix to the present chapter. Here we shall only state the results, which are as follows :

(1) the flow is unstable for $r < 1$.

(2) There are precisely n neutral solutions for $n < r < n + 1$.

(3) The solutions are neutral for $r = n$ ($n = 1, 2, ---$).

(4) There is another unstable, weakly amplifying, solution for all non-integral $r > 1$.

By a method of trial and error Kuo (1952 loc. cit.) calculated ξ_0 and therefore c for the range $0 < r < 1$. Green (1960, Stability Bibliography) calculated c numerically for an atmosphere bounded rigidly at $z = 1$ ($z = H$ dimensionally). He divided the interval $0 \leq z \leq 1$ into a large number of sub-intervals, expressing the amplitude equation (9.61) (with $\bar{u}$ a linear function of z and ε = constant), in finite-difference form, and solved the resultant algebraic equations for the eigenvalues c. Figure 9.2 shows his real and imaginary parts of c expressed as functions of the non-dimensional wave number α for $\gamma = \beta\varepsilon/m = 1$. The corresponding curves calculated by Kuo are also shown, together with the results for Eady's model. Figure 9.3 compares the growth rates as given by Kuo with those calculated by Green for the Boussinesq case, ρ_s = constant. In these figures c is non-dimensionalized with $U = mH$. Attention is called especially to the weak instabilities predicted by Green for $r > 1$. These correspond to those found analytically in the right neighborhood of integral values of r, although there is an apparent discrepancy : Green's growth-rate curve in Figure 9.2 has a finite tangent at $r = 1+$ whereas our analysis shows an $\omega^{3/2} = (r-1)^{3/2}$ dependency (cf. 9.90) giving a zero tangent.

The corresponding growth-rate curves for Eady's model are shown in Figure 9.4. Since here $c_i = F(\alpha)$ alone, both sides are divided by γ to permit comparison with the results of Kuo and Green.

The phases and amplitudes calculated by Kuo for the wave of maximum instability ($r = 0.5$) are shown in Figure 9.5 ; for comparison, those of Green for maximum instability of the strong type ($\gamma = 1$, $\alpha = 2.1$) and instability of the weak type ($\gamma = 1$, $\alpha = 1.0$) are shown in Figure 9.6.

1.0
0.5
0.25
0.15
0.1
r Eady
2.0
1.0
0.5
0.3
0.2
r Green Kuo
growth rate αc_i
0.3
0.2
0.1
0
1
2
3
α
Scaled wave number α

1.0
0.5
0.25
0.15
0.1
r Eady
2.0
1.0
0.5
0.3
0.2
r Green Kuo
wave phase velocity c_r
0.5
0.4
0.3
0.2
0.1
0
1
2
3
4
α
S.A. Green (full line) γ= 1
H. Lan Kuo (thin line) γ= 1
Eady (dashed line) γ= 0

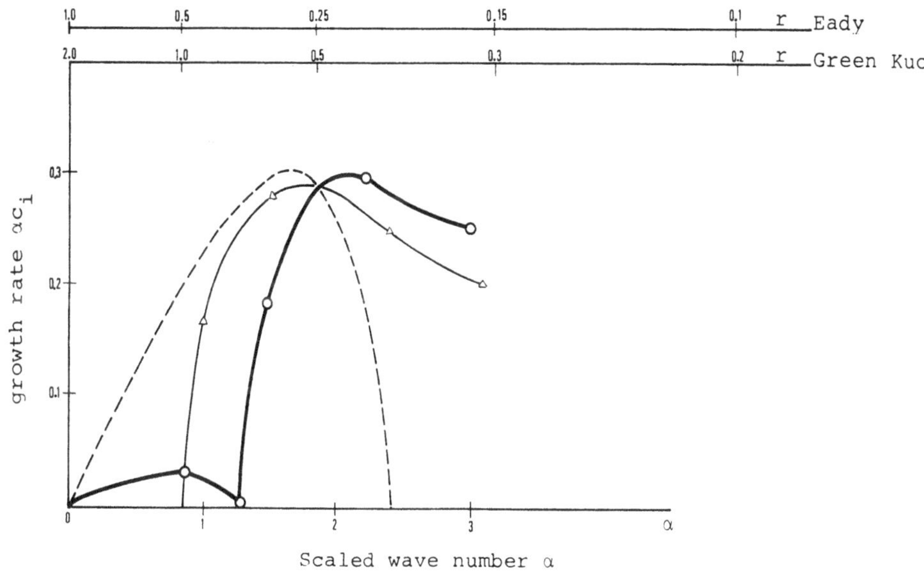

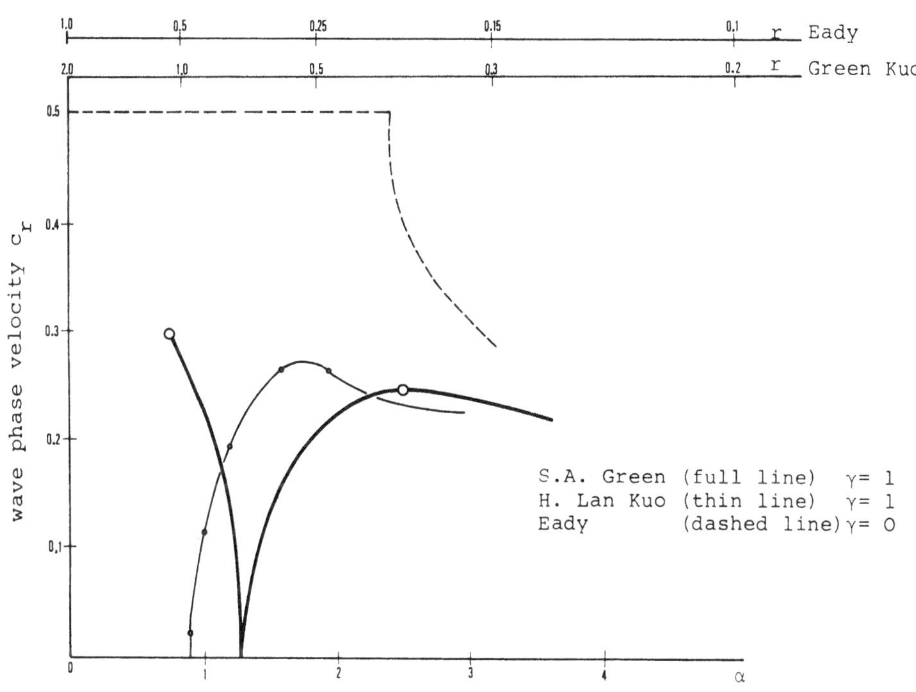

Figure 9.2

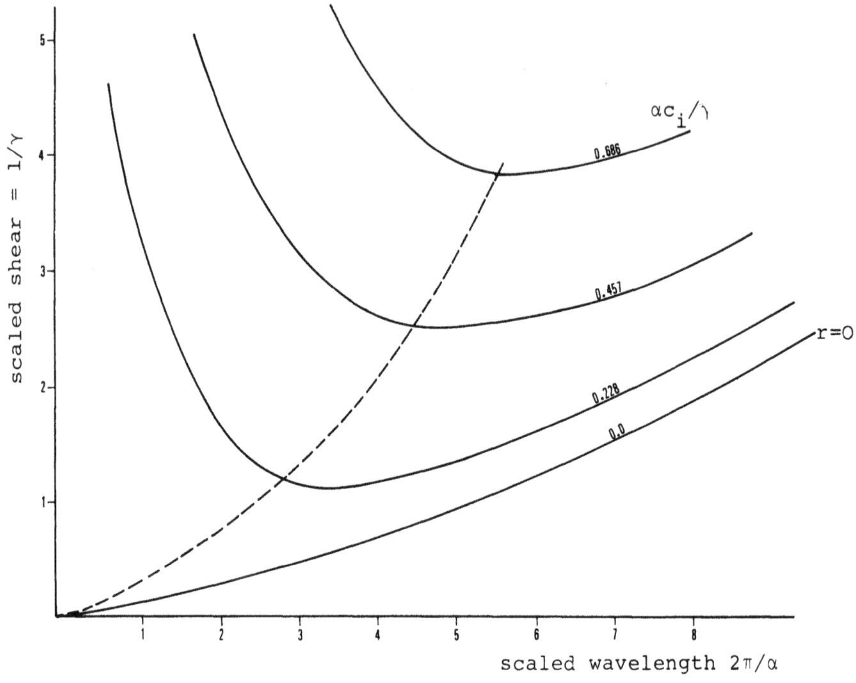

Hsiao-Lan Kuo

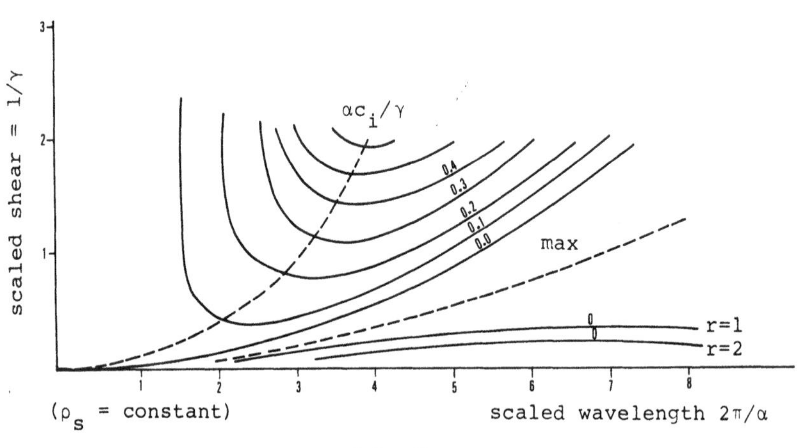

S.A. Green

Figure 9.3

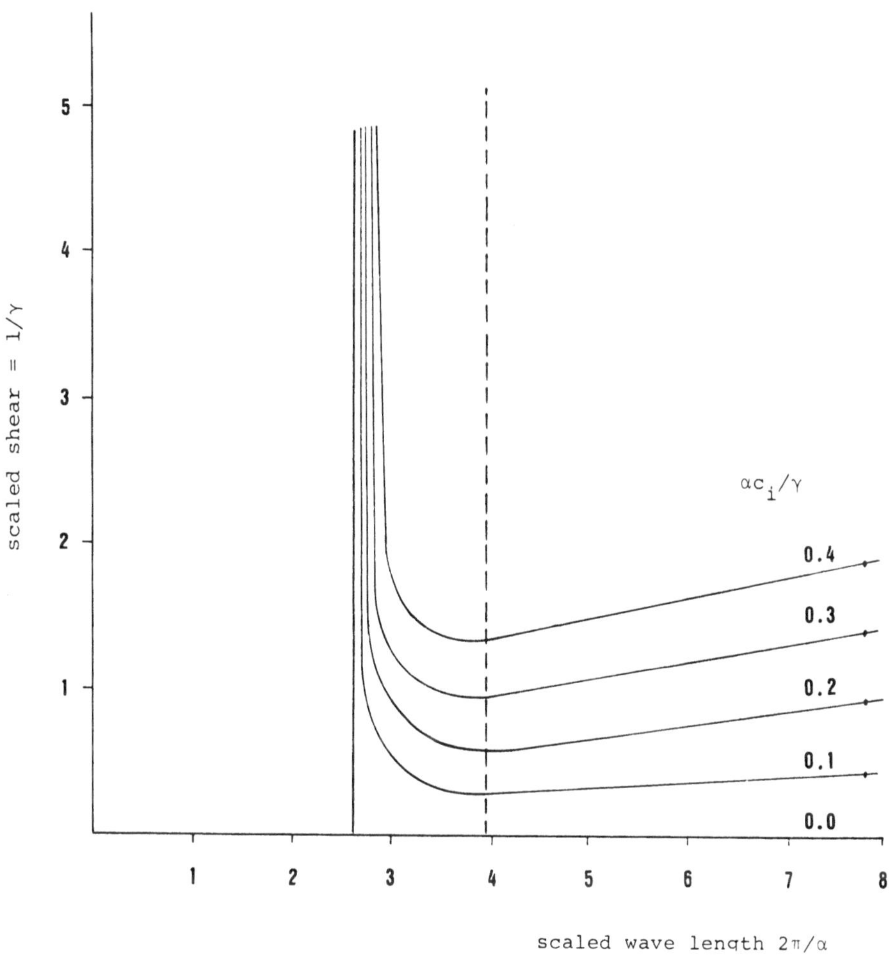

Eady (β = 0)

Figure 9.4

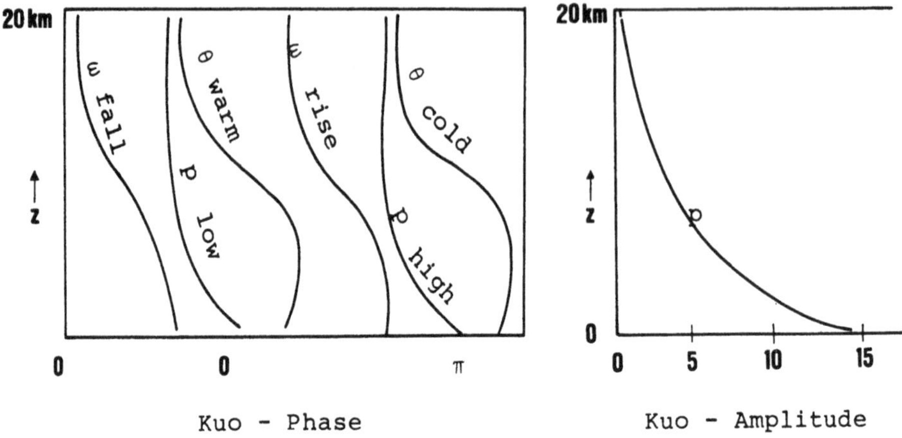

Figure 9.5

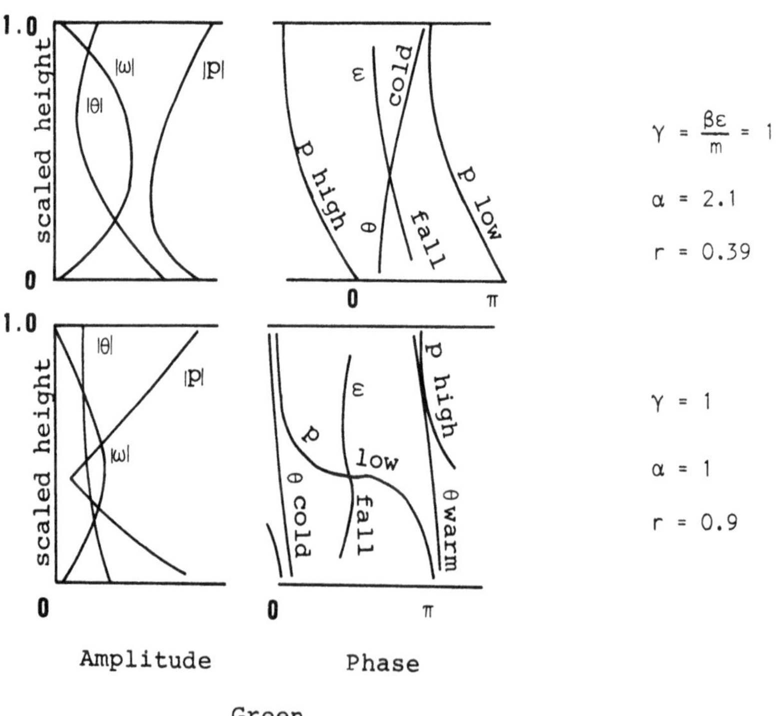

Figure 9.6

Appendix to Chapter IX

Equation (9.66) is a special case of the confluent hypergeometric equation

$$\xi \frac{d^2\chi}{d\xi^2} + (c - \xi) \frac{d\chi}{d\xi} - a\chi = 0 , \qquad (9.69)$$

one of whose solutions is the confluent hypergeometric function

$$\chi = M(a,c,\xi) \equiv 1 + \frac{a}{c}\frac{\xi}{1!} + \frac{a(a+1)}{c(c+1)}\frac{\xi^2}{2!} + \frac{a(a+1)(a+2)}{c(c+1)(c+2)}\frac{\xi^3}{3!} \cdot \cdot \cdot , \qquad (9.70)$$

and another

$$\chi = \xi^{1-c} M(a-c+1, 2-c,\xi) . \qquad (9.71)$$

In the present case the former is not defined, and the latter gives

$$\chi = \xi M(1-r, 2, \xi) . \qquad (9.72)$$

If χ_1 and χ_2 are two independent solutions, the general solution may be expressed as an arbitrary linear combination of the two. It may be shown by the method of Laplace* that two such solutions are

$$\chi_1 = \frac{1}{2\pi i} \int_{\lambda_1} (1 - \frac{\xi}{t})^r e^t \, dt ,$$

$$(9.73)$$

$$\chi_2 = \frac{1}{2\pi i} \int_{\lambda_2} (1 - \frac{\xi}{t})^r e^t \, dt$$

*See A. Erdelyi, ed. Bateman Manuscript Project : Higher transcendental functions I. McGraw-Hill , New York, 1953, 302 pp.

where the paths of integration are

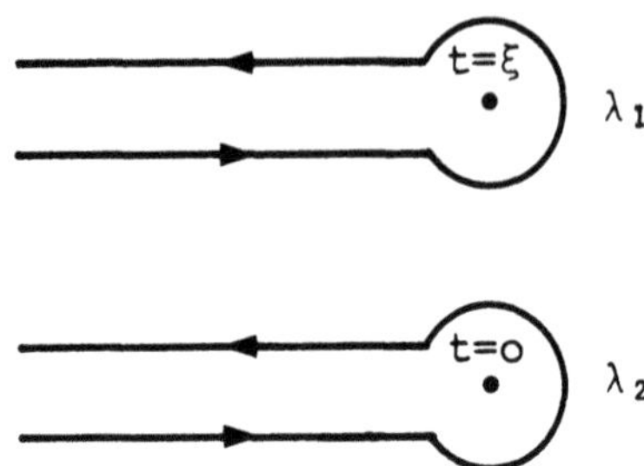

and that these solutions have the asymptotic expansions

$$\chi_1 \sim (-\xi)^r/\Gamma(r) \tag{9.74}$$

$$\chi_2 \sim \xi^{-r}\, e^{\xi}/\Gamma(-r) \tag{9.75}$$

It is evident from (9.68) that the second leads to infinite energies but that the first does not. The appropriate solution is therefore $\chi = \chi_1$.

The first integral in (9.73) can be expanded into the infinite series [See Charney (1947, loc. cit.)]:

$$\chi_1 = (-1)^{r+1} \frac{\sin \pi r}{\pi} \{-r\xi M(1-r,2,\xi) \left[\ln\xi + \frac{\Gamma'(-r)}{\Gamma(-r)} - \frac{\Gamma'(1)}{\Gamma(1)}\right]$$

$$+ 1 + \sum_{n=1}^{\infty} B_n \frac{-r(-r+1)\ldots.(-r+n-1)}{(n-1)!n!} \xi^n\} , \tag{9.76}$$

where

$$B_n = \sum_{\nu=0}^{n-1} \left(\frac{1}{\nu-r} - \frac{2}{\nu+1}\right) + \frac{1}{n} . \tag{9.77}$$

Making use of the relation for the Gamma function,

$$\Gamma(z+1) = z\Gamma(z)$$

we may also cast (9.76) in the form

$$\chi_1 = 1 + \sum_{n=1}^{\infty} \frac{\Gamma(n-r)p_n\xi^n}{\Gamma(-r)\Gamma(n)\Gamma(n+1)} \tag{9.78}$$

where

$$p_n = \ln\xi + \frac{\Gamma'(n-r)}{\Gamma(n-r)} - \frac{\Gamma'(n)}{\Gamma(n)} - \frac{\Gamma'(n+1)}{\Gamma(n+1)} . \tag{9.79}$$

If r is a positive integer n+1, (9.76) becomes undefined, and the appropriate solution is (9.72), which becomes

$$\xi M(-n, 2, \xi) = \xi(-1)^n \, n! \, L_n^1(\xi) \quad , \tag{9.80}$$

where $L_n^1(\xi)$ is the generalized Laguerre polynomial,

$$L_n^1(\xi) = \sum_{\nu=0}^{n} \binom{n-1}{n-\nu} \frac{(-\xi)^\nu}{\nu!} \quad , \tag{9.81}$$

$\binom{n-1}{n-\nu}$ being the binomial coefficient. The other solution behaves like $\xi^{-r}e^{\xi}$ for large ξ and must therefore be excluded.

From the nature of the solution χ_1 we see that (9.67) can have real roots, ξ_o, corresponding to neutral solutions, only if $\xi_o > 0$. Now it was shown, graphically by Charney (1947 loc. cit.) and analytically by Kuo, (1952, Stability Bibliography) that

$$\left(\frac{\chi_1'}{\chi_1}\right)_o < \frac{r}{\xi_o} \text{ for } 0 < r < 1 \quad , \tag{9.82}$$

so that (9.67) has no real roots for $r < 1$. If $n < r < n + 1$, we now show that (9.67) has just n real roots. First note that for $r = n + 1$, χ is given by (9.80), and that this expression has n + 1 real roots since the Laguerre polynomial of order n has n zeros (Erdelyi, loc. cit. vol. II). Putting $\chi_1 = ue^{\xi/2}$ in (9.66), we get

$$u_{\xi\xi} - \left(\frac{1}{4} + \frac{1-r}{\xi}\right)u = 0 \; , \tag{9.83}$$

and it follows from Sturm's oscillation theorem* that χ_1 has n zeros for $n < r < n + 1$. Now between two consecutive zeros of χ_1 there must be a zero of (9.67). This is because $\chi_1'(\xi_o)$ changes sign as ξ_o goes from one zero to the next so that χ_1'/χ_1 varies from plus infinity to minus infinity, or vice versa. Beyond the largest zero of χ_1 the asymptotic relation

$$\frac{\chi_1'}{\chi_1} \sim \frac{r}{\xi_o}$$

indicates that χ_1'/χ_1 varies from $+\infty$ to zero and consequently that (9.67) must have another root. Thus (9.67) has a total of n real (positive) roots.

Before considering the complex roots, we must determine the proper branch of the many-valued logarithm in (9.76). The indeterminacy disappears if we consider the eigenvalue problem as a limiting case of an initial value problem or allow the fluid to be diffusive and let the diffusivity tend toward zero. In the latter case the amplitude equation becomes of fourth order and the domain of validity of the solution, if it is to include the real z axis, can be determined by analogy to the corresponding two-dimensional viscous flow problem (Lin, 1955, footnote to page 155).

Since the source of the energy for the unstable disturbances is not kinetic but potential, one might expect that the removal of the indeterminacy requires the introduction of heat conduction rather than viscosity. Indeed, it is found that the introduction of the former raises the order of the differential equation whereas the latter does not . If viscosity is ignored, the non-dimensional perturbation form of (7.53) becomes

$$\mu(\Psi_{zzzz} - 2\Psi_{zzz} + \Psi_{zz}) = ik(\bar{u}-c)\left[\varepsilon(\Psi_{zz} - \Psi_z) - k^2\Psi\right] + \bar{q}_y ik\Psi \qquad (9.84)$$

---

*See for example E. L. Ince : Ordinary differential equations, Dover, New York, 1944, Chapter X.

where $\mu = \frac{\varepsilon KE}{\nu Ro}$. We wish to consider the asymptotic properties of (9.84) as μ approaches zero. When $\mu = 0$ the limiting equation has a singular point at $\bar{u} - c = 0$, or $\xi = \frac{2\delta}{m}(\bar{u} - c) = 0$, in our case, a logarithmic branch point. The behavior of the solutions of (9.84) in the vicinity of this singular point has been studied for the similar equation arising in the Orr-Sommerfeld problem (stability of parallel viscous flow) (see Lin, footnote page 155). Since c, and therefore ξ, may be complex, we must consider equation (9.84) in the complex ξ or z-plane to determine the domain of validity of the solution. This domain must include the real z-axis in the case of real c. Define z_c as the point at which $\bar{u} - c$ or ξ vanishes in the complex z-plane. Then it can be shown that the asymptotic solution is valid in the sector $-\frac{7}{6}\pi < \arg(z-z_c) < \frac{\pi}{6}$. If c is complex with a positive imaginary part, $z_c = (c-u_o)/m$ will lie in the upper half of the z-plane, and the sector includes the entire real axis. In this case the diffusion-free equation yields an approximation to the diffusive equation which is uniform at all heights. If z_c is below the real z-axis, i.e., if the solution is damped, a uniform approximation for all heights is not possible.

We see therefore that when z_c is close to the real axis and has a positive real part ($c_r > \bar{u}_o$), arg $(z-z_c)$ or arg ξ varies from $-\pi$ to 0 as z goes from 0 to ∞. In particular,

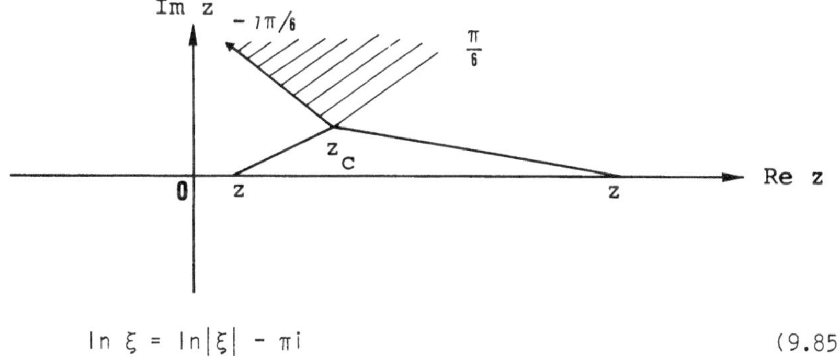

$$\ln \xi = \ln|\xi| - \pi i \tag{9.85}$$

for ξ real and negative.

Returning now to the eigenvalue problem, we show that there are complex roots in the vicinity of r = n with positive imaginary part for c.* First note that if n is an integer, χ_1'/χ_1 behaves like $1/\xi_o$ for small ξ_o, so that $\xi_o = 0$ is a root of (9.67). Consider then the behavior of the solution (9.76) in the vicinity of $r = n + 1$ ($n = 0, 1, \ldots$). The function $\Xi(x) = \Gamma'(x)/\Gamma(x)$ has poles for negative integral values of x and for $x = 0$, whose behavior may be inferred from the identity

$$\Xi(x) = \Xi(1 - x) - \pi \cot(\pi x) \; .$$

Following Miles**, we set $r = n + 1 + \omega$, so that

$$\Xi(-r) \simeq - \pi \cot(-\pi\omega) \simeq \frac{1}{\omega}$$

for $\chi_1 = \xi\phi$, (9.67) becomes

$$\phi' - \left(\frac{1}{2} - \frac{1}{4\delta}\right)\phi = 0 \; , \tag{9.86}$$

and the dominant terms in ϕ' and ϕ are

$$\phi' \simeq - \frac{1}{\xi_o^2} - \frac{r(1-r)}{2\omega} \; ,$$

$$- \phi \simeq - \frac{r}{\omega} \; ,$$

giving

$$- \frac{1}{\xi_o^2} - \frac{r(r - 1/2\delta)}{2} = 0 \; ,$$

or

$$\xi_o^2 = \frac{2\omega}{(n+1)(n+1-1/2\delta)} \equiv \frac{\omega}{q^2} \; . \tag{9.87}$$

$$\text{If } \omega > 0, \; \xi_o = \pm \frac{\sqrt{\omega}}{q}$$

$$\text{If } \omega < 0, \; \xi_o = \pm i \frac{\sqrt{-\omega}}{q} \tag{9.88}$$

---

* This was first shown by Burger (1962, Stability Bibliography).

**J. W. Miles : A note on Charney's model of zonal-wind instability, J.A.S., 21, 1964, pp. 451-452.

Thus there are pairs of both real and complex roots in the vicinity of integral r. However, we have seen that there cannot be negative real zeros of (9.67), corresponding to $\omega > 0$. We must therefore consider the second approximation to the negative real root of (9.87). Here we make use of the determination (9.85) of the proper branch of the logarithm in (9.76). To second approximation

$$\phi' - (\frac{1}{2} - \frac{1}{4\delta})\phi = -\frac{1}{\xi_0^2} - \frac{r(1-r)}{2}(-\pi i + \frac{1}{\omega}) + (\frac{1}{2} - \frac{1}{4\delta})r(-\pi i + \frac{1}{\omega}) = 0$$

or

$$\xi_0^2 = \frac{2\omega(1-\pi i\omega)^{-1}}{(n+1)(n+1-1/2\delta)} \simeq \frac{\omega(1+\pi i\omega)}{q} \quad , \tag{9.89}$$

giving

$$\xi_0 = \frac{\xi^{1/2}}{q} - \frac{\pi i\omega^{3/2}}{q} \; . \tag{9.90}$$

In first approximation, the roots corresponding to $\omega < 0$ are pure imaginary, so that the real part of the phase velocity is equal to the mean zonal velocity at the ground, $\bar{u}_0$. The roots corresponding to $\omega > 0$ give two kinds of solution : one a neutral solution, and the other a weakly amplifying one with phase velocity larger than $\bar{u}_0$, so that $c = \bar{u}$ at some intermediate point. It may be shown that the latter instability is associated with a critical layer in the vicinity of the height at which $c = \bar{u}$. The former corresponds to the overturning predicted by Eady. The latter has no counterpart in his theory.

X. FREE AND FORCED OSCILLATIONS*

Rossby waves in three-dimensions

In this section, we consider the free oscillations (normal modes) of a compressible atmosphere in uniform zonal motion on the β-plane. The oscillations are governed by equations (7.37) - (7.40). suppose that the atmosphere is isothermal in the undisturbed state. Then

$$\rho_s = \rho_{so} e^{-z/H} \quad ,$$

$$H = RT_s/g = \text{constant} \ , \qquad (10.1)$$

$$N^2 = g \frac{d}{dz} (\ln\theta) = g^2/c_p T_s = \text{constant}$$

If $\bar{u}$ is the undisturbed zonal velocity, we may write

$$\psi = - \bar{u}y + \psi'(x,y,z,t) \ .$$

Substitution in (7.37) and (7.38) and neglect of quadratic terms yields the perturbation equations

$$\left(\frac{\partial}{\partial t} + \bar{u}\frac{\partial}{\partial x}\right)\left[\nabla^2\psi' + \frac{f_o^2}{N^2}\left(\psi'_{zz} - \frac{\psi'_z}{H}\right)\right] + \beta\psi'_x = 0 \qquad (10.2)$$

---

* References : J.G. Charney and P. G. Drazin, J. Geophys. Res., 66, 1961 ; A. Eliassen, and E. Palm, Geofys. Publikasjoner, 22, 1961. A review of the theory of forced oscillations produced by airflow over mountain ranges and non-uniform convective heating is given by B. Saltzman, Meteorological Monographs, Vol. 8, No. 30, 4-19, 1968.

$$(\frac{\partial}{\partial t} + \bar{u}\frac{\partial}{\partial x})\psi'_z + \frac{N^2}{f_o} w' = 0 \qquad (10.3)$$

The plane wave solution

$$\psi' = e^{i(kx + \ell y + nz - \sigma t) + z/2H}$$

$$w' = \frac{1}{N}(\bar{n} - \frac{i}{\lambda})\frac{\beta k}{(k^2 + \ell^2 + \bar{n}^2 + \lambda^{-2})}\psi' \qquad (10.4)$$

$$\bar{n} = \frac{f_o}{N} n, \quad \lambda = \frac{2HN}{f_o},$$

satisfies (10.2) and (10.3) providing

$$c = -\frac{\sigma}{k} = \bar{u} - \frac{\beta}{k^2 + \ell^2 + \bar{n}^2 + \lambda^{-2}}, \qquad (10.5)$$

As in the two-dimensional case (cf. equation (6.45)), the component of the phase velocity in the x-direction is negative with respect to the zonal flow.

If n is pure imaginary, the expressions (10.4) represent a wave propagating horizontally with amplitude decreasing or increasing exponentially with height. If the wave is to be stationary,

$$\bar{n}^2 = \frac{\beta}{\bar{u}} - (k^2 + \ell^2 + \lambda^{-2}). \qquad (10.6)$$

Hence vertically propagating waves are possible only if the right hand side of (10.6) is positive ; otherwise the waves propagate horizontally only. In particular, if $\bar{u} < 0$, vertical propagation is impossible.

The group velocities in the x, y, and z direction are found from (10.5) to be

$$c_{gx} = -\frac{\partial\sigma}{\partial k} = \bar{u} + \frac{\beta(k^2 - \ell^2 - n^2 - \lambda^{-2})}{(k^2 + \ell^2 + n^2 + \lambda^{-2})^2},$$

$$c_{gy} = -\frac{\partial\sigma}{\partial\ell} = \frac{\beta(2\ell k)}{(k^2 + \ell^2 + \bar{n}^2 + \lambda^{-2})^2}, \tag{10.7}$$

$$c_{gz} = -\frac{\partial\sigma}{\partial n} = \left(\frac{f_o}{N}\right)\frac{\beta(2nk)}{(k^2 + \ell^2 + \bar{n}^2 + \lambda^{-2})^2},$$

In particular, if $n > 0$, the vertical group velocity is upward, and the waves move downward slantwise from east to west relative to $\bar{u}$. If $n < 0$, the vertical group velocity is downward, and the waves move slantwise upward from east to west relative to $\bar{u}$. If there are no energy sources at $z = \infty$, the waves must be generated within the atmosphere and must propagate energy upward above their region of generation. Hence, only waves with positive n, negative vertical phase velocity and positive vertical group velocity are permitted above the generation region. This may be verified by considering the vertical energy flow in the case

$$\psi' = \left[Ae^{i(kx + \ell y + \bar{n}z - \sigma t)} + Be^{i(kx + \ell y - \bar{n}z - \sigma t)}\right]e^{z/2H}.$$

We find

$$\overline{Re(p')Re(w')} = f_o\rho_s\,\overline{Re(\psi')Re(w')}$$

$$= \frac{f_o^2\rho_{so}}{4N^2}\,\frac{\beta k\bar{n}}{(k^2 + \ell^2 + \bar{n}^2 + \lambda^{-2})}(|A|^2 - |B|^2), \tag{10.8}$$

where the upper bar denotes an average over a wave length in both x and y directions. It is seen that the energy propagation for one wave component is independent of the other and that the energy flow for a given component is upward of the group velocity is upward.

Forced oscillations

Let us consider simultaneously the forced oscillations produced by flow over variable topography and by heating, with modifications introduced by boundary-layer friction. To illustrate the principles, it will suffice to study horizontally sinusoidal topography and heating. More general distributions can be obtained by Fourier superposition. The height of the ground is represented by $h' = h_o \exp[i(kx + \ell y)]$ and the heating rate by $Q' = Q_o \exp[i(kx + \ell y) - z/\delta]$, where h_o and Q_o may differ by a horizontal phase factor. For simplicity we suppose that H, N^2 and $\bar{u}$ are constant. Later we shall consider them as slowly varying functions of z.

Setting

$$\psi' = X(z)\, e^{i(kx + \ell y) + z/2H},$$

$$\zeta = \frac{Nz}{f_o}, \quad \lambda = \frac{2HN}{f_o} \tag{10.9}$$

and substituting in (7.43), we obtain

$$X_{\zeta\zeta} - (k^2 + \ell^2 + \lambda^{-2} - \frac{\beta}{\bar{u}})X = i\,\frac{2gS}{\lambda N Ro}\,(1 + \frac{H}{\delta})\,e^{-\frac{\zeta}{\lambda}(1 + \frac{2H}{\delta})} \tag{10.10}$$

where

$$Ro = \frac{k\bar{u}}{f_o}, \quad S = \frac{Q_o}{f_o c_p T_s}.$$

The vertical velocity at z = 0 due to flow over topography is $\bar{u}h'_x$, and that due to friction is $\frac{1}{2} D_E \nabla^2 \psi'(0)$. For small perturbations these are additive. Substitution into (7.44) then gives

$$(\frac{\partial}{\partial t} + \bar{u}\,\frac{\partial}{\partial x})\psi'_z + \frac{N^2}{f_o}(\bar{u}h'_x + \frac{D_E}{2}\nabla^2\psi') = \frac{g}{f_o c_p T_s}\,Q'(x,y,0) \text{ at } z = 0. \tag{10.11}$$

Or

$$X_\zeta + \frac{X}{\lambda}\left[1 + \gamma i\right] = - Nh_o - \frac{gS}{NRo}\, i \quad \text{at} \quad \zeta = 0 \ , \qquad (10.12)$$

where

$$\gamma = \frac{\sqrt{2}}{4}\frac{E^{1/2}}{Ro}\lambda^2(k^2 + \ell^2), \quad E = \frac{\nu_e}{f_o H^2} \ . \qquad (10.13)$$

We now seek solutions of the form $e^{m\zeta}$ and distinguish two cases

$$m^2 = k^2 + \ell^2 + \lambda^{-2} - \frac{\beta}{\bar{u}} \gtrless 0 \ . \qquad (10.14)$$

In the first case $(m^2 > 0)$, the solution to (10.10) consists of a forced oscillation satisfying the non-homogeneous equation plus an exponential function representing a stationary free oscillation which damps with height, i.e., a trapped wave. In the second case $(m^2 < 0)$, the free oscillation is sinusoidal in ζ and represents a vertically propagating wave. In the first case, the free component is $Ae^{m\zeta} + Be^{-m\zeta}$, and we must set $A = 0$ if the kinetic energy density $\frac{1}{2}\rho_s|\nabla\psi'|^2$ is to be bounded at $z = \infty$. In the second case, the free component is $Ae^{i\bar{m}\zeta} + Be^{-i\bar{m}\zeta}$ where $\bar{m}^2 = -m^2$, and if $\bar{m} > 0$ it follows from the previous section that $B = 0$. In either case, the solution of (10.10) subject to the boundary condition (10.12) is

$$\left(\frac{N}{g\lambda}\right)X = \frac{\left[2\mu\gamma - (1 - 4\mu H/\delta)i\right]\frac{S}{Ro} - \frac{N^2}{g}h_o}{1 - \lambda m + \gamma i}\, e^{-m\zeta} + 2\mu i \frac{S}{Ro}\, e^{-\frac{\zeta}{\lambda}\left(1 + \frac{2H}{\delta}\right)} \ , \qquad (10.15)$$

where

$$\mu = (1 \pm H/\delta)/\left(\left[1 + \frac{2H}{\delta}\right]^2 - m^2\lambda^2\right) \ ,$$

and m is to be interpreted as the positive root of (10.14) for the case $m^2 > 0$ and the <u>negative</u> imaginary root for the case $m^2 < 0$.

In order to compare the effects of heating and topography, we consider the limiting case $\delta \to \infty$, i.e., the case for which the heating per unit mass is uniform throughout the atmosphere. If friction is ignored, (10.15) becomes

$$\frac{N}{g\lambda} X = i \frac{S}{Ro} \left[-\frac{e^{-m\zeta}}{1-\lambda m} + \frac{2e^{-\zeta/\lambda}}{1-\lambda^2 m^2} \right] - \frac{N^2 h_o}{g} \frac{e^{-m\zeta}}{1-\lambda m} , \qquad (10.16)$$

where the factors S/Ro and $N^2 h_o/g$ may be interpreted as follows : Writing

$$\frac{S}{Ro} = \frac{Q_o}{c_p T_s k\bar{u}} = \frac{c_p \partial T/\partial t}{c_p T_s k\bar{u}} \sim \frac{\Delta T}{T_s} ,$$

we see that ΔT is the surface temperature change produced by heating during the time a particle with velocity $\bar{u}$ moves the distance k^{-1}; writing

$$\frac{N^2 h_o}{g} = \frac{h_o}{\theta_s} \frac{d\theta_s}{dz} \sim \frac{\Delta\theta_s}{\theta_s} ,$$

we see that $\Delta\theta_s$ is the change of θ_s in the height h_o. Let

$$k = \frac{2\pi}{2\pi a \cos\phi_o / s} = \frac{s}{a \cos\phi_o} ,$$

where s is the longitudinal wave number. Then we take s = 2, corresponding to two continents and two oceans, and set $\phi_o = 45°$, $\bar{u} = 15$m/sec, we obtain $\Delta T \sim 2.5°$C, whereas $\Delta\theta_s \sim 3°$C for $h_o = 1$ km. Thus the gross effects of heating and topography are comparable.

<u>$m^2 > 0$</u> (trapped waves)

If m is real and friction is neglected, resonance occurs for $\lambda^2 m^2 = 1$. Taking $\ell \sim k$, $\phi_o = 45°$ and $\bar{u} = 15$m/sec, we find the critical longitudinal wave number for resonance to be $s_r = (\beta/2\bar{u})^{1/2} a \cos\phi_0 \simeq 3.3$, and the critical longitudinal wave number for m to be 0 is $s = s_c = 2.3$.

The sea-level streamfunction is

$$\frac{N}{g\lambda} X(0) = i \frac{S/Ro}{1+\lambda m} - \frac{N^2 h_o/g}{1-\lambda\ m} , \tag{10.17}$$

and we see that : the sea-level topographic perturbation in the streamfunction is 180° out of phase with the surface elevation for $s_c < s < s_r$ and in phase for $s > s_r$; the sea-level heating perturbation is shifted 90° to the west for $s_c < s < s_r$ and 90° to the east for $s > s_r$.

In the present case ($m^2 > 0$), it follows from (10.16) and (10.17) that the coefficient of iS/Ro is positive at $\zeta = 0$ but becomes negative at the level ζ_c for which

$$e^{(1-\lambda m)\frac{\zeta_c}{\lambda}} = 1 + \lambda m \ .$$

At this level there is a 180° phase shift in the heating perturbation. No such shift occurs for the topographic perturbation. If friction is included in the latter, we see from (10.15) that there is an eastward shift through the phase angle $\tan^{-1} (\frac{\gamma}{1-\lambda m})$; hence if $s > s_r$ ($\lambda m > 1$) the shift is actually westward. This agrees with observation : the Rocky Mountains and Alps are characterized by rather high horizontal wave number and therefore by $\lambda m > 1$, they are accompanied by ridges in the streamfunction somewhat to the west of the mountain ridges.

The 180° phase shift at $\lambda m = 1$ is associated with the competing effects on changes in the relative vorticity component between horizontal advection of $\bar{q} = f = \beta(y - y_o)$ and horizontal advection of relative vorticity. If the wave length is large ($\lambda m < 1$), the β-effect dominates, and the decrease in the vertical component of relative vorticity due to the shrinking of the earth's vortex tubes when air flows up a mountain must be balanced by a flow to the north to increase $\beta(y - y_o)$ and decrease $\zeta = \nabla^2\psi$. If the wave-length is small ($\lambda m > 1$), the horizontal advection of relative vorticity dominates, the flow deviates anticyclonically to the south, and cyclonic vorticity is advected by $\bar{u}$ to compensate for the decrease of cyclonic vorticity due to vertical shrinking of the earth's vortex tubes.

$m^2 < 0$ (vertically propagating waves)

If $s < s_c \simeq 2.3$, m is pure imaginary. We note here only that the free oscillation now propagates vertically. It will later be shown that, owing to the variation of u with height, it is eventually damped.

It is difficult to apply the foregoing results to the actual distributions of heating and topography without considering vertical variations in $\bar{u}$. We call attention only to the following general conclusions : (a) the effects of heating and topography are roughly comparable ; (b) the heating perturbation shifts by 180° in phase in passing from sea-level to upper levels ; (c) the effects of topography and friction seem to account qualitatively for the western Pacific and Atlantic ridges in the streamfunction.

Figures 10.1 and 10.2, taken from an article by Charney and Eliassen (1949, Tellus, 1, 38-54) show the topographical profile at 45°N, the observed mean (normal) perturbation in the 500 mb streamfunction at 45°, and the calculated topographic perturbation for a single-layer model with and without friction. Since there are no vertical phase shifts for the topographic perturbation (even with increase of $\bar{u}$ with height), the single-layer model gives qualitatively correct results. (See Saltzman, loc. cit., on effects of heating).

Vertical propagation of energy

Suppose that $\bar{u}$ and N are slowly varying functions of z so that the amplitude equation (10.10) for the free oscillation may be written

$$X_{\zeta'\zeta'} + \nu^2 X = 0 \ , \tag{10.18}$$

where

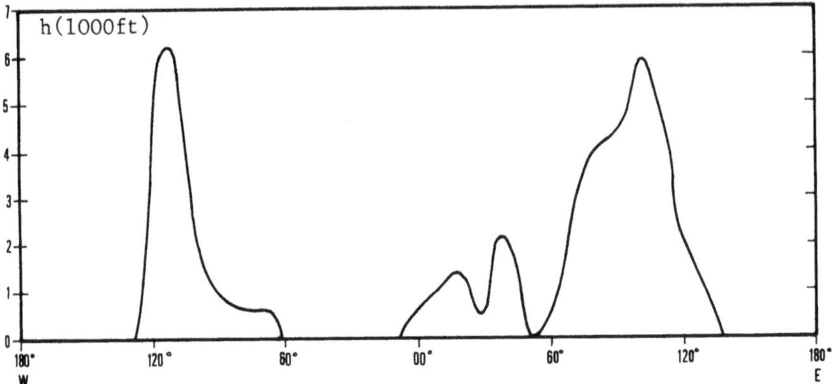

Fig. 10.1 - Topographical profile at 45° N. The elevations represent the mean height of the earth's surface averaged for the interval 40° N to 50° N, as a function of longitude.

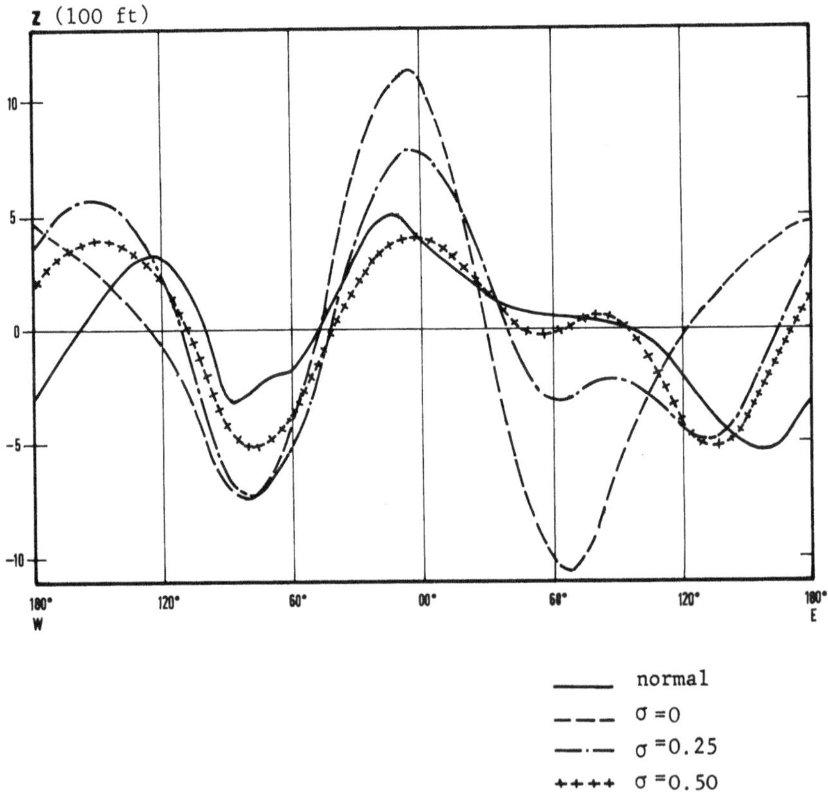

Fig. 10.2 - Normal height profile of the 500 mb surface at 45° N for January together with computed stationary profiles for $\sigma = 0$ (no friction), for $\sigma = 0.25$ (moderate friction) and for $\sigma = 0.50$ (strong friction). For purposes of comparison the heights are represented as deviations from their respective means.

$$\zeta' = \lambda\zeta$$

$$\nu^2 = \lambda^2\left[\frac{\beta}{\bar{u}} - (k^2 + \ell^2 + \lambda^{-2})\right] = - \lambda^2 m^2 . \qquad (10.19)$$

The quantity ν has the properties of an index of refraction. The solution of (10.18) will be of the form $e^{i\nu\zeta'}$, where $\zeta' = 1$ when $\frac{Nz}{f_o} / \frac{2HN}{f_o} = 1$, or $z = 2H \simeq 14$ km. Thus if $\nu^2 = -1$, the solution is an exponentially damped (trapped) wave whose amplitude is reduced by the factor e^{-1} in the height 2H. If $\nu^2 = 1$ the wave propagates vertically as an internal wave with wave-length $\frac{2\pi}{\nu} = 2\pi \times 2H$ in units of height.

We may now consider the vertical propagation of wave energy generated in the lower troposphere. Figures 10.3 and 10.4, taken from an article by Charney and Drazin (Geophys. Res., 66, 1, 83-109, 1961), show the non-dimensional index of refraction ν calculated from observed profiles of $\bar{u}$ and N^2 averaged between 30° and 60° for different seasons. The indices are plotted for different values of $L = 2\pi(k^{-2} + \ell^{-2})^{1/2}$. It will be seen that the waves are strongly trapped in the summer stratosphere above 20 km by the circumpolar atmospheric vortex. In this case $\bar{u} < 0$ and $\nu^2 = -\lambda^2 m^2 < 0$. In the winter the shorter-scale waves are also trapped because $\bar{u}$ is so large that again $m^2 > 0$. Only in the equinoctial periods does there seem to be a possibility of free propagation to high altitudes, and only in these periods are strong correlations observed between vortices at 10 mb and in the troposphere.

The trapping of planetary waves in the lower atmosphere in summer and of all but the longest planetary waves in winter accounts for the lack of correlation, as seen in Figures 1.2 - 1.4 and Figure 1.5, between tropospheric and upper stratospheric motions. There is a possibility that vertical propagation at extremely long wave lengths can account for the low wave-number asymmetric circulation at 10 mb in January seen in Figure 1.5 (see Figure 10.3).* Another possible factor which may influence the amplification of long planetary waves in the atmosphere is the instability of the so-called Polar-Night Jet which forms at high latitudes in the winter hemisphere due to the strong temperature gradient

* See T. Matsuno, J. Atmos. Sci., 27, 871-883, 1970 ; also 28, 1479-1494, 1971.

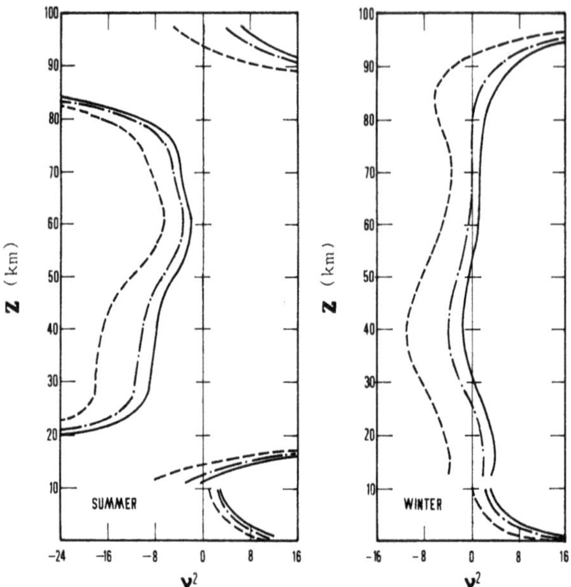

Fig. 10.3 - The square of the index of refraction for summer and winter, averaged between 30° and 60° N. The short-dashed lines correspond to L = 6,000 km, the long-dashed lines correspond to L = 10,000km and the solid lines correspond to L = 14,000 km.

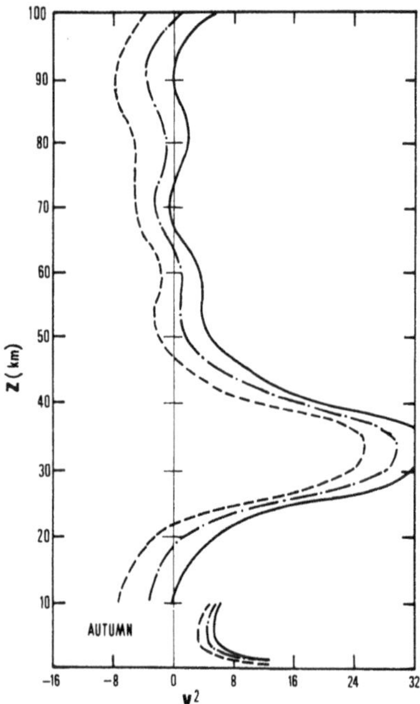

Fig. 10.4 - The square of the index of refraction for autumn, averaged for the period September 1 to December 1. See Figure 5 for meaning of lines.

induced by the strong latitudinal variation in oxygen absorption of solar radiation (see Charney and Stern, loc. cit.)

The trapping of vertically-propagating waves by an easterly zonal current also accounts for the fact that waves and vortices in the low-level easterly trade winds in the tropics are very often observed to be uncoupled to the waves and vortices of the upper level westerlies above the trades. Indeed, the reasoning which we have employed for vertical propagation applies also to horizontal meridional propagation, and we can account in this way for the fact that the waves and vortices in the mid-latitude troposphere arising from baroclinic instability do not propagate or propagate very little into the low-level trades. (See J.G. Charney, 1969, J.A.S. 26, pp. 182-185).

XI. ENERGY CASCADES, FRONTOGENESIS AND TURBULENCE IN THREE-DIMENSIONAL QUASI-GEOSTROPHIC FLOW

Energy cascades

Energy cascades from large to small scales in three-dimensional turbulent flow are due to eddy diffusive separation of fluid particles bringing about elongations of vortex tubes and increases of vorticity on ever smaller scales. Such elongations and vorticity increases are forbidden in two-dimensional incompressible flow, so that, as T. D. Lee* pointed out, one does not expect an unlimited energy cascade. The following simple but penetrating proof of this assertion was given by Fjørtoft.**

We consider two-dimensional non-viscous flow on the surface of a sphere. The motion is governed by the conservation of the component of vorticity normal to the surface of the sphere

$$(\frac{\partial}{\partial t} + \underline{V}\cdot\nabla)\nabla^2\psi = 0 \quad ,$$

where ψ is the streamfunction for the flow and $\nabla^2\psi$ is the normal component of the vorticity. It follows from this relationship that the kinetic energy is conserved, or if we write E for twice the kinetic energy, the quantity

$$E = \iint(\nabla\psi)^2 d\sigma = \text{constant (for the motion)},$$

where $d\sigma$ is an element of surface area. It also follows from the non-divergent property of the motion and multiplication of the vorticity equation through by $\nabla^2\psi$ that the integral of the square of the vorticity is a constant :

* T. D. Lee, 1951 : J. App. Phys., 22, 514.

**R. Fjørtoft, 1953 : Tellus, 5, 225.

$$F = \iint (\nabla^2\psi)^2 d\sigma = \text{constant (for the motion).}$$

Now let us express the streamfunction as an infinite sum of the eigenfunctions of the Laplace operator ∇^2, i.e.,

$$\psi = \sum_1^\infty a_n\psi_n \text{ , where } \nabla^2\psi_n = -\lambda_n\psi_n \ .$$

These eigenfunctions, one can show, are orthonormal in the sense that

$$\iint \psi_m\psi_n \, d\sigma = \delta_{mn} \ ,$$

where

$$\delta_{mn} = \begin{cases} 0, & m \neq n \\ 1, & m = n. \end{cases}$$

We then have

$$F = \sum_1^\infty \lambda_n^2 a_n^2 \ ,$$

and we can show by integration by parts that

$$E = -\iint \psi\nabla^2\psi d\sigma = \sum_1^\infty \lambda_n a_n^2 \ .$$

If we define

$$b_n = \lambda_n a_n^2 \ ,$$

then

$$E = \sum_1^\infty b_n = \text{constant}$$

and

$$F = \sum_1^\infty \lambda_n b_n = \text{constant} \ .$$

For the sphere

$$\lambda_n = \frac{n(n+1)}{a^2} ,$$

where a is the radius of the sphere.

We may interpret the relations E = constant, F = constant in the following mechanical way. Imagine that we have a weightless rod on which we suspend weights b_1, b_2, etc., at the distances $\sqrt{\lambda_1}$, $\sqrt{\lambda_2}$, etc., and that these weights are balanced by E. Since λ_n behaves like n^2 for large values of n, this says essentially that the mass E, together with the moment of inertia F, has to be preserved. Can we start with energy at the low wave numbers, represented by small values of λ_n, and transfer it out to high-wave numbers ? The answer is obviously no, for one cannot do so and still preserve the moment of inertia. Since energy can flow toward higher wave numbers providing more energy flows toward lower wave numbers to balance the moment of inertia, we see that at least three of the boxes in Figure 11.1 must be involved in any energy transaction, but that in any case the fraction of the energy that can flow to high wave numbers is clearly limited. The higher the wave number, the more it is limited.

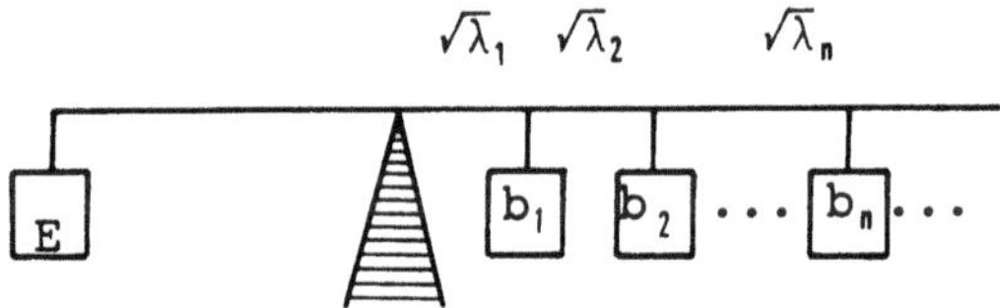

Figure 11.1

As we have seen, there is a close analogy between two-dimensional incompressible flow and three-dimensional adiabatic, non-viscous, quasi-geostrophic flow on the β-plane. Both are governed by a law of conservation of a single scalar invariant, the vorticity $\nabla^2\psi$ in the former case, and the quantity $q = L(\psi) + f_o + \beta(y - y_o)$ where

$$L(\psi) \equiv \nabla^2\psi + \frac{f_o^2}{\rho_s}\frac{\partial}{\partial z}\left(\frac{\rho_s}{N^2}\frac{\partial\psi}{\partial z}\right) ,$$

in the latter.

If $\underline{V}$ denotes the geostrophic velocity $\underline{k} \times \nabla\psi$, where $\underline{k}$ is a vertical unit vector, then the basic conservation equations are

$$\left(\frac{\partial}{\partial t} + \underline{V}\cdot\nabla\right)q = \left(\frac{\partial}{\partial t} + \underline{V}\cdot\nabla\right)L(\psi) + \beta\psi_x = 0$$

and

$$\left(\frac{\partial}{\partial t} + \underline{V}\cdot\nabla\right)\frac{\partial\psi}{\partial z} + \frac{N^2}{f}w = 0 .$$

Multiplying the first equation by $-\rho_s\psi$, integrating by parts and utilizing the second equation with $w = 0$ at $z = 0$, we find that the total energy, kinetic plus available potential, is conserved, i.e., if E denotes twice this energy.

$$\frac{dE}{dt} = \iiint \left[(\nabla\psi)^2 + \frac{f_o^2}{N^2}\left(\frac{\partial\psi}{\partial z}\right)^2\right]\rho_s d\tau = 0 .$$

Similarly, multiplication by $\rho_s L(\psi)$ gives

$$\frac{dF}{dt} = \frac{d}{dt}\iiint [L(\psi)]^2\rho_s d\tau = \beta\iint \left(\frac{f_o^2}{N^2}\psi_x\psi_z\rho_s\right)_{z=0} d\sigma ,$$

where $d\tau$ is the volume element, and all integrations are assumed to take place between rigid latitudinal walls at which $\psi_x = 0$.

Now L is a self-adjoint elliptic operator with a complete

orthonormal set of eigenfunctions ψ_n, and eigenvalues λ_n such that

$$L(\psi_n) = -\lambda_n\psi_n \quad .$$

As in the case of the Laplace operator, these eigenfunctions may be so ordered that λ_n is a positive, non-decreasing function of n which tends toward infinity like $(n^{1/3})^2$. Moreover, the nodal surfaces of the eigenfunctions divide the fluid domain into n subdomains whose volumes approach zero like $(n^{1/3})^{-2}$. Hence as before we set

$$\psi = \sum_1^\infty a_n\psi_n \ , \quad b_n = \lambda_n a_n^2$$

and get,

$$F = \sum_1^\infty \lambda_n b_n \quad .$$

To obtain an exact analogue of the two-dimensional energy cascade theorem, we would have to show that

$$E = -\iiint \psi L(\psi)\rho_s d\tau \quad ,$$

from which we would then obtain

$$E = \sum_1^\infty b_n = \text{constant},$$

and that F = constant. The energy cascade theorem would then follow. However, this is not usually the case : in general we find that

$$E + \iiint \psi L(\psi)\rho_s d\tau = -\iint \left(\frac{f_o^2}{N^2}\,\psi\,\frac{\partial\psi}{\partial z}\,\rho_s\right)_{z=0} d\sigma \ , \qquad (11.1)$$

and

$$\frac{dF}{dt} = \beta\iint\left(\frac{f_o^2}{N^2}\,\frac{\partial\psi}{\partial x}\,\frac{\partial\psi}{\partial z}\,\rho_s\right)_{z=0} d\sigma \ , \qquad (11.2)$$

where the right-hand integrals are taken at the earth's surface (z = 0).

The latter are zero only if the potential temperature at this surface is constant ($\partial\psi/\partial z = 0$). We thus obtain the result that an energy cascade is forbidden in quasi-geostrophic flow if the temperature gradient at the ground vanishes. This may be interpreted to mean that no frontal discontinuity can form, for a step function discontinuity in $\underline{V}$ or θ ($\nabla\psi$ or ψ_z) implies that b_n behaves like n^{-2} and $\Sigma\lambda_n b_n$ diverges, so that the "moment of inertia" cannot be conserved.

Conversely, if there is a surface potential temperature gradient, an energy cascade is possible and frontal discontinuities may form. Indeed, Stone* and Williams and Plotkin** have given simple analytic examples of the formation of a discontinuity. They utilize the fact that it is possible to find a quasi-geostrophic flow whose horizontal deformational field is time-invariant. In such a field, as Bergeron showed, frontal discontinuities must form - at least at the ground where the vertical velocity vanishes. One must bear in mind, however, that the motion cannot remain quasi-geostrophic within the frontal zone because of the large velocity gradients and high Rossby numbers, so that the quasi-geostrophic fronts are bound to be somewhat unrealistic. We might call them "pseudo-fronts". What has been shown by the present analysis is that frontal development requires surface temperature gradients, a fact well known to the synoptic meteorologist.

Charney and Stern (loc. cit.) prove a theorem to the effect that baroclinic instability in a zonal flow also requires the existence of a surface temperature gradient. This, together with the previous theorem, appears to account for the fact that frontogenesis takes place only in conjunction with a deepening baroclinic wave.

The analogy between two-dimensional and three-dimensional quasi-geostrophic flow may be exploited further. In the Charney-Stern article it was also shown that a surface temperature gradient in three-dimensional flow with uniform vertical shear plays the same role in regard to stability as the existence of a free surface in two-dimensional

---

* P. Stone, 1966 : J. Atm. Sci., 23, pp. 455-465.

** R. T. Williams and J. Plotkin, 1968 : J. Atm. Sci., 25, 201-206.

flow with uniform shear. Both act to destabilize the flow. In the proof of the cascade theorem for two dimensions, it was necessary to assume that the fluid covered the whole sphere or else was rigidly bounded. If there is a free surface, the theorem no longer holds ; a two-dimensional energy cascade becomes possible and discontinuities can form. Because of the simplicity of two-dimensional flow, it may be profitable to study the formation of such discontinuities in two-dimensional flow as a means of gaining greater insight into the process of frontogenesis in three dimensions.

Frontogenesis

A particularly simple example of frontogenesis has been given by Williams and Plotkin (loc. cit.). They seek solutions of the system (7.33), (7.35) and (7.36) with the property

$$\psi = \alpha xy + \chi(y,z,t) \tag{11.3}$$

i.e., a meridional flow superimposed on a pure deformation field. The scales L and U are prescribed so as to make $\alpha = \varepsilon = 1$, and the fluid is assumed to be Boussinesq and confined between the rigid horizontal planes $z = 0$ and $z = 1$. Setting $\sin\phi_o = 1$, which can be done without loss of generality, we obtain

$$\left(\frac{\partial}{\partial t} - x\frac{\partial}{\partial x}\right) q = 0 \quad , \tag{11.4}$$

$$\left(\frac{\partial}{\partial t} - x\frac{\partial}{\partial x}\right)\theta = 0 \qquad \text{at } z = 0,1 \ , \tag{11.5}$$

where

$$q = \chi_{xx} + \chi_{zz} \quad , \tag{11.6}$$

$$\theta = \chi_z \ . \tag{11.7}$$

Equations (11.4) and (11.5) assert that q and θ are conserved at a point which moves with the velocity

$$\frac{Dx}{Dt} = - x \ . \tag{11.8}$$

If the subscript "o" denotes a quantity in the initial configuration, (11.8) states that $x = x_o e^{-t}$. Hence

$$q(x,z,t) = q(x_o,z,0) = q(xe^t,z,0)$$
$$\chi_z(x,z,t) = \chi_z(x_o,z,0) = \chi_z(xe^t,z,0) \quad \text{at } z = 0,1 \tag{11.9}$$

Let the initial field of χ be

$$\chi(x,z,0) = zF(x) \quad ,$$

so that

$$\theta = \chi_z = F(x) \quad \text{at} \quad t = 0$$

is independent of z. Choosing $\theta = F(x) = -\frac{2}{\pi}$ arc tan x initially, we get

$$\theta = F(xe^t) = -\frac{2}{\pi} \text{ arc tan } (xe^t)$$
$$\lambda = \theta_x = e^t F'(xe^t) = -\frac{2}{\pi} \frac{e^t}{1 + x^2 e^{2t}} \quad \text{at } z = 0,1, \tag{11.10}$$

which shows that θ_x approaches zero with increasing t at all x except at x = 0, and at x = 0, θ_x approaches infinity, i.e., a front forms at x = 0. The interior solution is found by integration of (11.6), which becomes, after differentiation of (11.6) with respect to z

$$\theta_{xx} + \theta_{zz} = F''(e^t x) \ . \tag{11.11}$$

The solution for θ at t = 0 and t = ∞, satisfying (11.10) at z = 0,1 is shown in Figure 11.2. It will be seen that a discontinuity in θ forms only at the rigid horizontal boundaries and that the zone of strong gradient is vertical. Real fronts have a finite slope and can exist in the interior as well. However, as already mentioned, it is not to be expected that the quasi-geostrophic formalism will apply when the horizontal scale has been reduced so far that the Rossby number is no longer small.

Hoskins,* Hoskins and Bretherton,** and Williams*** have recently considered non-geotrophic modifications which give a finite slope and frontogenesis in a finite time, although no interior discontinuities form.

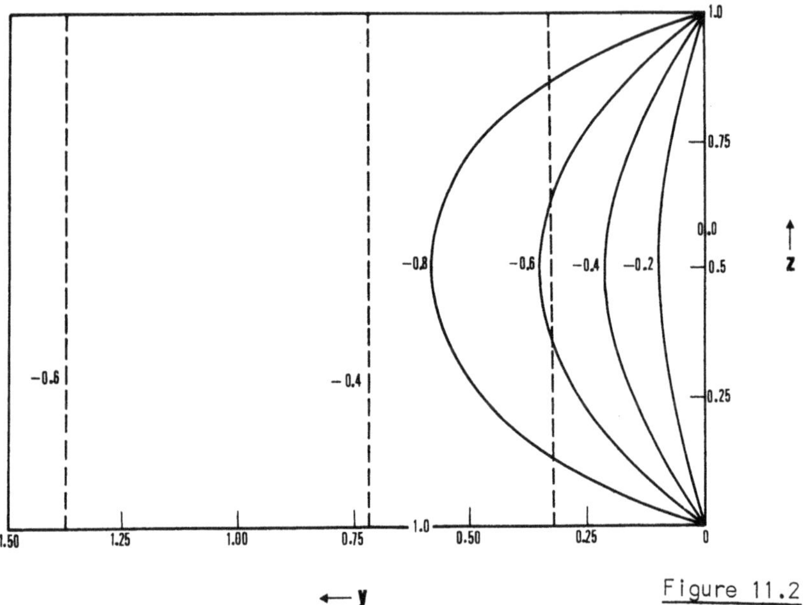

Figure 11.2

The field θ as a function of x and z. The dashed curves represent the initial state and the solid curves represent the final state ($t = \infty$).

The following is a brief presentation of Hoskins and Bretherton's (1972) results. They also seek solutions of the equations of motion in which the velocity field is the sum of a stationary pure deformation field independent of height and a field which varies only with x and z. The assumption of geostrophy is retained for the flow parallel to the axis of dilatation but is dropped for the normal horizontal component. The variation of f is ignored. The resulting equations of motion turn out to be a special case of the balance equations of Chapter VII. Indeed, if we set

---

* B.J. Hoskins, 1971 : Quart. J. Roy. Met. Soc., 97, 139-153

** B.J. Hoskins and F. Bretherton, 1972 : J. Atmos. Sci., 29, 11-37.

***R.T. Williams, 1972 J. Atmos. Sci., 29, 1-10.

$$\begin{aligned}
\psi &= \alpha xy + \chi(x,z,t) \\
\sigma &= \sigma(x,z,t) \\
\pi &= f\alpha xy - \frac{\alpha^2}{2}y^2 + \pi'(x,z,t) \\
u &= -\psi_y + \sigma_x = -\alpha x + u'(x,z,t) \\
v &= \psi_x + \sigma_y = \alpha y + v'(x,z,t) \\
w &= w(x,z,t) \\
\theta &= \theta(x,z,t)
\end{aligned} \tag{11.12}$$

in (7.60), we obtain, after some rearrangement of terms,

$$\frac{\partial}{\partial x}\left(fv' - \frac{\partial \pi'}{\partial x}\right) = 0 \ , \tag{11.13}$$

$$\frac{\partial}{\partial x}\left[\frac{D'v'}{Dt} + fu' + \alpha(v' + fx)\right] = 0 \ , \tag{11.14}$$

$$g(\ln\theta - \ln\theta_s) = \pi'_z \ , \tag{11.15}$$

$$(\rho_s u')_x + (\rho_s w)_z = 0 \ , \tag{11.16}$$

$$\frac{D'(\ln\theta)}{Dt} = 0 \ , \tag{11.17}$$

where

$$\frac{D'}{Dt} \equiv \frac{\partial}{\partial t} + u\frac{\partial}{\partial x} + w\frac{\partial}{\partial z} \equiv \frac{D}{Dt} - v\frac{\partial}{\partial y} \ . \tag{11.18}$$

The integral of (11.13) is just the geostrophic equation for v' ,

$$fv' = \frac{\partial \pi'}{\partial x} \ , \tag{11.19}$$

and that of (11.14) may be written

$$\frac{D'M}{Dt} = -\alpha M \ , \tag{11.20}$$

or

$$\frac{D'}{Dt}(Me^{\alpha t}) = 0 \ , \tag{11.21}$$

where

$$M = v' + fx \tag{11.22}$$

is proportional to the absolute angular momentum in a rotating plane at a large distance from the axis of rotation (see the latter part of Chapter V). Equations (11.15) - (11.20) are, with small modification, the equations employed by Hoskins and Bretherton.

To the present degree of approximation, the vorticity of the flow is $(-v'_z, u'_z, \zeta)$, where $\zeta = f + v'_x$; hence the potential vorticity, $\frac{1}{\rho_s}(-v'_z, u'_z, \zeta)\cdot(\theta_x, 0, \theta_z)$, is

$$\hat{q} = \frac{1}{\rho_s}\left[\theta_z(f + v'_x) - \theta_x v'_z\right] = \frac{1}{\rho_s}\frac{\partial(M,\theta)}{\partial(x,z)} \tag{11.23}$$

$$= \frac{\theta_s}{g\rho_s f}\left[(N^2 + \pi'_{zz})(f^2 + \pi'_{zz}) - (\pi'_{xz})^2\right].$$

Since $\hat{q}$ is a constant of the motion and is independent of y,

$$\frac{D'\hat{q}}{Dt} = 0 \quad . \tag{11.24}$$

Any two of the three conservation equations (11.17), (11.21) and (11.24) may be taken as governing the motion. However, their solution is complecated by their non-linear character. A considerable simplification may be brought about by replacing the independent variables x and z by $X = \frac{M}{f} = x + \frac{v'}{f}$ and $Z = z$. We get

$$\rho_s\hat{q} = \frac{\partial(M,\theta)}{\partial(x,z)} = \frac{\partial(M,\theta)}{\partial(X,Z)}\frac{\partial(X,Z)}{\partial(x,z)} = \zeta\theta_Z \ . \tag{11.25}$$

The transformation formulae

$$\frac{\partial}{\partial x} = \frac{\partial X}{\partial x}\frac{\partial}{\partial X} = \frac{\zeta}{f}\frac{\partial}{\partial X} \quad ,$$

$$\frac{\partial}{\partial z} = \frac{\partial X}{\partial z}\frac{\partial}{\partial X} + \frac{\partial}{\partial Z} = \frac{1}{f}\frac{\partial v'}{\partial z}\frac{\partial}{\partial X} + \frac{\partial}{\partial Z} \quad , \tag{11.26}$$

give

$$\frac{\partial v'}{\partial x} = \frac{\zeta}{f}\frac{\partial v'}{\partial X}$$

$$\frac{\partial v'}{\partial z} = \frac{\zeta}{f}\frac{\partial v'}{\partial Z}$$

$$\frac{\partial \theta}{\partial x} = \frac{\zeta}{f}\frac{\partial \theta}{\partial X} \qquad (11.27)$$

$$\frac{\partial \theta}{\partial z} = \frac{1}{f}\frac{\partial v'}{\partial X}\frac{\partial \theta}{\partial X} + \frac{\partial \theta}{\partial Z} \; .$$

Whence we obtain

$$\frac{\zeta}{f} = (1 - \frac{1}{f}\frac{\partial v'}{\partial X})^{-1} \qquad (11.28)$$

and

$$\rho_s \hat{q} = f\,\frac{\partial \theta}{\partial Z}\,(1 - \frac{1}{f}\frac{\partial v'}{\partial x})^{-1} \; . \qquad (11.29)$$

The thermal wind relationship

$$f\,\frac{\partial v'}{\partial z} = g\,\frac{\partial}{\partial x}\,(\ln\theta - \ln\theta_s) \qquad (11.30)$$

is found to have the same form in X and Z :

$$f\,\frac{\partial v'}{\partial Z} = g\,\frac{\partial}{\partial X}\,(\ln\theta - \ln\theta_s) \; , \qquad (11.31)$$

but this is the integrability condition for the equations

$$fv' = \frac{\partial \Pi}{\partial X} \; ,$$

$$g(\ln\theta - \ln\theta_s) = \frac{\partial \Pi}{\partial Z} \; . \qquad (11.32)$$

(It may be shown that $\Pi = \pi + v'^2/2$). Introducing the above expressions for v' and θ into (11.29), we get

$$\frac{1}{f^2}\frac{\partial^2 \Pi}{\partial X^2} + \frac{f\theta_s}{g\rho_s \hat{q}}\,(N^2 + \frac{\partial^2 \Pi}{\partial Z^2}) = 1 \; , \qquad (11.33)$$

which may be taken as determining the motion. If the initial $\hat{q}$ is expressed as a function $\hat{q}(X_o, \theta_o, 0)$ of the initial values of X and θ, then since $Xe^{\alpha t}$, θ and $\hat{q}$ are constants of the motion in the x,z plane.

$$\hat{q}(X,\theta,t) = \hat{q}(Xe^{\alpha t},\theta,0) , \tag{11.34}$$

so that $\hat{q}$ in (11.33) is a known function of X and of θ(or Π_Z).

At a rigid horizontal boundary θ(or Π_Z) is given initially as a function of X and is therefore determined at all times from

$$\theta(X,t) = \theta(Xe^{\alpha t},0) . \tag{11.35}$$

The solution of (11.33) subject to the given initial and boundary conditions can be found by numerical integration or by analysis in such simple cases as when the fluid is Boussinesq and $\hat{q}$ is constant. The variables $v(x,z)$ and $\theta(x,z)$ are obtained from (11.32) and the transformations

$$\begin{aligned} x &= X - \frac{v'(X,Z)}{f} \\ z &= Z . \end{aligned} \tag{11.36}$$

The streamlines at any given time may be found by using $M_o = Me^{\alpha t}$ and $\theta_o = \theta$ as Lagrangian tracers of the motion, or else, as in Chapter V, by eliminating the time derivatives of θ and M from their respective conservation equations. Introducing the stream-function expressions,

$$\rho_s u' = \frac{\partial\Psi}{\partial z} ,$$

$$\rho_s w = -\frac{\partial\Psi}{\partial x} ,$$

into (11.17) and (11.20) :

$$\begin{aligned} \frac{D'}{Dt}(\ln\theta) &= (\ln\theta)_t + (-\alpha x + \frac{\Psi_z}{\rho_s})(\ln\theta)_x - \frac{\Psi_x}{\rho_s}(\ln\theta)_z = 0 , \\ \frac{D'}{Dt} M &= v_t + (-\alpha x + \frac{\Psi_z}{\rho_s})M_x - \frac{\Psi_x}{\rho_s} M_z = -\alpha M , \end{aligned} \tag{11.37}$$

and eliminating v_t and $(\ln\theta)_t$ by means of (11.30), we obtain an elliptic equation in Ψ which is very similar to (5.18) and whose coefficients are determined from $M(x,z,t)$ and $\theta(x,z,t)$.

The relation of the present theory to the quasi-geostrophic theory may be brought out by substituting

$$\frac{g\rho_s \hat{q}}{f\theta_s} \equiv N^2(1 + q) \quad (N^2 = g\frac{\partial \ln\theta_s}{\partial z})$$

in (11.33). In view of the scale analysis of Chapter VII, $q = O(\text{Ro})$.

Multiplication by $f^2(1 + q)$ gives

$$(1 + q)\Pi_{XX} + \frac{f^2}{N^2}\Pi_{ZZ} = f^2 q(X,\theta,t) = f^2 q(Xe^{\alpha t},\theta,0) \ . \tag{11.38}$$

For a Boussinesq fluid in which N^2 is constant, this equation is seen to be formally equivalent to (11.6) if the geostrophic replacements $1 + q \to q$, $X \to x$, $Z \to z$, $\Pi \to \pi$ are made.

Hoskins and Bretherton present analytic solutions of (11.38) for the case $q \equiv 0$, i.e., for constant potential vorticity. The somewhat questionable assumption is made that the initial θ varies non-linearly with x but linearly with z. Such a state cannot satisfy $q = 0$. It could perhaps be argued that a modification of the initial θ distribution to satisfy $q \equiv 0$ would not affect the solution significantly. Their solution for the case where the initial distribution of θ at the upper and lower boundaries is the same as that assumed by Williams and Plotkin is shown in Figure 11.3 below.

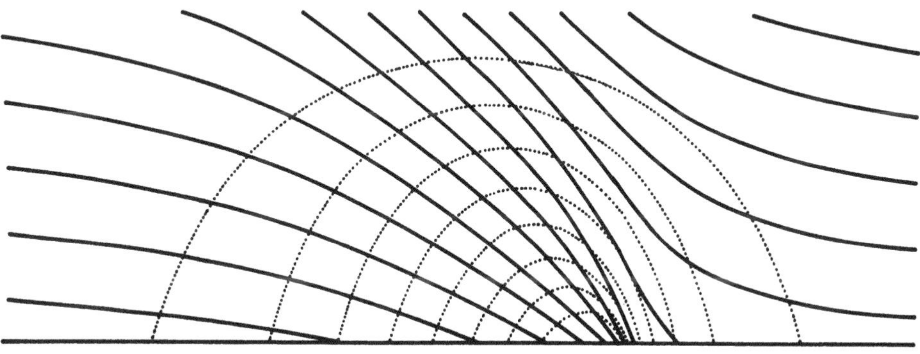

Figure 11.3

The solid curves are isotherms of the potential temperature θ, and the dashed curves are isolines of the long front velocity v'. It will be seen that the zone of strong gradient in θ is now inclined and that altogether the front is much more realistic.

Geostrophic turbulence

The large-scale waves and vortices in the atmosphere and oceans which arise from instabilities and thermal and topographic forcing are far from regular and periodic; they may be described as components of a turbulent motion. Through non-linear interactions these primary scales of motion give rise to secondary sum and difference scales and thus to a transfer of energy to both larger and smaller scales. Ultimately the scales become so small that the kinetic energy is dissipated by viscosity. In three-dimensional flow the mechanism of energy transfer to smaller scales is via the stretching of vortex tubes by the random motions. This process gives rise to vorticity and kinetic energy on ever smaller scales. In two-dimensional flow vorticity is either preserved or dissipated directly by viscosity; it is not increased; and no such energy cascade can take place.

Kolmogoroff derived a simple expression for the energy spectrum of three-dimensional turbulent flow in a homogeneous, incompressible fluid by postulating that : (a) at high wave numbers, away from boundaries, the turbulence is isotropic and homogeneous; (b) the non-linear interactions are local in wavenumber space, that is, the Fourier wave-components only interact with components of comparable wave number ; (c) in consequence the spectrum depends only on the viscosity and the energy dissipation per unit mass, ε, transferred from low to high wave numbers and then dissipated; (d) if the turbulent energy input is sufficiently high, there will be a subrange of wave numbers for which the local Reynolds numbers are so high that no dissipation takes place. In this case, the spectrum can depend only on ε, and a simple dimensional argument leads to the formula

$$E(k) = \alpha_1 \varepsilon^{2/3} k^{-5/3} , \qquad (11.39)$$

where k is the scalar wave number $(k_x^2 + k_y^2 + k_z^2)^{1/2}$, E is the kinetic energy per unit mass per unit wave number, and α_1 is a non-dimensional universal constant.

---

*J.G. Charney, J. Atmos. Sci., 28, 1087-1095, 1971.

No such energy cascade can take place in two-dimensional flow, but there is nothing preventing the generation of vorticity on ever smaller scales by non-linear interactions until finally it is dissipated by viscosity. Kraichnan,* Leith,** and Batchelor*** postulated an inertial subrange for two-dimensional flow based on the mean-square vorticity (enstrophy) transfer, η, toward higher wave numbers. A dimensional argument then gives

$$E(k) = \alpha_2 \eta^{2/3} k^{-3} , \qquad (11.40)$$

if the turbulence may be considered isotropic and homogeneous. In this case, the kinetic energy spectrum in any single direction will have the same functional dependence.

It is a matter of some interest, therefore, that the atmospheric horizontal kinetic energy spectrum taken along a latitude circle shows an approximate k^{-3} dependency for longitudinal wave-numbers k in the range k = 7 - 18, and computations with numerical models extend this range somewhat farther†. One does not expect such a dependency for lower wave numbers than k = 7 since excitation by baroclinic instability occurs at about k = 5. Figures 11.4 and 11.5, taken from an article by Julian, et. al.,**** show the observed spectra at 50°N for summer and winter at 500 mb.

However, one must ask if the appropriate k^{-3} behavior of the horizontal kinetic energy spectrum in the atmosphere can be ascribed merely to the fact that the flow is two-dimensional in any significant sense ? We assert that this is very unlikely, for, as we have shown, changes in the vertical vorticity component are as much due to stretching

---

* R.H. Kraichnan, 1967 : Phys. Fluids, 10, 1417

** C.E. Leith, 1968 : Phys. Fluids, 11, 671.

*** G.K. Batchelor, 1969 : Phys. Fluids, 12, II-233.

**** P.R. Julian, W.M. Washington, L. Hembree, and C. Ridley, 1970 : J. Atmos. Sci., 27, 376-387.

† It was shown by Charney (1971, loc. cit.) that a minus n power law for kinetic energy implies a minus n power law for the longitudinal spectrum.

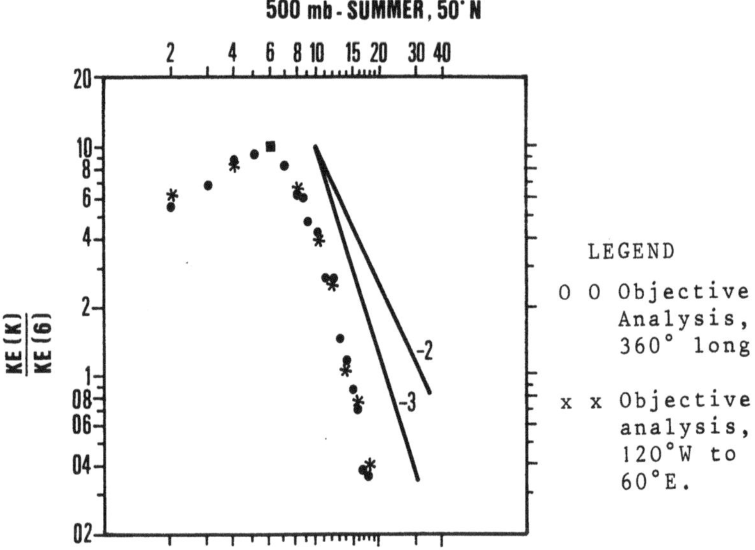

Fig.11.4 HEMISPHERIC WAVE NUMBER, K

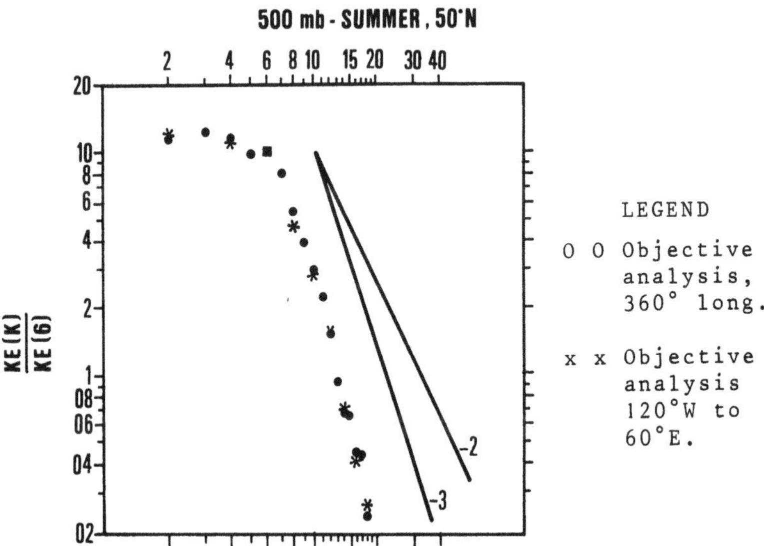

Fig.11.5 - HEMISPHERIC WAVE NUMBER, K

of the vortex tubes of the earth's rotation as to horizontal advection of vorticity. The real reason must be sought in the geostrophic character of the flow and the constraints that it imposes. We have seen than an energy cascade is forbidden in three-dimensional quasi-geostrophic flow, as well as two-dimensional flow, if the ground is isentropic. If we assure that small, synoptic scale motions within the atmosphere are not coupled with the ground (just as we assume that the small scales of three-dimensional turbulence in a box are independent of the wall constraints), an energy cascade is again forbidden. If we assume further that the vertical as well as horizontal scales are so small that ρ_s and N may be regarded as constant locally, we may set $\zeta = Nz/f_o$ and obtain for the equation of conservation of pseudo-potential vorticity

$$(\frac{\partial}{\partial t} + \underline{V}\cdot\nabla)\left[\psi_{xx} + \psi_{yy} + \psi_{\zeta\zeta} + \beta(y - y_o)\right] = 0 . \qquad (11.41)$$

We now postulate that the flow is isotropic and homogeneous in x, y, and ζ for sufficiently small k_x, k_y and k_ζ. This means that the total kinetic plus available potential energy per unit mass per unit scalar wave number k is a function only of η, the transfer of mean-squared pseudo-potential vorticity to higher wave numbers. We then obtain, as in the case of two-dimensional flow,

$$E(k) = \alpha_3 \eta^{2/3} k^{-3} , \qquad (11.42)$$

Because of the isotropy, one expects equipartition between the two components of horizontal kinetic energy and the available potential energy. Thus the components ψ_x, ψ_y, and ψ_ζ should have the same spectra in all directions, including the vertical.

One indication of the correctness of the foregoing hypothesis is the fact that the horizontal spectrum of temperature as well as kinetic energy has a k^{-3} dependence.*

The reasoning underlying the hypothesis of isotropy and homogeneity in the x,y,ζ space is as follows :

---

* S. K. Kao, 1970 : J.A.S. 27, 1000.

We assume first that the horizontal scales are so small and the non-linear interaction energies so large that the β-effect and the interaction with the mean flow via the horizontal Reynolds stresses may be ignored. We may then expect horizontal homogeneity and isotropy in $\nabla\psi$ and ψ_ζ separately. In this case turbulent energy can propagate in the vertical only through interaction with the horizontally-averaged mean flow, if we do not assume that the vertical scales are so small that the pairs $(\nabla\psi,\psi_\zeta)$ and (ψ,w) are also uncorrelated. Thus, if horizontally averaged values are denoted by a bar, the eddy energy equation in differential form (cf. equation 9.35) becomes

$$\frac{\partial}{\partial t}\left[\frac{\rho_s}{2}\cdot(\overline{(\nabla\psi')^2} + \overline{(\psi_\zeta')^2})\right] = \rho_s\,\overline{\psi_\zeta'\underline{V}'}\cdot\underline{\overline{V}}_\zeta - N\frac{\partial}{\partial\zeta}(\rho_s\,\overline{\psi' w'}) \qquad (11.43)$$

The first term on the right represents conversion of energy from the mean flow and the second term convergence of the vertical flow of wave energy.

We next suppose that $\underline{\overline{V}}_\zeta$ and N are uniform over the vertical scale of the small eddies and that surface boundary effects are not present. Then, at statistical equilibrium, the only parameters on which the eddy energy, the eddy heat transfer, and the vertical energy flow can depend are the vector $\underline{\overline{V}}_\zeta$ and the horizontal wave number $k_H = (k_x^2 + k_y^2)^{1/2}$, the parameters N and f having been eliminated in the vertical scaling of the equations of motion (cf. 11.41). Dimensional analysis then gives for the energy spectrum

$$E(k_H) = \alpha_4|\underline{\overline{V}}_\zeta|^2 k_H^{-3}\,, \qquad (11.44)$$

and similarly for the kinetic energy $K(k_H)$, the available potential energy $P(k_H)$, and the cospectrum $B(k_H)$ of ψ_ζ' and the component of $\underline{V}'$ in the direction of $\underline{\overline{V}}$:

$$K(k_H) = \alpha_5|\underline{\overline{V}}_\zeta|^2 k_H^{-3} \qquad (11.45)$$

$$P(k_H) = \alpha_6|\underline{\overline{V}}_\zeta|^2 k_H^{-3} \qquad (11.46)$$

$$B(k_H) = \alpha_7|\underline{\overline{V}}_\zeta|^2 k_H^{-3} \qquad (11.47)$$

We now observe that if $\underline{\overline{V}}_\zeta$ is uniform through several vertical scales, the energies $K(k_H)$ and $P(k_H)$ become independent of height and the turbulence becomes homogeneous in ζ as well as x and y. When this happens, ψ_ζ must become uncorrelated with $\nabla\psi$ for otherwise the thermal wind relationship, $(\nabla\psi)_\zeta = \nabla\psi_\zeta$, would prevent the horizontal velocities from remaining homogeneous in ζ. We also assume that $\nabla\psi_\zeta$ is not correlated with $\nabla\psi$. It then follows that $\nabla\psi$ is homogeneous in ζ as well as in x and y. Consequently ψ_ζ must also be homogeneous in ζ. We may therefore assume that there will be a suitable vertical scaling factor σ for which the three-dimensional vector $(\psi_x, \psi_y, \sigma\psi_\zeta)$ becomes isotropic. It is difficult to see how this scale can be other than unity because there are no other disposable parameters in the problem.

Thus we have shown that, under the somewhat restrictive condition that N and $\underline{\overline{V}}_\zeta$ are constant over a sufficiently large vertical stretch, geostrophic turbulence will be homogeneous and isotropic in x, y and ζ. In this case $\underline{V}$ and ψ_ζ will be uncorrelated and the turbulence will depend, not on $\underline{\overline{V}}_\zeta$, but on the dissipation of pseudo-potential enstrophy η as in (11.40). The most questionable feature of the foregoing analysis is the assumption that the small-scale internal motions are not coupled to the surface boundary. If the ground is not an isentropic surface, baroclinic instability and frontogenesis become possible. A front, being a discontinuity which extends into the interior, brings with it turbulent energy at small scales of motion. Although actual discontinuities do not occur in geostrophic dynamics, the zones of strong gradient accompanying the pseudo-fronts would be enough to alter the k^{-3} spectral dependency.

Let us idealize the front as a vertical discontinuity in $\underline{V}$ and ψ_ζ. If the discontinuities occur in random positions and orientations, they will give rise to a k_H^{-2} spectrum for E, K, and P.* If we assume that they are superimposed randomly over a class of motions having a k_H^{-3} spectrum, the combined spectrum would be of the form $Ak_H^{-3} + Bk_H^{-2}$. It might be be surmised from the fact that a k_H^{-3} behavior is observed in the intermediate range s = 7 to 18 that fronts occur so infrequently and at such small

*O. M. Phillips, J. Phys. Ocean. 1, 1-6, 1971.

amplitudes that the second term can become important only at quite high wave numbers. At still higher wave numbers one expects a Kolmogoroff $k_H^{-5/3}$ behavior. A possible form of the spectral law would therefore be $Ak_H^{-3} + Bk_H^{-2} + Ck_H^{-5/3}$, so that the slope in a log K - log k_H diagram would be observed to vary from -3 to -2 to -5/3 with increasing k_H. At the present time all that one can say is that observations do not contradict this possibility.

XII. THE GENERAL CIRCULATION OF THE ATMOSPHERE *

Introduction

A semi-quantitative theory of the general circulation of the atmosphere may be constructed as a synthesis of the physical ideas that have been established in earlier chapters. The theory is based on the principle that the changes in zonal momentum and density brought about by the real and eddy sources of momentum and heat must be such as to maintain a condition of balance in the zonally averaged flow (Chapter V). Since these changes are not usually compensating, meridional circulations are required to restore the balance.

The hydrodynamical problem may be stated as follows : a meridional distribution of heat sources decreasing in intensity from equator to pole produces a slow, axially-symmetric, convective circulation with a poleward temperature gradient and thereby a zonal wind field increasing in westerly intensity with height. This flow is unstable for small amplitude wavelike disturbances, which therefore grow to finite amplitude. We wish to determine the mean properties of the final state of motion and also, if possible, the nature of the perturbations.

One may distinguish two possible classes of motion : (1) the secondary wave disturbances may themselves be stable and approach a constant-amplitude equilibrium state or a regularly fluctuating state ; (2) several different disturbance modes may be excited by instability of the primary flow, by instability of the secondary disturbances, or by self-interaction of the secondary disturbances ; these modes will appear as a steady, a regularly fluctuating, or an aperiodic state of motion. Although the atmospheric circulation is better described by a fluctuating or turbulent state of motion, it may resemble the equilibrium state in important aspects. In this chapter no attempt will be made to deal directly with the regularly fluctuating or turbulent regimes. **

---

* Charney, J.G., 1959 : On the theory of the General Circulation of the Atmosphere, The Rossby Memorial Volume. See also P. Stone, J. Atmos. Sci., 29, April, 1972.

**The fluctuating state has been discussed by J. Pedlosky, 1970 : J.A.S. 27, pp. 15-27.

By confining attention to the steady-amplitude regime it becomes possible to introduce a number of simplifying approximations which greatly facilitate the mathematical analysis. We shall take into account only the non-linear interaction between the disturbance and the mean flow, disregarding self-interaction of the disturbance and the subsequent cascade of interactions. By the same token we shall permit changes in amplitude but no changes in the shape of the disturbance.

Within the limitations imposed by these approximations, one may envisage the process by which the finite-amplitude regime is established in the following way : Imagine a small wave disturbance superimposed on the steady symmetric flow produced by the symmetric heating. In the initial stages its interaction with the mean motion may be ignored. But as its amplitude grows, at first exponentially, it begins to distort the mean flow, directly through the action of the Reynolds stresses and eddy conduction of heat, and indirectly through the action of the forced meridional circulations. At length the mean flow becomes so modified that the rate at which the perturbation receives energy from the mean flow is just balanced by the rate at which it loses potential energy through radiative temperature equalization and kinetic energy through the action of the Reynolds stresses and frictional dissipation. The form of the disturbance is first calculated from the first-order instability theory : the modification of the mean flow as a function of an undetermined amplitude is then calculated by means of the mean momentum and heat equations ; and finally the amplitude is determined by making use of energy balance relationships.

<u>The mean momentum and heat equations</u>

In the interest of simplicity, the Boussinesq approximation will be made, and the geometry will be approximated by a periodic plane bounded to the north and south by walls whose distance apart will be taken to be the radius of the earth a, a distance somewhat less than the distance $\frac{\pi a}{2}$ from pole to equator. Because of the large meridional scale of the mean motion, quantities which are small for the eddies may no longer **be** small for the mean flow. For this reason the motion is supposed to satisfy the balance dynamics of Chapter VII, rather than geostrophic dynamics. Averaging equations (7.60 with respect to x and integrating the first two with respect to y gives

$$- f_o \bar{u}_\psi = \bar{\pi}_y - (\overline{v_\psi'^2})_y^{\ *}$$

$$\left(\frac{\partial}{\partial t} + \bar{v}_\sigma \frac{\partial}{\partial y} + \bar{w} \frac{\partial}{\partial z}\right)\bar{u}_\psi - f\bar{v}_\sigma = - \frac{\partial}{\partial y} \left(\overline{u_\psi' v_\psi'} + \overline{u_\psi' v_\sigma'}^{\ *} + \overline{u_\sigma' v_\psi'}^{\ *}\right) - \frac{\partial}{\partial z} \left(\overline{u_\psi' w'}\right)^{**}$$

$$g(\ln\bar{\theta} - \ln\theta_s) = \frac{\partial \bar{\pi}}{\partial z} \qquad (12.1)$$

$$\frac{\partial \bar{v}_\sigma}{\partial y} + \frac{\partial \bar{w}}{\partial z} = 0$$

$$\left(\frac{\partial}{\partial t} + \bar{v}_\sigma \frac{\partial}{\partial y} + \bar{w} \frac{\partial}{\partial z}\right)(\ln\bar{\theta}) = - \frac{\partial}{\partial y} \left(\frac{\overline{v_\psi'\theta'}}{\bar{\theta}} + \frac{\overline{v_\sigma'\theta'}}{\bar{\theta}}^{\ *}\right) - \frac{\partial}{\partial z} \left(\frac{\overline{w'\theta'}}{\bar{\theta}}\right)^{*} + \frac{\bar{Q}}{c_p \bar{T}}$$

where the subscripts "ψ" and "σ" denote components of $\underline{V}_\psi = \underline{k} \times \nabla\psi$ and $\underline{V}_\sigma = \nabla\sigma$ respectively, and we have added the heat source term to the thermal equation.

The single star terms are small, of order perturbation Rossby number, in comparison with the remaining terms. The double star terms would be small on the scale of the perturbation but are not small on the global scale. The former are therefore neglected but the latter are retained. The mean momentum and heat equations thus become

$$- \left(\frac{\partial}{\partial t} + \bar{v} \frac{\partial}{\partial y} + \bar{w} \frac{\partial}{\partial z}\right) \frac{\partial \bar{\psi}}{\partial y} - f\bar{v} = \frac{\partial}{\partial y} \left(\overline{\psi_x' \psi_y'}\right) + \frac{\partial}{\partial z} \left(\overline{\psi_y' w'}\right) \qquad (12.2)$$

$$\left(\frac{\partial}{\partial t} + \bar{v} \frac{\partial}{\partial y} + \bar{w} \frac{\partial}{\partial z}\right)\left(\frac{\partial \bar{\psi}}{\partial z} + \frac{g}{f_o} \ln\theta_s\right) = - \left[\frac{\partial}{\partial y} \left(\overline{\psi_x' \psi_z'}\right) + \frac{\partial}{\partial z} \left(\overline{w' \psi_z'}\right)\right] + \frac{gQ}{f_o c_p \bar{T}} . \qquad (12.3)$$

The heating function

A first estimate of the mean heating of the atmosphere by radiative transfer may be obtained by assuming the atmosphere to be an isothermal slab at temperature T_m transparent to solar radiation and "black" to terrestrial radiation. We shall adopt one of two models of the thermal properties of the earth's surface : (1) the ground is solid, its heat capacity is negligible, and the net flux of heat at the surface is zero ; (2) the earth is covered by an ocean whose heat capacity is so large that at the time scale of the planetary baroclinic wave motions its surface temperature may be regarded as a fixed function of latitude.

If S_o is the solar constant, the mean absorption of solar radiation at latitude ϕ is

$$S = \frac{(1 - A)}{\pi} S_o \cos \phi \ , \tag{12.4}$$

where A is the combined albedo of the atmosphere and ground. In the first model

$$\sigma T_o^4 = S + \sigma T_m^4 \ , \tag{12.5}$$

where T_o is the surface temperature, T_m is the temperature of the slab and $\sigma = 5.67 \times 10^{-5}$ erg cm^{-2}sec^{-1}deg^{-4} is the Stefan-Boltzmann constant. In the second model T_o is a prescribed function of y.

The net heating of a unit column in model (1) is

$$\int_0^\infty \rho_s \, Q dz = \sigma T_o^4 - 2\sigma T_m^4 = \frac{(1 - A)}{\pi} S_o \cos\phi - \sigma T_m^4 \ . \tag{12.6}$$

If we assume that the heating is distributed uniformly by mass over the entire unit column and define the radiative equilibrium temperature T_r by

$$T_r = \begin{cases} \left[\dfrac{(1 - A) S_o \cos\phi}{\sigma\pi}\right]^{1/4} = T_1 & \text{in } (1) \\ 2^{-1/4} \ T_o(y) = T_2 & \text{in } (2) \end{cases} \tag{12.7}$$

the heating per unit mass becomes

$$\frac{p_o Q}{g} = \begin{cases} \sigma(T_1^4 - T_m^4) & \text{in } (1) \\ 2\sigma(T_2^4 - T_m^4) & \text{in } (2) \end{cases}$$

where p_o is the mean surface pressure. Since $(T_r - T_m)/T_r \ll 1$, these expressions may be linearized to give

$$\frac{P_o Q}{g} = F[T_r(y) - T_m] \tag{12.8}$$

where

$$F = \begin{cases} 4\sigma T_1^3(\phi_o) & (1) \\ 8\sigma T_2^3(\phi_o) & (2) \,. \end{cases} \tag{12.9}$$

A similar linearization of $(\cos\phi)^{1/4} \simeq [\cos(\phi_o + \frac{y}{a})]^{1/4}$ gives

$$T_1 = T_1(\phi_o)[1 - \frac{y}{4a}\tan\phi_o] \quad ,$$

where $y = 0$ at $\phi = \phi_o$. Taking $\phi_o = 45°$, we get

$$T_1(y) = T_1(45°)\ (1 - \frac{y}{4a}) \ . \tag{12.10}$$

To handle both models simultaneously, we suppose that T_r has the same form in model 2. Then, in view of (12.9),

$$T_2(\phi_o) = 2^{-1/4}\, T_o(\phi_o) = 2^{-1/3} T_1(\phi_o) \quad . \tag{12.11}$$

Equation (12.8) is essentially a Newtonian cooling law, according to which the atmosphere heats or cools in proportion to its deviation from the radiative equilibrium temperature. We now allow for vertical temperature variation by postulating that (12.8) applies at all levels, not merely at a mean level, and assume that T_1, in addition to having the horizontal variation given by (12.10), also has a vertical variation corresponding to T_s. Such a variation would be determined not only by radiation but by dry and moist convection. Thus we assume that (12.8) can be replaced by the cooling law

$$\frac{P_o}{g} Q = FT^*\left(-\frac{y}{4a} + \frac{T_s - T}{T^*}\right) \quad , \tag{12.12}$$

where T^* is a mean of $T_s(z)$.

Linearizing the expression for ψ_z in (7.40) we obtain

$$\frac{f_o}{g}\psi_z \simeq \frac{\theta - \theta_s}{\theta_s} \simeq \frac{T - T_s}{T} \quad ; \qquad (12.13)$$

whence

$$\frac{p_o}{g} Q = FT^* \left(- \frac{y}{4a} - \frac{f_o \psi_z}{g} \right)$$

$$= \frac{4(1-A)S_o \cos\phi_o}{\pi} \left(- \frac{y}{4a} - \frac{f_o \psi_z}{g} \right) \qquad (12.14)$$

The symmetric Hadley circulation

In the absence of any circulation, $Q = 0$, and one obtains by differentiation of (12.14)

$$\bar{u}_z = - \bar{\psi}_{yz} = \frac{g}{4f_o a} = 4 \text{ m/sec/km} ,$$

giving the measure $U = \frac{1}{4} \frac{gD}{f_o a} = 40$ m/sec for the radiative-convective thermal wind for $D = 10$ km. Let us scale y by a, z by D, t by a/U, $\underline{V}$ by U, ψ by Ua, w by $\varepsilon RoUD/L$, T by T^*, so that

$$Ro = \frac{U}{f_o a} = \frac{gD}{4f_o^2 a^2} \simeq \frac{1}{16} ,$$

$$\varepsilon = \frac{f_o^2 a^2}{N^2 D^2} \simeq 40 ,$$

$$\Delta T \equiv \frac{(1-A)S_o \cos\phi_o}{c_p f_o T_s \pi} ; \quad \frac{p_o}{g} \simeq 4\times10^{-3} ,$$

where $N \simeq 10^{-2}\text{sec}^{-1}$, $f_o \simeq 10^{-4}\text{sec}^{-1}$, and ΔT may be interpreted as the mean non-dimensional temperature change in a vertical column produced by the net flux of solar radiation in the time f_o^{-1}. The non-dimensional forms of equations (12.2) and (12.3) then become

$$-\mathrm{Ro}\left(\frac{\partial}{\partial t} + \bar{v}\frac{\partial}{\partial y} + \varepsilon \mathrm{Ro}\bar{w}\frac{\partial}{\partial z}\right)\frac{\partial \bar{\psi}}{\partial y} - \bar{v} = \mathrm{Ro}\left[\frac{\partial}{\partial y}\overline{(\psi_x'\psi_y')} + \varepsilon \mathrm{Ro}\frac{\partial}{\partial z}\overline{(w'\psi_y')}\right] \quad (12.15)$$

$$\left(\frac{\partial}{\partial t} + \bar{v}\frac{\partial}{\partial y}\right)\frac{\partial \bar{\psi}}{\partial z} + \bar{w}(1 + \varepsilon \mathrm{Ro}\bar{\psi}_{zz}) = -\frac{\partial}{\partial y}\overline{(\psi_x'\psi_z')} - \varepsilon \mathrm{Ro}\frac{\partial}{\partial z}\overline{(w'\psi_z')} - 4\mathrm{Ro}^{-1}(\Delta T)(\bar{\psi}_z + y). \quad (12.16)$$

These, together with the continuity equation,

$$\frac{\partial \bar{v}}{\partial y} + \varepsilon \mathrm{Ro}\frac{\partial \bar{w}}{\partial z} = 0 \quad , \quad (12.17)$$

determine the mean flow. They are seen to be closely analogous to the balance equations, (5.12) and (5.13), for a symmetric vortex if the sources of heat and momentum are taken to include the eddy flux divergence terms.

Now it was shown in Chapter III that a symmetric, thermally-driven Hadley circulation in a fluid with a zero-stress upper boundary condition produces vertical Ekman pumping of order $E = \frac{\nu}{f_o D^2}$ in units of (D/L)U, or εRoE in the present units. Taking $\nu \approx 4\times10^5\,\mathrm{cm^2 sec^{-1}}$, we get $E \approx 4\times10^{-3}$ and $\bar{w} \sim \varepsilon\mathrm{RoE} \sim \bar{v} \sim 10^{-2}$. In contrast $4\mathrm{Ro}^{-1}(\Delta T) \sim .25$. Thus we see that the meridional transport terms are small in (12.15) and (12.16), and we may conclude that the symmetric circulation (in the absence of latent heat sources) can produce no significant change in radiative-convective equilibrium at middle latitudes. As in Chapter III the vanishing of the stress at the upper boundary also requires the interior $\bar{u}$ to vanish at the ground.

The mean circulation with eddies

The radiative-convective thermal gradient, which was found to be essentially unmodified by the frictionally forced symmetric circulation, produces a uniform zonal flow, which is zero at the ground and increases with height at the rate $g/4f_o a$ = 4m/sec/km. We now calculate the form of the eddies by solving the baroclinic stability problem for this flow. To simplify matters, we ignore the β-effect and assume that the atmosphere is a Boussinesq fluid with a rigid top at a mean tropopause height D of 10 km. The problem is then reduced to the one formulated by Eady (Chapter IX, pp. 9.15 - 9.20). Heating and friction will be ignored on the ground that their time constants are large in comparison with the amplification time of an unstable wave disturbance.

Since the distance between the latitudinal walls is taken to be the radius of the earth a, and a is also the scaling length, $y_2 - y_1 = 1$ and $\ell = n\pi$ ($n = 1, 2, \ldots$). Also $m = 1$ in the present non-dimensionalization. Substituting (9.49) into (9.48) and setting $z = 0$, we get

$$-\alpha c A - B = 0 ,$$

where, by (9.50),

$$c = \frac{1}{2}(1 + Ri) , \qquad \frac{R}{2} = \left(\frac{\coth\alpha}{\alpha} - \frac{1}{\alpha^2} - \frac{1}{4}\right)^{1/2} .$$

Hence

$$\Psi = A(\sinh \alpha z - \alpha c \cosh \alpha z),$$

and

$$\left.\begin{aligned}
\psi' &= Ae^{ik(x-ct)}(\sinh \alpha z - \alpha c \cosh \alpha z)\cos \ell y \\
u' &= \psi'_y = \ell\psi' \tan\ell y \\
v' &= \psi'_x = ik\psi' \\
\psi'_z &= \alpha A e^{ik(x-ct)}(\cosh \alpha z - \alpha c \sinh \alpha z)\cos \ell y \\
w' &= -\left(\frac{\partial}{\partial t} + \bar{u}\frac{\partial}{\partial x}\right)\psi'_z - v'\frac{\partial}{\partial y}(\bar{\Psi}_z) \\
&= -ik\left[(z - c)\psi'_z - \psi'\right] .
\end{aligned}\right\} \quad (12.18)$$

Utilizing these expressions, we obtain

$$\overline{\psi'_x\psi'_y} = \overline{\mathrm{Re}\psi'_x \mathrm{Re}\psi'_y} = -k\ell\tan\ell y\ \overline{\mathrm{Re}\ i\psi' \mathrm{Re}\psi'} = 0$$

$$\overline{w'\psi'_y} = -\ell\tan\ell y\ \overline{\mathrm{Re}\psi' \mathrm{Re}w'}$$

$$
\left.
\begin{aligned}
\overline{w'\psi'_y} &= k\ell\tan\ell y\left[(z-c_r)\overline{\mathrm{Re}\psi'\mathrm{Rei}\psi'_z} + c_i\,\overline{\mathrm{Re}\psi'\mathrm{Re}\psi'_z}\right] \\
&= \frac{k\ell\alpha^2A^2R}{8}\left[-(z-\frac{1}{2}) + \frac{1}{2}\,\frac{\sinh 2\alpha(z-\frac{1}{2})}{\sinh\alpha}\right]\sin 2\ell y \\
\overline{(w'\psi'_y)_z} &= \frac{k\ell\alpha^2A^2R}{8}\left[-1 + \frac{\alpha\cosh 2\alpha(z-\frac{1}{2})}{\sinh\alpha}\right]\sin 2\ell y \\
\overline{\psi'_x\psi'_y} &= k\,\overline{\mathrm{Rei}\psi'\mathrm{Re}\psi'_z} = \frac{k\alpha^2A^2R}{4}\cos^2\ell y \\
\overline{(\psi'_x\psi'_z)_y} &= -\frac{k\ell\alpha^2A^2R}{4}\sin 2\ell y \\
\overline{w'\psi'_z} &= k(\overline{\mathrm{Rei}\psi'\mathrm{Re}\psi'_z} - c_i\,\overline{\mathrm{Re}\psi'_z\mathrm{Re}\psi'_z}) \\
&= -\frac{k\alpha^3A^2R}{8\sinh\alpha}\left[\cosh 2\alpha(z-\frac{1}{2}) - \cosh\alpha\right]\cos^2\ell y \\
\overline{(w'\psi'_z)_z} &= -\frac{k\alpha^4A^2R}{4}\,\frac{\sinh 2\alpha(z-\frac{1}{2})}{\sinh\alpha}\cos^2\ell y
\end{aligned}
\right\} \qquad (12.19)
$$

$$
\overline{(u'^2 + v'^2)}_{z=0} = \frac{A^2}{2}\left(\alpha\coth\alpha - 1\right)\left[\ell^2\sin^2\ell y + k^2\cos^2\ell y\right] \qquad (12.20)
$$

$$
\begin{aligned}
\overline{(\psi'_z)^2} &= \frac{|\psi_z|^2}{2}\,\alpha^2\cos^2\ell y \\
&= \frac{\alpha^2A^2}{2}\left\{1 + \frac{\alpha}{2\sinh\alpha}\left[\cosh 2\alpha(z-\frac{1}{2}) - \cosh\alpha\right]\right\}\cos^2\ell y
\end{aligned}
\qquad (12.21)
$$

where the factor $\exp(kc_it)$ has been absorbed in A.

Energetics

In accordance with the principle announced at the outset of this chapter, we determine the amplitude factor A^2 from the assumption that the perturbation grows in amplitude, but without changing its shape, until the mean flow becomes so modified that the rate at which the perturbation receives energy from the mean flow is just balanced by the dissipations. In the present model it is necessary to add the dissipation of available potential energy through radiative temperature equalization.

If the perturbation term $-4Ro^{-1}(\Delta T)\psi'_z$, derived from the perturbation form of the non-dimensional heat equation (12.16), is added to the right-hand side of (9.34), the equilibrium perturbation energy equation (9.35) becomes, for a Boussinesq fluid,

$$\frac{d}{dt}\iint\frac{1}{2}\left[\overline{(\nabla\psi')^2} + \varepsilon\overline{(\psi'_z)^2}\right]dydz = \iint \overline{\psi'_x\psi'_y}\ \bar{u}_y\ dydz + \iint\varepsilon\ \overline{\psi'_x\psi'_z}\ \bar{u}_z\ dydz$$

$$- 4\varepsilon(\Delta T)Ro^{-1}\iint \overline{(\psi'_z)^2}\ dydz + \int \overline{(\psi'w')}_{z=0}\ dy = 0\ , \qquad (12.22)$$

where the limits of integration are $-\frac{1}{2}$ to $\frac{1}{2}$ for y and 0 to 1 for z. The last two integrals are the available potential energy dissipation and the kinetic energy dissipation respectively. The latter represents the pressure work by the fluid above the frictional boundary layer needed to supply the kinetic energy dissipated within the boundary layer. Thus, we obtain from (7.52) in non-dimensional form,

$$\int \overline{(\psi'w')}_{z=0}\ dy = \frac{Ro^{-1}E^{1/2}}{\sqrt{2}}\int \overline{\psi'\nabla^2\psi'}\ dy = -\frac{Ro^{-1}E^{1/2}}{\sqrt{2}}\int\overline{(\nabla\psi')^2}\ dy\ . \qquad (12.23)$$

In the present case $\overline{\psi'_x\psi'_y} = 0$, and we get

$$\varepsilon\iint\overline{\psi'_x\psi'_z}\ \bar{u}_z dydz - 4(\Delta T)\varepsilon Ro^{-1}\iint\overline{(\psi'_z)^2}dydz - \frac{Ro^{-1}E^{1/2}}{\sqrt{2}}\int\overline{(\nabla\psi')^2}_{z=0}\ dy = 0\ , \qquad (12.24)$$

which states that the generation of perturbation available potential energy is balanced by the dissipation of available potential energy by radiative equalization of temperature and by boundary-layer friction.

The mean flow

In the steady state it follows from (12.15) that $\bar{v} = O(Ro)$ and, since $\bar{w}$ is also $O(Ro)$, that the vertical transport terms in (12.16) are $O(Ro)^2$ and may therefore be ignored. The mean momentum and heat equations may therefore be approximated by

$$-\bar{v} = Ro\left[\frac{\partial}{\partial y}(\overline{\psi'_x\psi'_y}) + \varepsilon Ro\frac{\partial}{\partial z}(\overline{w'\psi'_y})\right] \qquad (12.25)$$

$$- \frac{\partial}{\partial y}\,(\overline{\psi'_x\psi'_z}) - \varepsilon Ro\,\frac{\partial}{\partial z}\,(\overline{w'\psi'_z}) - 4Ro^{-1}(\Delta T)(\psi_z + y) = 0\,. \quad (12.26)$$

Since the first term on the right-hand side of (12.25) vanishes, and since the second integrates to zero vertically, the integrated $\bar{v}$ transport in the interior vanishes to $O(Ro)^2$. Moreover, since the stress must vanish at the upper boundary, the upper boundary layer can contribute at most an order E transport. It follows that the surface mean zonal velocity must vanish, otherwise it would contribute an uncompensated order $E^{1/2}$ transport.

Differentiation of (12.26) and use of (12.19) gives

$$4Ro^{-1}(\Delta T)(\bar{u}_z - 1) = (\overline{\psi'_x\psi'_z})_{yy} + \varepsilon Ro(\overline{w'\psi'_z})_{zy} = -\frac{k\ell^2\alpha^2RA^2}{2}\cos 2\ell y + $$

$$+ \frac{\varepsilon Ro k\ell\alpha^4A^2R}{4\sinh\alpha}\sinh 2\alpha(z - \tfrac{1}{2})\sin 2\ell y\,, \quad (12.27)$$

whose integral, subject to the condition that $\bar{u}$ vanish at $z = 0$, is

$$\bar{u} = z(1 - B^2\cos 2\ell y) + \frac{\varepsilon Ro\alpha B^2}{4\ell\sinh\alpha}\left[\cosh 2\alpha(z - \tfrac{1}{2}) - \cosh\alpha\right]\sin 2\ell y \quad (12.28)$$

where

$$B^2 = \frac{k\ell^2\alpha^2RA^2}{8Ro^{-1}(\Delta T)}\,.$$

Finally, substituting from (12.19), (12.20), (12.21) and (12.27) into (12.24), and noting that the z-dependent term in $\bar{u}_z$ contributes nothing to the first integral, we obtain

$$\varepsilon\int \frac{k\alpha^2RA^2}{4}(1 - B^2\cos 2\ell y)\cos^2\ell y\,dy$$

$$- 2(\Delta T)\varepsilon Ro^{-1}\alpha^2A^2\iint\{1 + \frac{\alpha}{2\sinh\alpha}\left[\cosh 2\alpha(z - \tfrac{1}{2}) - \cosh\alpha\right]\}\cos^2\ell y\,dydz$$

$$- \frac{Ro^{-1}E^{1/2}}{\sqrt{2}}\,\frac{A^2}{2}\,(\alpha\coth\alpha - 1)\int\left[k^2\cos^2\ell y + \ell^2\sin^2\ell y\right]dy = 0\,, \quad (12.29)$$

which gives, upon integration,

$$0 = \varepsilon(\frac{\pi}{2\ell} - B^2 \frac{\pi}{4\ell}) \frac{k\alpha^2 RA^2}{4} - \frac{Ro^{-1}E^{1/2}}{\sqrt{2}} \frac{\varepsilon\alpha^2 A^2}{2} (\alpha\coth\alpha - 1) \frac{\pi}{2\ell}$$

$$- 2(\Delta T)\varepsilon Ro^{-1}\alpha^2 A^2 (\frac{\pi}{2\ell})(\frac{3}{2} - \frac{\alpha\cosh\alpha}{2\sinh\alpha})$$

or

$$1 - \frac{B^2}{2} = \frac{Ro^{-1}}{kR}[(2E)^{1/2}(\alpha\coth\alpha - 1) + 4(\Delta T)(3 - \alpha\coth\alpha)] \quad . \quad (12.30)$$

To estimate k, α and R, we choose $\alpha = 1.6$ as the value of α which gives the maximum growth rate, kc_i, for $\ell = 0$. In reality, the maximum growth rate occurs for $\ell = \pi$, which is the smallest permissible value of ℓ, and the value which gives maximum growth. However, since k^2/ℓ^2 will turn out to be large, the value $\alpha = 1.6$ is approximately correct for $\ell = \pi$ also. Substituting the numerical values $\ell = \pi$, $\alpha = 1.6$, $\varepsilon = 40$, $Ro = \frac{1}{16}$, $E = 4\times10^{-3}$, $\Delta T = 4\times10^{-3}$, $k = (\varepsilon\alpha^2-\ell^2)^{1/2} = 9.6$, $R = (\frac{4}{\alpha}\coth\alpha - \frac{4}{\alpha^2} - 1)^{1/2} = .375$ into (12.30) we obtain

$$\left.\begin{aligned} 1 - \frac{B^2}{2} &= \frac{16}{9.6\text{x}.375}(9\text{x}10^{-2}\text{x}.74 + 1.6\text{x}10^{-2}\text{x}1.27) = .30 + .09 \\ B^2 &= 1.22 \\ A^2 &= \frac{8Ro^{-1}(\Delta T)}{k\ell^2\alpha^2 R} B^2 = .68\text{x}10^{-2} \end{aligned}\right\} \quad (12.31)$$

The orders of magnitude of u' and v' are obtained from (12.18) :

$$u' \sim \ell A = .26 \quad ,$$

$$v' \sim kA = .80 \quad ,$$

and are thus the same as that of $\bar{u}$.

The mean zonal velocity, temperature, stability, and meridional velocities are given by (12.28), (12.26), (12.25), (12.17) and (12.19). We find

$$\bar{u} = z(1 - B^2\cos 2\ell y) + \frac{\varepsilon Ro\alpha B^2}{4\ell\sinh\alpha}\left[\cosh 2\alpha(z - \tfrac{1}{2}) - \cosh\alpha\right]\sin 2\ell y$$

$$= z(1 - 1.22\cos 2\ell y) + .164\left[\cosh 3.2(z-.5) - \cosh 1.6\right]\sin 2\pi y \qquad (12.32)$$

$$\left.\begin{aligned}\bar{\Psi}_z &= -y + \frac{B^2}{2\ell}\sin 2\ell y + \frac{\varepsilon Ro\alpha^2 B^2}{2\ell^2}\,\frac{\sinh 2\alpha(z-\frac{1}{2})}{\sinh\alpha}\cos^2\ell y \\ &= -y + 0.20\sin 2\pi y + 0.40\,\frac{\sinh 3.2(z-.5)}{\sinh 1.6}\cos^2\pi y\end{aligned}\right\} \qquad (12.33)$$

$$1 + \varepsilon Ro\bar{\Psi}_{zz} = 1 + \frac{\varepsilon^2 Ro^2\alpha^3 B^2}{\ell^2}\,\frac{\cosh 2\alpha(z-\frac{1}{2})}{\sinh\alpha}\cos^2\ell y = 1 + 3.2\,\frac{\cosh 3.2(z-\frac{1}{2})}{\sinh 1.6}\cos^2\pi y\,. \qquad (12.34)$$

$$\left.\begin{aligned}\tilde{v} &= \overbrace{\frac{\varepsilon Ro^2 k\ell\alpha^2 RA^2}{8}}^{.38\times 10^{-2}}\left[1 - \frac{\alpha\cosh 2\alpha(z-\frac{1}{2})}{\sinh\alpha}\right]\sin 2\ell y \\ \bar{w} &= \overbrace{\frac{Rok\ell^2\alpha RA^2}{4}}^{.60\times 10^{-2}}\left[-(z-\tfrac{1}{2}) + \frac{1}{2}\,\frac{\sinh 2\alpha(z-\frac{1}{2})}{\sinh\alpha}\right]\cos 2\ell y\,.\end{aligned}\right\} \qquad (12.35)$$

The relationships (12.31) - (12.34) may be given the following interpretations :

a) Temperature. Equation (12.33) states that the uniform northward decrease of temperature in the symmetric circulation, as represented by the first term (which is determined by radiative-convective equilibrium, since by (12.35) the meridional circulation is too weak to affect the temperature), is modified by the divergences of the horizontal eddy flux of heat, as represented by the third term. It may be seen from (12.19) that the northward eddy flux increases from zero at $y = -\frac{1}{2}$ to a maximum at $y = 0$, and then decreases to zero at $y = \frac{1}{2}$, thus giving an increase of temperature north of $y = 0$ and a decrease south of $y = 0$. It may also be seen that the vertical flux of heat is always upward, increasing from zero at $z = 0$ to a maximum at $z = \frac{1}{2}$ and then decreasing symmetrically back to zero at $z = 1$. This is because the release of mean flow available potential energy, the source of energy flux for the disturbance, entails rising warm (light) air and sinking cold (dense) air. Consequently, there is flux convergence above the mid-level, $z = \frac{1}{2}$, and divergence below, giving rise to heating above and cooling below. The left-hand side of (12.33) represents the balancing effect of radiation : a flux convergence of heat must be balanced by an increase in temperature to allow for compensating radiative cooling in the steady state.

(b) Stability. Since the horizontal eddy flux of heat does not vary with height in the present model, the modification of the stability produced by radiation and convection in the symmetric flow is due solely to the vertical eddy flux of heat. This, we have seen, has a maximum at $z = \frac{1}{2}$. The temperature is thereby decreased below $z = \frac{1}{2}$ and increased above $z = \frac{1}{2}$, resulting in an increase of stability everywhere. Since the perturbation has a maximum amplitude at $y = 0$, the effect is greatest there and decreases to zero both to the north and south. We note that the amplitude of the increase exceeds the radiative-convective value. This suggests that large-scale dynamical effects giving rise to vertical heat flow are as important in the troposphere as radiation and convection. Indeed, had we ignored moist convection, radiation and dry convection by themselves would have given rise to a near zero stability, and a model with zero stability in the symmetric state would then have been more appropriate for consideration. But in such a model the vertical heat flow induced dynamically by the disturbance would by itself have stabilized the lower atmosphere. We may conclude that moist convection is not essential to produce a gravitationally stable troposphere.

(c) Zonal velocity. The horizontal eddy convergence of heat north of $y = 0$ and the divergence south of this latitude tend to reverse the poleward temperature gradient at $y = 0$ and increase it at $y = \pm \frac{1}{4}$. The thermal wind $\bar{u}_z$ (and, since $\bar{u}(0) = 0$, the actual wind) are thus decreased in the region $|y| < \frac{1}{4}$ and increased in the region $|y| > \frac{1}{4}$ by this effect. The vertical eddy flow of heat, being a maximum at $z = \frac{1}{2}$ and $y = 0$, produces a temperature increase above $z = \frac{1}{2}$ and a decrease below. If $y < 0$, the vertical gradient of $\bar{u}$ is increased below $z = \frac{1}{2}$ and decreased above. If $y > 0$ the reverse is true. Because of the symmetry with respect to $z = \frac{1}{2}$, and the fact that $\bar{u}(0) = 0$, vertical heat flow gives rise to a wind increase south of $y = 0$, with a maximum at $z = \frac{1}{2}$, and a decrease north of $y = 0$, with a minimum at $z = \frac{1}{2}$. Figure 11.2 shows $\bar{u}$ at $z = \frac{1}{2}$

due to 1) the symmetric circulation, 2) the horizontal eddy flux of heat, and 3) the vertical eddy flux of heat. Curve 4) is the combined profile. We note that, in contrast to its influence on the static stability, vertical heat flow has little influence on the wind profile.

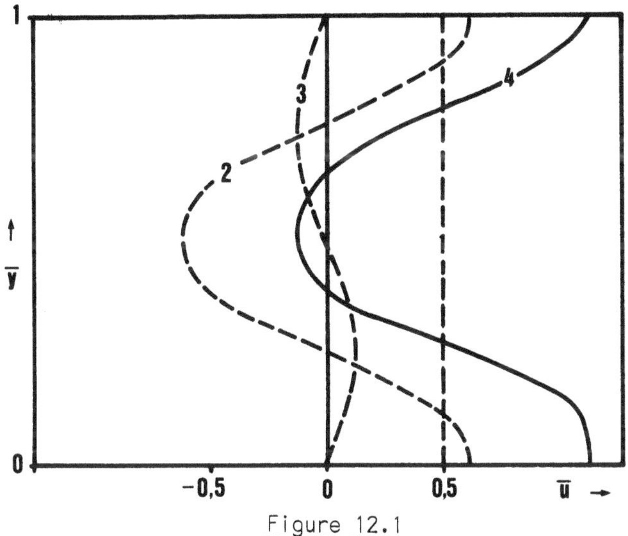

Figure 12.1

The maximum in the northward eddy flux of heat at the middle latitude gives rise to a double-jet structure with minimum winds near the middle latitude and maximum winds to the north and south. This property of the motion is not observed in the mean northern hemisphere zonal winds but is found in the southern hemisphere winter (See Fig. 1.8).

(d) Energy balance. The evaluation in (12.31) of the terms in (12.30) gives for the production and dissipation terms in (12.29) the proportions .39:.30:.09, i.e., the rates of conversion of mean flow available potential energy, of dissipation of kinetic energy by friction, and of dissipation of mean flow available potential energy are in the ratios .39 to .30 to .09. The balance is thus primarily between conversion of mean flow potential energy and dissipation by friction. The eddy heat fluxes have so modified the mean temperature field that the rate of conversion of available potential energy has been reduced in the ratio $1:1 - \frac{B^2}{2}$ or 1:.39.

Conclusion

The present model is of course too highly simplified to permit more than a qualitative comparison with reality - if that. Its primary objective has been to show how some of the important elements of the atmospheric general circulation are interrelated. A more precise treatment would have to take into consideration that : (1) the mean flow is so radically changed by the eddy fluxes of heat that the form of the perturbation must change in interaction with the mean flow ; (2) the use of Cartesian geometry and the linearization of the radiative heating exclude non-uniformity in the horizontal temperature gradients and therefore horizontal shear in the zonal wind and non-zero Reynolds stresses ; (3) the characteristic frictional relaxation time, $(f_o E^{1/2})^{-1}$, is about two days, and therefore not long compared with the e-folding time of the perturbation, $(kc_i)^{-1} = (\frac{kR}{2})^{-1} \frac{a}{U}$, about one day, and friction should not have been ignored in the calculation of the perturbation ; (4) β and the variation of ρ_s cannot be ignored in any quantitative analysis.

In the article on the general circulation by Charney (1958, loc. cit.) a non-trivial meridional circulation was introduced by postulating an appreciable internal viscosity, such as might exist in the presence of deep turbulent convection. This produced a zonal flow with horizontal shear. The stability problem for a two-layer model was solved numerically for this flow and the Reynolds stresses and eddy heat fluxes were calculated. The principal difference from the present analysis lay in the fact that the maximum heat flux did not occur at the mid-latitude but north and south of this latitude, thus producing a positive vertical wind shear at the mid-latitude. The results agreed qualitatively with earlier machine calculations made by Phillips* with essentially the same two-level model.

---

* Philipps, N. A., 1956 : Quart. J. Roy. Met. Soc., 82, pp. 123-164.

XIII. TROPICAL CYCLOGENESIS AND THE FORMATION OF THE INTERTROPICAL CONVERGENCE ZONE

The problem of tropical cyclogenesis

The great migratory waves and vortices of middle latitudes have now been tolerably well explained. As we have shown*, they are due to the instability of the mean, axially-symmetric, circumpolar vortex whose gravity and Coriolis forces balance the pressure forces associated with the radiatively induced pole to equator temperature gradient. In contrast, the origins of the migratory low pressure systems of the tropics (depressions and hurricanes) are far less well understood. An explanation analogous to that for the extra-tropical cyclones does not seem possible. Tropical temperature gradients are weak, and the associated potential energy supply is not by itself a likely source for the kinetic energy of the tropical disturbance. Indeed, tropical cyclones form most often over the tropical oceans whose surface temperatures are particularly uniform.

However, the oceans are also a source of moisture, and what is most characteristic of tropical depressions in oceanic and coastal regions is the copious rainfall they produce. It is very likely that the release of the latent heat of condensation is the primary driving mechanism. But tropical precipitation occurs for the most part in deep cumulus convection cells whose diameters are of the order of 1 to 10 km, and it is not easy to see how these cells are organized so as to supply energy to the depression whose scale is of the order of 1000 km.

Precipitation also occurs in extra-tropical cyclones, but here it is forced by the rising of the warm light air, which, in conjunction with sinking of the cold dense air, is required for the release of the potential energy associated with the strong horizontal temperature gradients. Since the hydrostatic pressure decreases with altitude, the rising air expands and cools adiabatically until it becomes saturated and is forced to release its moisture as cloud and precipitation. In the extratropical cyclone latent heat is an additional source of energy, not the primary source.

---

* See Chapter IX.

There is another mechanism, independent of horizontal temperature gradients, which is capable of organizing the release of heat of condensation. Surprisingly, it involves boundary friction, which one thinks of ordinarily as dissipative rather than generative. Although the mechanism is capable of operating alone, it probably never does. Asymmetries of the undisturbed flow may help or hinder the formation of the tropical depression. Indeed, most depressions form in the so-called Intertropical Convergence Zone (ITCZ), a narrow zone paralleling the equator but lying at some distance from it, in which air from one hemisphere converges against air from the other to produce cloud and precipitation. The ITCZ is characterized by low pressure and cyclonic shear vorticity in the surface layers. It will be suggested that the mechanism which leads to the formation of the tropical depression may also be responsible for the formation of the ITCZ itself.

Conditional instability

It will be necessary to consider motion on two scales, that of the cumulus cloud and that of the cyclone. While the detailed dynamics of the cumulus cloud are complicated, the explanation for its formation is simple. A combination of radiative, sensible and latent heat transfer maintains the tropical atmosphere in a state of hydrostatic equilibrium such that the temperature decreases with height at a rate of about 5.5°C per km in the lower part. Since a parcel of air rising adiabatically cools at a rate of 10°C per km, it finds itself colder and denser than its environment and is therefore acted upon by a restoring buoyancy force. If, however, the parcel is saturated, the moisture condenses, and the heat of condensation prevents it from cooling as rapidly as if it were not saturated. Since a rising, saturated parcel cools at a rate of about 4.5°C per km in the lower atmosphere, it finds itself warmer than its environment and is accelerated upward by the buoyancy force. The tropical atmosphere is therefore gravitationally stable for unsaturated air and gravitationally unstable for saturated air. Such an atmosphere is said to be conditionally unstable. Cumulus clouds are a manifestation of gravitational instability and convective overturning in an atmosphere whose lower part is kept moist by evaporation from the sea-surface.

Is it possible, then, that the tropical disturbance is simply a large-scale overturning in a conditionally unstable atmosphere ? We shall answer this question in the negative by showing that such an atmosphere is far more unstable to cumulus cloud scales than to disturbance scales, so that, were it not for friction and entrainment of dry air, the preferred convection cell would have an infinitely narrow ascending branch.

Competition or cooperation ?

To show that the cumulus-scale perturbation is favored over the cyclone-scale in a conditionally unstable atmosphere, we must consider the thermodynamics of unsaturated and saturated air.

Consider an air parcel which has reached saturation. If the parcel is lifted, the temperature will drop, moisture will condense, and latent heat will be liberated. We have $dQ = -Ldq_s$, and

$$ds = c_p d(\ln\theta) = -L(dq_s/T) , \qquad (13.1)$$

where q_s is the specific humidity and L is the latent heat of condensation per unit mass. Since fractional changes in q_s are far greater than those of T or L - the latter may be regarded as constant - Equation (13.1) may also be interpreted as stating that the entropy of the moist air

$$\tilde{s} = c_p \ln\theta + Lq_s/T = c_p \ln\theta_E = \text{constant} . \qquad (13.2)$$

The quantity $\theta_E = \theta\exp(Lq_s/c_pT)$ is called the "equivalent potential temperature" in analogy to "potential temperature". It is conserved for changes in state in which the latent heat of condensation is retained by the parcel. Such a process is called "moist-adiabatic". Integration of (13.1), after multiplication by T and use of the hydrostatic relation, also gives the result that the enthalpy

$$h = c_pT + gz + Lq_s = \text{constant}. \qquad (13.3)$$

Suppose now that a saturated air parcel is lifted from z to

z + dz. Again, the pressure at z + dz does not change, and we have from the gaz law that the restoring force per unit mass is

$$-N_m^2 dz = g(\delta\rho/\rho) = -g(\delta T/T) = -g(\Gamma_m - \Gamma)/T)dz \ , \qquad (13.4)$$

where Γ is the lapse rate of temperature in the environment, and $\Gamma_m = -(dT/dz)_s$ is the lapse rate of temperature for the moist adiabatic process. Thus if $\Gamma - \Gamma_m < 0$, or alternatively if $\partial\theta_E/\partial z < 0$, the atmosphere is conditionally unstable.

In reality, rising motion must be compensated by descending motion to preserve mass continuity. Let us schematically denote the area of rising motion at a fixed level by A_+ and the area of sinking motion by A_-. Let a typical upward displacement be ℓ_+ and a downward displacement ℓ_-. From mass continuity $\ell_+ A_+ \approx \ell_- A_-$. The work W_- done against gravity in the downward displacement is measured by $W_- = N^2 \ell_-^2 A_-$, and the work done by the buoyancy forces in the upward displacement is measured by $W_+ = N_m^2 \ell_+^2 A_+$. The ratio of the two is

$$\frac{W_+}{W_-} = \frac{N_m^2}{N^2} \frac{\ell_+^2}{\ell_-^2} \frac{A_+}{A_-} = \frac{N_m^2}{N^2} \frac{A_-}{A_+} \qquad (13.5)$$

If energy is to be released we must have

$$A_+/A_- < N_m^2/N^2 = (\Gamma - \Gamma_m)/(\Gamma_a - \Gamma) \sim \frac{1}{5} \ , \qquad (13.6)$$

where Γ_a is the dry-adiabatic lapse rate g/c_p. Thus the area of the rising branch of the circulation must be less than approximately one-fifth that of the descending branch. But it also follows from (13.5) that for a given amount of work done against gravity in the downward displacement, an unlimited amount can be realized through conditional instability in the upward displacement providing the ascending branch is of arbitrarily small cross-section. Of course, friction and entrainment of dry air will limit the width of the ascending branch, but it is known from the theory of Bénard convection that these limitations will not be effective until the width approaches that of the depth of the troposphere, i.e., the width of a cumulus cloud. It follows that conditional instability by itself will give rise to cumulus clouds, not depressions or hurricanes.

Nevertheless, conditional instability, by permitting cumulus convection must play a role in the formation of the tropical cyclone. As we have noted, the most striking characteristic of the depression or hurricane is its enormous rainfall ; the latent heat energy released is two orders of magnitude greater than the amount needed to maintain the kinetic energy against frictional dissipation. This suggests that we should look upon the depression and the cumulus cell not as competing for the same energy, for in this competition the cumulus cell must win ; rather we should consider the two as supporting one another, the cumulus cell by supplying the heat energy for driving the depression, and the depression by maintaining the moisture supply for driving the convection. The cumulus and cyclone-scale motions are thus to be regarded as cooperating rather than competing phenomena.

The frictional boundary layer

The clue to the cumulus-cyclone interaction lies in the character of the frictional boundary layer. It is here that most of the moisture is fed into the bases of the towering cumulonimbus clouds in the tropical depression. What causes the low-level convergence of mass and moisture ? The tentative answer is that it is caused by boundary friction. The vertical flow out of the frictional boundary layer in quasi-geostrophic motion (cf. Equation 7.52) is $D_E\zeta_g/2$ where D_E is the depth $(2\nu/f)^{1/2}$ of the boundary layer and ζ_g is the geostrophic wind above the boundary layer. Thus cyclonic vorticity produces horizontal mass convergence in the boundary layer and positive vertical mass flow out of the boundary layer.

Eliassen and Charney* have applied this result to show how the kinetic energy of a geostrophically balanced circular vortex in a

---

* J. G. Charney and A. Eliassen, 1949 : A numerical method for predicting the perturbations of the middle latitude westerlies. Tellus 1, 2, 38-54.

homogeneous fluid is destroyed by frictional interaction with the boundary. The frictional convergence in the boundary layer toward lower pressure must be compensated by divergence above the layer. The radial flow above the boundary layer is therefore directed outward toward high pressure, work is done against the inward directed pressure force, and the kinetic energy is reduced. Let V be the radial, and U the tangential, velocity above the boundary layer. Then by mass continuity, $V \sim UD_E/H$, and the **percentage** rate of reduction of U by outward expansion of rings of air with conservation of absolute angular momentum - or, alternatively, by action of the Coriolis force in relative flow on the radial velocity - is $fV/U \sim fD_E/H \sim fE^{1/2}$, where E is the nondimensional Ekman number $(\nu/fH^2)^{1/2}$. Thus the characteristic "spin-down" time for the vortex is $1/fE^{1/2}$. Another way of looking at it is to say that the cross-sectional areas of the vortex tubes in absolute motion, which are formed mainly by the vertical vortex lines of the rotating plane, expand at the rate w_E/H per unit area, so that the relative vertical vorticity component is reduced by the amount fw_E/H per unit time. The characteristic time for the destruction of the relative vorticity is again

$$\frac{\zeta_g}{fw_E/H} \sim \frac{\zeta_g}{f\zeta_g D_E/H} \sim \frac{1}{fE^{1/2}} \quad .$$

Thus we see that the adjustment time for the flow outside the boundary layer is longer than the boundary-layer adjustment time by the factor $E^{-1/2}$, which ordinarily is large, about 10 in the atmosphere. It is for this reason that the boundary layer itself can be considered to be in steady equilibrium, i.e., to reflect instantly the time-variable geostrophic flow above the boundary layer.

<u>Conditional instability of the second kind</u>

Consider now a conditionally unstable moist tropical atmosphere and suppose that a large-scale disturbance of small amplitude is somehow imparted. In regions of low-level cyclonic vorticity, there will be a boundary layer convergence, upward pumping of the moist surface air into the cumulus cells, and therefore enhancement of convection ; in regions of anti-cyclonic vorticity, there will be a boundary layer suction carrying down dryer air from aloft, and therefore quenching of cumulus convection. Moreover, as heat is liberated in the cyclonic region, the pressure will fall and the geostrophic vorticity will increase. Thus we have the possibility of a self-excited disturbance. It is this process that Charney and Eliassen (loc. cit.) have called "Conditional Instability of the Second Kind" (CISK).

The CISK process may be studied analytically for circular or slab symmetry. Other cases present problems because of the nonlinear character of the condensation process : the liberation of heat of condensation is proportional to $\frac{1}{2}(w_E + |w_E|)$. We will here consider a zonally- and slab-symmetric perturbation of a resting atmosphere on the rotating earth. Anticipating that the y-scale will be small, we ignore effects from the curvature of the earth and the variation with latitude of the Coriolis parameter, $f = 2\Omega\sin\phi$, where Ω is the angular speed of rotation of the earth and ϕ is the latitude. The perturbation equations of motion above the boundary layer become

$$\partial u'/\partial t' - fv' = 0 , \tag{13.7}$$

$$\partial v'/\partial t' + fu' = -(\partial/\partial y)(p'/\bar{\rho}), \tag{13.8}$$

$$0 = -\partial p'/\partial z' - \rho' g , \tag{13.9}$$

$$\partial \rho'/\partial t' + (\partial/\partial y')(\bar{\rho}v') + (\partial/\partial z')(\bar{\rho}w') = 0 , \tag{13.10}$$

$$\frac{1}{\bar{\theta}}\frac{\partial\theta'}{\partial t'} + \frac{N^2}{g}w' = \frac{Q'}{c_p\bar{T}} , \tag{13.11}$$

where primes denote perturbations, bars horizontally-averaged mean quantities, and the dimensional independent variables are also denoted by primes so that we may drop the primes for the nondimensional independent variables. In equation (13.11)

$$\frac{\theta'}{\bar{\theta}} = \frac{T'}{\bar{T}} - \frac{R}{c_p}\frac{p'}{\bar{p}} = \frac{c_v}{c_p}\frac{p'}{\bar{p}} - \frac{\rho'}{\bar{\rho}} \tag{13.12}$$

and

$$N^2 = g\partial(\ln\bar{\theta})/\partial z' .$$

From (13.9) and (13.12) we get

$$g\,\frac{\theta'}{\bar{\theta}} = \frac{\partial}{\partial z'}\left(\frac{p'}{\bar{p}}\right) - \frac{\partial \ln\bar{\theta}}{\partial z'}\,\frac{p'}{\bar{p}} \simeq \frac{\partial}{\partial z'}\left(\frac{p'}{\bar{p}}\right) , \qquad (13.13)$$

since $\ln\bar{\theta}$ is a slowly-varying function of z in the atmosphere.

From what has been said, we expect u to be geostrophic and v to be driven by friction. Hence we may scale the independent variables as follows : $t' = f^{-1}E^{-1/2}t$, $y' = By$, $z' = Hz$, where H is the scale height $R\bar{T}_m/g$ and $\bar{T}_m$ is a vertically averaged temperature. The dependent variables are scaled by

$$u',\ v',\ w',\ \frac{p'}{\bar{p}},\ \frac{\rho'}{\bar{\rho}} = Uu,\ E^{1/2}Uv,\ \frac{H}{B}\,UE^{1/2}w,\ fUB\psi,\ \frac{fUB}{gH}\,R. \qquad (13.14)$$

The equations of motion then become

$$\partial u/\partial t - v = 0 , \qquad (13.15)$$

$$E\partial v/\partial t + u = -\,\partial\psi/\partial y , \qquad (13.16)$$

$$\frac{f^2B^2}{gH}\,\frac{\partial R}{\partial t} + \frac{\partial v}{\partial y} + \frac{1}{\bar{\rho}}\,\frac{\partial(\bar{\rho}w)}{\partial z} = 0, \qquad (13.17)$$

$$\frac{\partial^2\phi}{\partial t\partial z} + \frac{N^2H^2}{f^2B^2}\,w = \frac{gHU}{f^2BE^{1/2}}\,\frac{Q}{c_p\bar{T}} . \qquad (13.18)$$

The Ekman number is of the order 10^{-2}, and since NH/f, Rossby's "radius of deformation", is the only important horizontal scale in the problem, we choose B so that N^2H^2/f^2B^2 is of order unity. Hence

$$f^2B^2/gH = (N^2H^2/f^2B^2)^{-1}(\partial\ln\bar{\theta}/\partial z) < O(1). \qquad (13.19)$$

Equations (13.16) and (13.17) therefore simplify to

$$u = -\ \partial\psi/\partial y\ , \qquad (13.16')$$

$$\bar{\rho}\ \partial v/\partial y + \partial(\bar{\rho}w)/\partial z = 0\ . \qquad (13.17')$$

The assumption is now made that all the moisture pumped out of the boundary layer condenses into heat and is distributed in the vertical according to a form function F whose vertical average is unity. Equation (13.18) then becomes

$$\partial^2\psi/\partial t\partial z + \lambda^2(w - F\eta w_E) = 0\ , \qquad (13.18')$$

where $\lambda = \varepsilon^{-1} = NH/fB$ is the nondimensional radius of deformation (it represents the horizontal influence for a spike function) and

$$\eta = \frac{Lq_s}{c_p\bar{T}}\left(\frac{\partial \ln\bar{\theta}}{z}\right)^{-1}\ , \qquad (13.20)$$

$$w_E = -\frac{1}{2^{1/2}}\ \frac{\partial u}{\partial y}\ (y,0) = \frac{1}{2^{1/2}}\ \frac{\partial^2\psi}{\partial y^2}\ (y,0), \qquad (13.21)$$

nondimensionally.

Since the statistical mechanics of cumulus convection is not understood well enough to permit a theoretical derivation of F, nor is it known empirically with any accuracy, nothing is gained by attempting to integrate the perturbation equations in full generality. We shall instead introduce the approximation of expressing vertical derivatives in finite-difference form and writing the equations of motion at two levels, centered at the midpoints by weight of the lower and upper halves of the atmosphere. Thus we divide the interval 0 to $\bar{p}(0)$ into four equal subintervals at the points $\bar{p} = \bar{p}(0)[p-\frac{1}{2}(n-\frac{1}{2})]$, ($n = 1, 1\frac{1}{2}, 2$) and write (13.15), (13.16') and (13.17')at the points n = 1,2, the last by setting

$$\frac{1}{\bar{\rho}}\ \frac{\partial(\bar{\rho}w)}{\partial z} = -\ gH\ \frac{\partial(\bar{\rho}w)}{\partial\bar{p}} = gH\ \frac{[(\bar{\rho}w)_{n+1/2} - (\bar{\rho}w)_{n-1/2}}{\frac{1}{2}\bar{p}(0)} \qquad (13.22)$$

and (13.18') at the point $n = 1\frac{1}{2}$ by setting

$$\partial\psi/\partial z = -\, gH\bar{\rho}(\partial\psi/\partial p) = gH\bar{\rho}_{3/2}(\psi_2 - \psi_1)/\tfrac{1}{2}\,\bar{p}(0)\ .$$

Utilizing the condition that $w = w_E$ at $\bar{p} = \bar{p}(0)$ and $\bar{\rho}w = 0$ at $\bar{p} = 0$ (no mass flux through the top of the atmosphere), and the approximation $\bar{p} = \bar{\rho}R\bar{T}_m = \bar{\rho}gH$, we get

$$\partial u_1/\partial t - v_1 = 0,\ \partial u_2/\partial t - v_2 = 0\ , \tag{13.23}$$

$$u_1 = -\,\partial\psi_1/\partial y\ ,\quad u_2 = -\,\partial\psi_2/\partial y\ , \tag{13.24}$$

$$\partial v_1/\partial y + w_{3/2} - 2w_E = 0,\ \partial v_2/\partial y - w_{3/2} = 0 \tag{13.25}$$

$$(\partial/\partial t)(\psi_2 - \psi_1) + \lambda^2(w_{3/2} - F\eta w_E) = 0\ . \tag{13.26}$$

To the present order of approximation, the lower level may be taken to be the top of the boundary layer. Hence

$$w_E = -\frac{1}{2^{1/2}}\frac{\partial u_1}{\partial y} = \frac{1}{2^{1/2}}\frac{\partial^2\psi_1}{y^2}\ . \tag{13.27}$$

Also

$$\begin{aligned} F &= 1, \quad w_E > 0\ , \\ F &= 0, \quad w_E < 0\ . \end{aligned} \tag{13.28}$$

We seek exponentially growing solutions, with rising motion for $|y| < a$ and sinking motion for $|y| > a$. Setting all dependent variables proportional to $\exp(\frac{1}{\sqrt{2}}\sigma t)$, we obtain by elimination :

$$\frac{\partial^2 v_2}{\partial y^2} - \frac{1}{\lambda^2}\left(\frac{2 + 2\sigma}{2 + \sigma - \eta}\right) v_2 = 0 \quad \left(\frac{\partial v_2}{\partial y}\right) > 0\ \ , \tag{13.29}$$

$$\frac{\partial^2 v_2}{\partial y^2} - \frac{1}{\lambda^2}\left(\frac{2 + 2\sigma}{2 + \sigma}\right) v_2 = 0\left(\frac{\partial v_2}{\partial y} < 0\right) \quad . \tag{13.30}$$

Mass continuity at $y = a$ gives the jump condition $[v_2] = 0$. Continuity of pressure, and therefore of temperature, gives

$$(\sigma + 2 - \eta)(\partial v_2/\partial y)(a-) = (\sigma + 2)(\partial v_2/\partial y)(a+). \tag{13.31}$$

The conditions at $y = - a$ follow from the anti-symmetry of v.

Equations (13.23)-(13.31) are solved by

$$v_2 = A(\sin y/\lambda_+)/(\sin a/\lambda_+), \ |y| < a \quad , \tag{13.32}$$

$$= A \exp(- |y - a|/\lambda_-), \ |y| > a \quad ,$$

$$\lambda_+ = \left(\frac{\eta - 2 - \sigma}{2 + 2\sigma}\right)^{1/2} \lambda \quad , \quad \lambda_- = \left(\frac{2 + \sigma}{2 + 2\sigma}\right)^{1/2} \lambda$$

providing σ satisfies the **eigenvalue** equation,

$$\frac{1}{\lambda_+} \tan \frac{a}{\lambda_+} = \frac{1}{\lambda_-} \tag{13.33}$$

It may be shown from (13.33) that σ increases monotonically from zero at $a/\lambda = (\eta/2 - 1)^{1/2}\tan^{-1}(\eta/2 - 1)^{1/2}$ to $\eta/2 - 1$ at $a/\lambda = 0$. Hence the criteria for instability are $\eta/2 > 1$ and $a/\lambda < (\eta/2 - 1)^{1/2}\tan^{-1}(\eta/2 - 1)^{1/2}$.

In the rain areas of the oceanic tropics $\eta = 2.14$. Thus $\sigma_{max} \simeq 0.07$, and the cut-off $a/\lambda = (0.07)^{1/2}\tan^{-1}(0.07)^{1/2} \simeq 0.07$. Taking $\phi = 15°$, we get $\lambda \sim 3000$ km, and the cut-off $a \sim 210$ km. This is indeed the right order of magnitude for the width of the rain area in a tropical depression or hurricane.

A schematic graph of u_1, v_1 and w_E is shown in Figure 13.1.

We see that the low-level shear vorticity is positive in the region of ascent ($w_E > 0$), and negative in the region of descent.

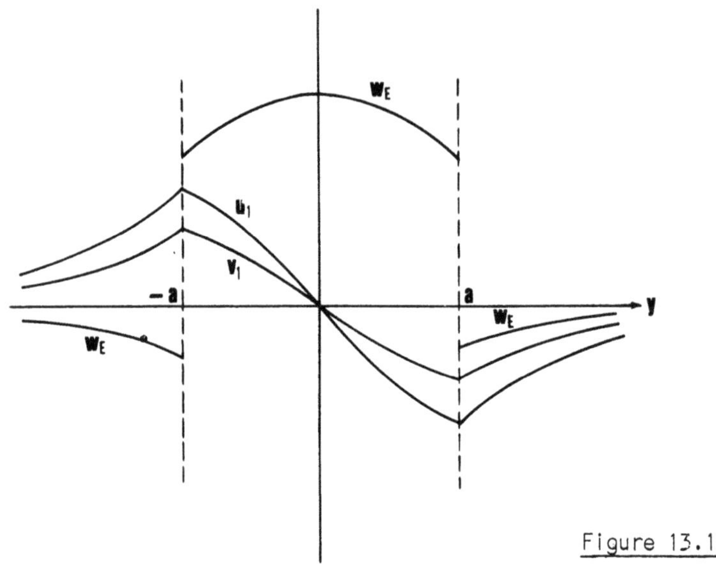

Figure 13.1

The Intertropical Convergence Zone

It was mentioned in the introduction that the ITCZ is a zone of cyclonic shear vorticity which paralleles the equator but lies at some distance from it. It was partly for this reason that the stability analysis was given for a zonally symmetric flow, the shear in the flow being expected to determine the symmetry of the disturbances which form in it. The existence of easterly winds in the tropics decreasing to near zero at the equator - and therefore of cyclonic shear vorticity - is a necessary property of the thermally-driven, zonally-symmetric circumpolar vortex. Disturbances in this region should therefore also have zonal symmetry, and the CISK process should be expected to intensify the cyclonic shear vorticity into a narrow zone, i.e., CISK should produce an ITCZ within the region of weak cyclonic vorticity. According to our analysis, its width should be approximately 300 km, and this is what is observed.

The process that we have described seems plausible, but many questions remain. Why is it that, contrary to popular notion, the zone of maximum tropical rainfall is not at the equator - at least not over the oceans where, unlike over continents, convection requires dynamical pumping of moisture ? A possible explanation comes from the nature of the CISK process itself. In dimensional form, $\sigma_{max} = f(\eta - 1)E^{1/2}$, and if $f = 2\Omega\sin\phi$ is now considered to vary parametrically, we get

$$\sigma_{max} \simeq (2\Omega\nu/H^2)^{1/2}(\eta/2 - 1)(\sin\phi)^{1/2}.$$

Thus σ_{max} vanishes at the equator because the efficiency of the Ekman pumping goes to zero. At the same time, q_s, and therefore η, strongly depends on temperature and decreases with increasing latitude until $\eta/2 - 1$ eventually becomes negative and CISK no longer possible. It appears to be for this reason that the ITCZ is close to, but not at, the equator.

A numerical calculation of the radiatively-driven, zonally-symmetric circulation for a finite amplitude two-level model of the kind used in the present analysis did indeed produce a sharp ITCZ near to, but not at, the equator.*

Cloud bands

Observations reveal that cumulus convection in the rain areas of tropical disturbances is not uniform but tends to be concentrated in bands of the order 20 - 50 km in width. The ITCZ itself usually consists of one or more such bands paralleling the equator. A remaining problem is to explain the origin of these bands. Here the clue may lie precisely in the fact that the CISK analysis gives no lower limit to the size of the region of convection. Do the geostrophic and Ekman dynamics we have assumed apply on scales as small as the band widths ? To answer this question, let us again consider a zonally symmetric perturbation but this time perturb a horizontally shearing flow, $\bar{u} = \bar{u}(y)$, such as we know exists in the ITCZ and is produced by the simple CISK process. The perturbation equations (13.7) - (13.11) remain the same as before, but (13.7) becomes

$$\partial u'/\partial t' - (f - \partial\bar{u}/\partial y)v' = 0 ,$$

* J. G. Charney, 1969 : The Intertropical Convergence Zone and the Hadley circulation of the atmosphere. Proc. WMO/IUGG Symp. on Num. Wea. Pred. in Tokyo, 1968, pp. III-73 - III-80, (published by Japan Meteorological Agency).

and we must solve an additional problem : to find the boundary-layer structure for the mean flow when the shear vorticity $-\partial\bar{u}/\partial y$ is no longer small relative to f. In general, the boundary layer equations are non-linear and parabolic (second-order in z and first-order in y) and cannot be solved except by numerical methods. However, in the case of uniform horizontal shear, the y-variation can be removed, and the ordinary differential equations in z are similar to those obtained by von Karman in the analysis of the boundary layer flow over a rotating plate. Solutions have been found which justify qualitatively the approximations to be made below.

We must ask : does the time-scale for the adjustment of the boundary layer remain small compared to the time scale of the modified CISK process, and does the zonal flow remain quasi-geostrophic ? A naive Oseen type of approximation which simply replaces f by $Z = f-\partial\bar{u}/\partial y$ in the x-momentum equation, both above and within the frictional boundary layer, can be justified qualitatively. The new space and time scales are then obtained by replacing f by $(fZ)^{1/2}$. The modified Ekman depth becomes $\sim \nu^{1/2}(fZ)^{-1/4}$, the boundary layer adjustment time $(fZ)^{-1/2}$, and the spin-up time $(fZ)^{-1/2}E^{-1/2}$, where E is now $\nu(fZ)^{-1/2}H^{-2}$. The ratio of the boundary-layer adjustment time to the spin-up time thus becomes even smaller since $Z > f$. The ratio of the relative acceleration to Coriolis force in the y-momentum equation is O(E) as before, and therefore also smaller, so the zonal flow remains quasi-geostrophic. Thus the CISK process continues to apply, and it is plausible that the cloud bands themselves are a manifestation of a modified CISK process. The ultimate limitation in size may then be due to limitations on geostrophy which have not been considered, to shear instabilities in the bands, or, as seems more likely, to the fact that the quasi-Ekman dynamics no longer operates in the presence of concentrated cumulus convection.

BIBLIOGRAPHY OF HYDRODYNAMICAL STABILITY STUDIES RELEVANT TO GEOPHYSICAL FLUID MECHANICS

Arnason, G., 1963 : The stability of nongeostrophic perturbations in a baroclinic zonal flow. Tellus XV, 3, 205-229.

Barcilon, V., 1964 : Role of the Ekman layers in the stability of the symmetric regime obtained in a rotating annulus. J. Atmos. Sci., 21, 291-299.

Bates, J. R., 1970 : Dynamics of disturbances of the Intertropical Convergence Zone. Quart. J. Roy. Meteor. Soc., 96, 677-701.

Bénard, M., 1901 : Les tourbillons cellulaires dans une nappe liquide transportant de la chaleur par convection en régime permanent. Ann. Chim. (Phys.), 23, 62-144.

Bezold, W. von, 1888 : Zür Thermodynamik der Atmosphäre. Pts. I, II, Sitz. K. Preuss. Akad. Wissensch. pp. 485-522, 1189-1206 ; Gesammelte Abhandlungen, 91-144. Also in the Mechanics of the Earth's Atmosphere (see reference under "Helmholtz, 1889").

Bjerknes, J., 1938 : Saturated-adiabatic ascent of air through dry-adiabatically descending environment. Quart. J. R. Meteor. Soc., 64, 275, 325-330.

Bretherton, F. P., 1966a : Critical layer instability in baroclinic flows. Quart. J. R. Meteor. Soc., 92, 325-334.

Bretherton, F. P., 1966b : Baroclinic instability and the short wavelength cut-off in terms of potential vorticity. Quart. J. R. Meteor. Soc., 92, 335-345.

Brown, J. A., 1968 : A numerical investigation of hydrodynamic instability and energy conversions in the quasi-geostrophic atmosphere. Ph. D. thesis. University of Colorado.

Burger, A. P., 1962 : On the non-existence of critical wavelengths in a continuous baroclinic stability problem. J. Atmos. Sci., 19, 31-38.

Burger, A. P., 1966 : Instability associated with the continuous spectrum in a baroclinic flow. J. Atmos. Sci., 23, 272-277.

Chandrasekhar, S., 1953 : The instability of a layer of fluid heated below and subject to Coriolis forces. Proc. Roy. Soc., A 217, 306-327.

Chang, C.-P., 1971 : On the stability of low-latitude quasi-geostrophic flow in a conditionally unstable atmosphere. J. Atm. Sci., 28, 270-274.

Charney, J. G., 1947 : The dynamics of long waves in a baroclinic westerly current. J. Meteor., 4, 5, 135-162.

Charney, J. G. and Stern M. E., 1962 : On the stability of internal baroclinic jets in a rotating atmosphere. J. Atmos. Sci., 19, 159-172.

Charney, J. G. and Eliassen A., 1964 : On the growth of the hurricane depression. J. Atmos. Sci., 21, 68-75.

Derome, J. and Dolph C. L., 1970 : Three-dimensional non-geostrophic disturbances in a baroclinic zonal flow. Geo. Fluid Dyn. 1, 1, 91-122.

Eady, E. J., 1949 : Long waves and cyclone waves. Tellus, 1, 3, 33-52.

Eliassen, E., 1960 : On the initial development of frontal waves, Publ. Danske Met. Inst., 13, 107 pp.

Faller, A. J. and Kaylor R. E., 1966 : A numerical study of the instability of the laminar Ekman boundary layer. J. Atmos. Sci., 23, 466-80.

Fjørtoft, R., 1944 : On the frontogenesis and cyclogenesis in the atmosphere. Pt. 1. On the stability of the stationary circular vortex. Geof. Publ., 16, 5, 27 pp.

Fjørtoft, R., 1950 : Application of integral theorems in deriving criteria of stability for laminar flows and for the baroclinic circular vortex. Geof. Publ., 17, 6, 52 pp.

Goldstein, S., 1931 : On the stability of superposed streams of fluids different densities. Proc. Roy. Soc., A132, 524-548.

Green, J. S. A., 1960 : A problem in baroclinic stability. Quart. J. R. Meteor. Soc., 86, pp. 237-251.

Halley, E., 1686 : An historical account of the trade winds and monsoons by E. Halley. Phil. Trans. Roy. Soc., 26, 153-168.

Haque, S. M. A., 1952 : The initiation of cyclonic circulation in a vertically unstable stagnant air mass. Quart. J. R. Meteor. Soc., 78, 337, 394-406.

Heisenberg, W., 1924 : Über Stabilität und Turbulenz von Flussigkeitsströmen. Ann. Phys., Lpz., 4, 74, 577-627.

Helmholtz, H. von, 1888 : Uber atmosphärische Bewegungen. I. Sitzungsber. Gesselsch. Wissensch. Berl. 647-663. Also in The Mechanics of the Earth's Atmosphere. A collection of translations by C. Abbe published by the Smithsonian Institution, in Smithsonian Miscellaneous Collections (2 vols.), 78-93.

Helmholtz, H. von, 1889 : Über atmosphärische Bewegungen. II. Sitzungsber. Gessellsch. Wissensch. Berl. 761-780. Also in The Mechanics of the Earth's Atmosphere. A collection of translations by C. Abbe published by the Smithsonian Institution, in Smithsonial Miscellaneous Collections (2 vols.), 94-129.

Höiland, E., 1939 : On the interpretation and application of the circulation theorems of V. Bjerknes. Arch. Math. og Naturv., 42, 5, 68 pp.

Höiland, E., 1941 : On the stability of the circular vortex. Avh. Norske Videnskaps-akad. Oslo, Nat.-naturv. Kl., 1, 11, 24 pp.

Howard, L. N., 1961 : Note on a paper of John W. Miles. J. Fluid Mech., 10, 509-512.

Jeffreys, H. 1926 : The stability of a layer of fluid heated below. Phil. Mag. 7, 2, 833-44.

Kelvin, Lord, 1971 : Influence of wind and capillarity on waves in water supposed frictionless. Mathematical and Physical Papers, iv, 76-85, Cambridge University Press, (1910).

Kochin, N., 1932 : Über die Stabilität von Margulesschen Diskontinuitatsflachen Beitr. Phys. Frei. Atmos., 18, 129-164.

Kuo, H-L., 1949 : Dynamic instability of two dimensional non-divergent flow in a barotropic atmosphere. J. Meteor., 6, 2, 105-122.

Kuo, H-L., 1952 : Three-dimensional disturbances in a baroclinic zonal current. J. Meteor., 9, 260-278.

Lilly, D. K., 1960 : On the theory of disturbances in a conditionally unstable atmosphere. Mon. Wea. Rev., 88, 1, 1-18.

Lilly, D. K., 1966 : On the instability of the Ekman boundary layer, J. Atmos. Sci., 23, 481-494.

Lin, C. C., 1945 : On the stability of two-dimensional parallel flows. Parts I, II, III. Q. Appl. Math., 3, 117-42, 218-34, 277-301.

McIntyre, M. E., 1967 : Convection and baroclinic instability in rotating fluids. Ph.D. thesis, University of Cambridge.

McIntyre, M. E., 1970a : Diffusive destabilization of the baroclinic circular vortex. Geophys. Fluid Dyn. 1, 19-52.

McIntyre, M. E., 1970b : On the non-separable baroclinic parallel flow instability problem. J. Fluid Mech., 40, 2, 273-306.

McIntyre, M. E., 1970c : Baroclinic instability of Murray's continuous model of the polar night jet. (Unpublished).

Miles, J. W., 1957 : On the generation of surface waves by shear flows. J. Fluid Mech., 3, 185-204.

Miles, J. W., 1959 : On the generation of surface waves by shear flows. J. Fluid Mech., 6, 583-598.

Miles, J. W., 1960 : On the generation of surface waves by turbulent shear flows. J. Fluid Mech., 7, 459-480.

Miles, J. W., 1961 : On the stability of heteorogeneous shear flows. J. Fluid Mech., 10, 496-508.

Miles, J. W., 1964a : Baroclinic instability of the zonal wind. Rev. of Geophys., 2, 1, 155-176.

Miles, J. W., 1964b : A note on Charney's model of zonal-wind instability. J. Atmos. Sci., 21, 4, 451-452.

Miles, J. W., 1964c : Baroclinic instability of the zonal wind : Part II. J. Atmos. Sci., 21, 5, 500-506.

Miles, J. W., 1964d : Baroclinic instability of the zonal wind : Part III. J. Atmos. Sci., 21, 6, 603-609.

Miles, J. W., 1965 : Effects of diffusion on baroclinic instability of the zonal wind. J. Atmos. Sci., 22, 2, 146-151.

Orlanski, I., 1968 : Instability of frontal waves. J. Atmos. Sci., 25, 2, 178-200.

Ooyama, K., 1966 : On the stability of the baroclinic circular vortex. J. Atmos. Sci., 23, 6, 43-53.

Orr, W. McF., 1906 : The stability or instability of the steady motions of a liquid. Proc. R. Irish Acad. A 27, 9-27, 69-138.

Pedlosky, J., 1964a : The stability of currents in the atmosphere and the ocean : Part I. J. Atmos. Sci., 21, 2, 201-219.

Pedlosky, J., 1964b : The stability of currents in the atmosphere and the ocean : Part II. J. Atmos. Sci, 21, 4, 342-353.

Pedlosky, J., 1964c : An initial value problem in the theory of baroclinic instability. Tellus, 16, 12-17.

Pedlosky, J., 1965 : On the stability of baroclinic flows as a functional of the velocity profile. J. Atmos. Sci., 22, 2, 137-145.

Rayleigh, Lord, 1880 : On the stability,or instability, of certain fluid motions. Scientific Papers, 1, 474-487, Cambridge University Press.

Rayleigh, Lord, 1882 : On the question of the stability of the flow of fluids. Scientific Papers. 3, 575-584, Cambridge University Press.

Rayleigh, Lord, 1916a : On convection currents in a horizontal layer of fluid when the higher temperature is on the under side. Scientific Papers. 6, 432-446, Cambridge University Press.

Rayleigh, Lord, 1916b : On the dynamics of revolving fluids. Scientific Papers. 6, 447-453, Cambridge University Press.

Richardson, L. F., 1920 : The supply of energy from and to atmospheric eddies. Proc. Roy. Soc., A 97, 686, 354-373.

Rumford, Count, 1797 : Of the propagation of heat in fluids. In Complete Works, 1, p. 239, Amer. Acad. Arts and Sciences, Boston, 1870.

Solberg, H., 1928 : Integrationen der Atmosphärischen Störengsgleichungen, Geof. Publ., 5, 9, 120 pp.

Solberg, H., 1936 : Le mouvement d'inertie de l'atmosphère stable et son rôle dans la théorie des cyclones. Proc. -verb. Assoc. météor. U.G.G.I. (Edinburgh), pt. II (Mém.), 66-82.

Sommerfeld, A. 1908 : Ein Beitrag zur hydrodynamischen Erklärung der turbulent Flüssigkeitsbewegung. Proc. 4th Int. Congr. Math. Rome, 116-124.

Stone, P. H., 1966 : On non-geostrophic baroclinic stability. J. Atmos. Sci., 23, 4, 390-400.

Stone, P. H., 1969 : The meridional structure of baroclinic waves. J. Atmos. Sci., 26, 3, 376-389.

Stone, P. H., 1970 : On non-geostrophic baroclinic stability : Part II. J. Atmos. Sci., 27, 5, 721-726.

Taylor, G. I., 1915 : Eddy motion in the atmosphere. Phil. Trans. Roy. Soc., A 215, 1-26.

Taylor, G. I., 1923 : Stability of a viscous liquid contained between two rotating cylinders. Phil. Trans. Roy. Soc., A 223, 289-343.

Taylor, G. I., 1931 : Effect of variation in density on the stability of superposed streams of fluid. Proc. Roy. Soc., A 132, 499-523.

Tollmien, W., 1935 : Ein allgemeines Kriterium der Instabilität laminarer Geschwindigkeitsverteilungen. Machr. Ges. Wiss. Gootingen, Math.-phys. Klasse, 50, 79-114.

Yamasaki, M., 1969 : Large-scale disturbances in the conditionally unstable atmosphere at low-latitudes. Papers in Meteor. and Geophys. 290-336. Meteor. Res. Inst., Tokyo.

LECTURES IN SUB-SYNOPTIC SCALES OF MOTION AND TWO-DIMENSIONAL TURBULENCE

D.K. LILLY

National Center for Atmospheric Research
BOULDER, Colorado (U.S.A.)

CONTENTS

LECTURES IN SUB-SYNOPTIC SCALES OF MOTION AND TWO-DIMENSIONAL TURBULENCE

By

D. K. LILLY

National Center for Atmospheric Research
Boulder, Colorado 80302 - U.S.A.

I. INTRODUCTION

It has long been recognized that the atmosphere shares certain attributes with those of a turbulent fluid. Its apparent randomness and unpredictability on many scales combined with long term statistical order are just about what one would expect to observe if he were, say, an ant dwelling amongst the eddies of a turbulent pipe flow. Until recently, however, little use has been made of this sort of analogy except in the case of the surface boundary layer. The principal reason was first (before about World War II) the non-recognition by meteorologists of the special characteristics of the large scale atmosphere as a quasi-two-dimensional fluid and later the non-recognition by fluid dynamicists of the meaningfulness and general characteristics of two-dimensional turbulence.

At one time it was thought that the large scale disturbances of the atmosphere were shearing instability phenomena deriving at least some of their energy from the mean westerlies and easterlies. Jeffreys (1926) pointed out, however, that the very existence of mid-latitude and polar westerlies at the surface required that non-symmetric circulations exist capable of transporting angular momentum poleward, i.e. toward regions of westerly flow. Although he did not carry the argument further, it can be shown that his analysis implies a transfer of kinetic energy from the disturbances to the mean flow.Much later, after Fjørtoft's (1953)explanation of the general tendency of two-dimensional flows to transfer energy from small scales to large, it was assumed that the large scale atmosphere was not turbulent in any proper sense. And yet it is certainly not steady-state or periodic and it seems to be often unstable to perturbations. When Benton and Kahn (1958) and Ogura (1958) looked at the energy spectrum of the large scale atmosphere they found that it apparently followed a power law in wave number (near -3) but not the one predicted for and observed in three-dimensional turbulence.

A major part of this series of lectures is devoted to describing the recent developments in two-dimensional turbulence theory and its atmospheric applications which are bringing to atmospheric dynamics a higher degree of integration with other areas of fluid mechanics.

A principal recent thrust in atmospheric dynamics is the growing use of numerical simulation methods, not only in practical and experimental forecasting but as an aid in solving fundamental problems in fluid dynamics. Because most atmospheric scales of motion involve turbulence the simulation modelers must attempt to deal with the closure problem of turbulence theory, but in a somewhat different form from that familiar in laboratory-oriented fluid mechanics. Some current developments in this area will be reviewed.

Finally, the last lecture of this series will review the current problems and prospects in certain mesoscale phenomena, usually involving turbulence, which constitute some of the most serious blocks to further development of large scale atmospheric simulation models.

2. HOMOGENEOUS TURBULENCE THEORY - CONCEPTS AND NOTATION

In the first few lectures of this series we will be dealing with aspects of the theory of turbulence in two and three dimensions. For this purpose it is convenient to introduce some standard notation and concepts from the theory of homogeneous turbulence. Much of this material, especially for the three-dimensional case, is quoted or directly derived from Batchelor (1956). As does Batchelor we will use mixed tensor and vector expressions. Many of the definitions and derivations can also be found in texts by Hinze (1959) and Lumley and Panofsky (1964).

We first define a velocity field as a function of space and time, i.e.

$$u_i = u_i(\vec{x},t), \quad \vec{x} = \{x_1, x_2, x_3\} \quad , \quad i = 1,2,3. \qquad (2.1)$$

In general this velocity field, or at least parts of it, is expected to be somewhat random in nature. It is therefore desirable to deal with statistical moments and their Fourier transforms. In particular we define the two-point tensor velocity correlation as

$$R_{ij}(\vec{x},\vec{r}) = \overline{u_i(\vec{x}-\vec{r}/2)\, u_j(\vec{x}+\vec{r}/2)} \ , \qquad (2.2)$$

where the over-bar represents an average over all x-space. If the turbulence can be considered spatially homogeneous the above correlation is a function of the separation vector only, that is

$$R_{ij}(\vec{x},\vec{r}) = R_{ij}(\vec{r}) = \overline{u_i(\vec{x})\, u_j(\vec{x}+\vec{r})} \ . \qquad (2.3)$$

We deal generally with incompressible flows, so that $\partial u_i/\partial x_i = 0$. From this it can be easily verified that

$$\frac{\partial R_{ij}(\vec{r})}{\partial r_i} = \frac{\partial R_{ij}(\vec{r})}{\partial r_j} = 0 \ . \qquad (2.4)$$

From the general theory of tensors it can be shown that if R_{ij} is isotropic (i.e. invariant to an arbitrary rotation and to reflection), it can be expressed in terms of two functions, F and G, of the scalar separation distance, i.e.

$$R_{ij}(\vec{r}) = F(r)\, r_i r_j + G(r)\, \delta_{ij} \ . \qquad (2.5)$$

Upon differentiating Eqn. (2.5) and applying (2.4) we can show that the two functions are not independent in an incompressible flow, i.e.

$$\frac{\partial R_{ij}(\vec{r})}{\partial r_i} = r_j \left[(n+1)F + r\frac{dF}{dr} + \frac{1}{r}\frac{dG}{dr}\right] = 0 \ , \qquad (2.6)$$

where n is the number of dimensions present in the physical framework (2 or 3). The most convenient combination of F and G for many purposes is that which appears when i and j are equal and summed, that is

$$R(r) = \tfrac{1}{2}R_{ii}(\vec{r}) = \tfrac{1}{2}(r^2F + nG) \tag{2.7}$$

The quantity R(r) is called the scalar velocity correlation function.

It is also convenient to define a vorticity correlation function $Z_{ij}(\vec{r})$ as

$$Z_{ij}(\vec{r}) = \overline{\omega_i(\vec{x})\,\omega_j(\vec{x}+\vec{r})} \tag{2.8}$$

where ω_i is the ith component of vorticity. Batchelor (1956) shows that $Z_{ij}(\vec{r})$ is related to $R_{ij}(\vec{r})$ by the equation

$$Z_{ij}(\vec{r}) = \nabla^2 R_{ji}(\vec{r}) + \left[\frac{\partial^2}{\partial r_i \partial r_j} - \delta_{ij}\nabla^2\right] R_{kk}(\vec{r}) , \tag{2.9}$$

where $\nabla^2 = \partial^2/\partial r_k^{\,2}$. A scalar vorticity correlation function, Z(r), can also be defined analogously to the scalar velocity correlation function as

$$Z(r) = \tfrac{1}{2}Z_{ii}(\vec{r}) = -\nabla^2 R(r) . \tag{2.10}$$

From a system of hydrodynamic equations, such as the Navier-Stokes equations, it is possible to derive expressions for the rates of change of the above correlation functions. Certain terms involving the pressure become rather difficult to evaluate, however, and this difficulty has led to widespread use of the Fourier transforms of the correlations, the equations for which do not present such difficulties. In Batchelor's notation the Fourier transform of the tensor velocity correlation function is given by

$$\Phi_{ij}(\vec{k}) = \frac{1}{(2\pi)^n} \int R_{ij}(\vec{r})\, e^{-i\vec{k}\cdot\vec{r}}\, d\vec{r} , \tag{2.11}$$

with the inverse transform

$$R_{ij}(\vec{r}) = \int \Phi_{ij}(\vec{k})\ e^{i\vec{k}\cdot\vec{r}}\ d\vec{k}\ , \tag{2.12}$$

where $d\vec{r} = dr_1\ dr_2\ dr_3$, etc. Since R_{ij} is real, $\Phi_{ij}(-\vec{k}) = \Phi_{ij}{}^*(\vec{k})$. We may note that the velocity correlation for zero separation is given by the integral of Φ_{ij}, that is

$$R_{ij}(0) = \overline{u_i(\vec{x})u_j(\vec{x})} = \int \Phi_{ij}(\vec{k})\ d\vec{k}\ , \tag{2.13}$$

and therefore that the kinetic energy is expressed as

$$\overline{u_i^2}/2 = {}^1\!/_2 \int \Phi_{ii}(\vec{k})\ d\vec{k}\ . \tag{2.14}$$

In the isotropic case Φ_{ii} is a function only of the amplitude of its argument, so that

$$\overline{u_i^2} = \begin{cases} 4\pi \int k^2\Phi_{ii}(k)\ dk\ , & \text{3 dimensions} \\ 2\pi \int k\Phi_{ii}(k)\ dk\ , & \text{2 dimensions} \end{cases} . \tag{2.15}$$

We then may define the scalar energy spectrum function E(k) as the energy on a circle (two-dimensional) or spherical shell (three-dimensional) in wave vector space. Thus in isotropic turbulence

$$E(k) = \begin{cases} 2\pi k^2 \Phi_{ii}(k), & \text{three-dimensional} \\ \pi k\Phi_{ii}(k)\ , & \text{two-dimensional} \end{cases} \text{ and } \overline{u_i^2}/2 = \int E(k)\ dk\ . \tag{2.16}$$

Equations may also be derived directly relating the scalar velocity correlation function R(r) and the scalar energy spectrum function E(k). From Eqns. (2.7), (2.12), and (2.16) one may write for the three-dimensional case

$$R(r) = \int \frac{E(k)}{4\pi k^2}\ e^{i\vec{k}\cdot\vec{r}}\ d\vec{k} = \int_0^\infty \int_0^\pi \tfrac{1}{2}E(k)\ e^{ikr\cos\theta}\ \sin\theta\ d\theta\ dk \tag{2.17}$$

c

where the integration element $d\vec{k}$ has been taken as $2\pi k^2 \sin\theta \, d\theta \, dk$ in spherical coordinates with θ the angle between $\vec{r}$ and $\vec{k}$, measured from the direction of $\vec{r}$. Integration of (2.17) over θ leads to the expression

$$R(r) = \int_0^\infty \frac{\sin kr}{kr} E(k) \, dk \text{ , three dimensions} \tag{2.18}$$

A similar derivation in the two-dimensional case yields

$$R(r) = \int_0^\infty J_o(kr) \, E(k) \, dk \text{ , two dimensions} \tag{2.19}$$

where J_o is the zeroth order Bessel function. The inverse expressions are given by

$$E(k) = \frac{2}{\pi} \int_0^\infty kr \sin kr \, R(r) \quad dr\text{, three dimensions} \tag{2.20}$$

$$E(k) = \int_0^\infty kr \, J_o(kr) \, R(r) \quad dr\text{, two dimensions .} \tag{2.21}$$

A tensor vorticity spectrum function may be defined analogously to the tensor velocity correlation function as

$$\Omega_{ij}(\vec{k}) = \frac{1}{(2\pi)^n} \int Z_{ij}(\vec{r}) \, e^{-i\vec{k}\cdot\vec{r}} \, d\vec{r} \quad , \tag{2.22}$$

where $Z_{ij}(\vec{r})$ is the vorticity correlation function. Upon substitution of (2.9) and two successive integrations by parts (and assuming that $R_{ij}(\vec{r}) \to 0$ as $r \to \infty$) we find that $\Omega_{ij}(\vec{k})$ is related to $\Phi_{ij}(\vec{k})$ by the equation

$$\Omega_{ij}(\vec{k}) = (k^2\delta_{ij} - k_i k_j) \, \Phi_{\ell\ell}(\vec{k}) - k^2\Phi_{ji}(\vec{k}) \tag{2.23}$$

In the summation case we obtain

$$\Omega_{ii}(\vec{k}) = k^2 \Phi_{ii}(\vec{k}) \; . \tag{2.24}$$

The function Ω_{ii} is related to enstrophy (half the squared vorticity) similarly to the relationship between Φ_{ii} and energy, i.e.

$$\overline{\omega_i^2}/2 = \tfrac{1}{2} \int \Omega_{ii}(\vec{k}) \quad d\vec{k} \quad . \tag{2.25}$$

Thus in isotropic turbulence we can define the scalar enstrophy spectrum function $\Omega(k)$ as the contribution to enstrophy on a circle or spherical shell in wave space. But from (2.23) and (2.24) it is evident that

$$\Omega(k) = k^2 E(k) \quad \text{and} \quad \overline{\omega_i^{\,2}}/2 = \int k^2 E(k) \; dk \; . \tag{2.26}$$

Equation (2.26) is also generalizable to higher moments, so that, e.g.

$$\overline{(\partial\omega_i/\partial x_i)^2}/2 = \int k^4 E(k) \; dk \; . \tag{2.27}$$

In all the above relations an infinite spatial and wave number domain has been assumed, with some corresponding special mathematical difficulties discussed by Batchelor. In the case of a finite volume (L^n) and resolution (N^n Fourier amplitudes or mesh points), as is usual in numerical simulation studies, similar definitions hold but with the integrals replaced by finite sums, i.e.

$$\Phi_{ij}(\vec{k}) = (L/2\pi N)^n \sum_{r \le L} R_{ij}(\vec{r}) \, e^{-i\vec{k}\cdot\vec{r}} \tag{2.28}$$

$$R_{ij}(\vec{r}) = (\pi/L)^n \sum_{k \le \pi N/L} \Phi_{ij}(\vec{k}) \, e^{i\vec{k}\cdot\vec{r}} \; . \tag{2.29}$$

In this case the practical way to obtain the spectrum functions is by direct Fourier analysis, i.e.

$$u_i(\vec{k}) = (L/2\pi N)^n \sum_{x \le L} u_i(\vec{x}) \, e^{-i\vec{k}\cdot\vec{x}}, \text{ where } u_i(-\vec{k}) = u_i^*(\vec{k}) \tag{2.30}$$

and the converse

$$u_i(\vec{x}) = (\pi/L)^n \sum_{k \le \pi N/L} u_i(\vec{k}) \, e^{i\vec{k}\cdot\vec{x}} \; . \tag{2.31}$$

We then find that

$$\Phi_{ij}(\vec{k}) = (\pi/L)^n \, u_i^*(\vec{k}) \, u_j(\vec{k}) \; . \tag{2.32}$$

The scalar energy spectrum function $E(k)$ is half the integral of $\Phi_{ii}(\vec{k})$ around a shell or circle in wave space. Since an arbitrarily defined shell or circle will not coincide with the ends of more than a few wave vectors in a discrete system it is necessary to estimate E(k) by its average over a finite band width. Sometimes it is more convenient to arbitrarily define E(k) to be the integral around a square or cube in k-space rather than a circle.

3. THE MICRO-SCALES OF TURBULENCE

An important aspect of turbulence in a viscous two- or three-dimensional fluid is its character at the smallest spatial scales. We confine ourselves to incompressible constant density fluids, but the principal results hold in slightly altered form for a barotropic gas at low Mach number.

We will assume the usual Navier-Stokes equations of motion and incompressibility in the form

$$\frac{\partial u_i}{\partial t} + u_j \frac{\partial u_i}{\partial x_j} + \frac{\partial (p/\rho)}{\partial x_i} = \nu \frac{\partial^2 u_i}{\partial x_j{}^2} , \qquad (3.1)$$

$$\frac{\partial u_i}{\partial x_i} = 0 , \qquad (3.2)$$

where density ρ is assumed to be constant. The vorticity equation is obtained by taking the curl of (3.1) in the form

$$\frac{\partial \omega_i}{\partial t} + u_j \frac{\partial \omega_i}{\partial x_j} - \omega_j \frac{\partial u_i}{\partial x_j} = \nu \frac{\partial^2 \omega_i}{\partial x_j{}^2} \qquad (3.3)$$

where $\omega_i = \varepsilon_{ijk} \frac{\partial u_k}{\partial x_j}$, with ε_{ijk} the unit alternating tensor, and other notations are standard. By multiplying (3.1) and (3.3) respectively by u_i and ω_i and applying (3.2) to each, we may obtain equations for kinetic energy and enstrophy as follows :

$$\frac{\partial (u_i{}^2/2)}{\partial t} = - \frac{\partial}{\partial x_j} (u_j u_i{}^2/2) - \frac{\partial}{\partial x_i} (u_i p/\rho) + \nu \frac{\partial^2 (u_i{}^2/2)}{\partial x_j^2} - \nu \left[\frac{\partial u_i}{\partial x_j}\right]^2 \qquad (3.4)$$

$$\frac{\partial(\omega_i^2/2)}{\partial t} = -\frac{\partial}{\partial x_j}(u_j\omega_i^2/2) + \omega_i\omega_j\frac{\partial u_i}{\partial x_j} + \nu\frac{\partial^2(\omega_i^2/2)}{\partial x_j^2} - \nu\left[\frac{\partial\omega_i}{\partial x_j}\right]^2 \qquad (3.5)$$

We note that the kinetic energy is <u>conserved</u> except for viscous terms, i.e. if Eqn. (3.4) is integrated over a closed or periodic or infinite volume, all terms on the right vanish except the last viscous dissipation term, which is always negative. The enstrophy is not so conserved because of the presence of the second term on the right of (3.5). If the flow is confined to two dimensions, however, this term disappears and the enstrophy is also conserved except for viscous dissipation. We symbolize these two important dissipation parameters by

$$\text{energy dissipation}: \quad \varepsilon = \overline{\nu(\partial u_i/\partial x_j)^2} = \overline{\nu\omega_i^2} \quad , \qquad (3.6)$$

$$\text{enstrophy dissipation}: \quad \eta = \overline{\nu(\partial\omega_i/\partial x_j)^2} \; , \qquad (3.7)$$

where the over-bar represents time, space and/or ensemble averaging. The second equality in (3.6) is a kinematic identity for homogeneous incompressible turbulence. In applying definition (3.7) we will usually assume that the motion field is confined to a plane and ω_i reduced to a scalar, ω.

It has long been recognized that the viscous terms of Eqns.(3.1) and (3.3), because they are of the highest differential order, should be most important at the smaller spatial scales. A more quantitative appraisal may be obtained by performing a formal scale analysis of these equations. We let L, U, Z, P and T represent scaling parameters for distance, velocity, vorticity, pressure and time respectively. If L is to represent the scale at which the nonlinear and viscous terms of (3.1) and (3.3) are of the same magnitude, we then find that

$$UL/\nu = 1 \; , \; P = \rho U^2, \quad T = L/U. \qquad (3.8)$$

It is then convenient to introduce ε and η as parameters by making (3.6) and (3.7) dimensionless, so that

$$\nu U^2/L^2 = \varepsilon = \nu Z^2 , \tag{3.9}$$

and
$$\nu Z^2/L^2 = \eta . \tag{3.10}$$

From (3.8) and (3.9) we then deduce the usual three-dimensional Kolmogoroff micro-scales (dissipation scales) of velocity, v_λ, and length, λ, and the corresponding pressure and time scales, as

$$\begin{aligned} \lambda &= L = (\nu^3/\varepsilon)^{1/4}, \quad v_\lambda = U = (\nu\varepsilon)^{1/4} , \\ P/\rho &= (\nu\varepsilon)^{1/2} , \quad T = (\nu/\varepsilon)^{1/2} . \end{aligned} \tag{3.11}$$

From (3.9) we also have

$$Z = v_\lambda/\lambda = (\varepsilon/\nu)^{1/2} . \tag{3.12}$$

Then (3.10), (3.11) and (3.12) combine to form alternative two-dimensional micro-scale definitions, i.e.

$$\lambda = (\nu^3/\eta)^{1/6} , \quad v_\lambda = (\nu^3\eta)^{1/6} . \tag{3.13}$$

The micro-scale definitions (3.11) and (3.13) are completely equivalent, but the former is more useful in three-dimensional turbulence and the latter more useful in two dimensions. The reason lies in the interpretation and measurement of ε and η with respect to the total flow field.

In three dimensions kinetic energy is conservative in the absence of forcing and viscosity, so that ε represents a net energy loss from the flow. Enstrophy is not conserved, however, being continually generated by vortex stretching processes, so that its dissipation does not necessarily represent a net loss from the flow and is not measurable as a large scale property. In two dimensions enstrophy and energy are both conserved except for the viscous dissipation. But if there are many scales of motion present, i.e. if the energy spectrum is broad, then we will show that the fractional rate of dissipation of energy is small compared to that of enstrophy, and the latter is thus the more significant parameter. To show this we define moments of the scalar energy spectrum as follows

$$k_2{}^2 = \frac{\overline{\omega^2}/2}{\overline{E}} = \frac{\int k^2\, E(k)\, dk}{\int E(k)\, dk}\ , \quad k_4{}^4 = \frac{\overline{(\nabla\omega)^2}}{\overline{E}} = \frac{\int k^4\, E(k)\, dk}{\int E(k)\, dk}\ . \qquad (3.14)$$

From Equations (3.4) to (3.7) we then write the mean energy and enstrophy equations in two dimensions as

$$\begin{aligned} d\overline{E}/dt &= -\nu k_2{}^2\overline{E} \\ \frac{d}{dt}(k_2{}^2\overline{E}) &= -\nu k_4{}^4\overline{E}\ . \end{aligned} \qquad (3.15)$$

From these we see that

$$\frac{d(\ln \overline{E})}{dt} = -\nu k_2{}^2\ , \quad \frac{d(\ln \overline{\omega^2}/2)}{dt} = \frac{\nu k_4{}^4}{k_2{}^2}\ . \qquad (3.16)$$

By Schwarz's inequality, $k_4{}^2 \geq k_2{}^2$ in any case, but if $E(k)$ has significant amplitude over a wide range of k (for example any power law) then $k_4{}^2 >> k_2{}^2$ and the second expression in (3.16) is much larger in magnitude.

4. THE INERTIAL RANGES

It is commonly believed that in flow of sufficiently large Reynolds number there exists a range of scales of motion which are intermediate between and not directly affected by either large scale forcing (or initial large scale state of flow) or viscous decay processes. In this range, the "inertial range", the flow is expected to be random and isotropic, and its principal function is to transfer energy (in three dimensions) and enstrophy (in two dimensions) from the large scales down to those of viscous dissipation. It is postulated that certain gross statistics of the flow (correlation functions, energy spectra) have a form which depends only on the relevant dissipation parameter. In particular the energy spectra are predicted to conform to the relations

$$E(k) = \alpha\varepsilon^{2/3} k^{-5/3} \qquad \text{(three dimensions)} \qquad (4.1)$$

$$E(k) = \beta\eta^{2/3} k^{-3} \qquad \text{(two dimensions)} \qquad (4.2)$$

where α and β are supposedly universal dimensionless constants. The qualitative and semi-quantitative arguments used to justify Eqn. (4.1) can be found in Batchelor (1956) and those for Eqn. (4.2) in papers by Kraichnan (1967) and Batchelor (1969).

A great deal of work has been done in attempting to derive (4.1), (4.2), and the constants from basic principles. The basic justification for their survival, however, remains that they -especially (4.1)- fit observational data remarkably well (Lilly and Panofsky, 1967). Many atmospheric and oceanographic scientists in small scale turbulence research tend now to test the calibration of new instruments and techniques on how well they produce a -5/3 power law. The -3 power law is apparently rather well satisfied by the large scale atmosphere, for planetary wave numbers greater than about 5-6 (Wiin-Nielsen, 1967 ; Julian, et al., 1970) which suggests that the large scale atmosphere may be regarded as a quasi-two-dimensional turbulent fluid. A number of numerical simulation experiments have been conducted for the purpose of testing the -3 power law prediction, and these have produced for the most part positive verification.

Probably the simplest of these simulation tests to interpret are those done on decaying two-dimensional flow (Lilly, 1971). These are started by introducing forcing for a short time to produce a quasi-periodic flow field. The forcing is then removed and Eqn. (3.3) (without the third term on the left) is integrated numerically for over 1000 time steps. The results are satisfyingly random and exhibit (see Figure 4.1) several octaves of a -3 power law spectrum. The constant β is found to be of order 2, which compares to values of $\alpha \sim 1.5$ from geophysical turbulence data analysis.

The most unique feature of two-dimensional turbulence is its energy trapping or blocking characteristic. From Eqn. (3.16) we saw that energy dissipation tends to be negligible compared to enstrophy dissipation for large Reynolds number flow and that the scale defined by $2\overline{E}/\overline{\omega^2}$

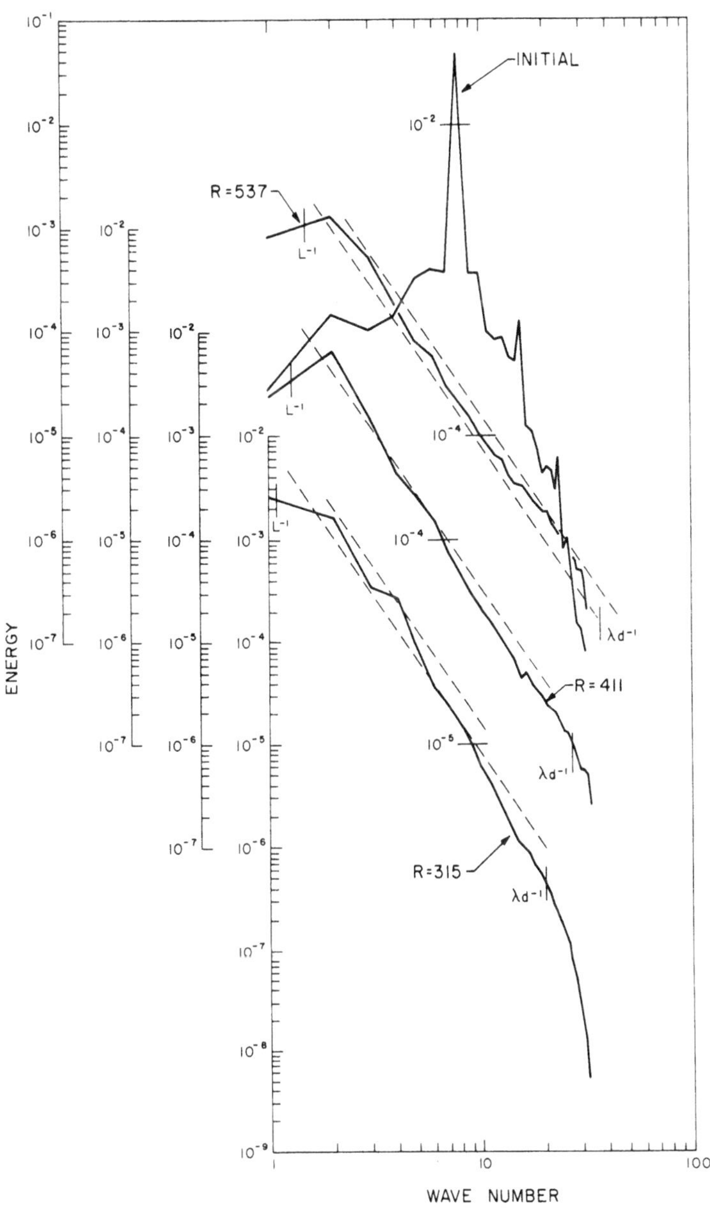

Fig. 4.1

(From Lilly, 1971) - The development of the scalar energy spectrum of decaying two-dimensional turbulence from a numerical simulation experiment. The "initial" spectrum is produced from a few time steps of forcing at wave number 8. The three lower curves are for different values of the final Reynolds number, R, where $R = \overline{E}^{1/2} L/\nu$ and $L = \overline{E}^{1/2}/\eta^{1/3}$. L^{-1} and λ_d^{-1} (from Eqn. 3.13) are indicated by vertical hatch marks. The dashed curves have -3 slope.

therefore increases with time. From Fig. 4.1 we see, however, that the development of the -3 spectrum has proceeded both up and down the scale range and that most of the kinetic energy is ultimately associated with larger scales than that of the initial forcing. From a statistical point of view this result is an essential consequence of the -3 spectrum and the quasi-conservation of energy. If Eqn (4.2) is valid for $k > k_c$, say, then the total energy in the flow may be obtained by integration, and is

$$\overline{E} = \tfrac{1}{2}\beta\eta^{2/3}\, k_c^{-2} .$$

As the total enstrophy decays, its dissipation rate decreases and since $\overline{E}$ is nearly conserved k_c must also decrease. Somewhat more fundamentally Fjørtoft (1953), in a paper much pre-dating the development of two-dimensional turbulence theory, pointed out that the nonlinear interactions of the equations of motion occur in wave number triads typically consisting of a small, an intermediate, and a larger wave number. He showed that if enstrophy is transferred from the intermediate to the larger wave number (as we now know is necessary for maintenance of the enstrophy inertial range) then energy must be transferred to the smaller wave number.

The above arguments also hold in general for turbulence maintained by continuous forcing at large or intermediate scales. In this case Kraichnan (1967) predicts a second inertial range at scales larger than that of the forcing, in which energy flows "backwards" from the generating scale to larger scales. The energy spectrum is given by

$$E(k) = \alpha' \left[\frac{d\overline{E}}{dt}\right]^{2/3} k^{-5/3} \qquad (4.3)$$

where $d\overline{E}/dt$ is the rate of energy increase produced by forcing and α' is another dimensionless constant (not necessarily equal to α or β). The high frequency limit on the validity of (4.3) is the forcing frequency, but the low frequency limit continues to move toward smaller k as the total energy increases until the entire available spectrum (defined by the size of the physical regime) is filled. After that, the energy continues to pile up indefinitely at the lowest frequency. We entitle this prediction the "Infrared Catastrophe in Flatland Fluid Dynamics". Kraichnan's predictions have been partially tested by numerical simulation (Lilly, 1969) and found in agreement with the results.

5. APPLICATION OF TWO-DIMENSIONAL TURBULENCE TO THE ATMOSPHERE

Real geophysical fluids, in particular the atmosphere, do not experience the I-R catastrophe in spite of their gross two-dimensional behavior because of the existence of surface drag. Surface friction, although it involves turbulence, is not sensitive to the horizontal scale of motion, but only to its amplitude. Thus it tends to remove energy from the peak of the energy spectrum and prevent the "backwards" inertial range from piling up energy indefinitely at the largest scales.

A simple model of a forced two-dimensional fluid has been constructed which reproduces several of the gross statistical features of the atmosphere. It is produced by numerical time integration of the following differential equations :

$$\frac{\partial\omega}{\partial t} + u\frac{\partial\omega}{\partial x} + v\frac{\partial\omega}{\partial y} = F + \nu\nabla^2\omega - K\omega\ ,$$

$$u = -\frac{\partial\psi}{\partial y}\ ,\quad v = \frac{\partial\psi}{\partial x}\ ,\omega = \nabla^2\psi\ , \tag{5.1}$$

where ω is the z-component of vorticity, ψ is a stream function, F is a forcing function and K is a surface drag coefficient. This is equivalent to a barotropic model of the atmosphere but with generation and dissipation terms. In the real atmosphere kinetic energy is obtained from the release of potential energy, either by direct heating from fixed and moving heat sources, or through baroclinic instability. In the model two kinds of forcing have been applied, both of which are somewhat arbitrary and artificial but analogous to the above two kinds of real forcing. The first kind, the "uncorrelated forcing" is an arbitrary distribution of vorticity sinks and sources. The distribution is fixed in its amplitude and scale characteristics, but may vary with time in a manner uncorrelated with the flow. The second kind, "negative viscosity forcing", consists of a positive amplification of a portion of the existing vorticity field. It is thus analogous to a flow instability, since in the absence of the nonlinear and dissipative terms, the flow amplitude would grow exponentially. In most cases both kinds of forcing have been restricted to a narrow scale range, for example, all the Fourier modes lying on the rim of a box in wave number space.

Figures (5.1) and (5.2) show representative maps of stream function and vorticity at a time when the simulated turbulence is fully developed and roughly steady-state in its statistical structure. This case is for negative viscosity forcing at wave number 8 with the two damping constants chosen so as to produce an energy spectrum with a peak near wave number 2 and a k^{-3} inertial tail. Figure 5.3 is a map of the forcing function F at the same time, showing its characteristic lack of strong correlation with the flow field it is generating. Figure 5.4 shows the scalar energy spectrum E(k), defined as the sum around boxes in wave space. Figures 5.5, 5.6 and 5.7 are time plots of kinetic energy, enstrophy, and enstrophy dissipation rate during 4000 time steps of the experiment. The rather large oscillations observed are typical of the experimental simulation results. From these figures and the coefficients of viscosity and surface drag it is possible to evaluate the energy and enstrophy budgets. In the next section we will compare these results with a simplified model intended to predict them and other aspects of forced two-dimensional flow with surface drag.

6. A PROBLEM IN TURBULENCE THEORY AND PLANETARY ATMOSPHERES

In this section we introduce a problem relevant to planetary and (possibly) solar atmospheres. The problem, which was inspired by Monin's lectures on planetary atmospheres, can be solved in an approximate way by application of some concepts of the inertial ranges of two-dimensional turbulence.

We assume that an atmosphere is known or postulated to be quasi-two-dimensional and that it has a source of kinetic energy and enstrophy with a characteristic wave length L_F, or a characteristic wave number $k_F = 2\pi/L_F$. This energy source could consist, for example, of either a fluid instability or a distribution of heat sources and sinks.

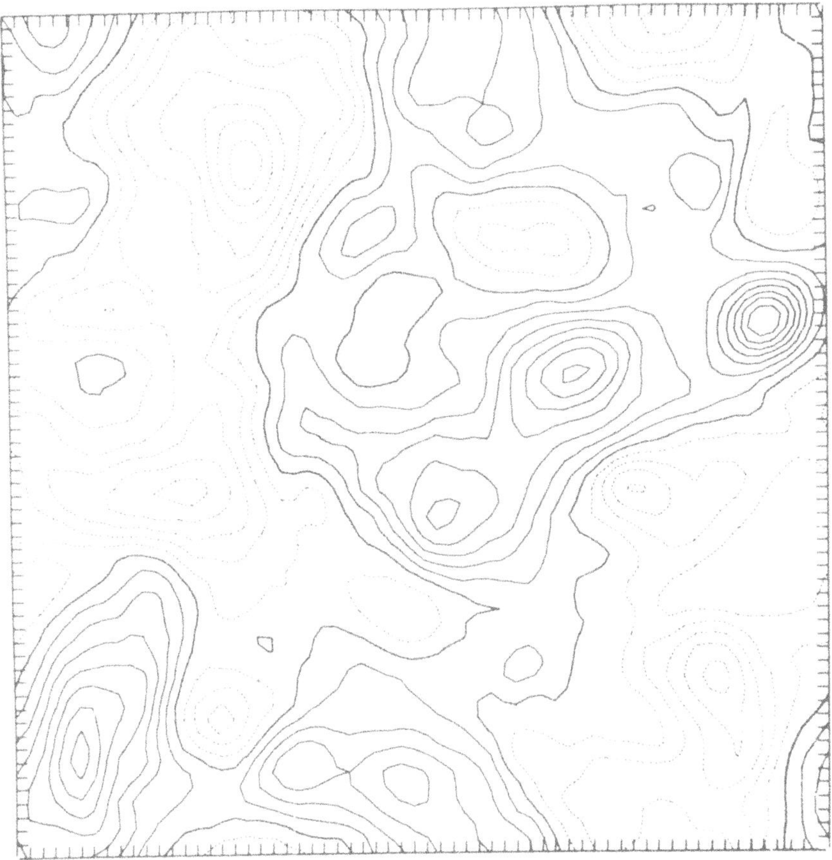

Fig. 5.1

Stream function map developed at the 2200th time step (22.0 time units) of a numerical simulation of two-dimensional turbulence with surface drag. The mesh interval is shown by hatch marks along the boundaries. Positive contours are solid, negative dashed, with contour interval 0.095. The mesh interval is 0.125, $\nu = 5.0 \times 10^{-4}$, and K = 0.02. The forcing is of the negative viscosity type, i.e. $F = \omega_F/\tau$, where ω_F is the portion of the vorticity field associated with wave number 8. The time constant $\tau = 2.0$. All units are dimensionless. The isoline analysis was performed by the computer, using a two-dimensional linear interpolation scheme, and photographed from a cathode-ray-tube presentation.

Fig. 5.2

Vorticity map at the same time as Fig. 5.1. Contour interval = 2.24.

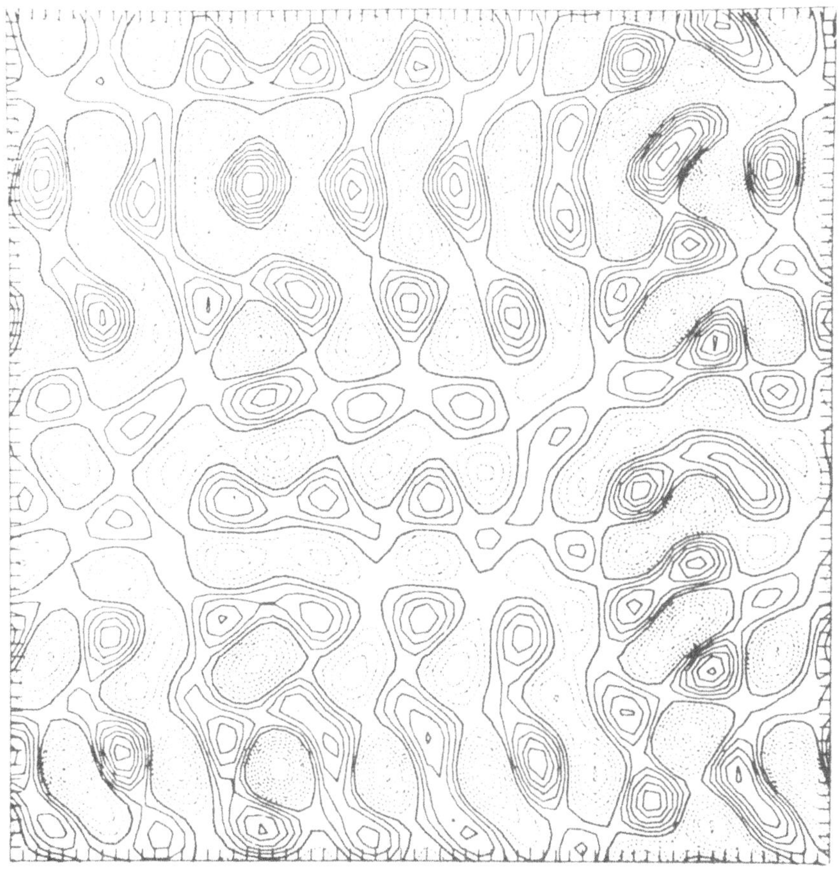

Fig. 5.3

Forcing function map at the same time as Fig. 5.1.
Contour interval = .196.

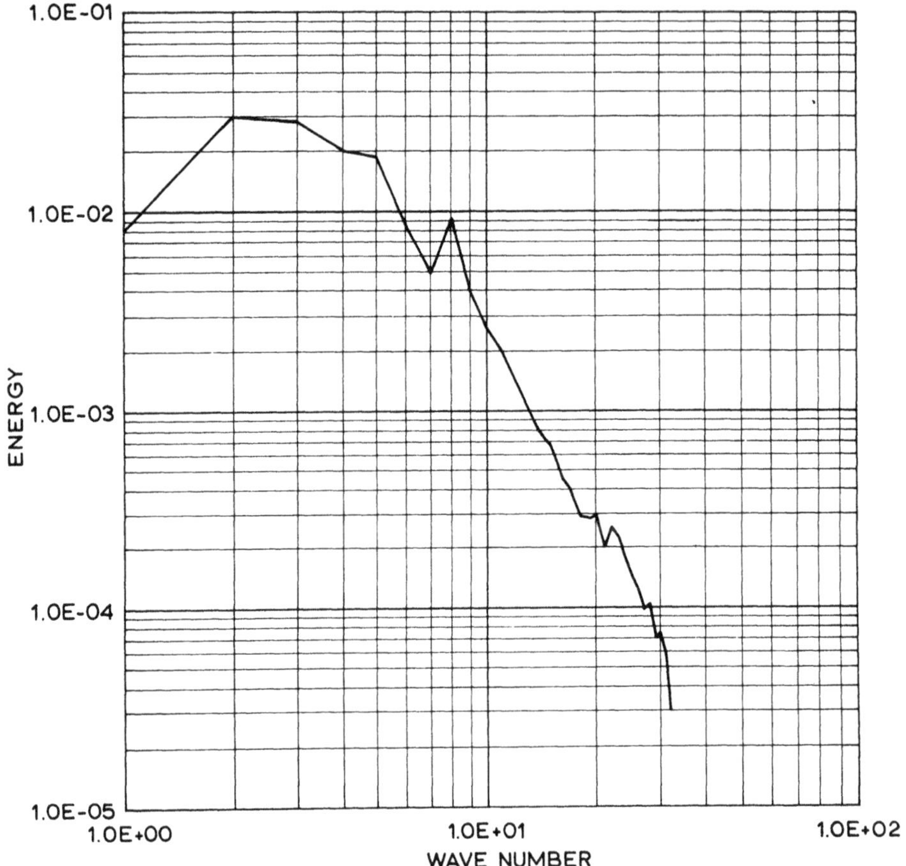

Fig. 5.4

The scalar energy spectrum for the same time as Fig. 5.1. The plotted values represent the energy associated with the surface of a square in wave number space rather than the circle normally assumed in turbulence theory.

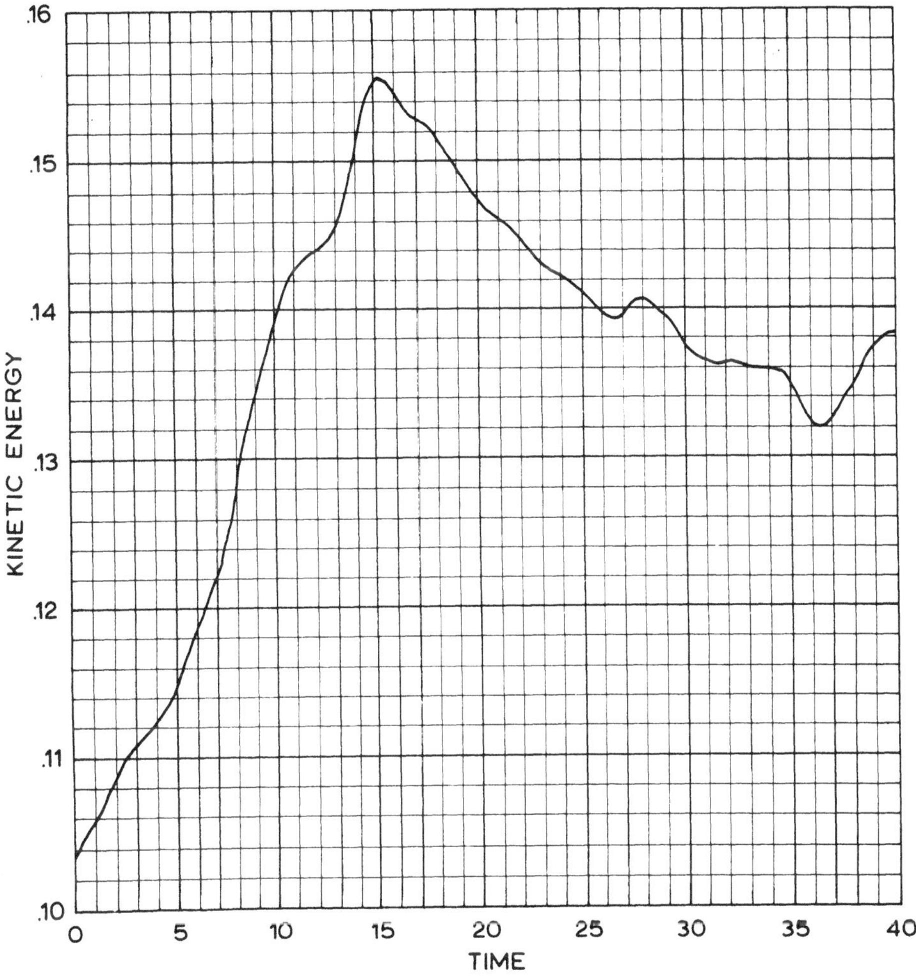

Fig. 5.5

Kinetic energy as a function of time for 4000 time steps of the simulation experiment described in the text. Figures 5.1-5.4 present data corresponding to time = 22.0 units on this figure.

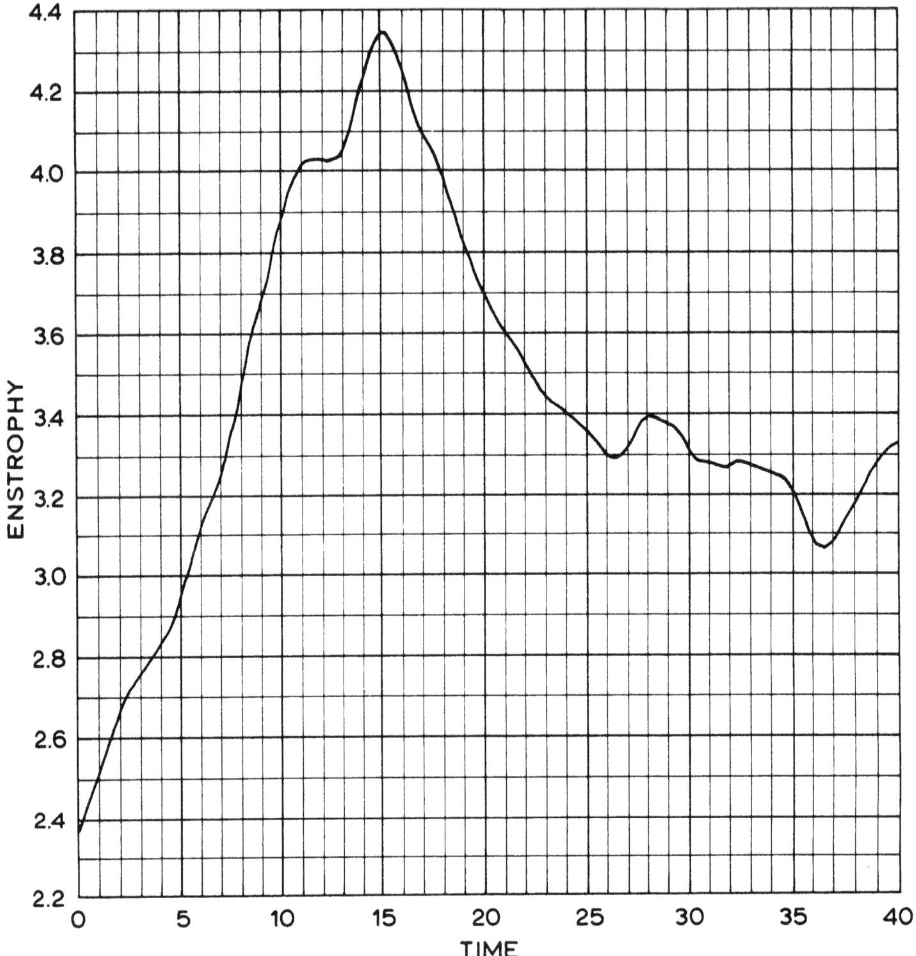

Fig. 5.6

Enstrophy vs. time for the same experiment.

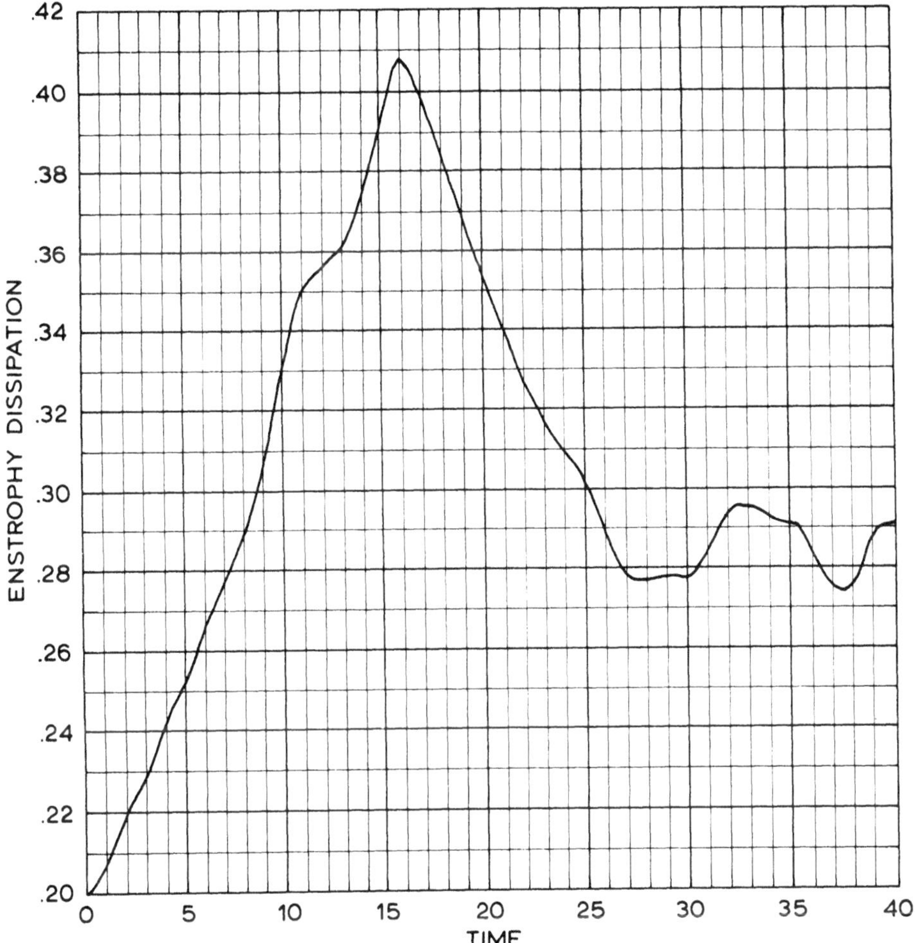

Fig. 5.7

Viscous dissipation of enstrophy vs. time for the same experiment.

We will assume the former. If L_F is less than the circumference of the planet then it is possible that motions of scales larger than L_F will be generated and maintained by Kraichnan's backwards energy cascade. We ask what is the characteristic large scale, if different from L_F, and what is its corresponding velocity amplitude. In order to obtain a definite result we will have to assume a value of the surface drag coefficient (which is mainly a function of the surface roughness) and the existence of molecular (or eddy) viscosity with an enstrophy dissipation scale much smaller than L_F. This last assumption can be tested a posteriori after the solution is obtained.

We now assume that Equation (5.1) adequately describes the dynamics of the atmosphere. The forcing function F is of the negative viscosity type, representing a fluid instability, and given by

$$F = \omega_F/\tau \quad , \tag{6.1}$$

where ω_F is the part of the vorticity field associated with the Fourier modes of wave number k_F and τ is a characteristic time scale of growth. Thus

$$\nabla^2 F = - k_F{}^2 F \quad . \tag{6.2}$$

Upon multiplication by $-\psi$ and ω and averaging over all space we respectively obtain the mean energy and enstrophy equations as

$$\frac{d\overline{E}}{dt} = - \overline{\psi F} - \varepsilon \tag{6.3}$$

$$\frac{d\overline{\omega^2}/2}{dt} = \overline{\omega F} - \eta \tag{6.4}$$

where ε and η are the dissipation rates of energy and enstrophy respectively. Each of these terms is produced partly by surface drag and partly by viscous dissipation, i.e.

$$\varepsilon = 2K\overline{E} + \nu\overline{\omega^2} \tag{6.5}$$

$$\eta = K\overline{\omega^2} + \nu\overline{(\nabla\omega)^2} \,. \tag{6.6}$$

The atmosphere will be assumed to be statistically steady-state, so that the left sides of(6.3) and (6.4) vanish and the generation and dissipation rates are equal. In addition we can deduce immediately from (6.1) and (6.2) that η and ε are proportional to each other through the forcing frequency, k_F, i.e.

$$\frac{\eta}{\varepsilon} = -\frac{\overline{\omega F}}{\overline{\psi F}} = -\frac{\overline{(\nabla^2\psi)F}}{\overline{\psi F}} = -\frac{\overline{\psi(\nabla^2 F)}}{\overline{\psi F}} = {k_F}^2 \,. \tag{6.7}$$

If planetary rotation is an important dynamic influence the surface drag terms should be altered to account for the fact that drag is relative to the moving surface. We will neglect these alterations and our results will thereby be in error with regard to the existence and characteristics of a mean zonal flow.

The questions posed above can be answered from the form and amplitude of the scalar energy spectrum, E(k), as defined in Eqn. (2.13), if this is known. One might assume, for example, that a model spectrum like that proposed by Kraichnan for maintained two-dimensional turbulence would hold for this case with $d\overline{E}/dt$ replaced by the energy dissipation rate ε, i.e.

$$E(k) = \alpha' \varepsilon^{2/3} \quad k^{-5/3} \,, \; k_c < k < k_F \,, \tag{6.8}$$

$$E(k) = \beta \eta^{2/3} \quad k^{-3} \,, \; k_F < k < k_d \,, \tag{6.9}$$

where k_c would be determined somehow from the magnitude of the drag coefficient K, and k_d would be inversely proportional to λ from Eqn. (3.13). Because it assumes that the energy and enstrophy transfers are constant within the respective inertial ranges this solution would only be valid if the energy dissipation were all at very large scales and the enstrophy dissipation all at very small scales. That requirement means that the following conditions must be assumed :

a) the spectral contributions to (6.5) are essentially all associated with wave numbers much smaller than k_F ;

b) the spectral contributions to (6.6) are essentially all associated with wave numbers much greater than k_F .

As we shall show, conditions (a) and (b) are only fulfilled for special circumstances, in particular a very low drag coefficient, that are probably not common in planetary atmospheres.

Our approach will be to alter (6.8) and (6.9) in a somewhat intuitive way to account for the non-fulfillment of conditions (a) and (b) and then to test these new spectral predictions in simple ways for internal self-consistency. The tests ultimately lead to a prediction for k_c as a function of k_F, K and ε, which is then compared to the results of simulation experiments. The drag factor K, which is actually an inverse time scale of surface frictional losses, is then re-evaluated in terms of the normal dimensionless boundary layer drag coefficient C_D. Finally the ratio k_c/k_F is shown to be a simple function of the dimensionless combination $H/C_D L_F$, where H is the characteristic depth (usually scale height) of the atmosphere.

Kraichnan's derivation of Eqns. (6.8) and (6.9) depended on having a constant rate of energy transfer to smaller wave numbers (equal to ε in our case) and a constant rate of enstrophy transfer to larger wave numbers equal to η. We shall preserve the form of the equations but replace the dissipation rate arguments by the local rate of spectral transfer of energy or enstrophy. Thus we replace ε in (6.8) by the total dissipation of energy at wave numbers smaller than k, since all that energy must be transferred across κ, and similarly we replace η in (6.9) by the total dissipation of enstrophy at wave numbers smaller than k. The result is a pair of integral equations, i.e.

$$E(k) = \alpha' \left[2K \int_{k_c}^{k} E(k) \, dk\right]^{2/3} k^{-5/3} , \quad k_c < k < k_F \qquad (6.10)$$

$$E(k) = \beta \left[2K \int_{k}^{k_d} \underline{k}^2 E(k) \, dk + \eta_V\right]^{2/3} k^{-3}, \quad k_F < k < k_d \qquad (6.11)$$

The term in brackets in (6.10) is 2K multiplied by the energy associated with wave numbers less than k and the first term in brackets in (6.11) is

2K times the enstrophy associated with wave numbers greater than k. The second term in brackets in (6.11), η_V, is the enstrophy lost by molecular viscosity. It is assumed to occur at wave numbers $k \sim k_d \gg k_F$.

The above equations may be transformed by differentiation into differential equations and easily solved. With the use of the boundary conditions $E(k) = 0$ at k_c, k_d the results may be written as

$$E(k) = \alpha'^3 K^2 k^{-5/3} (k_c^{-2/3} - k^{-2/3})^2, \quad k_c < k < k_F \qquad (6.12)$$

$$E(k) = \frac{4}{9} \beta^3 K^2 k^{-4} (\ln k_*/k)^2, \quad k_F < k < k_d \qquad (6.13)$$

where $\ln k_*/k_d = 3\eta_V^{1/3}/(2K\beta)$. These expressions look reasonable and promising, since they seem to be similar to (6.8) and (6.9) when $k_c \ll k \ll k_d$. If we evaluate (6.12) at $k = k_F$, the factor multiplying $\alpha' k_F^{-5/3}$ must be the 2/3 power of the total energy dissipation at wave numbers smaller than k_F - compare (6.10)-, and evaluation of (6.13) similarly yields the total enstrophy dissipation for wave numbers larger than k_d. Thus we obtain

$$2K \int_{k_c}^{k_F} E(k)\, dk = \frac{\alpha'^3 K^3}{k_F^2}\left[\left(\frac{k_F}{k_c}\right)^{2/3} - 1\right]^3 , \qquad (6.14)$$

$$2K \int_{k_F}^{k_d} k^2 E(k)\, dk + \eta_V = \left[\frac{2}{3} \beta K \ln k_*/k_F\right]^3 . \qquad (6.15)$$

Also, equating the two expressions (6.10) and (6.11) at $k = k_F$ yields a relation between β and α', i.e.

$$\beta/\alpha' = \left[\frac{3}{2} \frac{(k_F/k_c)^{2/3} - 1}{\ln k_*/k_F}\right]^{2/3} . \qquad (6.16)$$

Next we evaluate the energy dissipation occuring at wave numbers greater than k_F and thereby determine how well condition (a) above is fulfilled. This quantity is the integral

$$2K \int_{k_F}^{k_d} E(k)\, dk = \frac{8}{9} \beta^3 K^3 \int_{k_F}^{k_d} k^{-3}(\ln k_*/k)^2\, dk \tag{6.17}$$

in which we have utilized (6.13). Somewhat surprisingly this expression can also be readily integrated with the result

$$2K \int_{k_F}^{k_d} E(k)\, dk = \frac{4}{9} \frac{\beta^3 K^3}{k_F^{\,2}} \left\{ \left[\ln \frac{k_*}{k_F}\right]^2 - \ln \frac{k_*}{k_F} + \frac{1}{2} - \left[\frac{k_F}{k_d}\right]^2 \left[\left(\ln \frac{k_*}{k_d}\right)^2 \quad \ln \frac{k_*}{k_d} + \frac{1}{2}\right] \right\}. \tag{6.18}$$

when $k_d >> k_F$ the first term inside the brackets dominates. Upon neglecting the remaining terms and substituting (6.16) to remove β we obtain

$$2K \int_{k_F}^{k_d} E(k)\, dk \approx \frac{\alpha'^3 K^3}{k_F^{\,2}} \left[\left(\frac{k_F}{k_c}\right)^{2/3} - 1 \right]^2 . \tag{6.19}$$

Upon comparing (6.19) with (6.14), we see that the former is much smaller if $k_c << k_F$. Thus condition (a) is fulfilled only for those circumstances.

There have been no obvious self-inconsistencies in the model and we now may construct an equation for predicting the scale k_c. The sum of the left sides of (6.14) and (6.19) is the total energy dissipation rate. Upon adding the right sides, also, we obtain

$$\varepsilon = \frac{\alpha'^3 K^3}{k_F^{\,2}} \left[\left(\frac{k_F}{k_c}\right)^{2/3} - 1 \right]^2 \left[\frac{k_F}{k_c}\right]^{2/3} . \tag{6.20}$$

We may similarly evaluate the total enstrophy dissipation from the assumed spectral forms and compare it with the simulation results to obtain an estimate of k_*/k_F. Neglecting some small contributions from the low wave number end, η is given essentially by the term in brackets in (6.11) evaluated at $k = k_F$. Using (6.11) and (6.13) and (6.16) we obtain

$$\eta \approx \left[\beta^{-1} k_F^{\,3}\, E(k_F)\right]^{3/2} = \frac{3}{2}\, \alpha'^3 K^3 \left[\left(\frac{k_F}{k_c}\right)^{2/3} - 1 \right] \ln k_*/k_F \tag{6.21}$$

Solving for k_*/k_F and making use of (6.7) and (6.20) we obtain the rather simple result

$$\ln k_*/k_F \simeq \frac{3}{2}\left[\frac{k_F}{k_C}\right]^{2/3} \qquad (6.22)$$

We now may compare the predictions of this model with the results of the simulation experiment described in the previous section. Figure 6.1 shows the theoretically predicted spectra of Equations (6.12) and (6.13), with β/α' evaluated from (6.16) and k_*/k_F from (6.22), plotted for various values of k_C/k_F. The simulated spectrum is shown by the solid points and is a time average of the spectra for the simulation illustrated in Section 5. The time averaged spectrum does not differ greatly from that of Fig. 5.4. Except for the sharp spike at the forcing frequency it is evidently well represented at both low and high k by the theoretical prediction for $k_C/k_F \sim 0.09$. The constant α' can now be estimated from Eqn. (6.20)

$$\alpha'^3 = \frac{k_C^2\,\varepsilon}{K^3\left[1-(k_C/k_F)^{2/3}\right]^2} = \frac{\left[K\overline{\omega^2} + \nu\overline{(\nabla\omega)^2}\right](k_C/k_F)^2}{K^3\left[1-(k_C/k_F)^{2/3}\right]^2} \qquad (6.23)$$

where Eqns. (6.6) and (6.7) have been introduced into the second equality The constant β can be evaluated from (6.16), which with the aid of (6.22) becomes

$$\beta/\alpha' = \left[1-(k_C/k_F)^{2/3}\right]^{2/3} \quad . \qquad (6.24)$$

Evaluation of $\overline{E}$, $\overline{\omega^2}/2$, and $\nu\overline{(\nabla\omega)^2}$ from Figs. 5.5, 5.6, and 5.8 respectively, lead to the estimates

$$\alpha' = 7.4 \quad , \qquad \beta = 6.4 \quad .$$

Now that we have tested the theoretical predictions against numerical simulation results it seems appropriate to proceed to a more complete atmospheric interpretation. We therefore proceed to evaluate the parameter K in more meteorologically suitable terms.

The energy dissipation in the surface boundary layer is given (Monin, Eqn. 8) by

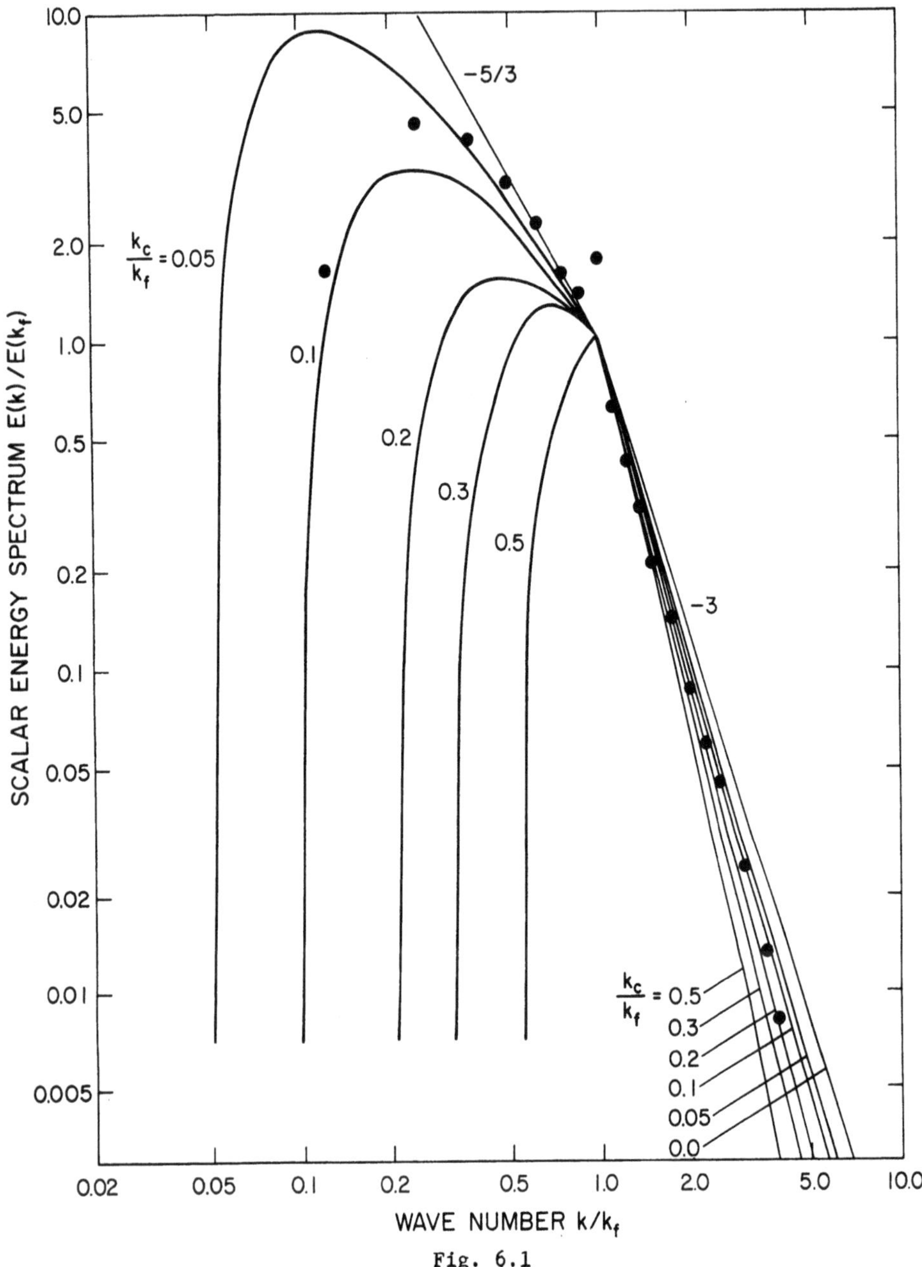

Fig. 6.1

Theoretical predictions of the scalar energy spectrum of forced two-dimensional turbulence with viscosity and surface drag. The different curves are for differing values of the ratio of cut-off wave number to forcing wave number, as specified by Eqns. (6.20), (6.24), and (6.31). The plotted points define the time averaged spectrum of the simulation experiment described in Section 5.

$$\varepsilon = \frac{\tau}{\rho} \frac{\partial \bar{u}}{\partial z} , \tag{6.25}$$

where τ is the stress. If we integrate this expression over the mass of the boundary layer, we obtain

$$\int_0^h \rho\varepsilon \, dz = \int_0^h \tau \frac{\partial \bar{u}}{\partial z} \sim \tau_0 \bar{u}_h \tag{6.26}$$

where $\bar{u}_h$ is the value of $\bar{u}$ at $z = h$. The last approximation can be made because most of the shear occurs in the constant stress layer. The drag coefficient, C_D, may be defined by the relation

$$\tau_0 = \rho_0 C_D \bar{u}_h^2 \tag{6.27}$$

so that (6.26) becomes

$$\int_0^h \rho\varepsilon \, dz \sim \rho_0 C_D \bar{u}_h^3 . \tag{6.28}$$

We wish to use a value of dissipation equivalent to an average over the total mass of the atmosphere. Thus we divide (6.28) by $\rho_0 H$, where H is something like an atmospheric scale height, and write

$$\varepsilon = C_D (2\bar{E})^{3/2} / H \tag{6.29}$$

where we have replaced $\bar{u}_h$ by $(2\bar{E})^{1/2}$, a characteristic velocity for the system.

We have assumed, however, that the dissipation is also given by the first term of Eqn. (6.5). Upon eliminating $\bar{E}$ from that equation and Eqn. (6.29) we obtain

$$K^3 = C_D^2 \bar{E}/H^2 . \tag{6.30}$$

Upon inserting the above two relations into (6.20) we find that the rate of dissipation is eliminated from the equation as an explicit parameter, and it becomes

$$\left[\left(\frac{k_F}{k_c} \right)^{2/3} - 1 \right]^2 \left[\frac{k_F}{k_c} \right]^{2/3} = \left[\frac{k_F H}{C_D \alpha'^{3/2}} \right]^2 . \tag{6.31}$$

This is a very interesting result. It states that the largest length scale that can be simulated by forcing at a much smaller scale is limited by the ratio of atmospheric scale depth to drag coefficient. Thus when the parameter on the right of (6.31) is large we have

$$L_c = \frac{2\pi}{k_c} \sim \frac{2\pi H}{C_D \alpha'^{3/2}} \quad . \qquad (6.32)$$

For the earth's atmosphere $C_D \sim 10^{-3}$ and $H \sim 10^4$m. Using $\alpha' = 7.4$ (from evaluations of the simulation experiments) we thus find $L_c \sim 3500$ km, a surprisingly small value about equal to planetary wave number 8-10. Of course the forcing in the earth's atmosphere arises from both geographic and baroclinic instability sources and therefore has components from wave numbers 1-10 at least. Under these circumstances Equation (6.31) simply says that the response scales of the atmosphere will be significantly larger than the forcing scales, i.e. $k_F/k_c > 1$. For example, with forcing at L_F = 15000 km, $L_c \sim$ 20000 km is predicted.

For application to planetary atmospheres it is necessary to know or estimate what sort of driving scales are present, as well as to know the character of the underlying surface well enough to guess the drag coefficient. A particularly interesting case arises with the solar atmosphere. It is known that a general circulation exists in the outer layers of the sun, as indicated by the observational evidence of a differential rotation rate as a function of latitude. Evidence also exists (Ward, 1964; 1965) that this mean zonal shear is driven by at least somewhat smaller scales of quasi-two-dimensional motion, but the only motions for which a definite driving force is known are associated with the small eddies by means of the reverse cascade process. Equation (3.2) predicts that this process could be effective if H/C_D is of the order of the solar diameter. The scale depth of the chromosphere is of order 300 km and increases rapidly downwards, while the solar diameter is 1.4×10^6 km. The drag mechanism between the convective and lower layers of the sun is not well understood but may be similar to the interactions between a convective boundary layer in the atmosphere and a stable layer above it with shear across the interface. An effective drag coefficient of order 10^{-3} therefore seems reasonable. This then implies that the reverse cascade mechanism is plausible.

7. THE FORMAL CLOSURE PROBLEM AND THE PREDICTABILITY PROBLEM FOR TWO-DIMENSIONAL ATMOSPHERIC SIMULATION

Assume a two-dimensional atmosphere described by Eqn. (3.1), re-written in component form as

$$\frac{\partial u}{\partial t} + \frac{\partial}{\partial x}(u^2) + \frac{\partial}{\partial y}(uv) + \frac{\partial(p/\rho)}{\partial x} = \nu\left[\frac{\partial^2 u}{\partial x^2} + \frac{\partial^2 u}{\partial y^2}\right] \tag{7.1}$$

$$\frac{\partial v}{\partial t} + \frac{\partial}{\partial x}(uv) + \frac{\partial}{\partial y}(v^2) + \frac{\partial(p/\rho)}{\partial y} = \nu\left[\frac{\partial^2 v}{\partial x^2} + \frac{\partial^2 v}{\partial y^2}\right] \tag{7.2}$$

$$\frac{\partial u}{\partial x} + \frac{\partial v}{\partial y} = 0 \quad . \tag{7.3}$$

We neglect rotation, surface drag and forcing terms because they are not involved in present considerations. The restriction to two dimensions is also not essential here. It has been made largely for simplicity. The three-dimensional case has been considered from the same viewpoint by Lilly (1967). In a simulation problem or numerical forecast we commonly define variables at centers of mesh boxes of side Δ and think of them as representing the spatial average in that box. Using an over-bar to represent this averaging process we define for any dependent variable $F(x,y,t)$

$$\overline{F}(x,y,t) = \frac{1}{\Delta^2}\int_{-\Delta/2}^{\Delta/2}\int_{-\Delta/2}^{\Delta/2} F(x + x', y + y', t)\, dx'\, dy' \; . \tag{7.4}$$

Examination of (7.4) shows that the averaging operator commutes freely with space and time differentiation. We may therefore apply this operator to Eqns. (7.1)-(7.3) to obtain predictive equations for $\overline{u}$, $\overline{v}$. The difficulty in making use of such equations is in defining $\overline{u^2}$, $\overline{uv}$, and $\overline{v^2}$ in terms of $\overline{u}$ and $\overline{v}$. The problem is more clearly defined by averaging (7.1) and (7.2) and rearranging the terms as follows

$$\frac{\partial \overline{u}}{\partial t} + \frac{\partial}{\partial x}(\overline{u}^2) + \frac{\partial}{\partial y}(\overline{u}\overline{v}) + \frac{\partial(\overline{p}/\rho)}{\partial x} = \frac{\partial \tau_{xx}}{\partial x} + \frac{\partial \tau_{xy}}{\partial y} \tag{7.5}$$

$$\frac{\partial \overline{v}}{\partial t} + \frac{\partial}{\partial x}(\overline{u}\overline{v}) + \frac{\partial}{\partial y}(\overline{v}^2) + \frac{\partial(\overline{p}/\rho)}{\partial y} = \frac{\partial \tau_{yx}}{\partial x} + \frac{\partial \tau_{yy}}{\partial y} \tag{7.6}$$

where

$$\tau_{xx} = -\tau_{yy} = \frac{\overline{u^2} - \bar{u}^2 - \overline{v^2} + \bar{v}^2}{2} + \nu\left[\frac{\partial\bar{u}}{\partial x} - \frac{\partial\bar{v}}{\partial y}\right]$$

$$\tau_{xy} = \tau_{yx} = \bar{u}\,\bar{v} - \overline{uv} + \nu\left[\frac{\partial\bar{u}}{\partial x} + \frac{\partial\bar{v}}{\partial y}\right] \qquad (7.7)$$

$$p = \bar{p} + \rho\left[\frac{\overline{u^2} - \bar{u}^2 + \overline{v^2} - \bar{v}^2}{2}\right]$$

The nonviscous parts of the τ_{ij}'s are the Reynolds stresses defined in such a way that they all vanish in isotropic turbulence. The diagonal part of the usual Reynolds stress tensor, which does not vanish in isotropic turbulence, has been incorporated into the pressure term and pressure is thus augmented by the small scale kinetic energy to form a "macro-pressure". The usual Reynolds postulates cannot be applied here because the over-bar is a running average, and in general $\bar{\bar{F}} \neq \bar{F}$, etc. Nevertheless, the principal contributions to the Reynolds stresses come from the integrated effects of motion on scales less than the mesh interval. It is formally possible to derive equations for the rates of change of the τ_{ij}'s, but these equations then contain triple-product correlations on the right. The generalization of this process leads to the usual closure problem of turbulence. There are always more terms required than there are equations to relate them.

The central problem of turbulence theory (still unsolved) is to truncate the infinite series of moment equations by means of a suitable set of assumptions relating higher order moments to those of lower order. It must be recognized, however, that no matter how successful such a theory may be in this respect it can make no more than a statement of probability. For example, if we have a relation giving τ_{ij} as a function of $\bar{u}$, $\bar{v}$ and the τ_{ij}'s at all previous times, the relation will at best give a prediction of the most probable value with perhaps some measure of the expected statistical deviation of that value. Errors in the predicted τ_{ij}'s cannot be expected to be self-correcting. On the contrary they will immediately begin to affect the $\bar{u}$, $\bar{v}$ fields and ultimately lead to complete invalidation of the detailed flow forecast. This is because the equations of motion of large Reynolds number flow are unstable to small perturbations. This is the central problem of predictability and will be discussed further in subsequent lectures.

There is another framework of equations with which to look at the closure and predictability problems that has always been useful in turbulence theory and is now becoming more popular in numerical simulation and forecasting. That is the Fourier mode system. If we define, as in (2.30),

$$u(\vec{x},t) = (\pi/L)^2 \,\Sigma\, u(\vec{k},t)e^{i\vec{k}\cdot\vec{x}} \; , \; v(\vec{x},t) = (\pi/L)^2 \,\Sigma\, v(\vec{k},t)e^{i\vec{k}\cdot\vec{x}} \; , \qquad (7.8)$$

and recognizing from the continuity equation that

$$k_1 \; u(\vec{k}) + k_2 \; v(\vec{k}) = 0 \; , \qquad (7.9)$$

then Equations (7.1) and (7.2) may be written in Fourier space as

$$\frac{du(\vec{k})}{dt} + \nu k^2 u(\vec{k}) = k_2 \sum_{\vec{p}+\vec{q}=\vec{k}} K(k,p,q) \frac{u(\vec{p})u(\vec{q})}{p_2 q_2} \; ,$$

$$\frac{dv(\vec{k})}{dt} + \nu k^2 v(\vec{k}) = -k_1 \sum_{\vec{p}+\vec{q}=\vec{k}} K(k,p,q) \frac{v(\vec{p})v(\vec{q})}{p_1 q_1} \; , \qquad (7.10)$$

where

$$K(k,p,q) = i \; (L/2\pi)^2 \frac{(k_1 p_1 - k_2 q_2)\,(p_2 q_1 - p_1 q_2)}{k^2} \qquad (7.11)$$

By comparison with (7.1) and (7.2), expressed in finite difference form, this system has the advantage that all numerical errors associated with spatial differences are removed. Orszag (1970) has shown that with proper organization of the calculations a high resolution numerical simulation can be done with better accuracy with Fourier modes in a comparable time to that required for (7.1) and (7.2) cast into finite difference form.

The equivalent to the Reynolds stress problem appears here in a different form. If we limit the resolution by confining k, p and q to a finite area in Fourier space (say $k, p, q < K$) and remove all interaction terms on the right leading to $k > K$ the energy and enstrophy conservation principles remain equivalent to those for an unbounded series. In doing this we have removed any way for energy or enstrophy to be exchanged across $k = K$. What is required for correct dynamic behavior is a statistical expression for the right hand side of (7.13) and (7.14) when p and/ or q > K. In general the expression must have the effect of decreasing the amplitude of the Fourier modes near k = K.

8. CLOSURE HYPOTHESES

Zero'th order.

The simplest solution to the problem of Reynolds stresses is to ignore them. This has, in fact, been a fairly common approach in simulation studies, especially when the numerical discretization and integration methods are stable. If the viscosity is sufficiently large that $\lambda > \Delta$, then there is no need to include those terms, but this is not ordinarily the case in atmospheric or oceanographic flow problems. The results of neglecting the Reynolds stresses in many cases are :

1. If the numerical procedures are accurate then little or no energy or enstrophy can be removed from the system at high Reynolds number.

2. If, again, the numerical methods are very good (like Fourier mode systems) it is believed (Lee, 1952) that there will be a tendency to develop an equipartition spectrum, in which energy is randomly and equally distributed through all modes. This is a highly unrealistic solution for most practical problems.

3. More commonly the numerical scheme will either develop a form of computational instability or, if the scheme is a strongly damped one, all the high frequency modes will disappear because of a very small effective Reynolds number.

First order. Non-linear eddy viscosity

In an attempt to remove or reduce the above difficulties Smagorinsky (1963) adopted and generalized a form of variable viscosity deduced earlier (Von Neumann and Richtmyer, 1950) by workers in shock-wave dynamics simulation. It was soon realized that the form chosen was related to and consistent with both the standard mixing length approximation of boundary layer theory and the Kolmogoroff inertial range predictions. Those relationships have since been quantified (Lilly, 1967 ; Deardoff, 1970) and the form also extended (Leith, 1969) to two-dimensional turbulence. At the present time one or another version is being used by several groups doing general circulation or fluid dynamics simulation studies. Further work needs to be done, however, in adapting the concepts to Fourier space and in several other areas of generalization or improvement.

The stress-strain relationships used are the following :

$$\left.\begin{aligned} \tau_{ij} &= \nu_e \overline{D}_{ij} \\ \nu_e &= \ell^2 \overline{D} \end{aligned}\right\} \quad \text{(three dimensions)} \qquad (8.1)$$

$$\left.\begin{aligned} \frac{\partial \tau_{xx}}{\partial x} + \frac{\partial \tau_{xy}}{\partial y} &= -\nu_e \frac{\partial \bar{\omega}}{\partial y} = \nu_e \nabla^2 \bar{u} \\ \frac{\partial \tau_{yx}}{\partial x} \quad \frac{\partial \tau_{yy}}{\partial y} &= \nu_e \frac{\partial \bar{\omega}}{\partial x} = \nu_e \nabla^2 \bar{v} \\ \nu_e &= \ell^3 \ |\nabla \bar{\omega}| \end{aligned}\right\} \quad \text{(two dimensions)} \qquad (8.2)$$

$\overline{D}_{ij}$ is the deformation tensor of the mesh-box-averaged flow, given by

$$\overline{D}_{ij} = \frac{\partial \bar{u}_i}{\partial x_j} + \frac{\partial \bar{u}_j}{\partial x_i} \qquad (8.3)$$

and $\overline{D} = \left[\frac{\partial \bar{u}_i}{\partial x_j} \ \overline{D}_{ij} \right]^{1/2}$. The scale length ℓ is proportional to the mesh-box averaging length, Δ, with a proportionality constant which has been estimated theoretically and by experience to be of order 0.2.

We now look at the properties of the above forms and try to see why they are plausible. We notice first that the stresses represent down-gradient transfer of velocity in the three-dimensional system and vorticity in the two-dimensional case. This means that they are all direct energy or enstrophy removal mechanisms. Thus the mean kinetic energy equation for the three-dimensional case is

$$\frac{\partial \left\langle \bar{u}_i^{\,2}/2 \right\rangle}{\partial t} = -\left\langle \nu_e \bar{D}^2 \right\rangle = -\varepsilon \tag{8.4}$$

where the brackets $\langle\,\rangle$ represent an integral or average over all space. Now $\bar{D}^2$ is the mean squared deformation of a velocity field which has been smoothed over a length scale Δ, and its magnitude must increase with decreasing Δ. Thus ν_e must typically decrease with decreasing Δ. If we assume that motions of scale Δ lie within the inertial range we can estimate the characteristic magnitude of $\nu_e(\Delta)$ from the Kolmogoroff scaling parameters to be

$$\nu_e \sim \Delta^{4/3} \quad \varepsilon^{1/3} \quad . \tag{8.5}$$

From this and (8.4) we may eliminate ε to obtain a local estimate of eddy viscosity as

$$\nu_e \sim \Delta^2 \, \bar{D} \quad , \tag{8.6}$$

as was assumed in (8.1).

In two dimensions we start with the mean enstrophy equation in the form

$$\frac{d}{dt} \left\langle \bar{\omega}^2/2 \right\rangle = -\left\langle \nu_e (\nabla\bar{\omega})^2 \right\rangle = -\eta \tag{8.7}$$

By the two-dimensional inertial scaling we assume that

$$\nu_e \sim \Delta^2 \, \eta^{1/3} \tag{8.8}$$

and obtain

$$\nu_e \sim \Delta^3 \, |\nabla\bar{\omega}| \tag{8.9}$$

as was assumed in (8.2).

The proportionality constant ℓ/Δ must be derived from a rather complex integration involving either the energy spectrum or the correlation function to obtain an estimate of $[\overline{D}(\Delta)]^2$. An attempt was made to do this (Lilly, 1966) in the three-dimensional case with a resulting estimate $\ell/D \sim 0.2$, but because of various uncertainties this should only be considered valid to ±25%.

The dimensional consistency of the variable eddy viscosity formulations with appropriate inertial ranges is further borne out by the experience (Deardoff, 1970) that numerical simulations using these forms tend to produce the proper inertial range tails in their spectra. Another desirable feature (really another aspect of the same thing) is that these forms have less damping effect on the larger scales than a constant viscosity which produces the same rate of dissipation. The apparent reason is that, since the coefficients are themselves proportional to the local intensity of turbulence, the damping is strongest "where it needs to be". Finally, experience has shown simulations performed using the variable viscosity formulations tend to be more stable than those with laminar viscosity.

Higher order schemes.

The closure problem described in Section 7 is a special example of the more general closure problem of turbulence theory. It presumes that the motions in the lower frequency ranges are known or calculated explicitely and only the higher frequencies are to be treated statistically. Many turbulence specialists attempt to treat the entire field of motion (with the exception of a time-averaged mean) as random and seek only statistical measures of its characteristics. Thus most closure attempts have been cast within that framework. There is no particular reason (except a shortage of dedicated and competent manpower in this rather esoteric field of applied mathematics) why the more successful of these could not be applied to the simulation problem.

Along the lines of the eddy viscosity hypotheses a higher order formulation (the visco-elastic model) has been suggested by Lilly (1967) and Crow (1968). This consists of a closed system of equations for rates of change of the τ_{ij}'s and the turbulent energy, E. The eddy viscosity formulation is preserved as a special steady-state solution, but the more general equations allow nonequilibrium effects to be more accurately treated. In addition the stresses are not in general exactly parallel with the strains, as they are in the eddy viscosity forms. One disadvantage is that there are one or two additional constants that have to be evaluated from integrals of the presumed inertial range structure. Another is that the additional equations (6 in the case of constant density flow) impose strong limitations on computer speed and memory. Up to now the system has not been tried in practice.

Most other closure models have been formulated in Fourier space, which is a mathematical convenience but sometimes leads to difficulties in physical interpretation. The most well known (or notorious) of these is the quasi-normal model. Although this model turned out to be grossly unsuccessful, several more successful and/or more sophisticated models incorporate some of its features. It can be rather simply derived for the two-dimensional case for Eqns. (7.1)-(7.3).

The basic assumption of the quasi-normal model is that fourth order correlations of velocity components of their Fourier transforms can be derivable for those of second order as if the variable components or Fourier modes were distributed as Gaussian variables with zero mean. Thus if a, b, c, d are four such variables, then

$$\overline{abcd} = \overline{ab}\;\overline{cd} + \overline{ac}\;\overline{bd} + \overline{ad}\;\overline{bc} \quad . \tag{8.10}$$

In distinction to a <u>fully</u> Gaussian model, however, one does <u>not</u> assume that $\overline{abc} = \overline{a}\;\overline{bc} + \overline{ab}\;\overline{c} + \overline{ac}\;\overline{b} = 0$. To apply this assumption we construct equations for the second moments of the Fourier velocity component transforms from Eqn. (7.10) as follows :

$$\frac{d}{dt}\left[\overline{u(\vec{k})u(\vec{\ell})}\right] + \nu(k^2+\ell^2)\,\overline{u(\vec{k})u(\vec{\ell})} = k_2 \sum_{\vec{p}+\vec{q}=\vec{k}} K(k,p,q)\,\frac{\overline{u(\vec{p})u(\vec{q})u(\vec{\ell})}}{p_2 q_2}$$

$$+ \ell_2 \sum_{\vec{p}+\vec{q}=\vec{\ell}} K(\ell,p,q)\,\frac{\overline{u(\vec{p})u(\vec{q})u(\vec{\ell})}}{p_2 q_2} \quad . \tag{8.11}$$

Similarly, equations for the triple correlations may be derived as

$$\frac{d}{dt}\left[\overline{u(\vec{p})u(\vec{q})u(\vec{\ell})}\right] + \nu(p^2+q^2+\ell^2)\ \overline{u(\vec{p})u(\vec{q})u(\vec{\ell})} = p_2 \sum_{\vec{r}+\vec{s}=\vec{p}} K(p,r,s)\frac{\overline{u(\vec{r})u(\vec{s})u(\vec{q})u(\vec{\ell})}}{r_2 s_2}$$

$$+\ q_2 \sum_{\vec{r}+\vec{s}=\vec{q}} K(q,r,s,)\frac{\overline{u(\vec{r})u(\vec{s})u(\vec{p})u(\vec{\ell})}}{r_2\, s_2} + \ell_2 \sum_{\vec{r}+\vec{s}=\ell} K(\ell,r,s)\frac{\underline{u(\vec{r})u(\vec{s})u(\vec{p})u(\vec{q})}}{r_2\, s_2} \tag{8.12}$$

But by the quasi-normal hypothesis,

$$\overline{u(\vec{r})u(\vec{s})u(\vec{q})u(\vec{\ell})} = \overline{u(\vec{r})u(\vec{s})}\ \overline{u(\vec{q})u(\vec{\ell})} + \overline{u(\vec{r})u(\vec{q})}\ \overline{u(\vec{s})u(\vec{\ell})} + \overline{u(\vec{r})u(\vec{\ell})}\ \overline{u(\vec{s})u(\vec{q})} \tag{8.13}$$

etc. so that the system can be closed, provided the wave numbers k, ℓ, p, q, r, s are limited to some finite range.

Numerical integrations of these equations (reduced in volume by use of the isotropic assumption) were performed by Ogura (1963). The results were disastrous. For large Reynolds numbers (small ν) some of the energy correlations $\left[\ \overline{u(\vec{k})u(-\vec{k})}\ \right]$ became negative, i.e. the predicted flow could not be real. Attempts at understanding the meaning of this result and avoiding it in the future have occupied the attention of fluid dynamicists for much of the time since Ogura's work became known. The field is in a rapid state of change at the moment, but some of the simplest and most promising results have come from adding an eddy damping term to Eqn. (8.12). The result can be shown to be very nearly equivalent to the variable eddy viscosity models discussed above.

The informed student will note the absence of any explanation or description of the prodigious, almost solo efforts at producing sophisticated closure models by R. Kraichnan (1964, 1965, 1966, 1968 among numerous other papers). The lecturer feels that to do justice to Kraichnan's work would require at least the entire two-week period of these lectures. There is still much doubt as to how practical the models are in terms of their application to a wide variety of turbulence problems or (as given special emphasis here) as a recipe for specifying the statistical behavior of the submesh scales of numerical simulation models.

9. THE THEORETICAL PREDICTABILITY PROBLEM

It is now widely recognized that operational forecasts of at least the large scale aspects of the atmosphere have been considerably improved by the introduction of numerical simulation. Experimental forecasts utilizing sophisticated models (Miyakoda, et al., 1969) appear to verify considerably better. There is reason to expect, therefore, that continued improvement of the models and computing facilities, together with a major enhancement in the quality and quantity of global meteorological observational data will produce useful predictions for time periods at least twice as long as the current limit. But what should be our ultimate expectation ? Also, what can we expect in the way of shorter term small scale predictions by numerical simulation, assuming that practical methods of observation can be developed ?

The evidence is accumulating that a large share of the answers to the above must come from turbulence theory, or alternatively that the theoretical predictability problem and the turbulence closure problem will be solved together if at all. The greatest amount of evidence has come from the simulation experiments themselves, but other lines of research have been converging in this direction also.

The basic aspect of turbulence that is relevant here is that it is apparently totally unpredictable beyond some time limit. Turbulence consists of a continuing series of flow instabilities. Although the total variance of the various fields will remain bounded, any introduced uncertainty continues to grow, so that two initial states differing by some infinitesimal amount will diverge until they differ as much as two randomly chosen states having the same statistical properties. These properties had been predicted by Lorenz (1963, 1969) and have been largely verified in numerical simulations of two-dimensional turbulence (Lilly, unpublished). At the same time these simulation results are apparently similar in important respects to the earlier predictability tests of numerical general circulation models.

Lorenz's hypothesis for error growth and its tests by numerical simulation

Lorenz (1969) recently proposed a theory for the growth of observational errors leading to deterioration of predictions based on numerical simulation. We will describe his theory and some numerical experiments designed in part to test it.

Lorenz begins by assuming that the larger scales of motion are accurately observed and specified as initial conditions but that beyond some small scale limit there is effectively no knowledge of the initial state. This seems intuitively reasonable. Beyond that it can be shown that if the basic observational field consisted of pressure sensors at uniform space intervals, the scalar error energy spectrum in a quasi-geostrophic atmosphere would be proportional to k^3 for scales larger than the separation of data points, while for smaller scales there would be essentially no knowledge of the initial data.

Lorenz then proposes that uncertainty propagates and infects larger scales, through the nonlinear terms of the dynamic equations. This may appear to conflict with the principle of transfer of energy or enstrophy to higher frequencies in the inertial ranges of turbulence. There is no real conflict, however. These transfers occur in their preferred directions basically because of the existence of a viscous sink at high frequencies and a source at low frequencies of otherwise conservative properties. The nonlinear dynamics of fluids are basically a "scrambling" mechanism that would tend in the absence of sources and sinks to distribute amplitude uniformly among all Fourier modes. The "uncertainty energy" is a property with an initial source at small scales, which is therefore subject to transfer to large scales. Further, as we shall see, there is no conservation property for uncertainty energy until it becomes as large as the total flow energy, so the total amplitude of this property may grow rapidly.

Lorenz assumes that the time it takes before complete uncertainty at wave number 2k infects wave number k, so it also becomes completely uncertain, is proportional to the time scale of turbulent turn-over at scale k, $\tau(k)$. This time might be defined by k^{-1} divided by the characteristic velocity scale at wave number k, or in terms of energy spectra in an isotropic fluid, $\tau(k) = k^{-3/2}[E(k)]^{-1/2}$. Thus the predictability time for uncertainty to propagate N octaves, from say $2^N k$ to k is given by

$$\sum_{n=0}^{N-1} \tau(2^n k) = k^{-3/2}[E(k)]^{-1/2} + (2k)^{-3/2}[E(2K)]^{-1/2} + \ldots + (2^{N-1}k)^{-3/2}[E(2^{N-1}k)]^{-1/2} \quad (9.1)$$

It is interesting to consider this series when E(k) can be approximated by a power law. In the case of a three-dimensional inertial range, where $E(k) = \alpha\varepsilon^{2/3}k^{-5/3}$, the series becomes

$$\sum_{n=0}^{N-1} \tau(2^n k) = \alpha^{-1/2}\ \varepsilon^{-1/3}\ k^{-2/3}\left[1 + 2^{-2/3} + 2^{-4/3} + \ldots + 2^{-\frac{2}{3}(N-1)}\right]. \quad (9.2)$$

Remarkably, this series has a finite sum for $N \to \infty$, i.e.

$$\sum_{n=0}^{\infty} \tau(2^n k) = \frac{\alpha^{-1/2}\ \varepsilon^{-1/3}\ k^{-2/3}}{1 - 2^{-2/3}} \sim 2.7\alpha^{-1/2}\ \varepsilon^{-1/3}\ k^{-2/3} = 2.7\tau(k) \quad . \quad (9.3)$$

This says that the predictability period for the larger scales of three-dimensional turbulence is only a few times the turn-over time for those scales, essentially regardless of how well the initial state is known. This could be an extremely discouraging result for those interested in mesoscale forecasting by simulation methods.

For the two-dimensional enstrophy inertial range, $E(k) = \beta\eta^{2/3}\ k^{-3}$, and the predictability time series becomes

$$\sum_{n=0}^{N-1} \tau(2^n k) = \beta^{-1/2}\ \eta^{-1/3}(1+1+1+\ldots) = N\ \beta^{-1/2}\ \eta^{-1/3} \quad . \quad (9.4)$$

Here we see that the sum diverges as $N \to \infty$, so that in theory very long predictions would be possible with sufficiently good initial data. In the case of the atmosphere, where perhaps only 7 or 8 octaves of information are likely to be available as initial data, the difference is only a factor of 3 for even the planetary scale. Nevertheless, acceptance of (9.3) would lead us to believe that we are already near the optimum. Fortunately all evidence available indicates that if either of (9.3) or (9.4) is valid, the latter is based on a more realistic state.

Some two-dimensional turbulence simulation experiments have been conducted in an attempt to test Lorenz's predictions. The model used was that described in Eqn. (5.1) with surface drag, viscosity, and an

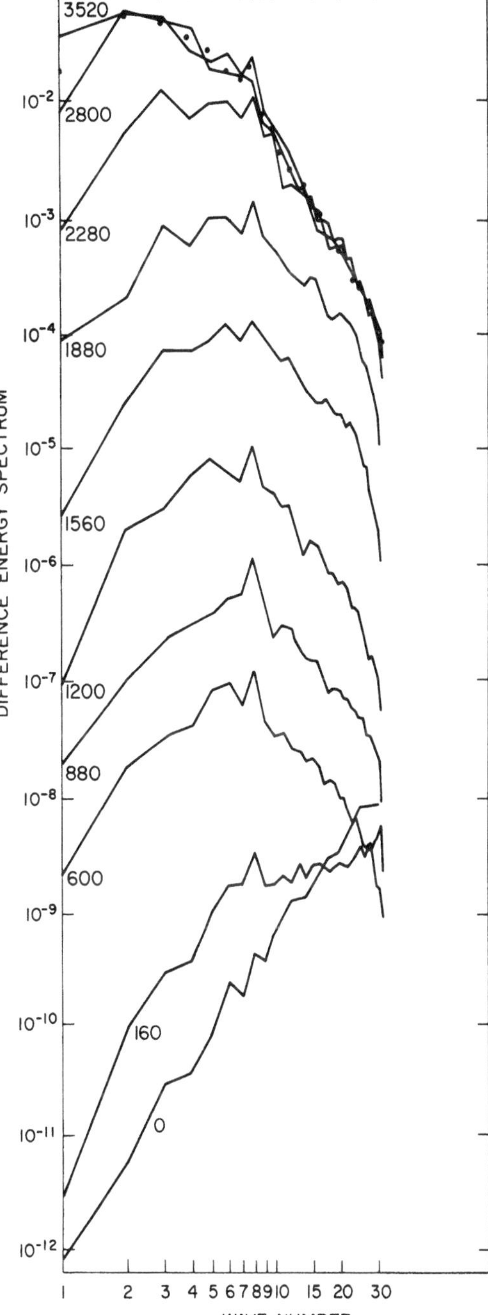

Fig. 9.1

The growth of the difference energy spectrum in a numerical predictability experiment. The various spectra are labelled by the time step to which they pertain (Δt = .01 dimensionless units). The solid points at the top are values of twice the energy spectrum of one of the realizations averaged over 4000 time steps. They also corresponds to the plotted points on Fig. 6.1.

energy source term. Initially the model was run long enough with a given set of parameters that a statistically steady-state was believed to exist. At the beginning of the predictability experiment two computer runs were made. Run #1 was the continuation of the initialization run, while for run 2 the initial conditions were altered by a small white noise perturbation in the stream function. The two runs were continued with the same time step and forcing function for several thousand time steps. Analysis was made of both the individual runs and the difference (uncertainty) and the experiment was terminated when the latter became as large in amplitude as runs #1 and #2 themselves.

Figure 9.1 shows a plot of the difference energy spectrum as it grew with time until becoming essentially identical with twice the energy spectrum of run 1. The most notable feature is that the difference energy spectrum apparently reached a self-preserving shape rather early in the experiment and then grew exponentially. The shape of this difference spectrum is not greatly different from the total energy spectrum of the separate runs, although somewhat skewed toward higher frequencies.

As a test of Lorenz's theory in detail, these experiments are a bit inconclusive. If one plots the time at which each wave number peak becomes unpredictable, it appears that essentially all the frequencies in the inertial range (above about wave number 8) lose predictability almost simultaneously, although for those of larger scale the uncertainty does seem to proceed toward larger scales. This result may be largely caused by insufficient resolution and therefore a poorly developed inertial range. In recent unpublished computations from a newly developed closure theory of turbulence Leith has essentially reproduced the above results. But when resolution is increased greatly, his model predicts that the difference spectrum slopes downward more shallowly than that of the total flow, so that uncertainty proceeds toward larger scale as Lorenz predicts.

The growth of uncertainty can be considered a linear perturbation problem up to the point when the difference amplitude is becoming nearly as large as the mean. If we define for any variable $\tilde{F} = (F_1 + F_2)/2$ and $F' = (F_1 - F_2)/2$ as the mean and difference, respectively, of two realizations, then Eqn. (5.1) can be expanded as

$$\frac{\partial \tilde{\omega}}{\partial t} + \tilde{u}\,\frac{\partial \tilde{\omega}}{\partial x} + \tilde{v}\,\frac{\partial \tilde{\omega}}{\partial y} + u'\,\frac{\partial \omega'}{\partial x} + v'\,\frac{\partial \omega'}{\partial y} = \tilde{F} + \nu\nabla^2\tilde{\omega} - K\tilde{\omega} \qquad (9.5)$$

$$\frac{\partial\omega'}{\partial t} + \tilde{u}\,\frac{\partial\omega'}{\partial x} + \tilde{v}\,\frac{\partial\omega'}{\partial y} + u'\,\frac{\partial\tilde{\omega}}{\partial x} + v'\,\frac{\partial\tilde{\omega}}{\partial y} = F' + \nu\nabla^2\omega' - K\omega' \quad . \tag{9.6}$$

When $|\omega'| << |\tilde{\omega}|$ everywhere, then the solution of Eqn. (9.6) is obviously identical with that of the linear stability problem of a mean flow which is two-dimensionally complicated and nonsteady.

Let us consider the case where F is the uncorrelated forcing function, so that $F' = 0$. By analogy with the known stability results for steady recti-linear flows we expect that growth of ω' will be mainly confined to regions of strong shear with a point of inflection. That some such effect is operating is indicated by inspection of the fields of ω' in the predictability simulation experiments. Figure (9.2) shows a typical example of ω, with $\tilde{\psi}$ and $\tilde{\omega}$ also shown for comparison on Figures 9.3 and 9.4. Obviously the difference vorticity is much more intermittent and inhomogeneous than the mean, and comparison with earlier and later times shows that the high amplitude regions are mostly those which are growing rapidly. However, the mean flow itself changes continually, and the regions of rapid growth do not stay long in the same place.

It is possible to solve Eqn. (9.6) directly with $\tilde{\psi}$ artificially held constant with time and thus look at the mode of instability which would grow most rapidly if the mean flow were fixed. Figure 9.5 shows the results of such a solution for ω', the difference of instability vorticity obtained for the mean stream field shown in Figure 9.6. The exponential rate of growth of the disturbance on the fixed mean flow is about twice as large as that of the difference field growing with respect to a changing mean flow and the intermittency is even more striking. Qualitatively this sort of result should be expected since in the perturbation experiment the disturbance has time to adjust its shape to correspond to the most rapidly growing mode.

10. THE PROBLEM OF VERTICAL EDDY FLUXES

We have thus far considered the effects of turbulence only in either fully three-dimensional or two-dimensional fluids, thus purposefully

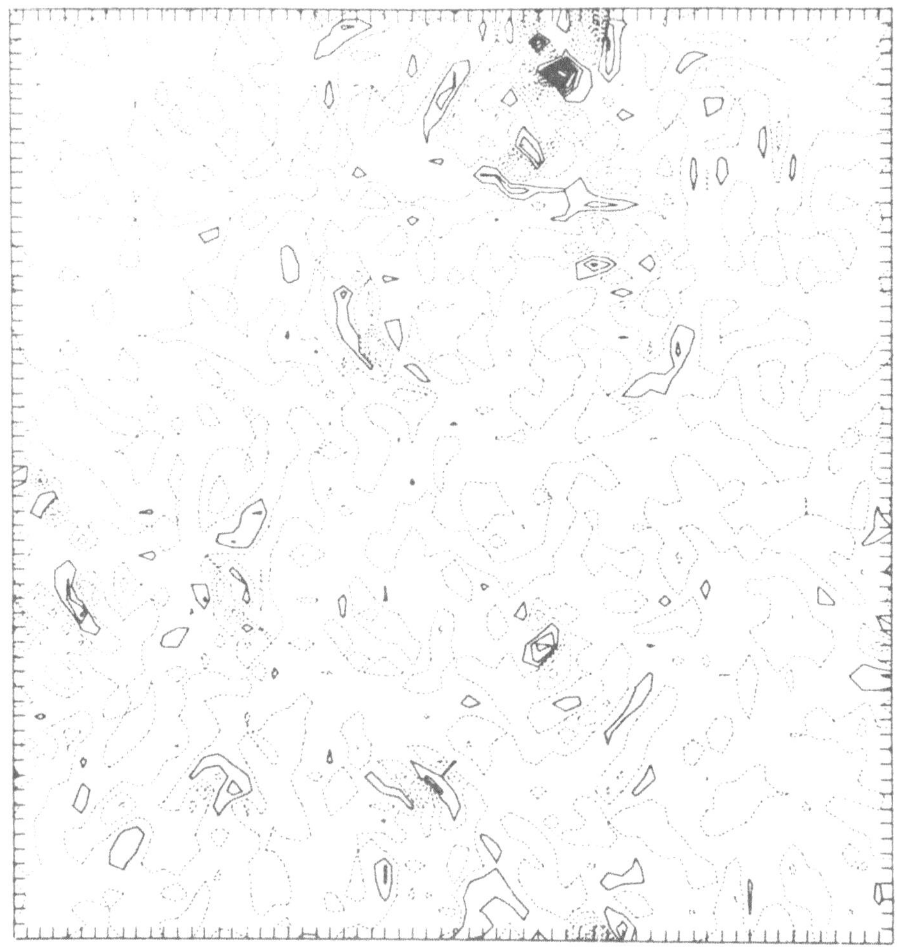

Fig. 9.2

Map of the difference vorticity generated during a numerical predictability experiment. Contour interval = 0.309.

Fig. 9.3

Map of the total vorticity for one of the realizations of the predictability experiment at the same time as Fig. 9.2. Contour interval = 1.33.

Fig. 9.4

Map of the stream field corresponding to Fig. 9.3.
Contour interval = 0.0526.

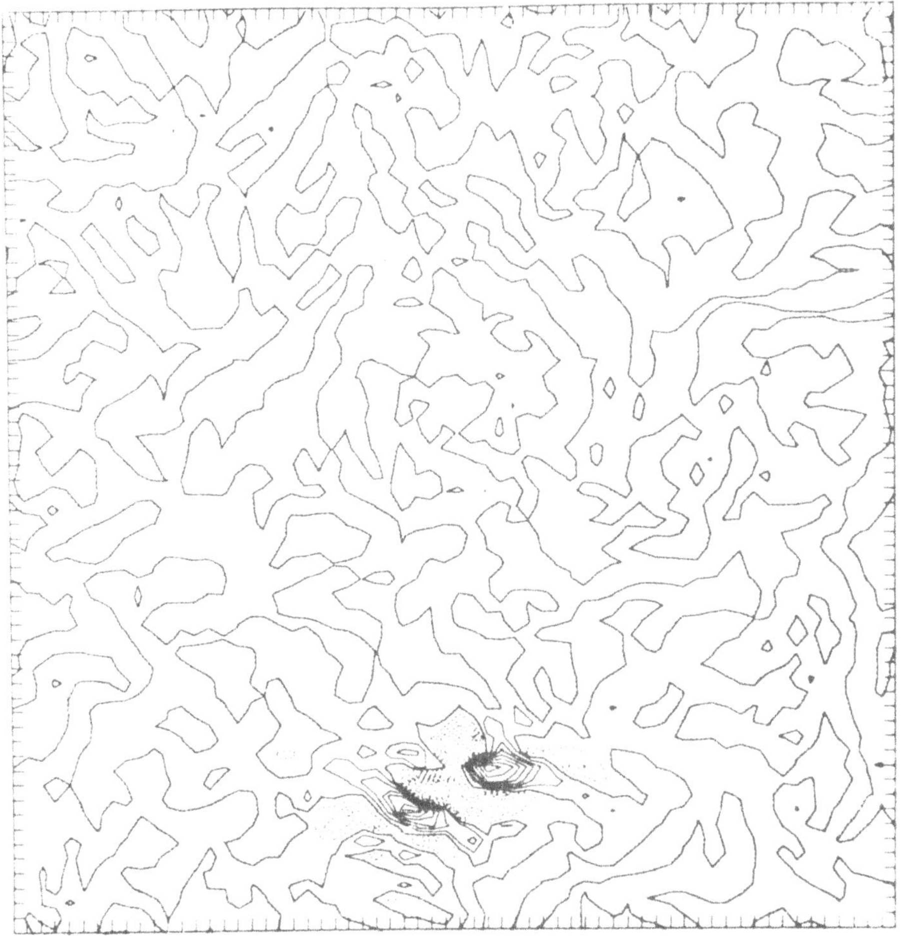

Fig. 9.5

Map of the perturbation vorticity after 2000 steps of integration of Eqn. 9.6. Only the region in lower center has significant amplitude.

Fig. 9.6

Map of the mean stream field on which the perturbation of Fig. 9.5 grows.

ignoring some of the most difficult aspects of the atmosphere as a quasi-two-dimensional fluid. These involve its ability to generate three-dimensional systems of motion in limited regions which can vertically transport large amounts of momentum, heat and moisture and then return to nearly two-dimensional flow. The resulting vertical eddy fluxes are crucial to some aspects of the general circulation and also are related to important small scale weather events, e.g. cloud convection including severe local storms, mountain winds, and clear air turbulence. Only in the case of the planetary boundary layer has much order been imposed on the chaotic structure of complex observational models and theoretical imagination. In some other cases the significant "events", as best they can be defined, are both rather high intensity and rare, so that probability estimates are not very useful and direct extrapolative forecasting suffers severely from inadequate data. Inertial range concepts are not useful because the largest scales are important.

Cloud convection systems.

This subject covers an enormous array of phenomena, ranging from radiatively active stratocumulus decks and the small cumuli tops of mostly invisible thermals to middle latitude and tropical squall systems. One area which has been studied with some success by means of laboratory (Deardorff, Willis, Lilly, 1969 ; Turner and Kraus, 1967) and theoretical models (Kraus and Turner, 1967 ; Ball, 1960) is penetrative convection under an inversion. A relatively simple situation arises when a stratocumulus deck exists over water in the presence of weak but sustained large scale subsidence, e.g. off the subtropical west coasts of large continental areas (Lilly, 1968). The combination of radiative cooling off the cloud tops and evaporation from the water surface allows development of a nearly steady-state and very strong inversion (up to 20°C) which contains the cloud in the lowest 1000 m or less below very dry middle tropospheric subsiding air. This regime is important to the general circulation for its special radiative characteristics and as the upwind source of the tropical trade wind inversion. It appears to be adequate to assume that the cloud region is in nearly reversible adiabatic equilibrium and that a fraction (probably small) of the kinetic energy made available from thermal convection goes into mixing some of the upper dry air downward and thus maintaining the cloud layer against subsidence. These assumptions plus suitable models of radiative cooling rates and boundary layer fluxes make possible predictions of the gross properties of the cloud layer from upper and lower boundary conditions and large scale subsidence only.

When the moist layer becomes deeper, due to weakened subsidence or surface heating, it becomes more inhomogeneous, containing a small fraction of active ascending cloud thermals, up to 90% decaying "dead" cumulus and large areas of drier subsiding air. This regime has not yet been modelled adequately in a way that allows prediction of its contributions to large scale vertical fluxes as a function of boundary layer and large scale motion field properties. Clearly, however, it is very sensitive to small changes in the large scale field of vertical motion. In particular a net upward vertical motion of $\lesssim$ 1 cm/sec maintained for 12 hours or so will apparently promote development of a tropical disturbance containing numerous tall Cb's from an initially undisturbed trade wind moist layer of ~ 2-3 km depth (Krishnamurti, 1968). Among the rather puzzling aspects of this type of development is the fact that neither before, during, or afterwards is there any very substantial change in the thermal stability of the air, which is nearly always conditionally unstable in the tropics. The strong convective instability events of middle latitudes and some tropical continents, the squall lines, are preceded by a greater than normal degree of thermal instability. Still, large scale convergence would probably be the best forecasting parameter, but one which is itself difficult to observe or predict.

Thermal convection is principally responsible for conveying upward the heat necessary to balance the net radiative cooling in the tropics and the middle latitudes in summer, with the relatively rare Cb systems doing most of the work in the upper troposphere (Riehl and Malkus, 1958). The momentum exchange produced by these systems is often incidental to their essential dynamics but may be important in the large scale dynamic budgets. In particular Newton (1969) has estimated that a U.S. squall line may develop a rate of downward eddy flux of westerly momentum during its 6-12 hours of life equal to about one-third of that required for the entire extratropical northern hemisphere.

The present state of understanding and utilization of cloud convection with respect to numerical simulation models of the large scale atmosphere is very crude indeed. Basically only two models have been applied. Manabe and associates (Manabe and Strickler, 1964 ; Smagorinsky, Manabe and Holloway, 1965 ; Manabe and Smagorinsky, 1967 ; Manabe, Holloway and Stone, 1970) used the "convective adjustment" model. In this it is taken that when a portion of the atmosphere becomes saturated in the presence of upward vertical motion, its vertical temperature gradient immediately adjusts to moist adiabatic. Unsaturated or descending motion regions are similarly constrained to be more stable than dry adiabatic. The effect of this model is to produce a simulated atmosphere in the tropics and other convectively active regions which is, as observed, conditionally unstable. It is usually unrealistically moist, however, since the adjustment and accompanying precipitation only is allowed when the atmosphere reaches saturation. In nature the mean humidity over a mesh-box area (200 km square) is almost never completely saturated through any substantial depth. In the region of simulated saturation, the convective adjustment process typically cools the lower levels of the atmosphere and warms the upper troposphere, thus producing a "cold-core" disturbance qualitatively resembling some of those found in the wet-season tropics. Usually the amplitude of the real disturbances (in winds and pressure drop) are several times smaller than those simulated, however. In an attempt to alleviate this discrepancy and to decrease the mean humidity of the simulated atmosphere Miyakoda, et al. (1969) have decreased the relative humidity required for moist adiabatic adjustment from 100% to 80%. This value or any other is, of course, arbitrary and difficult to justify on fundamental grounds.

The other principal model of convective-large scale interaction is that proposed by Charney and Eliassen (1964), Kuo (1965), and Ooyama (1969) and applied by Ooyama, Rosenthal (1970) and others to tropical cyclone simulation models. It has become commonly known as "Conditional Instability of the Second Kind" (CISK). Although the philosophy is different from that of the convective adjustment model the principal effective difference lies in how the heat of condensation is distributed. In the CISK models when saturation occurs somewhere in the troposphere the latent heat released is vertically distributed according to some predetermined formula, being usually proportional to the difference between the temperature of the atmosphere and that of a saturated undiluted parcel rising moist-adiabatically from cloud base. In contrast with the convective adjustment model there is no cooling at the bottom. Thus large scale disturbances produced by this mechanism are always warmer at all levels than their environment. It is, therefore, not possible to simulate the cooling produced by convective downdrafts with this model but tropical cyclones are easily generated.

Clear air turbulence, gravity and mountain lee waves.

In the above subjects several previously rather isolated lines of research are now becoming more connected through improved theoretical and observational understanding of their inter-relationships. First, the work of Kung (1966a, 1966b, 1967, 1969) has shown that a major fraction of the kinetic energy dissipation of the atmosphere occurs in the upper troposphere, probably associated with the sporadic regions of clear air turbulence which occur there. The dissipation is accompanied by degradation of large scale energy through an inertial range, and thus is given mainly by

$$\varepsilon \quad -(\rho\ \overline{\vec{V}'w'})\cdot(\partial\bar{\vec{V}}/\partial z) \qquad (10.1)$$

where the over-bar represents averaging over space and time scales large compared to the turbulence elements but small compared to the depth of the atmosphere and synoptic scales. The eddy momentum flux term in (10.1), $\rho\overline{\vec{V}'w'}$, is also likely to be of significance to both the general circulation and its component meteorological systems.

With regard to the generation of clear air turbulence it is now widely believed that the most commonly available machinery is unstable Kelvin-Helmholtz waves, whose instability depends almost entirely on the local Richardson number, R_i, where

$$R_i = \frac{\frac{g}{\theta}\frac{\partial\theta}{\partial z}}{\left[\frac{\partial\bar{V}}{\partial z}\right]^2} \qquad (10.2)$$

where θ is potential temperature. Drazin and Howard (1966) have shown that a necessary condition for instability of infinitesimal perturbations in a rectilinear inviscid flow is that $R_i < 1/4$, and in a number of kinds of flow profiles this is also a sufficient condition. Laboratory experiments (Clark, et al., 1969 ; Scotti and Corcos, 1969 ; Thorpe, 1969) tend to confirm the theory.

Some (mostly unpublished) data from aircraft observations support the existence of wave-like perturbations associated with CAT outbreaks, but a full observational test of the theory has not yet been attained. It is very difficult to make observations of Ri on the proper scales (~ 100's meters) and at times and places just prior to turbulence outbreaks. What is rather clear from aircraft observations (Panofsky, et al., 1968 ; Reiter and Burns, 1966) is that strong turbulence is usually associated with both strong shears and thermal stability. It is, of course, easier to reach the critical stability criterion $R_i = 1/4$ for weak stability, but then not much energy is available from the mean shear to be transformed into turbulence. But when the stratification is sufficiently large to allow the shear to become large (either by large scale or mesoscale mechanisms) then the breakdown, when it occurs, will be more violent. Thus, we see, for example, quite strong turbulence in the stratosphere on occasion, especially when produced by mountain waves. (Lilly and Toutenhoofd, 1969 ; Panofsky, et al., 1968 ; Prophet, 1970).

The Future.

It is probably safe to say that there is a considerable insufficiency of definitive data relevant to the interaction of the large scale atmosphere and the mesoscales which produce most of the vertical turbulent fluxes above the planetary boundary layer and that this insufficiency seriously handicaps the formulation and use of appropriate dynamical equations. It has always been difficult to make measurements on the mesoscale. The operational weather data is too coarse in resolution and small scale fixed data networks are too expensive to maintain for a long enough time to obtain data on relatively rare phenomena. Aircraft are commonly the logical choice for instrument platforms but until recently the navigational capabilities of aircraft were inadequate to measure accurately the fields of motion on the 10 km scale.

At the present time there are available, at high but not impossible costs, stabilized platforms for mounting on aircraft which allow measurement of horizontal velocities of ~.3 m/sec accuracy and vertical velocities to accuracies of .1 m/sec or better over scales from meters to hundreds of km. With a few such aircraft made available it should be possible to make direct measurements of both the mesh-box averaged flow and the turbulent stresses and fluxes in such programs as the GARP tropical experiment. The dynamicists can expect such data to be of great use to them in their future attempts to come to grips with the mesoscale and therefore should be prepared to spend much time and careful thought on the planning of the observational facilities and programs.

REFERENCES

BALL, F. K., 1960 : Control of inversion height by surface heating. Quart. J. Roy. Meteor. Soc., 86, 483-494.

BATCHELOR, G. K., 1956 : The Theory of Homogeneous Turbulence. Cambridge, England, Cambridge University Press, 197 pp.

BATCHELOR, 1969 : Computation of the energy spectrum in homogeneous two-dimensional turbulence. High Speed Computing in Fluid Dynamics, Phys. Fluids Suppl. II, 233-239.

BENTON, GEORGE S., and Arthur B. KAHN, 1958 : Spectra of large-scale atmospheric flow at 300 millibars. J. Meteor., 15, 404-410.

CHARNEY, J. G., and A. ELIASSEN, 1964 : On the growth of the hurricane depression. J. Atmos. Sci., 21, 68-75.

CLARK, J. W., R. C. STOEFFLER, and P. G. VOGT, 1969 : Research on Instabilities in Atmospheric Flow Systems Associated with Clear Air Turbulence. United Aircraft Research Laboratories, Rept. H910563-9 (Nasa Contract NASW - 1582), East Hartford, Connecticut, 66pp.

CROW, S. C., 1968 : Viscoelastic properties of fine-grained incompressible turbulence. J. Fluid Mechs., 33, 1-20.

DEARDORFF, J. W., G. E. WILLIS, and D. K. LILLY, 1969 : Laboratory investigation of non-steady penetrative convection. J. Fluid Mechs., 37, 7-31.

DEARDORFF, J. W., 1970 : A numerical study of three-dimensional turbulent channel flow at large Reynolds numbers. J. Fluid Mechs., 41, 453-480.

DRAZIN, P. G., and L. N. HOWARD, 1966 : Hydrodynamic stability of parallel flow of inviscid fluid. Adv. Appl. Mech., 9, 1-89.

FJØRTOFT, R., 1953 : On the changes in the spectral distribution of kinetic energy for two-dimensional non-divergent flow. Tellus, 5, 225-230.

HINZE, J. O., 1959 : Turbulence. New York, McGraw-Hill.

JEFFREYS, Harold, 1926 : On the dynamics of geostrophic winds. Quart. J. Roy. Meteor. Soc., 52, 85-104.

JULIAN, Paul R., W. M. WASHINGTON, L. HEMBREE, and C. RIDLEY, 1970 : On the spectral distribution of large-scale atmospheric kinetic energy. J. Atmos. Sci., 27, 376-387.

KRAICHNAN, Robert H., 1964 : Decay of isotropic turbulence in the direct-interaction approximation. Phys. Fluids, 7, 1030-1048.

KRAICHNAN, Robert H., 1965 : Lagrangian-history closure approximation for turbulence. Phys. Fluids, 8, 575-598.

KRAICHNAN, Robert H., 1966 : Isotropic turbulence and inertial-range structure. Phys. Fluids, 9, 1728-1752.

KRAICHNAN, Robert H., 1967 : Inertial ranges in two-dimensional turbulence. Phys. Fluids, 10, 1417-1423.

KRAICHNAN, Robert H., 1968 : Convergents to infinite series in turbulence theory. Physical Rev., 174, 240-246.

KRAUS, E. B., and J. S. TURNER, 1967 : A one-dimensional model of the seasonal thermocline. Part II. The general theory and its consequences. Tellus, 19, 98-106.

KRISHNAMURTI, T. N., 1968 : A calculation of percentage area covered by convective clouds from moisture convergence. J. Appl. Meteor., 7, 184-195.

KUNG, E. C., 1966a : Kinetic energy generation and dissipation in the large-scale atmospheric circulation. Mon. Weather Rev., 94, 67-82.

KUNG, E. C., 1966b : Large-scale balance of kinetic energy in the atmosphere. _Mon. Weather Rev._, 94, 627-640.

KUNG, E. C., 1967 : Diurnal and long-term of the kinetic energy generation and dissipation for a five-year period. Mon. Weather Rev., 95, 593-606.

KUNG, E. C., 1969 : Further study on the kinetic energy balance. _Mon. Weather Rev._, 97, 573-581.

KUO, H. L., 1965 : On formation and intensification of tropical cyclones through latent heat release by cumulus convection. _J. Atmos. Sci._, 22, 40-63.

LEE, T. D., 1952 : On some statistical properties of hydrodynamical and magneto-hydrodynamical fields. _Quart. Appl. Math._, 10, 69-74.

LEITH, C. E., 1969 : Numerical simulation of turbulent flow. In _Properties of Matter under Unusual Conditions_. New York, John Wiley, 267-271.

LILLY, D. K., 1966 : On the application of the eddy viscosity concept in the inertial sub-range of turbulence. NCAR Ms. No. 123 (January), 19 pp.

LILLY, D. K., 1967 : The representation of small-scale turbulence in numerical simulation experiments. In _Proceedings of the IBM Scientific Computing Symposium on Environmental Sciences_, IBM Data Processing Division, White Plains, N.Y., 195-210.

LILLY, D.K., and Hans A. PANOFSKY, 1967 : Summary of progress in research on atmospheric turbulence and diffusion. _Trans. Amer. Geophy. Union_, 48, 449-453.

LILLY, D. K., 1968 : Models of cloud-topped mixed layers under a strong inversion. _Quart. J. Roy. Meteor. Soc._, 94, 292-309.

LILLY, D. K., 1969 : Numerical simulation of two-dimensional turbulence. _Phys. Fluids Suppl. II_, 240-249.

LILLY, D. K., and W. TOUTENHOOFD, 1969 : The Colorado lee wave program. In Clear Air Turbulence and Its Detection. New York, Plenum Press, 232-245.

LILLY, D. K., 1971 : Numerical simulation of developing and decaying two-dimensional turbulence. Accepted for publication in J. Fluid Mechs.

LORENZ, Edward N., 1963 : Deterministic nonperiodic flow. J. Atmos. Sci. 20, 130-141.

LORENZ, Edward N., 1969 : The predictability of a flow which possesses many scales of motion. Tellus, 21, 289-307.

LUMLEY, J. L., and H. A. PANOFSKY, 1964 : The Structure of Atmospheric Turbulence. New York, John Wiley.

MANABE, S., and STRICKLER, R., 1964 : Thermal equilibrium of the atmosphere with a convective adjustment. J. Atmos. Sci., 21, 361-385.

MANABE, S., and J. SMAGORINSKY, 1967 : Simulated climatology of a general circulation model with a hydrologic cycle. II. Analysis of the tropical atmosphere. Mon. Weather Rev., 95, 155-169.

MANABE, S., J. L. HOLLOWAY, and H. M. STONE, 1970 : Tropical circulation in a time-integration of a global model of the atmosphere. J. Atmos. Sci., 27, 580-613.

MIYAKODA, K., J. SMAGORINSKY, R. F. STRICKLER, and G. D. HEMBREE, 1969 : Experimental extended predictions with a nine-level hemispheric model. Mon. Weather Rev.,93, 1-76.

NEWTON, C. W., 1969 : The role of extratropical disturbances in the global atmosphere. In The Global Circulation of the Atmosphere, ed. G. A. Corby, London, Salisbury Press, 137-158.

OGURA, Y., 1958 : On the isotrophy of large-scale disturbances in the upper troposphere. J. Meteor., 15, 375-382.

OGURA, Y., 1963 : A consequence of the zero-fourth-cumulant approximation in the decay of isotropic turbulence. *J. Fluid Mechs.*, 16, 33-40.

OOYAMA, K., 1969 : Numerical simulation of the life cycle of tropical cyclones. *J. Atmos. Sci.*, 26, 3-40.

ORSZAG, Steven A., 1970 : Transform method for the calculation of vector-coupled sums : Application to the spectral form of the vorticity equation. *J. Atmos. Sci.*, 27, 890-895.

PANOFSKY, H. A., J. A. DUTTON, K. H. HEMMERICH, G. McCREARY, and N. V. LOVING, 1968 : Case studies of the distribution of CAT in the troposphere and stratosphere. *J. Appl. Meteor.*, 7, 384-389.

PROPHET, David T., 1970 : High altitude clear air turbulence probability based on temperature profiles and rawinsonde ascensional rates. *Mon. Weather Rev.*, 98, 704-707.

REITER, E. R., and A. BURNS, 1966 : The structure of clear air turbulence derived from "TOPCAT" aircraft measurements. *J. Atmos. Sci.*,23, 206-212.

RIEHL, H., and J. S. MALKUS, 1958 : On the heat balance in the equatorial trough zone. *Geophysica*, 6, 503-538.

ROSENTHAL, S. L., 1970 : A circularly symmetric primitive equation model of tropical cyclone development containing an explicit water vapor cycle. *Mon. Weather Rev.*, 98, 643-663.

SCOTTI, R. S., and G. M. CORCOS, 1969 : Measurements on the growth of small disturbances in a stratified shear layer. *Radio Sci.*, 4, 1309-1313.

SMAGORINSKY, J., 1963 : General circulation experiments with the primitive equations. I. The basic experiment. *Mon. Weather Rev.*, 91, 99-164.

SMAGORINSKY, J., S. MANABE, and J. L. HOLLOWAY, 1965 : Numerical results from a nine-level general circulation model of the atmosphere. *Mon. Weather Rev.*, 93, 727-768.

THORPE, S. A., 1969 : Experiments on the stability of stratified shear flows. Radio Sci., 1327-1331.

TURNER, J. S., and KRAUS, E. B., 1967 : A one-dimensional model of the seasonal thermocline. Part I. A laboratory experiment and its interpretation. Tellus, 19, 88-97.

VON NEUMANN, J., and R. D. RICHTMYER, 1950 : A method for the numerical calculation of hydrodynamic shocks. J. Appl. Phys., 21, 232-237.

WARD, Fred, 1964 : General circulation of the solar atmosphere from observational evidence. Pure Appl. Geophys., 58, 157-186.

WARD, Fred, 1965 : The general circulation of the solar atmosphere and the maintenance of the equatorial acceleration. Astrophys. J., 141, 534-547.

WINN-NIELSEN, A., 1967 : On the annual variation and spectral distribution of atmospheric energy. Tellus, 19, 540-559.

BOUNDARY LAYERS IN PLANETARY ATMOSPHERES

A.S. MONIN

Institute of Oceanology
Academy of Sciences
MOSCOW, (U.S.S.R.)

CONTENTS

BOUNDARY LAYERS IN PLANETARY ATMOSPHERES

By

A.S. MONIN

Institute of Oceanology, Academy of Sciences, moscow, USSR

1. DEFINITION OF THE ATMOSPHERIC BOUNDARY LAYER

In large-scale air currents, near the surface of a planet, the combined action of the pressure gradient, turbulent friction and Coriolis force results in the formation of the atmospheric boundary layer. Unlike most boundary layers dealt with in aerodynamical engineering, the atmospheric boundary layer is characterized by the influence of the Coriolis force (i.e., the planet's rotation) and the density stratification of air (affecting turbulence through buoyancy forces). Thus the atmospheric boundary layer is a turbulent boundary layer in a rotating stratified fluid.

Only in the equatorial zone and inside rapidly rotating tropical storms is the Coriolis force negligible. Then, the thickness of the atmospheric boundary layer can change only along the wind direction, growing with its fetch in a manner similar to the thickness of a boundary layer near a solid body. To define this boundary layer we introduce characteristic scales : the scale of length L and the scale of velocity U. Using molecular kinematic viscosity ν we determine the Reynolds number $Re = \frac{UL}{\nu}$ and then the boundary layer thickness is defined by $\delta \sim \frac{L}{\sqrt{Re}}$. One of the most important characteristics of the boundary layer is the surface stress τ which is defined by

$$\tau \sim \rho\nu \frac{U}{\delta} \sim \rho\sqrt{\frac{\nu U^3}{L}} \qquad (1)$$

Its ratio to ρU^2 is called the friction coefficient C_f :

$$C_f = \frac{\tau}{\rho U^2} \sim \frac{1}{\sqrt{Re}} \tag{2}$$

In a boundary layer near a flat plate we introduce the distance X from the upwind edge of the plate and define the scale of length as L = X. Then $Re = \frac{UX}{\nu}$, and

$$\delta \sim \sqrt{\frac{\nu x}{U}}\ ;\ \tau \sim \rho\sqrt{\frac{\nu U^3}{x}} \tag{3}$$

One could try to apply this theory to atmospheric boundary layers in typhoons and in the equatorial zone but observational data necessary to confirm theoretical predictions are rather poor up to now. Beyond the particular regions mentioned above one must explicitely take the Coriolis force into account and the thickness of the boundary layer can be determined by the equation h = c(G/f), where G is the wind velocity at the upper boundary of the layer, $f = 2\omega\sin\phi$ is the vertical projection of the vorticity of the planet's rotation, called the Coriolis parameter (ω being the angular velocity of the planet's rotation and ϕ the latitude) and c is a nondimensional factor. Such an atmospheric boundary layer is called an Ekman layer in honour of Ekman (1) (1905), who was the first to construct a theoretical model of the boundary layer in a rotating fluid. We shall henceforth confine ourselves to the consideration of the Ekman layer alone.

According to Charney (2), an atmospheric boundary layer is hydrodynamically stable (and may be steady) only as long as its Reynolds number Re = (Gh/K) (where K is the effective value of turbulent viscosity) does not exceed some critical value Re_{cr}. Taking : $K \approx h^2 f$ we obtain the criterion $\frac{G}{hf} \lesssim Re_{cr}$ or $h \gtrsim (G/Re_{cr} f)$ for stability of an Ekman layer. In their experiments on the stability of the laminar Ekman layer Faller (3) Tatro and Mollo-Christiansen (4) & Green (5) obtained $Re_{cr} \approx 100$. If this estimate is extended into the turbulent Ekman layer the minimum thickness of a stable layer for the Earth, with G = 10 m/sec and $f = 10^{-4} sec^{-1}$, appears to be 1 km, i.e., one order of magnitude less than the effective thickness of the Earth's atmosphere, $h_a \approx 10$ km. In other words, the Ekman layer on the Earth is thin, and from this point of view the Earth should be considered a rapidly rotating planet.

The scales of inhomogeneities of meteorological fields much larger and much smaller than the effective thickness of the atmosphere h_a (say, the so-called "thickness of a homogeneous atmosphere" $h_a = \frac{p_0}{\rho_0 g}$), should reasonably be called "large" and "small" respectively (Kolesnikova & Monin, (6, 7)). The large-scale inhomogeneities are evidently quasi-two-dimensional (quasi-horizontal), whereas the small-scales ones, on the contrary, are essentially three-dimensional (quasi-isotropic) ; the latter are the object of micrometeorology. Thus the atmospheric boundary layer on the Earth is a micrometeorological formation.

Let us define the time scale (or the characteristic period) as $T = \frac{L}{U}$ and corresponding frequency as $f = \frac{2\pi}{T} \approx \frac{U}{L}$. Then the spectrum of scales L of atmospheric motions is equivalent to spectra of their periods T or frequencies f. This spectrum is shown on Fig. 1.

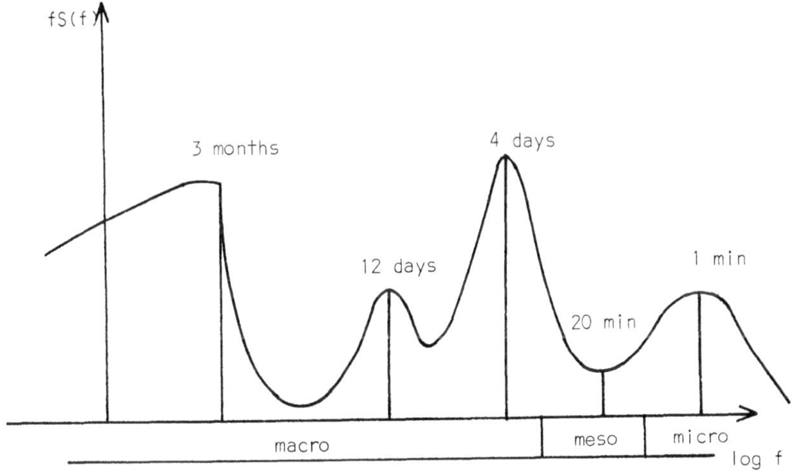

Fig. 1. Spectrum of periods of atmospheric motions ; f = frequency, S(f) = spectral density of energy.

There is a pronounced micrometeorological maximum in the spectrum near the period $T \sim 1$ min, then a mesometeorological gap in region 1 day $> T > 1$ min, then a very well pronounced synoptic maximum near the period $T \sim 4$ days, then a secondary maximum at $T \sim 12$ days (corresponding to the so-called "index-cycle" or cycle of the zonal circulation oscillations). Some recent estimates of spectral density in the range $T > 12$ days (by V. N. Kolesnikova in Moscow) seem to show that there exists another gap in spectrum, a very sharp rise near the period $T \sim 2{,}7$ months and then a rather slow fall of the energy spectral density. If this behavior of the spectrum is confirmed by future data it could be of utmost importance for the long-term weather prediction.

Turbulence is the most important of the micrometeorological phenomena ; its presence is an essential feature of the atmospheric boundary layer. Therefore a hydromechanical description of the layer requires the use of methods from the theory of turbulence. This description should be first of all statistical, based on a certain operation of averaging and a consideration of the resulting mean hydromechanical quantities.

Practically, averaging in time is used in spite of the fact that the spectral power density of atmospheric motions is continuous, so that mean values are not quite stationary but rather do change, depending upon the period of averaging. Yet, this difficulty is overcome because of the presence of a deep mesometeorological minimum (discovered for the first time in the wind-velocity spectrum by van der Hoven (8) (1957))in the spectrum of motions between the large-scale and micrometeorological domains, in the range of periods from minutes to hours, centered near the period $h_a/G \simeq 20$ min. In (6.7), a proposed explanation of this minimum is based on the strong damping action of turbulent viscosity in the mesometeorological range. On account of this minimum, averaging over periods of the order of 10-20 minutes yields statistically stable results.

Large scale processes or "weather conditions" are external parameters determining the internal structure of the atmospheric boundary layer and, first of all, the regime of turbulence. The latter, in turn, exerts a reverse influence on the large-scale processes as a sink of large scale kinetic energy and a source of upwelling fluxes (with a vertical velocity determined in a simple model of the boundary layer, by Dyubyuk's (9) equation :

$$w \sim h\Delta p/f\rho_a$$

where Δp is the horizontal pressure Laplacian and ρ_a the air density. The boundary layer is in effect the agency which produces the exchange of momentum, heat, humidity, and other pollutants between the underlying surface and the free atmosphere. This transfer to the free atmosphere circulation is particularly efficient in the regions of cumulus convection and in the regions where the layer becomes separated from the surface i.e. at atmospheric fronts and in typhoons (see Priestley (10)).

Thus a knowledge of the atmospheric boundary layer is necessary for computing the evolution of large-scale atmospheric processes for long-term weather forecasts and for the development of a theory of climate. Besides, turbulence in the atmospheric boundary layer is of paramount importance in determining the wind loading of constructions (buildings, bridges, towers, electric lines), the generation of wind waves and drift currents in the ocean, the buffeting of aircrafts, the diffusion of atmospheric pollutants, the fluctuations of light and radiowaves propagation in the atmosphere from land and space sources, the scattering of short radiowaves creating conditions for over-the-horizon UHF telecommunications, and many other phenomena of practical importance.

2. MODEL OF THE ATMOSPHERIC BOUNDARY LAYER

Observation shows that the structure of the atmospheric boundary layer and particularly, the vertical wind profile, do not by any mean always follow similarity rules and are often very irregular. For instance, a wind-velocity maximum is observed sometimes at relatively small heights of 100 to 300 m, and assumes, in the most pronounced cases, the form of a so-called "low level jet". Irregular changes of wind direction with height with no resemblance whatsoever to the "Ekman spiral" are also observed rather often. Such irregularities may be a consequence of a number of complicating factors : density stratification of the atmosphere, horizontal inhomogeneity of the temperature field (creating a so-called thermal wind (see for instance, Utina (11)), irregularities of the underlying surface (see Laikhtman (12) and (13-17)), irregular density and curvature of isobars (cyclonicity increases and anticyclonicity decreases the wind veering in the layer (see for instance (18)), orographic effects, nonstationarity effects, etc.

Yet even in the absence of these factors of inhomogeneity, the structure of the boundary layer proves to be rather complicated. Therefore we shall confine ourselves mainly to discussion of the simplest case, that of a stationary atmospheric boundary layer above a plane and homogeneous underlying surface with rectilinear and evenly distributed isobars, i.e., with a pressure field :

$$p(x,y,z) = P(z) + \rho_0 fG(x \sin\alpha - y \cos\alpha) \qquad (4)$$

Here x, y, z are Cartesian coordinates, the z-axis is along the vertical and the x-axis coincides with the direction of the surface wind; α is the angle between the isobars and the surface wind, ρ_0 is the standard average air density in the layer; G is the geostrophic wind velocity. With the pressure field (4), the Reynolds equations for the mean motion read :

$$\begin{aligned} f(\bar{v} - G \sin\alpha) + \frac{\partial}{\partial z}\left(\frac{\tau_{xz}}{\rho_0} + \nu\frac{\partial \bar{u}}{\partial z}\right) &= 0 \\ -f(\bar{u} - G \cos\alpha) + \frac{\partial}{\partial z}\left(\frac{\tau_{yz}}{\rho_0} + \nu\frac{\partial \bar{v}}{\partial z}\right) &= 0 \end{aligned} \qquad (5)$$

Here $\bar{u}$, $\bar{v}$ are the components of the mean wind velocity, ν is the coefficient of molecular viscosity of air, and $\tau_{xz} = -\rho_0\overline{u'w'}$, $\tau_{yz} = -\rho_0\overline{v'w'}$ are the components of the vertical turbulent momentum flux (Reynolds stresses), w' being the fluctuation of the vertical velocity; here and below, an over-bar means averaging and primes are used to identify fluctuations, i.e. deviations from average values. In view of our choice of direction for the x-axis $\frac{\tau_{yz}}{\rho_0} + \nu\frac{\partial \bar{v}}{\partial z}$ vanishes at the surface and the value of $\frac{\tau_{xz}}{\rho_0} + \nu\frac{\partial \bar{u}}{\partial z}$ approaches some positive limit u_*^2; the parameter u_* is called the friction velocity.

In order to describe the density stratification in the atmospheric boundary layer in term of the vertical temperature and humidity profiles, let us consider the equations for the mean heat and humidity transfer which, with our present assumption of statistical stationarity and horizontal homogeneity, and in the absence of phase transformations of water vapor, reduce to :

$$\rho_0 T_0 \overline{s'w'} - C_p\rho_0\chi\frac{\partial \bar{T}}{\partial z} + H_R = \text{const} \qquad (6)$$

$$\frac{H_L}{\mathcal{L}} - \rho_0 D\frac{\partial \bar{q}}{\partial z} = E = \text{const} \qquad (7)$$

Here S is the entropy, T the temperature (T_0 is the standard average temperature in the layer), q the specific humidity. $C_p \simeq 1$ Joule/g.deg and $\mathcal{L} \approx 2500$ Joule/g are the specific heat capacity of air under constant pressure and latent heat of evaporation. $C_p \rho_0 \chi$ and $\rho_0 D$ are the coefficients of molecular conductivity and water vapor diffusivity. H_R is the vertical radiative heat flux, $H_L = \mathcal{L}\rho_0 \overline{q'w'}$ is the vertical latent heat flux and E is the vertical humidity flux which, being constant, is equal to the rate of evaporation at the surface. According to reference (19), the first term in Equation (6) should be written in the form :

$$\rho_0 T_0 \overline{s'w'} = H_T + \left(\frac{S_v - S_d}{C_p}\right)\left(\frac{C_p T_0}{\mathcal{L}}\right) H_L$$

where $H_T = C_p \rho_0 \overline{T'w'}$ is the vertical turbulent heat flux, and S_v and S_d are the specific entropies of water vapor and dry air respectively. If vertical changes of the radiative heat flux H_R (usually resulting in slow evolution) are neglected, equation (6) reduced to :

$$H_T - C_p \rho_0 \chi \frac{\partial \bar{T}}{\partial z} = H = \text{const} \tag{6a}$$

The principle of the similarity theory of the boundary layer consists in assuming that the structure of the layer depends solely upon these internal parameters H, E and u_*.

3. ENERGY EQUATION

The influence of density stratification on turbulence is clearly explained with the aid of the turbulent energy equation. For a stationary and horizontally homogeneous atmospheric boundary layer (see §§ 6.2 - 6.6 of the book by Monin and Yaglom (20)), this equation reads :

$$\left(\frac{\tau_{xz}}{\rho_0}\frac{\partial \bar{u}}{\partial z} + \frac{\tau_{yz}}{\rho_0}\frac{\partial \bar{v}}{\partial z}\right) - \frac{g}{\rho_0}\overline{\rho'w'} - \frac{1}{\rho_0}\frac{\partial Q}{\partial z} = \varepsilon \tag{8}$$

where g is the acceleration of gravity, Q is the vertical flux of turbulent energy, and ε is the rate of energy dissipation per unit mass. The first term in parentheses on the left side of (8) describes the generation of turbulent energy caused by the work of Reynolds stresses. The next term is the gain or loss of turbulent energy caused by the work of buoyancy forces under stable or unstable stratification. According to (19), the buoyancy term can be written :

$$-\frac{g}{\rho_0}\overline{\rho' w'} = \frac{\beta(H_T + bH_L)}{C_p \rho_0} \tag{9}$$

where β is the buoyancy parameter, i.e. the product of g by the thermal expansion coefficient, equal to $1/T_0$ for air and $b = (R_v/R_d - 1)C_p T_0/\mathcal{L} \approx 0.07$. R_v = 0.461 Joule/g.deg and R_d = 0.287 Joule/g.deg are the gas constants for water vapor and dry air respectively ($B_0 = H_T/H_L$ is called the Bowen ratio). Thus if $H_T + bH_L \neq 0$ the presence of heat and humidity fluxes H_T and E affects turbulent energy so that neither heat nor humidity can be considered passive substances. Often (but not always) the contribution of humidity to density stratification is small and for simplicity we shall neglect this contribution further on. The relative contribution of density stratification to the generation of turbulent energy may be characterized by the ratio of the work of the buoyancy forces to the work of Reynolds stresses :

$$Rf = -\frac{\beta H_T}{C_p\left(\tau_{xz}\frac{\partial \bar{u}}{\partial z} + \tau_{yz}\frac{\partial \bar{v}}{\partial z}\right)} \tag{10}$$

which is called the flux Richardson number.

The requirement of stationarity seems to be the most limiting simplification accepted in our model (Equations 5, 6, 7). Under natural conditions it is often disturbed by both the synoptic evolution of the pressure field (through usually slow) and the diurnal changes of radiative heat influx leading to changes of thermal stratification, thus turbulence, and finally wind profile. In the model, such changes can be described by accepting the hypothesis of quasi-stationarity, i.e., by assuming all parameters of the atmospheric boundary layer to be dependent on time only through the internal parameters u_*, H, and E or the external parameters G, $\delta T = T_h - T_s$ and $\delta q = q_h - q_s$ which determine them (here and below, s and h denote values at the surface and at the upper boundary of the layer, respectively).The behavior of quasi-stationary and non-stationary models of the diurnal changes of the atmospheric boundary layer will be compared later.

4. STRUCTURE OF THE ATMOSPHERIC BOUNDARY LAYER (SURFACE LAYER, DYNAMIC SUBLAYER, VISCOUS SUBLAYER)

The lower part of the atmospheric boundary layer, where only small changes of the vertical heat and humidity fluxes (equations 6a and 7) and also of the vertical momentum flux components :

$$\frac{\tau_{xz}}{\rho_0} + \nu \frac{\partial \bar{u}}{\partial z} \approx u_*^2 \quad , \quad \frac{\tau_{yz}}{\rho_0} + \nu \frac{\partial \bar{v}}{\partial z} \approx 0 \tag{11}$$

are observed, is called the surface layer. In other words, the surface layer is the lower part of the atmospheric boundary layer in which the action of the Coriolis force can be neglected : the surface layer should resemble the wall region of boundary layers in nonrotating stratified fluid. The thickness of the surface layer is measured in tens of meters. In fact, we can show (see for instance, (20) § 6.6) that the thickness of the layer in which the conditions (11) are fulfilled within one percent does not exceed au_*^2/fG, where a is a numerical factor of the order of 0.2. Now $u_*/G \approx 0.05$ usually. Thus for G = 10 m/sec and $f = 10^{-4}$, the thickness of the surface layer is 50m. A similar estimate confirms rather well the quasi-stationarity of the layer with respect to diurnal and synoptic changes of the external parameters.

It is shown below that the effect of the density stratification on turbulence in the atmospheric boundary layer decreases as the surface is approached, and that there is a sublayer near the surface where the influence of stratification can be neglected ; the dynamic of this sublayer should resemble the wall region of a boundary layer in a homogeneous fluid). Consequently, turbulence in this sublayer is determined only by dynamic factors, which is why it is called the dynamic sublayer. We shall see later that the thickness of the dynamic sublayer is of the order of $C_p\rho_0 u_*^3/\beta H_T$. In the atmospheric boundary layer on the Earth, this thickness may change from several meters in the case of very strong hydrostatic instability or stability, to very large values under neutral stratification when H_T is close to zero (in which case the whole of the surface layer can be dynamic).

All properties of the dynamic sublayer are well determined by the two parameters ν and u_* and the roughness of the underlying surface, first of all by the mean height of roughness h_s. If $h_s \lesssim (\nu/u_*)$, the roughness does not affect the structure of the dynamic sublayer so that the surface is called dynamically smooth. Very near a smooth surface, a viscous sublayer is formed where Reynolds stresses are small compared with viscous stresses. By omitting the term τ_{xz}/ρ_0 of expression (11), we obtain a linear velocity profile :

$$\bar{u}(z) = \frac{u_*^2 z}{\nu}$$

in the viscous sublayer ; this law proves to apply up to height $z \lesssim 5(\nu/u_*)$ (about one millemeter, in the atmosphere). If $h_s >> \nu/u_*$ on the other hand, the flow near the surface consists of vortices formed around roughness elements, and no viscous sublayer exists ; in this case the surface is called dynamically rough. The land surface is always so.

5. DYNAMIC SUBLAYER AND AIR-SURFACE INTERACTION

For $z >> (\nu/u_*)$ or h_s (depending upon roughness) all dynamic quantities of the dynamic sublayer are determined by reasonably large turbulent motions, the regime of which depends very little on molecular viscosity. Consequently, these dynamic quantities can be functions of one dimensional parameter u_* only (and of course the height z). Such quantities include evidently the mean velocity gradient, which, from dimensional considerations, must be expressed by the following expression :

$$\frac{\partial \bar{u}}{\partial z} = \frac{u_*}{\kappa z}$$

(where κ is the so-called von Kàrmàn constant, for which measurements in the dynamic sublayer yield the value 0.4). The well-known logarithmic law follows :

$$\bar{u}(z_1) - \bar{u}(z_2) = \frac{u_*}{\kappa} \log \frac{z_1}{z_2} \tag{12}$$

The upper part of the dynamic sublayer in which this law obtains, is sometimes called the logarithmic sublayer. If the drag coefficient $C_f(z)$ is defined as the ratio of the turbulent stress $\rho_0 u_*^2$ to $\rho_0 \bar{u}^2$ and $\bar{u} = u_*/\sqrt{C_f}$ is introduced into equation (12), the resulting equation for C_f indicates that the value $z_0 = z \exp\{\kappa/\sqrt{C_f(z)}\}$ does not depend on the height z in the logarithmic sublayer. This parameter of the logarithmic sublayer can depend only on the size and form of mean roughness elements of the underlying surface and is therefore called the roughness parameter,(because the logarithmic sublayer is part of the dynamic sublayer. z_0 cannot depend on the density stratification). From this definition of z_0, it follows that $C_f = \kappa\ x^2/(\log z/z_0)^2$.Returning to $\bar{u} = u_*/\sqrt{C_f}$ we obtain finally :

$$\bar{u}(z) = \frac{u_*}{\kappa} \log \frac{z}{z_0} \tag{12a}$$

Above land, the values of z_0 are only a small fraction of the height h_s of the actual roughness elements. For example z_0 = 0.001 cm above a smooth snow surface, 0.03 cm above the sandy surface of deserts, 0.2 to 0.7 cm above mowed grass (h_s = 1.5 to 3 cm), 4 to 9 cm above grass 60 to 70 cm high (for wind velocity of 1.5 to 6 m/sec), 10 cm above shrubs and trees, 5 to 10 cm above forests ($h_s \sim 10$ m), and some 100 cm above cities (see for instance, Laikhtman (12) and Priestley (21) and one of the latest paper by Lettau (22)). The above-mentioned dependence of z_0 on wind velocity above high grass is explained by the bending of grass stems to the ground under the action of wind !

The logarithmic law (Eq. 12a) has so far been substantiated only on condition that $z >> h_s$. If z/h_s is not too large, h_s can affect the wind velocity profile and then we have :

$$\frac{\partial \bar{u}}{\partial z} = \frac{u_*}{\kappa z} \phi_1\left(\frac{h_s}{z}\right)$$

where ϕ_1 is some unknown function with $\phi_1(0) = 1$. Yet, if heights are counted from the level $z_1 = \phi_1'(0) \cdot h_s$, we still have a logarithmic law :

$$\frac{\partial \bar{u}}{\partial z} \simeq \frac{u_*}{\kappa(z - z_1)}$$

to the second order of the small quantity $h_s/(z-z_1)$ when z is replaced by $z - z_1$. The new reference level z_1 can be called the displacement height (by analogy with the displacement height from boundary-layer theory). For high vegetation z_1 usually lies between $h_s/2$ and h_s.

Above the sea surface, z_0 depends on a number of factors, and first of all, on the local wind velocity. Heights of roughness elements of the sea surface can be measured by the scale $h_s = u_*^2/g$. When the wind is weak :

$$u_* \lesssim (g\nu)^{1/3} \text{ and } h_s \lesssim (\nu/u_*)$$

the sea surface is dynamically smooth and $z_0 = m_0(\nu/u_*)$, with an empirical constant $m_0 \approx 0.1$. Under moderate or strong wind conditions, the sea-surface drag varies depending on wind duration or fetch. For $h_s \gg (\nu/u_*)$ and fully developed waves (when the average wave height exceeds 100 h_s) we may assume that $z_0 = m_1(u_*^2/g)$ (Charnock (23)), with a different empirical constant $m_1 \approx 0.035$. The drag coefficient C_f appears to depend on wind velocity, linearly at moderate winds, as found by Munk (24), and confirmed by some direct measurements of the momentum flux $u_*^2 = -\overline{u'w'}$ and wind profile $\bar{u}(z)$ (Zubkovsky et al. (25.26)). See an approach to the theoretical computation of the sea surface drag in the book of Phillips (27), and reviews of the experimental data in Roll (28) and Kitaigorodsky and Volkov (29,30).

Wind action on the sea surface results in the formation of the so-called drift current in the upper sea layer ; the direction of the drift current differs generally from the direction of the tangential wind stress (i.e., according to (Eq. 11) from the x-axis). The velocity of the current is usually one order of magnitude less than the friction velocity u_*. Let u_s, v_s be the components of this current which makes then an angle $\alpha_s = \tan^{-1}(v_s/u_s)$ with the x-axis. According to Faller's measurements (31) $\alpha_s \approx 13°$ in the temperate latitudes of the Northern Hemisphere. Laikhtman's theoretical computations (32) yield $\alpha_s \approx 5°$ to 15° in the Northern Hemisphere (and $\alpha_s < 0$ in the Southern Hemisphere).

In the dynamic sublayer, heat and humidity are passive substances, i.e. they do not exert a dynamic influence on turbulence and the buoyancy parameter β is not significant. At the same time, the hydrodynamic quantities associated with heat and humidity transfer through the dynamic sublayer should depend respectively on the parameters H and E determined by equations 6a and 7. Using these parameters and the friction velocity u_*, we may define scales of vertical change of temperature and specific humidity :

$$T_* = -\frac{H}{\kappa C_p \rho_0 u_*}, \quad q_* = -\frac{E}{\kappa \rho_0 u_*} \tag{13}$$

where the numerical factor κ has been introduced into the denominators for convenience. Signs were chosen so that $T_* < 0$ and $q_* < 0$ when temperature and humidity decrease with height (because then $H > 0$ and $E > 0$), and conversely, $T_* > 0$ and $q_* > 0$ when temperature and humidity increase with height (because then $H < 0$ and $E < 0$).

Because in the logarithmic sublayer, all statistical quantities of the temperature and humidity fields are determined by turbulent motions of not too small scale, they can depend only on the three constant dimensional parameters u_*, T_* and q_* (and on the height z). Hence for the mean temperature and humidity gradients, the following equations obtain :

$$\frac{\partial \bar{T}}{\partial z} = \frac{T_*}{\alpha_H{}^0 z}, \quad \frac{\partial \bar{q}}{\partial z} = \frac{q_*}{\alpha_q{}^0 z}$$

where $\alpha_H{}^0$ and $\alpha_q{}^0$ are numerical constants (if the exchange coefficients for momentum, heat and humidity :

$$K_M = \frac{u_*^2}{\partial \bar{u}/\partial z}, \quad K_H = -\frac{H}{C_p \rho_0 \partial \bar{T}/\partial z}, \quad K_q = \frac{-E}{\rho_0 \partial \bar{q}/\partial z} \tag{14}$$

are introduced, the constants $\alpha_H{}^0$ and $\alpha_q{}^0$ are just the ratios $\alpha_H = K_H/K_M$ and $\alpha_q = K_q/K_M$). Consequently, the mean temperature and humidity profiles in the logarithmic sublayer are given by expressions similar to equation (12) :

$$\bar{T}(z)-T_s = \delta T_s + \frac{T_*}{\alpha_H{}^0} \text{Log} \frac{z}{z_0} , \quad \bar{q}(z) - q_s = \delta q_s + \frac{q_*}{\alpha_q{}^0} \text{Log} \frac{z}{z_0} \tag{15}$$

where δT_s and δq_s are parameters (independent on z), the values of which determine the coefficients of heat and humidity transfer :

$$C_H = \frac{H}{C_p \rho_0 \bar{u}(T_s - \bar{T})} , \quad C_q = \frac{E}{\rho_0 \bar{u}(q_s - \bar{q})} \tag{16}$$

also called sometimes the Stanton and Dalton numbers. The quantities $\delta T_s/T_*$ and $\delta q_s/q_*$ should be regarded as functions, primarily of the roughness Reynolds number $u_* h_s/\nu$. They are apparently close to each other ; above land, laboratory measurements yield values of the order of $0.2\ (u_* h_s/\nu)^{1/2}$ (Owen & Thomson (33)).

Above the sea, the saturation humidity at temperature T_s is taken for q_s, and u_*^2/g can be used, as above, for h_s. The available experimental data on values of C_H and C_q (obtained under weak and moderate winds) show that these coefficients change by two orders of magnitude depending on a number of factors, and that the considerable scatter of the observed values can be significantly reduced by introducing functions of $u_* h_s/\nu$ or $u_* z_0/\nu$ for C_H and C_q (see Kitaigorodsky & Volkov (34) and Fig. 1 taken from this paper, as well as the theoretical work by Bortkovsky & Byutner (35)).

The air-surface interaction is mainly a matter of exchange of momentum, heat and humidity between the atmosphere and the solid or fluid surface. There are exchanges of other substances : sea salt which contribute to condensation nuclei in the atmosphere, some gases like oxygen and CO_2 which are important for biological processes, aerosols which contributes to the sedimentation processes in the ocean, etc ... But more than 2/3 of the surface of our planet are covered by oceans and therefore the air-surface interaction is mainly the ocean-atmosphere interaction. Let us then consider atmospheric influences on the ocean and oceanic influences on the atmosphere separately. Beginning with the ocean, one could say that almost all the oceanic motions (except tidal motions, some local currents generated by rivers and rather rare motions of tectonic origin like tsunamis) are the consequences of atmospheric influences. These are

(1) Wind waves
(2) Drift currents
(3) Internal waves
(4) Oceanic turbulence
(5) Main oceanic currents.

As for the wind waves, one could say perhaps that the real theory of their growth does not exist up to now. We have only linearized theories for the first stages of wind waves development - the J. Miles theory of hydrodynamic instability of the sea surface and atmospheric surface layer system and the O. Phillips theory taking into account the resonance between air pressure fluctuations and the sea surface oscillations. We have also the O. Phillips spectrum of developed waves of the form $S(f) \sim g^2 f^{-5}$, determined by the acceleration of gravity as the only parameter. Finally we know that the development of waves depends upon the wind duration (or upon its fetch) and there were attempts to describe this dependence in the frame of some similarity theory.

Drift currents are the immediate results of the wind drag on the sea surface. They appear similar to the flow in the atmospheric boundary layer and in particular show an Ekman layer structure near the ocean surface. Let us take note however that the problem of the partition of the total momentum flux $\rho_0 u_*^2$ between drift currents and wind waves on the sea surface is outstanding.

Wave motions are much more pronounced in the ocean than they are in the atmosphere. Sometimes we find internal waves in the ocean with amplitudes of several tens of meters. This phenomenon may be explained by the very stable density stratification in the ocean which is heated from above, at variance with the atmosphere which is heated from below. The density stratification can be described in three different ways : by the density profile $\rho(z)$ (which is nearly constant in the upper homogeneous layer of the ocean, nearly discontinuous at the lower boundary of this layer and then increases slightly below);by the natural frequency of internal waves $N = \left(\frac{g}{\rho} \frac{\partial \rho_p}{\partial z} \right)^{1/2}$ where ρ_p is the so-called potential density.

(N goes through maximum corresponding to periods of 10 - 15 min just below the upper homogeneous layer and decreases at greater depth); and, from the point of view of oceanic turbulence, by the profile of the mass flux $M = \overline{\rho' w'}$ which seems to be very small in the bulk of the ocean.

The problem of internal waves generation is also outstanding. They may be of tidal origin or generated by bottom topography, and there is a possibility of internal waves generation by non-linear interaction of surface waves and large-scale turbulent motions in the upper layer of the ocean. The instability of internal waves is also a problem and their decay may be the main source of turbulence in the interior of the ocean.

In the study of oceanic turbulence, one encounters immediately one difficulty : how to distinguish between the turbulence and a random wave field. At first glance one could say that waves are potential motions and turbulent motions include vorticity. However, the finite-amplitude waves have vorticity as well, of the order of a square of their slope. Therefore one usually defines oceanic turbulence as the part of velocity field which is incoherent with surface waves and with pronounced internal waves. Oceans provide a very important example of turbulence in a stably stratified fluid. Under these conditions, fluid tends to develop a layer structure and very many thick or thin layers are usually observed within the ocean. Let us mention the additional difficulty of having to deal with different molecular diffusivities of momentum, heat and salts in sea water.

One usually believes that the main oceanic currents like the Gulf-Stream and the Kuroshivo are wind-driven circulations. Recent measurements of the currents by means of currentmeter, show a very complicated picture however. It seems that all the main currents are rather narrow jets produced by non-linear fluid dynamics or by negative large-scale viscosity (which may be the same). Moreover there usually are countercurrents near the currents as if some tendency existed for local compensation of the loss of water, the overall compensation by giant oceanic gyres being insufficient. Finally there are slow moving meanders on the currents. All these peculiarities want explanation.

Let us consider the heating and cooling of the ocean which are of course controlled by the atmosphere. The main heating mechanism seems to be the absorption of the direct solar radiation in the upper tens of meters of the ocean. Therefore the screening action of cloudiness seems to be very essential and one could imagine the development of positive feedback oscillations in the ocean-atmosphere system, starting from a cloudless sky and intensive heating of the ocean, then the development of convection in the atmosphere above the warm oceanic regions, associated with cloudiness and less solar energy, the subsequent cooling of the ocean, the decay of cloudiness, and so on.

Due to the large heat capacity of the sea water, heating and cooling of the ocean should be slow and the periods of the above-mentioned feedback oscillations should be large. Theoretical models have indicated periods of the order of several months. That is why the ocean-atmosphere interactions should necessarily be taken into account in long-range weather prediction.

Another mechanism of intensive cooling of the ocean is the evaporation from the sea surface since the latent heat of evaporation $L \approx 2500$j/g is so large. Calculations show that the mean value of the latent heat flux over oceans is of the order of 1/3 of the available solar heat flux $\frac{1}{4} S_0(1 - A)$ (S_0 = solar constant, A = albedo of the Earth). If the Dalton number C_q increases linearly with the wind velocity, say, the evaporation increases quadratically and the regions of strong winds (storms) are the main sources of heat and humidity for the atmosphere.

Because of their large thermal inertia, oceans influence the atmosphere in the long period range of the spectrum, way beyond the periods of atmospheric Rossby waves. This supports the view that some kind of adjustment of the atmosphere to the thermal state of the ocean should exist, the Rossby waves being only the transient processes of the adjustment. Then the adjusted atmospheric and oceanic fields will be the main object of long-range weather prediction and one could try to develop a theory of quasi-stationary potential vorticity and entropy fields in the atmosphere with the sea-surface temperature as the only time-dependent field.

The last item in our list of the atmosphere-surface interaction consequences, is the formation of climate. It is a vast problem and the only question I would like to mention here is the Bjerknes theory of climatic oscillations in 17-19 centuries, (the so-called "little ice age") by air-sea interaction over the Atlantic.

Some authors support the possibility of using standard values of the coefficients C_H and C_q equal to 0,002 to compute turbulent heat and humidity fluxes above the oceans (see for instance, Robinson (36)). Data in Fig. 2 show that the error involved in such computations may be as large as many hundred percent. These errors are particularly dangerous in attempts to use equations (16) for the estimation of the climatic mean value H and E from climatic maps of $\bar{u}, \bar{q}$ and $\bar{\bar{T}}$ as done by Jakobs (37) and his co-workers : as C_H and C_q increase rapidly with the wind velocity, storms contribute significantly to the climatic mean values H and E, and yet do not appear on monthly climatic average maps.

6. SIMILARITY THEORY OF THE SURFACE LAYER

Above the dynamic sublayer, heat and humidity cannot be regarded as passive substances and the number of parameters determining the turbulent regime should be enlarged to include the buoyancy parameter β. Then we can accept the following similarity hypothesis (developed by Obukhov and Monin (38-41) ; the relevant bibliography is in reference (20), as a generalization of the similarity hypotheses for the logarithmic sublayer discussed above. In the surface layer with $z \gg (\nu/u_*)$ or h_s, the laws governing vertical variations of the mean hydrodynamic fields determined by the components of turbulence of not too small scale, may depend only on the four dimensional parameters u_*, $H/C_p\rho_0$, E/ρ_0 and β .

As in the logarithmic sublayer, we can use the parameter u_* for a velocity scale and determine the temperature and humidity scales T_* and q_* from equations (13). But, whereas in the logarithmic sublayer the only reference length was the height z, the dimensional parameters available in the surface layer can be used to construct a new length scale :

$$L = \frac{C_p \rho_0 u_*^3}{\kappa \beta H_T} \tag{17}$$

(the sign of L is chosen so that $L < 0$ for $H_T > 0$, when the density stratification is unstable, and conversely $L > 0$ when $H_T > 0$ and the stratification is stable ; the numerical factor κ is introduced into the denominator for convenience). According to the proposed similarity hypothesis, the nondimensional local statistical quantities obtained when using scales L, u_*, T_* and q_* should be universal functions of the nondimensional height $\zeta = z/L$. In particular,

$$\frac{\kappa z}{u_*}\frac{\partial \bar{u}}{\partial z} = \phi(\zeta) \; ; \; \frac{z}{T_*}\frac{\partial \bar{T}}{\partial z} = \frac{\phi(\zeta)}{\alpha_H(\zeta)} \; , \; \frac{z}{q_*}\frac{\partial \bar{q}}{\partial z} = \frac{\phi(\zeta)}{\alpha_q(\zeta)} \tag{18}$$

where $\phi(\zeta)$ is some universal function, and α_H and α_q as above, are the ratios of exchange coefficients K_H/K_M and K_q/K_M. By substituting the first of equations (18) into (10) we obtain the relation $Rf = \zeta/\phi(\zeta)$ for the flux Richardson number, which shows that ζ and Rf are equivalent parameters for describing the density stratification. With the aid of equations (14) and (10), the following equations are derived :

$$Rf = \alpha_H Ri, \quad Ri = \beta \frac{\partial \bar{T}}{\partial z} \left(\frac{\partial \bar{u}}{\partial z} \right)^{-2} \tag{19}$$

where Ri is the gradient Richardson number. By integrating equations (18), we obtain :

$$\begin{aligned} \bar{u}(z_1) - \bar{u}(z_2) &= \frac{u_*}{\kappa}\left[f\left(\frac{z_1}{L}\right) - f\left(\frac{z_2}{L}\right) \right] \\ \bar{T}(z_1) - \bar{T}(z_2) &= T_*\left[f_T\left(\frac{z_1}{L}\right) - f_T\left(\frac{z_2}{L}\right) \right] \end{aligned} \tag{20}$$

and a similar equation for the humidity profile q(z). The available data are consistent with $\alpha_q = \alpha_H$ and $f_q = f_T$; thus we shall not discuss further humidity profiles.

As neutral stratification is approached, i.e., for $H_T \to 0$, or $|L| \to \infty$ or $\zeta = z/L \to 0$, equations (20) should reduce to the logarithmic law (Eqs. 12a, 15) and $\phi(0) = 1$. Then we should have :

$$f(\zeta) \simeq \log \zeta + \text{const}, \; f_T(\zeta) \simeq \frac{1}{\alpha_H 0} \log \zeta + \text{const} \; (|\zeta| \ll 1) \tag{21}$$

In this case $Rf \simeq \zeta$ and $Ri \simeq \zeta/\alpha_H 0$. It should be noted that the condition $|\zeta| \ll 1$ obtains also for fixed H_T when $z \to 0$; this both proves the existence of the dynamic sublayer and yields an estimate of the order of $|L|$ for its thickness. Somewhat above this dynamic sublayer, where $|\zeta| \lesssim 1$, we expand the right-hand sides of equation (18) in power series of ζ and retain only the linear terms. Thus :

$$f(\zeta) \approx \log\zeta + \beta_u\zeta + \text{const} \;;\; f_T(\zeta) \approx \frac{1}{\alpha_H^0}\log\zeta + \beta_T\zeta + \text{const}(|\zeta| \lesssim 1) \quad (21a)$$

The numbers β_u, β_T are different for $\zeta>0$ and $\zeta<0$ but remain positive to yield the flattening of wind and temperature profiles caused by intensification of turbulent mixing under an increasingly unstable stratification. Equations (21a) constitute the "log-plus-linear law" for wind and temperature profiles suggested in (40, 41).

Under very strong instability, i.e., for large positive H_T, the values of L prove to be small and negative, and $\zeta = z/L$ is large and negative. Such values of ζ are reached for fixed $H_T > 0$ as well, if $u_* \to 0$ so that the limiting case of strong instability is free convection. For free convection, the second equation (20) should not contain u_* , which is possible only if $f_T(\zeta) \sim \zeta^{-1/3}$. Note, however, that the exclusion of the parameter u_* does not demand the absence of wind : it is only required that momentum become, so to say, a passive quantity. Therefore the limit $\alpha_H^{-\infty}$ of the ratio $\alpha_H = K_H/K_M$ as $\zeta \to -\infty$ can be finite. But in this case $f(\zeta) \sim \zeta^{-1/3}$ as well, and we have

$$f(\zeta) \approx C\zeta^{-1/3} + \text{const} \;;\; f_T(\zeta) \approx \frac{C}{\alpha_H^{-\infty}}\zeta^{-1/3} + \text{const}\ (\zeta<<-1) \qquad (22)$$

where C is a dimensionless constant. (Then $Rf = \alpha_H^{-\infty}\,Ri = -3\zeta^{4/3}/C$). This "minus one-third law" was proposed in (38, 39) ; it was later suggested by Priestley (42) for $f_T(\zeta)$ from other considerations (Priestley, in addition, established empirically that it is already applicable when $\zeta < -0.03$) ; see also the paper by Kazansky & Monin (43)).

Under very strong stability, i.e., at large negative H_T, the values of L prove to be small and positive and $\zeta = z/L$ is large and positive. Buoyancy forces interfere with the development of turbulence, and the energy equation (Eq. 8) shows that turbulence can be maintained only for not too large Rf. Since Rf evidently increases with ζ, the existence of a finite limiting value R for Rf at large $\zeta > 0$ follows. Then $\phi(\zeta) \simeq (\zeta/R)$. As stability increases, the value of α_H apparently decreases. As a model of infinitely large stability we can use the free surface of an incrompressible fluid through which turbulent heat transfer is impossible yet momentum can be transferred by pressure fluctuations, and $\alpha_H = 0$ (see Stewart (44)).If α_H has a nonzero limit α_H^∞ at large positive ζ, it follows from equation (18) that :

$$f(\zeta) \simeq \frac{\zeta}{R} + \text{const}, \quad f_T(\zeta) \simeq \frac{\zeta}{\alpha_H^\infty R} + \text{const} \quad (\zeta >> 1) \qquad (23)$$

If $\alpha_H^\infty = 0$, $f_T(\zeta)$ will increase with its argument more rapidly than according to the linear law (23).

The theoretical predictions contained in equations (20) to (23) agree well experimental data (summed up in Chapter 4 of (20) and in the book by Lumley & Panofsky (45)). As an example, Figure 3 shows an empirical graph of the function $f(\zeta)$, constructed for the first time in (40) and clearly showing the validity of the asumptotic laws, equations (21)to (23). The numerical constants in these laws were most carefully evaluated by Zilitinkevich & Chalikov (46). For rough estimates, it is recommended in (46) to assume that both f and f_T are given by :

$$f(\zeta) \simeq f_T(\zeta) \simeq \begin{cases} \frac{1}{4}(1 + 5\zeta^{-1/3}) & \text{for } \zeta < -0.07 \\ \text{Log } |\zeta| & \text{for } -0.07 < \zeta < 0 \\ \text{Log } |\zeta| + 10\,\zeta & \text{for } \zeta > 0 \end{cases}$$

It should be noted, however, that according to measurements in Australia including cases of stronger instability (up to $\zeta = -4{,}5$), $\alpha_H^{-\infty}$ appears to be much larger - about 3.5 (see Charnock (47)). Moreover, according to some data (summed up in § 8.2 of (20), see also (48)), for $\zeta < -1$ the "minus one-third law" of equation (22) for temperature profiles is replaced by a still quicker flattening of the temperature profile with height, corresponding to $\alpha_H(\zeta)$ increasing like a power law of ζ (the so-called "windless convection"). A discrepancy is also found in the case of very strong stability ; if most authors obtained values of the order of 0.1 for $R(=1/\beta_u)$, very small values, of the order of 0.01, were found in some measurements of $\alpha_H^\infty (=\beta_u/\beta_T)$ so that $Ri = Rf/\alpha_H$ appears to be of the order of 10 (see (20) and (49)).

The similarity hypothesis formulated in the preceding section can be applied to the statistical quantities derived from turbulent velocity, temperature, and humidity fluctuations in the surface layer : one-point moments, correlations, structure functions and spectra (50-53). Nondimensional RMS values of these fluctuations

$$\sigma_u/u_*,$$

$$\sigma_v/u_*,$$

$$\sigma_w/u_*,$$

$$\sigma_T/|T_*|,$$

$$\sigma_q/|q_*|.$$

should be functions of $\zeta = z/L$. According to experimental data (see § 5.3 and § 8.5 of (20)), the first three are respectively equal to 2.3, 1.7, and 0.9 in the logarithmic sublayer (some authors have obtained somewhat larger values for σ_w/u_* in the atmosphere). As the stratification changes from stable to unstable, these three functions tend to increase ; for strong instability, the situation tends toward horizontal isotropy and an excess of σ_w over $\sigma_u \approx \sigma_v$.

Temperature fluctuations are found to be dependent on stratification in a peculiar way. Under adiabatic temperature stratification they are very small, since mixing of an adiabatically stratified layer cannot result in temperature fluctuations ; therefore in this case the heat flux tends to zero (54). However, in both unstable and stable stratification the temperature fluctuations increase, in the latter case, in spite of the fact that stable stratification suppresses turbulence.

The u' and w' correlation coefficient is negative and close to − 0.5 under neutral stratification according to experimental data ; its absolute value decreases as instability increases. The T' and w' correlation coefficient is positive, close to 0.5 under neutral stratification and apparently increasing with increasing instability . Hence it can be seen that u' and T' should mostly have opposite signs so that there should be a horizontal turbulent heat flux $H_1 = C_p\rho_0 \overline{u'T'}$ with the opposite sign to that of the vertical heat flux H_T and small only under neutral stratification and strong instability (the same should be true for humidity fluxes). Direct measurements of H_1 made by Zubkovsky & Zvang (55) yielded values of H_1/H_T ranging from -3 for weak instability to -1 for moderate instability.

The vertical turbulent fluxes of momentum $\rho_0 u_*^2 = -\rho_0 \overline{u'w'}$, heat $H_T = C_p \rho_0 \overline{T'w'}$, and humidity $E = \rho_0 \overline{q'w'}$, can be measured directly from simultaneous recordings of u', w', T' and q' fluctuations or may be computed from mean profiles $\bar{u}(z)$, $\bar{T}(z)$ and $\bar{q}(z)$. One would, for instance, measure the wind velocity $\bar{u}$ at a fixed height z and the temperature and humidity differences $\overline{\delta T}$ and $\overline{\delta q}$ between heights z/2 and 2z. Then $u_*/\bar{u}$, $H_T/C_p\rho_0\bar{u}\delta T$ and $E/\rho_0\bar{u}\delta q$ (the latter two apparently coinciding) are functions of the empirical Richardson number $\beta z \delta T \bar{u}^{-2}$ and non-dimensional roughness parameter z_0/z. These functions can be expressed in term of the functions $f(\zeta)$ and $f_T(\zeta)$ of equations (20) and knowing the latter, one may construct the corresponding nomograms. Such nomograms were constructed by Kazansky & Monin (43, 56, 57) ; they were recently improved by Zilitinkevich & Chalikov (58).

The nomograms for altitude z = 1m show that, as the stratification changes from strong stability ($Ri \sim 0{,}3$) to strong instability ($Ri \sim -0.05$), the value of $u_*/\bar{u}$ increases from 0.06 or 0.07 to 0.11 or 0.12, and $H_T/C_p\rho_0 u\delta T$ becomes 25 times larger. Since strong stability ($\delta T < 0$) is observed, as a rule, under weak wind and since, on the other hand, strong instability is usually associated with active weather, the product $|u\delta T|$ also increases (by a large factor), and H_T changes by two orders of magnitude : stratification influences turbulent heat and humidity transfer much more strongly than it affects momentum transfer.

When turbulent fluctuations are measured in term of scales u_*, T_*, q_* and wave numbers k in term of the scale 1/z, the nondimensional spatial spectra of horizontal turbulent fields in the range of not too large k will be functions only of the nondimensional wave number k_z and the stratification parameter $\zeta = z/L$ or Ri. From spatial spectra of turbulent fluctuations along a direction parallel to the wind, one may derive frequency spectra of fluctuations at a fixed point of space on the basis of the "frozen turbulence" hypothesis of G. I. Taylor, i.e. by assuming

$k = f/\bar{u}$. For instance, the spectral density of vertical velocity fluctuations will have the form

$$S_w(f) = \frac{u_*^2 z}{\bar{u}} \tilde{S}_w \left(\frac{fz}{\bar{u}}, Ri \right) \qquad (24)$$

where $\tilde{S}_w$ is some universal function. The results of measurements of spectra and co-spectra of the fluctuations of u', v', w', T' and q' are discussed in (52), (45) and in § 23 of (20) (see also a recent paper by Zubkovsky & Koprov (59) on the spectra of momentum and heat fluxes). Measurements show that within a wide range of nondimensional wave numbers the spectra prove to be proportional to $(fz/\bar{u})^{-5/3}$, i.e. the Kolmogorov-Obukhov "minus-five-thirds law" obtains. The lower frequency limit of this inertial range appears to be dependent on stratification : for $S_w(f)$ the limit is $fz/\bar{u} \simeq 2.5$ at Ri = -0.76 ; 4.5 at Ri = 0 and 12 at Ri = 0.26. With a further decrease of f, sometimes at sufficiently large altitudes under unstable stratification, velocity fluctuations spectra grow more slowly and temperature spectra more rapidly than the "minus-five-third law". Under stable stratification, on the contrary, velocity fluctuations spectra grow more rapidly and temperature spectra more slowly (see, for instance, the theoretical computation made by Monin (60) and § 21.7 of (20)). Further down the frequency scale, the growth of spectra ceases and maxima are observed for periods of the order of one minute. This behavior of velocity and temperature spectra is shown on Fig. 4a,b.

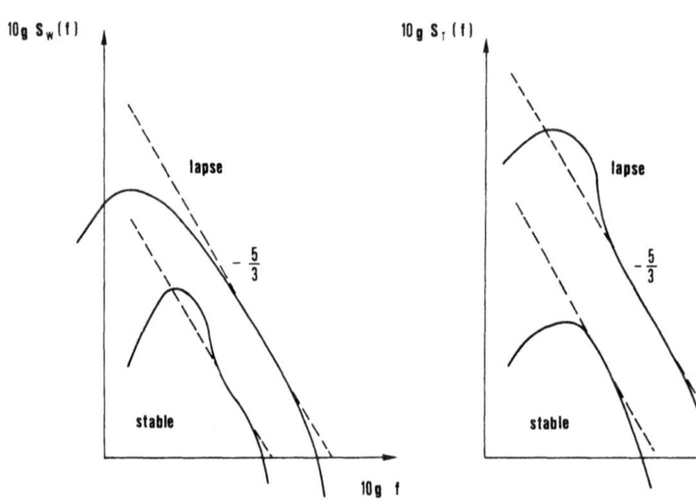

Fig. 4a. The shape of velocity spectra

Fig. 4b. The shape of temperature spectra

Attempts at theoretical determinations of the universal functions postulated by the similarity theory (a review of such attempts is given in §§ 7.4 to 7.5 of (20)), have been based primarily on the use of the energy equation (8). In this equation it is convenient to introduce, instead of ε, the turbulence scale $\ell = (K_M{}^3/\varepsilon)^{1/4}$, which must be essentially proportional to the height z in the logarithmic sublayer and at greater heights assumes the form $\kappa z\lambda(\zeta)$, where $\lambda(\zeta)$ is some universal function. Then the energy equation in the surface layer will have the form :

$$u_*^2 \frac{\partial \bar{u}}{\partial z} + \frac{\beta H_T}{C_p \rho_0} - \frac{1}{\rho_0}\frac{\partial Q}{\partial z} = \frac{K^3}{\ell^4} \qquad (8a)$$

where K is related to $\partial\bar{u}/\partial z$ by $K \frac{\partial \bar{u}}{\partial z} = u_*^2$

or to $\phi(\zeta) = \frac{uz}{u_*}\frac{\partial u}{\partial z}$ by $K = \frac{\kappa u_* z}{\phi}$

Let (σ = 1) be the ratio of the diffusivity term in Equation (8a) to the work of buoyancy forces ; in this case Equation (8a) will be

$$\phi^4 - \sigma\zeta\phi^3 - \lambda^{-4} = 0 \qquad (8b)$$

In the simple semi-empirical theory, it is assumed that $\lambda = 1$ and σ = const. (and α_H = const.) ; for $\sigma \approx 12$ to 14 such a theory agrees well with measurements. In a more detailed theory, the vertical flux of turbulent energy was taken equal to $Q = -\rho_0 K_Q(\partial B^2/\partial z)$, where B^2 is the kinetic energy of turbulence per unit mass, related to K_M by $K_M \sim B\ell$ (Monin 39)). More detailed expressions were also suggested for the turbulence scale ℓ; for instance, by generalizing the well-known von Kàrmàn equation $\ell = -\kappa \frac{\partial\bar{u}/\partial z}{\partial^2\bar{u}/\partial z^2}$ Zilitinkevich & Laikhtman (61) arrived at the expression $\ell = -\kappa\psi/(\partial\psi/\partial z)$ where $\psi = B/\ell$ considered together with Equation 8.

The most detailed semi-empirical theory of the surface layer published so far is that developed by Monin (62). Instead of one energy equation (Eq. 8), he examined the dynamic equations for all one-point second moments of the fluctuations u', v', w' and T' (neglecting the third moments of these fluctuations, i.e., diffusivity terms).

7. SIMILARITY THEORY FOR THE EKMAN BOUNDARY LAYER

Above the surface layer, the influence of the Coriolis force becomes significant ; therefore the Coriolis parameter should be added to the four parameters u_*, $H/C_p\rho_0$, E/ρ_0 and β taken into consideration in the previous similarity theory. Then in addition to L , one more length scale $h = \kappa u_*/f$ can be introduced and interpreted as the thickness of the atmospheric (Ekman) boundary layer. The factor ε in the equation $h = \varepsilon(G/f)$ in our first section will be $\varepsilon = \kappa u_*/L$ (of the order of 10^{-2}). Consequently, we have, in addition to the nondimensional parameter $\zeta = z/L$ depending on height, another constant parameter also representing the vertical stratification :

$$\mu = \frac{h}{L} = -\frac{\kappa^2\beta H_T}{C_p\rho_0 u_*^2 f} \tag{25}$$

(introduced by Kazansky & Monin (63)). Thus all nondimensional functions of ζ discussed above will also depend on the parameter μ in the atmospheric boundary layer. It is more convenient to regard them as functions of $\eta = z/h = \zeta/\mu$ and μ; these functions of η and μ depend only on the product of their arguments $\eta\mu = \zeta$. Such a similarity theory for the atmospheric boundary layer is formulated in (63) ; for neutral stratification ($\mu = 0$) the theory was used by Monin (64) as far back as 1950.

By writing relations of the type of equation (20) (with the nondimensional functions depending also on μ) for the velocity components $u_h = G\cos\alpha$ and $v_h = G\sin\alpha$ at the upper limit of the atmospheric boundary layer and the differences $\delta T_0 = T_h - \bar{T}(z_0)$ and $\delta q_0 = q_h - q(z_0)$ we shall see that G/u_*, $\delta T_0/T_*$ and $\delta q_0/q_*$ are functions of z_0/h and μ. These three relations can be used to express the internal parameters u_* , $H/C_p\rho_0$ and E/ρ_0 in terms of the external parameters G, δT_0, δq_0, z_0 and the constant parameters β and f (the true external parameters are $\delta T = T_h - T_s$ and $\delta q = q_h - q_s$, but the differences $\delta T_s = \bar{T}(z_0) - T_s$ and $\delta q_s = q(z_0) - q_s$ should be determined from equations 15 and 16). The latter six parameters can be used to construct two nondimensional combinations, by choosing the Rossby number Ro and the external parameter of stratification S defined for this purpose by :

$$Ro = \frac{G}{z_0 f}, \quad S = \frac{\beta \delta T_0}{Gf} \tag{26}$$

Thus for the atmospheric boundary layer, the following similarity hypothesis can be accepted : for $z >> \nu/u_*$ or h_s the nondimensional mean quantities derived from the hydrodynamical fields, determined by turbulent motions of not too small scales and obtained using the velocity scale G, length G/f, temperature δT_0 and humidity δq_0, may depend only on the two constant nondimensional parameters Ro and S (Zilitinkevich & Monin (65, 66)).

The relations connecting the internal and external parameters (the laws of momentum, heat and humidity exchanges for the atmospheric boundary layer) are derived using the asymptotic logarithmic laws in the dynamic sublayer. They have the form

$$\text{Log Ro} = B(\mu) - \text{Log}\frac{u_*}{G} + \sqrt{\kappa^2\left(\frac{u_*}{G}\right)^{-2} - A^2(\mu)}$$

$$\text{Sin}\,\alpha = -\frac{A(\mu)}{\kappa}\,\frac{u_*}{G}, \quad \frac{\delta T_0}{T_*} = \frac{1}{\alpha_H^{\,0}}\left[\text{Log}\left(\frac{u_*}{G}\,\text{Ro}\right) - C(\mu)\right] \tag{27}$$

where A, B, C are unknown functions of μ. The equation for $\delta q_0/q_*$ is entirely analogous to equation (27) for $\delta T_0/T_*$. The first of these equations (for $\mu = 0$) was obtained for the first time by Kazansky & Monin (67). The case $\mu = 0$ is also discussed by Gill (68), Charnock & Ellison (69), and Blackadar (70).

Reviews of empirical data on the dependence of the geostrophic drag coefficient u_*/G and wind turning angle α upon Ro (without taking into account stratification) are given by Lettau (71, 72) and Blackadar (70, 73). According to these data, u_*/G decreases roughly from 0.05 to 0.02 and α from 35° to 13° when Ro increases from 10^5 to 10^{10}. Empirical measurements of u_*/G obtained by Byzova & Mashkova (74) and measurements of α by Kurpakova & Orlenko (75) show that u_*/G decreases (and for fixed G the thickness of the atmospheric boundary layer decreases as well), and α grows , when the stratification changes from lapse to stable.

In (68) estimates of $A(0) \simeq 4.2$ to 4.7 and $B(0) \approx 1$ to 2 are given;

more precise estimates ($2 < A(0) < 3$ and $B(0) \approx 2$) are given in (69). The empirical dependence of A, B, C on μ was derived by Zilitinkevich & Chalikov (76) (they are presented also in (65)) ; according to their data, A increases, B and C decrease to large negative values when the stability increases. Proceeding from these results, Chalikov (77) constructed nomograms for the determination of u_*/G, α and $H/C_p\rho_0 G T_0 \approx E/\rho_0 G\delta q_0$ as functions of Log Ro and Log $|S|$(Fig. 5-7). Values of Log Ro in these nomograms are written on the curves. These nomograms can be used for describing the atmospheric boundary layer in mathematical models of the general atmospheric circulation.

The nondimensional wind velocity profiles $\sqrt{u^2 + v^2}/u_*$ versus $\eta = z/h$ for different values of μ were constructed by Kazansky & Monin (63) and Byzova & Mashkova (74). Values of $\sqrt{u^2 + v^2}/u_*$ at each fixed η proved to be regularly increasing as μ increases, and profiles (63) under stable stratification ($\mu > 0$) had well pronounced maxima at heights of 0.2 to 0.3 η ; since for large positive μ , the atmospheric boundary layer is thin, these maxima look like "low-level jet". The empirical dependence of $(G \cos \alpha - u)/u_*$, $(G \sin \alpha - v)/u_*$ upon η for $\mu = 0$ were derived by Gill (68), and the nondimensional hodographs (spirals) of the wind vector (u/G, v/G) as a function of η under lapse, neutral and stable stratification were constructed by Kurpakova & Orlenko (75). Examples of the nondimensional temperature profiles $(T - T_1)/|T_*|$ as a function of η at different μ are presented by Mashkova (78).

8. SEMI-EMPIRICAL THEORIES OF THE EKMAN BOUNDARY LAYER

For a long time the theory of the atmospheric boundary layer consisted only in the solution of Reynolds equations (5), in which τ_{xz}, τ_{yz}, were taken to be proportional to $\partial\bar{u}/\partial z$ and $\partial\bar{v}/\partial z$, the proportionality coefficient K_M being given as some simple function of z (a review of these theories can be found, for instance, in (79)). It would be more correct however, not to assume a value of K_M a priori but rather derive it together with $\bar{u}(z)$ and $\bar{v}(z)$. For this purpose, one can use Reynolds equations and the energy equation in the following form :

$$f(\bar{v} - G\sin\alpha) + \frac{\partial}{\partial z} K \frac{\partial \bar{u}}{\partial z} = 0$$

$$-f(\bar{u} - G\cos\alpha) + \frac{\partial}{\partial z} K \frac{\partial \bar{v}}{\partial z} = 0 \qquad (28)$$

$$K\left[\left(\frac{\partial \bar{u}}{\partial z}\right)^2 + \left(\frac{\partial \bar{v}}{\partial z}\right)^2\right] + \frac{\beta H_T}{C_p \rho_0} - \frac{1}{\rho_0}\frac{\partial Q}{\partial z} = \frac{K^3}{\ell^4}$$

Monin (64) used these equations in the case of neutral stratification, yet taking into consideration the diffusivity term $Q = -\rho_0 K_M(\partial B^2/\partial z)$ where $K_M \sim B\ell$, and solved these equations numerically, in term of the turbulence scale $\ell = \kappa z$ as sole parameter. For the first time, the dependence of u_*/G and α upon Ro were indeed computed.

During recent years, this approach has attracted the attention of various authors who have made similar computations with more detailed equations for the turbulence scale (yet without consideration of the diffusivity term) :

- Blackadar (73) : $\ell = \kappa z(1 + Cfz/G)^{-1}$
- Lettau (72) : $\ell = \kappa z\left[1 + C(z/h)^{5/4}\right]^{-1}$
- Appleby & Ohmsted (80) : $\ell = \ell_\infty\left[1 - \exp(-\kappa z/\ell_\infty)\right]$
- Zilitinkevich, Laikhtman & Tseitin (81) : $\ell = -\kappa\psi/(\partial\psi/\partial z)$ where $\psi = B/\ell$.

Finally, Bobyleva, Zilitinkevich & Laikhtman (82) used the latter equation for computing a stratified atmospheric boundary layer (with the diffusivity term). Their computation appears to be the most complete so far. It includes the derivation of u_*/G and α in term of Ro and μ , as well as the nondimensional coefficients of turbulent viscosity $K_M/\kappa u_* h$ and energy dissipation $\kappa h\varepsilon/u_*^3$, and the "velocity deficiency" components (κ/u_*) $(u - G\cos\alpha)$ and (κ/u_*) $(v - G\sin\alpha)$ as functions of z/h at different μ (these results are presented also in the review (79) ; note the clearly defined "low-level jet" under stable stratification in the two latter graphs.

Zilitinkevich & Chalikov (83) found that results similar to (82) can be obtained if $K/\kappa u_*|L|$ is taken equal to $|\zeta|$ for $|\zeta| \leqslant m$ and $m(|\zeta|/m)^{1/3}$ for $|\zeta| \geqslant m$ under unstable stratification, and equal to ζ for $\zeta \leqslant n$ and n for $\zeta \geqslant n$ under stable stable stratification, where $\zeta = z/L$, L being the scale of the dynamic sublayer (equation 17), and m = 0.064, n = 0.1 are numerical constants. All quantities, including the functions $A(\mu)$, $B(\mu)$, $C(\mu)$ of equations (27) can be obtained then in analytic form without difficulty.

The method of computing wind and temperature profiles and second moments of turbulent fluctuations in the surface layer suggested in (62), was extended by Monin to the neutral (84) and the stratified atmospheric boundary layer (85). In the course of this theory, the following particular relations were established :

$$\frac{\tau_{xz}}{\rho_0 \partial\bar{u}/\partial z} = \frac{\tau_{yz}}{\rho_0 \partial\bar{v}/\partial z} \quad (= K_M)$$

and

$$\frac{\tau_{yz}}{\tau_{xz}} = \frac{\overline{v'T'}}{\overline{u'T'}} \quad (= \tan\psi, \text{ say})$$

Also the following expressions were derived for the coefficients of velocity fluctuation anisotropy :

$$\frac{\sigma_u^2}{B^2} = \frac{\sigma_{u0}^2}{B_0^2} + \frac{M\cos^2\psi}{1 - Rf},$$

$$\frac{\sigma_v^2}{B^2} = \frac{\sigma_{v0}^2}{B_0^2} + \frac{M\sin^2\psi}{1 - Rf}, \qquad (29)$$

$$\frac{\sigma_w^2}{B^2} = \frac{\sigma_{w0}^2}{B_0^2} - \frac{M\,Rf}{1 - Rf},$$

where $M = (\sigma_{u0}^2 - \sigma_{v0}^2)/B_0^2$ and the subscript zero denotes values under neutral stratification. Finally, K_M and the nondimensional momentum flux $\tau = (\tau_{xz} + i\tau_{yz})/\rho_0 u_*^2$ are given by the following equations :

$$K_M = \kappa u_* \cdot L \cdot Rf \cdot |\tau|^2 \ , \ Rf \cdot |\tau|^2 \cdot \frac{\partial^2 \tau}{\partial \zeta^2} = i\mu\tau \qquad (30)$$

Now, if the hypothesis of quasistationarity of the atmospheric boundary layer formulated in section 1 is abandoned, the terms $\partial\bar{u}/\partial t$ and $\partial\bar{v}/\partial t$ of the Reynolds equations (28) must be explicitely included for describing the evolution of the layer at time t and the equations of heat and humidity transfers should be written :

$$\rho_0 T_0 (\partial\bar{s}/\partial t) = - \partial Q_s/\partial z, \rho_0(\partial\bar{q}/\partial t) = - \partial Q_q/\partial z$$

where Q_s and Q_q are vertical entropy and humidity fluxes. In papers devoted to diurnal changes of the atmospheric boundary layer, such equations have been used for a long time in a very simple form : turbulent fluxes of momentum, heat and humidity were determined from equations of the type of equations (14), with coefficients K_M, K_H, K_q given in the form of products of simple functions of z by periodic functions of t.

In the last few years, some authors dealing with diurnal changes of the atmospheric boundary layer began to express the exchange coefficients in terms of wind velocity and temperature gradients through relations similar to the stationary energy equation (Estoque (86), Sharon (87), Krishna (88)). Finally, Vager & Zilitinkevich (89) used the non-stationary form of energy equation (28) with the term $\partial B^2/\partial t$ to the right hand side and solved it together with the complete equations of motion and heat transfer with periodic values of temperature $\bar{T}(z_0, t)$ being given as a lower boundary condition :

$$\bar{T}(z_0, t) = T_m \sin \omega t.$$

In particular, they computed diurnal changes of the nondimensional parameters u_*/G, α and $H/C_p\rho_0 GT_m$ and found from the latter the diurnal changes of the parameter $\mu = \mu(t)$. By excluding the time t, Vager & Zilitinkevich could construct curves of the variations of u_*/G, α and $H/C_p\rho_0 GT_m$ on μ in the form of "hysteresis loops" which have two ordinates, for each value of μ, one corresponding to increasing μ and the other to decreasing μ. For u_*/G and particularly for $H/C_p\rho_0GT_m$ these loops turned out to be rather narrow, i.e., nonstationarity affects but little the computed values of these important quantities. This can justify the use of the hypothesis of quasistationarity of the atmospheric boundary layer.

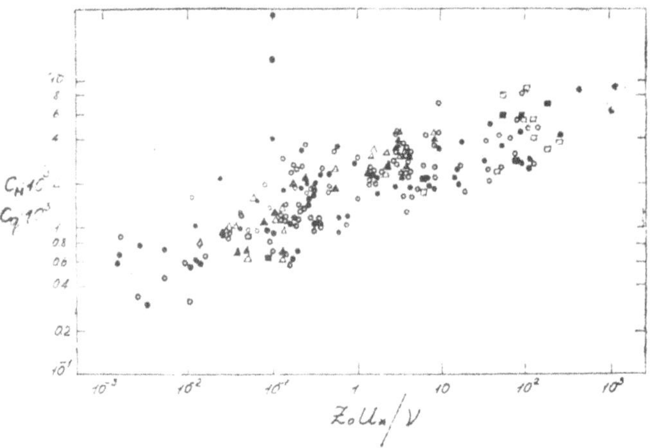

Fig. 2

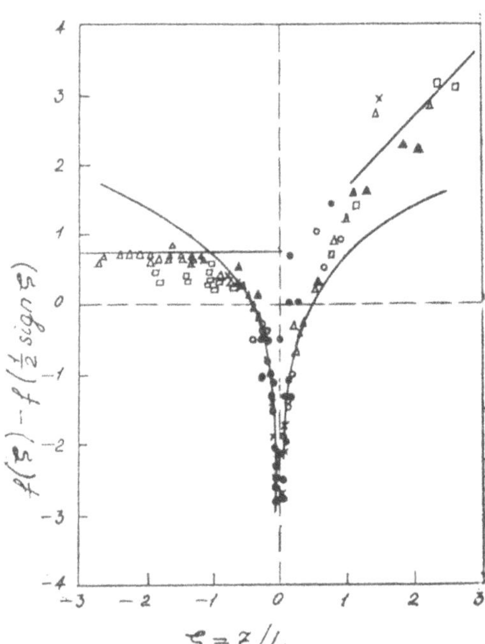

Fig. 3

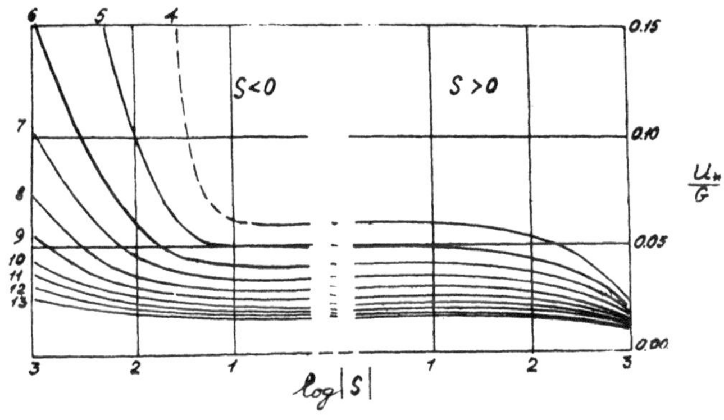

Fig. 5

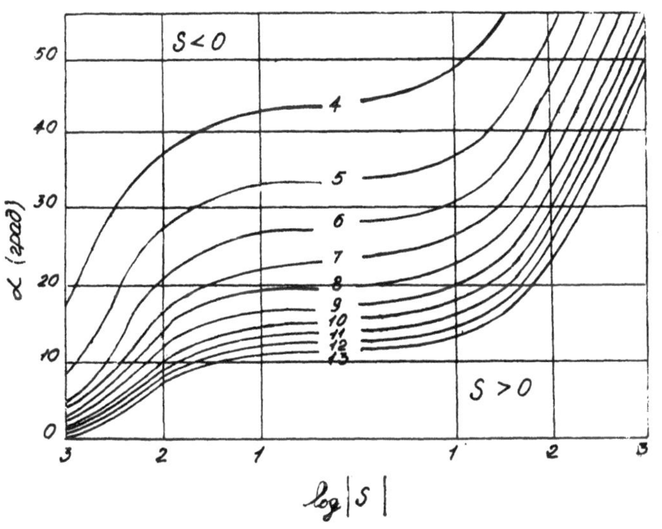

Fig. 6

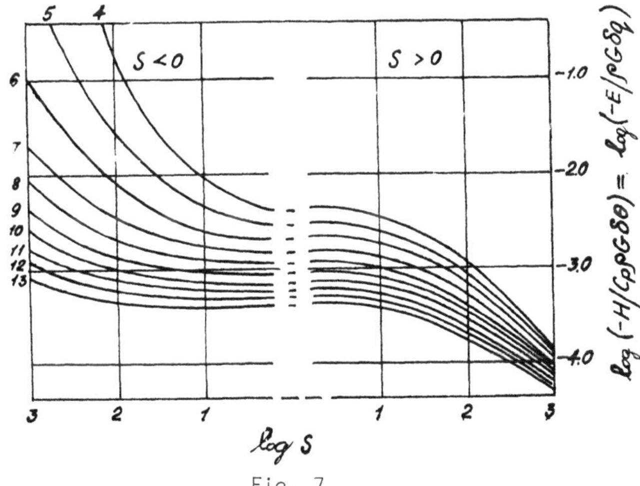

Fig. 7

9. ATMOSPHERES OF OTHER PLANETS

Let us restrict ourselves to planet atmospheric circulations driven by solar radiation i.e. Mercury, Venus, the Earth and Mars. Some characteristic parameters of their atmospheres are listed in Table 1 :

q = solar constant in cal/cm^2 min ;
A = albedo (the fraction of reflected solar radiation) ;
r = radius of the planet in km ;
M = total mass of the atmosphere ($M = 1$ for the Earth) :
T_1 and T_2 = maximum and minimum surface temperatures,
$\eta = \frac{T_1 - T_2}{T_1}$ = thermodynamic efficiency of a perfect heat engine ;
t = period of rotation in Earth days.

The estimates of M, T_1, T_2 and η here should be considered as preliminary ones.

Table 1	q	A	r	M	T_1	T_2	η	t
Mercury	13,4	0,09	2430	4.10^{-4}	600	150	0,75	59
Venus	3,8	0,76	6060	80	700	650	0,07	- 243
Earth	1,9	0,29	6370	1	300	240	0,20	1
Mars	0,87	0,26	3400	5.10^{-3}	280	220	0,21	1,02

One can estimate the mean rate of kinetic energy generation per unit mass in the planetary atmosphere by the following expression :

$$\varepsilon = C\eta \frac{q(1-A)\pi \cdot r^2}{M} , \qquad (31)$$

where C is some empirical numerical factor. For the Earth ε was estimated by a number of independent methods empirical or theoretical and is of the order of 4 $cm^2\ sec^{-2}$.

From this we get $C \approx 0,1$. Then one could estimate the characteristic period of large-scale atmospheric motions t_0, the characteristic velocity U and the large-scale horizontal mixing coefficient K_h by means of the equations :

$$t_0 \approx \varepsilon^{-1/3} L_0^{2/3}$$

$$U \approx (\varepsilon L_0)^{1/3} \qquad (32)$$

$$K_h \approx 0,1\, \varepsilon^{1/3} L_0^{4/3}$$

where L_0 is the characteristic horizontal scale of length of atmospheric motions defined by $L_0 = \frac{C_0}{2\omega}$ (C_0 = sound velocity ; 2ω = vorticity of planet rotation), if this value is less than, say, the diameter 2r of the planet and L_0 = 2r otherwise. After that, one could estimate the atmospheric boundary layer parameters u_*, H_T, L (equation 17) and T_* (equation 13) using the relations

$$u_* \approx 0,03\ U \qquad (33)$$

$$H_T \approx 0,1\ q\ (1-A) \text{ in daytime only.}$$

The results are given in Table 2 (after Golitzyn (90, 91) with some modifications) :

Table 2	ε	L_0	t_0	U	K_h	$g\beta$	u_*	$H_T/C_p\rho_0$	- L	T_*
Mercury	5.10^4	5	0,2	300	150	1	900	1.10^4	1800	25
Venus	1.10^{-2}	12	60	2,4	3	1,2	7,2	5.10^{-2}	500	0,01
Earth	4	3	3	10	3	3,3	30	7	20	1
Mars	120	2	1	30	6	2	90	6.10^2	25	15

Here ε is measured in $cm^2\ sec^{-3}$, L_0 in 10^3 km, t_0 in days, U in m sec^{-1}, K_h in $10^6\ m^2\ sec^{-1}$, $g\beta$ in cm $sec^{-2}(°k)^{-1}$, u_* in cm sec^{-1}, $H_T/C_p\rho_0$ in deg.cm.sec^{-1}, L in m, T_* in degree Kelvin.

We considered so far the temperature difference $\delta T = T_1 - T_0$ as a quantity given a priori. It seems desirable however to estimate this quantity from external parameters. Let us introduce the new notations

$$M_0 = \frac{M}{4\pi r^2} \left(= \frac{p_0}{g} \right) \tag{34}$$

for the mass of an air column of unit cross-section and

$$q_A = \frac{1}{4} q\ (1 - A) \quad (= \sigma T_0{}^4) \tag{35}$$

for the available solar energy flux averaged over the surface of the planet ($\sigma = 5{,}67\ 10^{-5}$ g sec^{-3} deg^{-4} is the Stephan-Boltzmann constant and T_0 is the effective radiation temperature). The effective thermodynamic efficiency of the atmosphere can be rewritten in the form

$$C\eta = C_1 \frac{\delta T}{T_0}\ ; \quad C_1 = C \frac{T_0}{T_1} \tag{36}$$

where C_1 is a renormalized empirical factor. Following Golitzyn's reasoning (92) let us consider the balance between the vertically integrated heat convection by air currents and the radiation

$$C_p M_0 u \cdot \nabla T = \sigma T_0{}^4 \tag{37}$$

Equations (31) and (37) can be rewritten in the form

$$\frac{U^3}{L_0} = C_1 \frac{\delta T}{T_0} \frac{q_A}{M_0} \tag{31a}$$

$$C_p M_0 U \frac{\delta T}{(\pi r/2)} = \sigma T_0{}^4 \tag{37a}$$

Eliminating U from these equations and using the definition of T_0 we obtain the expression for δT :

$$\delta T = \left(\frac{\pi}{2}\right)^{1/4} \cdot C_1^{-1/4} \cdot \sigma^{-1/16} \cdot C_p^{-3/4} \cdot q_A^{9/16} \cdot \left(\frac{r}{M_0}\right)^{1/2} \cdot \left(\frac{\pi r}{2L_0}\right)^{1/4} \qquad (38)$$

The last factor of (38) describes the influence of planet rotation. In the case of a rapidly rotating planet $L_0 \sim \omega^{-1}$ and $\delta T \sim \omega^{1/4}$ (the size of atmospheric vortices L_0 being rather small, the equator-to-poles heat transfer by these vortices is not efficient enough, and the equator-to-poles temperature difference δT becomes rather large). In the case of a slowly rotating planet, on the other hand, $L_0 \sim 2r$, $\frac{\pi r}{2L_0} \sim 1$ and the rotation does not influence the temperature difference δT. Note that equation (38) gives an estimate δT for Venus of the order of 2°K i.e. much less than adopted in Table 1. Therefore all estimates for Venus need recalculation.

Using equation (38) we get the following dependence of atmospheric circulation parameters upon the mass M_0 :

$$\delta T \approx M_0^{-1/2} \; ; \; c\eta \approx M_0^{-1/2} \; ; \; \varepsilon \approx M_0^{-3/2}$$

$$t_0 \approx M_0^{1/2} \; ; \; U \approx M_0^{-1/2} \; ; \; K_h \approx M_0^{-1/2} \qquad (39)$$

$$u_* \approx M_0^{-1/2} \; ; \; \frac{H_T}{C_p \rho_0} \approx M_0^{-1} \; ; \; L \approx M_0^{-1/2} \; ; \; T \approx M_0^{-1/2}$$

Note that the total kinetic energy of atmospheric circulation $E \approx M_0 U^2$ is independent of the mass of the atmosphere, a possibly fundamental conclusion.

10. SIMILARITY THEORY FOR PLANETARY ATMOSPHERE CIRCULATIONS

One is tempted to draw more general conclusions from the material of the previous section. All the equations written there are expressed in term of 6 truly "external parameters" : the Stephan-Boltzmann constant σ(g sec^{-3} deg^{-4}), the specific heat capacity of air (C_p cm^2 sec^{-2} deg^{-1}), the available solar energy flux q_A(g sec^{-3}), the radius of the planet r(cm), the mass of the atmospheric column of unit cross section M_0(g cm^{-2}) and the angular velocity of planet rotation ω(sec^{-1}). Let then assume a general similarity hypothesis proposed by Golitzyn (92), that these six external parameters determine entirely the planetary atmosphere circulation.

We have six parameters and only four dimensions : length, time, mass and temperature. Therefore there exist two independent dimensionless combinations of the parameters. To write them in a convenient way, let us introduce the radiation temperature $T_0 = (q_A/\sigma)^{1/4}$ and constant volume specific heat of the atmosphere $C_v = C_p - \frac{R}{\mu}$ (R = universal gas constant and μ = mean molecular weight), $\chi = \frac{C_p}{C_v}$ and finally define the sound velocity

$$C_0 = \sqrt{\chi \frac{R}{\mu} T_0} = (\chi - 1)^{1/2} \cdot \sigma^{-1/8} \cdot C_p^{1/2} \cdot q_A^{1/8} \tag{40}$$

Then we have the length scale $L_0 = \frac{C_0}{2\omega}$ and our first dimensionless combination will be r/L_0 . We shall use the parameter ω further on only in this combination. There exists only one independent dimensionless combination of remaining 5 parameters σ, C_p, q_A, r and M which is

$$M = \sigma^{3/8} \cdot C_p^{-3/2} \cdot q^{5/8} \cdot r \cdot M_0^{-1} \tag{41}$$

Physical meaning of the parameter M will be explained later.

According to Golitzyn's similarity hypothesis any characteristic quantity F of the atmospheric circulation is a properly chosen combination F_0 of the 4 parameters σ, C_p, q_A and r multiplied by some universal function of r_0/L_0 and M to be determined by further physical reasoning :

$$F = F_0(\sigma, C_p, q_A, r) \cdot \psi_F(\frac{r}{L_0}, M) \tag{42}$$

The dimensionless numbers r/L_0 and M play therefore the role of similarity criteria for atmospheric circulations. Now M is of the order of 1 for Mercury but only $1{,}3 \cdot 10^{-5}$ for Venus, $1{,}1 \cdot 10^{-3}$ for Earth, and $3{,}4 \cdot 10^{-2}$ for Mars. Leaving Mercury alone, we conclude that M is small and our first guess on the asymptotic behaviour of the function ψ_F for small M will be a power law :

$$\psi_F\left(\frac{r}{L_0},M\right) \sim M^{n_F}\cdot\Psi\left(\frac{r}{L_0}\right), \text{ if } M << 1 \qquad (43)$$

(the exponent n_F is given in (39) for a variety of quantities F). We have already seen that the total kinetic energy of the atmosphere E is independent on M_0, so that $n_\Sigma = 0$, and

$$E = \sigma^{1/8}\cdot C_p^{-1/2}\cdot q_A^{7/8}\cdot r^3\cdot\Psi\left(\frac{r}{L_0}\right) \qquad (44)$$

This equation can be rewritten in the form

$$E = \frac{(\chi-1)^{1/2}\cdot\Psi}{\pi}\cdot\pi\cdot r^2\cdot q_A\,\frac{r}{C_0} \qquad (44a)$$

so that E is proportional to the product of the total available solar radiation flux $\pi r^2 q_A$ and the time constant r/C_0 for the relaxation of pressure or density disturbances.

For the Earth E/Ψ is equal to $1{,}12\cdot10^{27}$ erg while empirical estimates give $E = 6$ to $9\cdot10^{27}$ erg (including seasonal changes). For Mars E/Ψ equals $0{,}97\cdot10^{26}$ erg while a numerical experiment by Leovy and Mintz has given the value $E = 1{,}2$ to $1{,}6\cdot10^{26}$ erg.

If E is known we can define the typical wind velocity U by means of the equation $E = \frac{1}{2}MU^2 = 2\pi r^2 M_0 U^2$:

$$U = \left(\frac{\Psi}{2\pi}\right)^{1/2}\cdot\sigma^{1/16}\cdot C_p^{-1/4}\cdot q_A^{7/16}\cdot r^{1/2}\cdot M_0^{-1/2} \qquad (45)$$

and then define the Mach number $Ma = \frac{U}{L_0}$

$$Ma^2 = \frac{\Psi}{2\pi(\chi-1)}\cdot\sigma^{3/8}\cdot C_p^{-3/2}\cdot q_A^{5/8}\cdot r\cdot M_0^{-1} = \frac{\Psi}{2\pi(\chi-1)}\cdot M \qquad (46)$$

This provides the physical meaning of the dimensionless parameter M which proves to be proportional to the square of the Mach number.

Let us finally define the time scale of atmospheric motions as $t_0 = \frac{L_0}{U}$ and the rate of kinetic energy generation (per unit mass) as :

$$\varepsilon = \frac{E/4\pi \cdot r^2 M_0}{t_0} \approx \frac{1}{2}\frac{U^3}{L_0} = \left[\frac{(\chi-1)^{1/2}}{4\pi}\cdot\Psi\cdot\frac{r}{L_0}\ Ma\right]\frac{q_A}{M_0} \quad (47)$$

The factor in square brackets gives the expression for the thermodynamic efficiency of the atmosphere $\left(C\eta = C_1\frac{\delta T}{T_0}\right)$ which proves to be proportional to the Mach number. From this we derive the following equation for δT :

$$\delta T = \frac{(\chi-1)^{1/2}}{4\pi\ C_1}\cdot\Psi\cdot\frac{r}{L_0}\cdot Ma\cdot T_0 \quad (48)$$

Comparing this result with equation (38) we obtain an expression of the function $\Psi\left(\frac{r}{L_0}\right)$:

$$\Psi\left(\frac{r}{L_0}\right) = (2\pi)^{4/3}\cdot C_1^{1/2}\cdot\left(\frac{r}{L_0}\right)^{-1/2} \quad \text{if} \quad \frac{r}{L_0} > 1 \quad (49)$$

Now all the equations of the theory contain only one empirical numerical factor C_1 , and the theory is ready for applications. Using equation (49) we obtain the final expressions for the total kinetic energy :

$$E = (2\pi)^{4/3}\cdot C_1^{1/2}\cdot\sigma^{1/8}\cdot C_p^{-1/2}\cdot q_A^{7/8}\cdot r^3\cdot\left(\frac{r}{L_0}\right)^{-1/2}, \quad (44b)$$

the characteristic wind velocity :

$$U = (2\pi)^{1/6}\cdot C_1^{1/4}\cdot\sigma^{1/16}\cdot C_p^{-1/4}\cdot q_A^{7/16}\cdot r^{1/2}\cdot M_0^{-1/2}\cdot\left(\frac{r}{L_0}\right)^{-1/4} \quad (45a)$$

the rate of kinetic energy generation :

$$\varepsilon = \frac{1}{2}(2\pi)^{1/2}\cdot C_1^{3/4}\cdot\sigma^{3/16}\cdot C_p^{-3/4}\cdot q_A^{21/16}\cdot r^{1/2}\cdot M_0^{-3/2}\left(\frac{r}{L_0}\right)^{1/4} \quad (47a)$$

the thermodynamic efficiency :

$$C\eta = \frac{1}{2}(2\pi)^{1/2}\cdot C_1^{3/4}\cdot\sigma^{3/16}\cdot C_p^{-3/4}\cdot q_A^{5/16}\cdot r^{1/2}\cdot M_0^{-1/2}\cdot\left(\frac{r}{L_0}\right)^{1/4} \quad (49)$$

and the temperature difference between equator and pole :

$$\delta T = \frac{1}{2}(2\pi)^{1/2} \cdot C_1^{-1/4} \cdot \sigma^{-1/16} \cdot C_p^{-3/4} \cdot q_A^{9/16} \cdot r^{1/2} \cdot M_0^{-1/2} \cdot \left(\frac{r}{L_0}\right)^{1/4} \qquad (48a)$$

In actual applications of the theory one should be careful not to forget that $C_1 = C \frac{T_0}{T_1}$ and that $L_0 \sim 2r$ for slowly rotating planets.

*

* *

"This text is an adaptation of A.S. Monin's paper : "The Atmospheric Boundary Layer" in Ann. Rev. of Fluid Mech., Vol. 2, 225-250, 170, with some additions".

BIBLIOGRAPHY

1. EKMAN, V. W., Arkiv. Mat. Astron. Fiz., 2, No. 11, 1-52 (1905-1906)

2. CHARNEY, J. G., Okeanologiya, 9, No. 1, 143-45 (1969)

3. FALLER, A. J., J. Fluid Mech., 15, No. 4, 560-76 (1963)

4. TATRO, P. R., MOLLO-CHRISTIANSEN, E. L., J. Fluid Mech., 28, No. 3, (1967)

5. GREEN, A. W., An Experimental study of the interactions between non-steady Ekman layers and an annular vortex, (Doctoral Thesis M I T, Cambridge, Mass., 1968)

6. KOLESNIKOVA, V. N., MONIN, A. S., Izv. Akad. Nauk SSSR, Atmospheric Oceanic Physics, 1, No. 7, 653-69 (1965)

7. KOLESNIKOVA, V. N., MONIN, A. S., Meteorol. Res., 16, 30-56 (1968)

8. VAN DER HOVEN, J., J. Meteorol., 14, No. 2, 160-64 (1957)

9. DYŬBYUK, A. F., Tr. Res. Inst. GUGMS, 2, No. 4, 18-51 (1947)

10. PRIESTLEY, C. H. B., Phys. Fluids, Suppl., 38-46 (1967)

11. UTINA, Z. M., Tr. Gl. Geofiz. Observ., No. 187, (in Russian) 146-48 (1966)

12. LAIKHTMAN, D. L., Physics of the Atmospheric Boundary Layer, 142-94 (Gidrometizdat, 1961)

13. PANOFSKY, H. A., TOWNSEND, A. A., Quart. J. Roy. Meteorol. Soc., 90, No. 384, 147-55 (1964)

14. OLIPHANT, J. E., PANOFSKY, H. A., Final Rep. Contract No. AF (604)-6641, Pennsylvania State Univ. (1965)

15. TOWNSEND, A. A., J. Fluid Mech., 22, No. 4, 799-822 (1965)

16. TOWNSEND, A. A., J. Fluid Mech., 23, No. 4, 767-78 (1965)

17. BLACKADAR, A. A., PANOFSKY, H. A., GLASS, P. E., BOOGARD, J. F., Phys. Fluids, Suppl., 209-11 (1967)

18. MONIN, A. S., Izv. Akad. Nauk SSSR, Ser Geograph. Geofiz., 3, No. 3, 220-37 (1949)

19. MONIN, A. S., Dokl. Akad. Nauk SSSR, 175, No. 4, 819-22 (1967)

20. MONIN, A. S., YAGLOM A. M., Statistical Hydromechanics (in Russian) (Nauka Publishers, Moscow, Part 1, 1965; Part 2, 1967)

21. PRIESTLEY, C. H. B., Turbulent Transfer in the Lower Atmosphere (Chicago Univ. Press, 1959)

22. LETTAU, H. H., In The collection and processing of field data, 3-40 (Interscience, New York, 1967)

23. CHARNOCK, H., Quart. J. Roy. Meteorol. Soc. 81, No. 350, 639-40 (1955)

24. MUNK, W. H., Quart. J. Roy. Meteorol. Soc. 81, No. 350, 639-40 (1955)

25. ZUBKOVSKY, S. L., TIMANOVSKY, D. F., Izv. Akad. Nauk SSSR, Atmospheric and Oceanic Physics, 1, No. 10, 1005-13 (1965)

26. ZUBKOVSKY, S. L., KRAVCHENKO, T. K., Izv. Akad, Nauk SSSR, Atmospheric and Oceanic Physics, 3, No. 2, 127-35 (1967)

27. PHILLIPS, O. M., The Dynamics of the Upper Ocean (Cambridge Univ. Press, London, 1966)

28. ROLL, H. U., Physics of the Marine Atmosphere (Academic, New York-London, 1965)

29. KITAIGORODSKY, S. A., VOLKOV, Yu. A., Izv. Akad. Nauk SSSR, Atmospheric and Oceanic Physics, 1, No. 9, 973-88 (1965)

30. KITAIGORODSKY, S. A., Izv. Akad. Nauk SSSR, Atmospheric and Oceanic Physics, 4, No. 8, 870-78 (1968)

31. FALLER, A. J., Tellus, 16, No. 3, 363-70 (1964)

32. LAIKHTMAN, D. L., Izv. Akad. Nauk SSSR, Atmospheric and Oceanic Physics, 3, No. 10, 1017-25 (1966)

33. OWEN, P. R., THOMSON, W. H., J. Fluid Mech., 15, No. 3, 321-34 (1963)

34. KITAIGORODSKY, S. A. VOLKOV, Yu. A., Izv. Akad. Nauk SSSR, Atmospheric and Oceanic Physics, 1, No. 12, 1317-36 (1965)

35. BORTKOVSKY, R. S., BYUTNER, E. K., Izv. Akad. Nauk SSSR, Atmospheric and Oceanic Physics, 5, No. 5, 494-503 (1969)

36. ROBINSON, G. D., Quart, J. Roy, Meteorol. Soc., 92, No. 394, 451-65 (1966)

37. JAKOBS, W. C., Compendium Meteorol., 1057-70 (Am. Meteorol. Soc., (Boston, Mass., 1951)

38. OBUKHOV, A. M., Tr. Inst. Teor. Geofiz. Akad. Nauk SSSR, 1, 95-115 (1946)

39. MONIN, A. S. Inform. Bull. Chief. Administr. Hydrometeorol. Serv., No. 1, 13-27 (1950)

40. MONIN, A. S., OBUKHOV, A. M., Dokl. Akad. Nauk SSSR, 93, No. 2, 223-26 (1953)

41. MONIN, A. S., OBUKHOV, A. M., Tr. Geofiz. Inst. Akad. Nauk SSSR, No. 24 (151), 163-87 (1954)

42. PRIESTLEY, C. H. B., Quart. J. Roy. Meteorol. Soc., 81, No. 348, 139-43 (1955)

43. KAZANSKY, A. B., MONIN, A. S., Izv. Akad. Nauk SSSR, Ser. Geofiz., No. 6, 741-51 (1958)

44. STEWART, R. W., Advan. Geophys., 6, 303-11 (1959)

45. LUMLEY, J. L., PANOFSKY, H. A., The Structure of Atmospheric Turbulence (Interscience, New York, 1964)

46. ZILITINKEVICH, S. S., CHALIKOV, D. V., Izv. Akad. Nauk SSSR, Atmospheric Oceanic Physics, 4, No. 3, 294-302 (1968)

47. CHARNOCK, H., Quart. J. Roy. Meteorol. Soc., 93, No. 395, 97-100 (1967)

48. DEARDORFF, J. W., WILLIS, G. E., Quart. J. Roy. Meteorol. Soc., 93, No. 396, 166-75 (1967)

49. MONIN, A. S., Atmospheric Turbulence and Radiowave Propagation, (in Russian) 113-20 (Nauka Publishers, Moscow, 1967)

50. MONIN, A. S., Theory of probability and its application, 3, No. 3, 285-317 (1958)

51. MONIN, A. S., J. Geophys. Res., 64, No. 12, 2196-97 (1959)

52. MONIN, A. S., J. Geophys. Res., 67, No. 8, 3103-09 (1962)

53. MONIN, A. S., Tr. Inst. Atmospheric Physics, Akad. Nauk SSSR, No. 4, 5-20 (1962)

54. MONIN, A. S., Izv. Akad. Nauk SSSR, Mechanics of Fluid and Gas, No. 1, 37-43 (1966)

55. ZUBLOVSKY, S. L., ZVANG, L. R., Izv. Akad. Nauk SSSR, Atmospheric and Oceanic Physics, 2, No. 12, 1307-10 (1965)

56. KAZANSKY, A. B., MONIN, A. S., Izv. Akad. Nauk SSSR, Ser. Geofiz., No. 1, 79-86 (1956)

57. KAZANSKY, A. B., MONIN, A. S., Meteorol. Hydrol, No. 12, 3-8 (1962)

58. ZILITINKEVICH, S. S., CHALIKOV, D. V., Izv. Akad. Nauk SSSR, Atmospheric and Oceanic Physics, 4, No. 9, 915-29 (1968)

59. ZUBKOVSKY, S. L., KOPROV, B. M., Izv. Akad. Nauk SSSR, Atmospheric and Oceanic Physics, 5, No. 4, 323-31 (1969)

60. MONIN, A. S., Izv. Akad. Nauk SSSR, Atmospheric and Oceanic Physics, No. 3, 397-407 (1962)

61. ZILITINKEVICH, S. S., LAIKHTMAN, D. L., Tr. Gl. Geofiz. Observ. No, 167, 44-8 (1965)

62. MONIN, A. S., Izv. Akad. Nauk SSSR, Atmospheric and Oceanic Physics, 1, No. 1, 45-54 (1965)

63. KAZANSKY, A. B., MONIN, A. S., Izv. Akad. Nauk SSSR, Ser. Geofiz. No. 1, 165-8 (1960)

64. MONIN, A. S., Izv. Akad. Nauk SSSR, Ser. Geograph. Geofiz., 14, No. 3, 232-54 (1950)

65. ZILITINKEVICH, S. S., MONIN, A. S., "Global Atmospheric Research Programme", Rep. Study Conf. Stockholm 28 June - 11 July, 1967, 5, 1-37 (Stockholm, 1967)

66. MONIN, A. S., Boundary Layers and Turbulence, 31-7 (Phys. Fluids, Suppl. 1967)

67. KAZANSKY, A. B., MONIN, A. S., Izv. Akad. Nauk SSSR, Ser. Geofiz., No. 5, 786-8 (1961)

68. GILL, A. E., The Turbulent Ekman Layer, (Dept. Appl. Math. Theor. Phys. Cambridge Univ., 1967)

69. CHARNOCK, H., ELLISON, T. H., Global Atmospheric Research Programme, Rept. Study Conf. Stockholm 28 June - 11 July 1967, 3, 1-16 (Stockholm, 1967)

70. BLACKADAR, A. K., Global Atmospheric Research Programme, Rept. Study Conf. Stockholm 28 June - 11 July 1967, 4, 1-11 (Stockholm, 1967)

71. LETTAU, H. H., Advan. Geophys., 6, 241-57 (1959)

72. LETTAU, H. H., Beitr. Phys. Atmosph., 35, No. 3-4, 195-212 (1962)

73. BLACKADAR, A. K., J. Geophys. Res., 67, No. 8, 3095-102 (1962)

74. BYZOVA, N. L., MASHKOVA, G. B., Izv. Akad. Nauk SSSR, Atmospheric and Oceanic Physics, 1, No. 11, 1209-11 (1964) ; 2, No. 7, 681)87 (1966)

75. KURPAKOVA, T. A., ORLENKO, L. R., Tr. Gl. Geofiz. Observ., No. 205, 13-24 (1967)

76. ZILITINKEVICH, S. S., CHALIKOV, D. V., Izv. Akad. Nauk SSSR, Atmospheric and Oceanic Physics, 4, No. 7, 765-72 (1968)

77. CHALIKOV, D. V., Meteorol. Hydrol., No. 8, 10-19 (1968)

78. MASHKOVA, G. B., Tr. Inst. Appl. Geophys. No. 2, 44-56 (1955)

79. ZILITINKEVICH, S. S., LAIKHTMAN, D. L., MONIN, A. S., Izv. Akad. Nauk SSSR, Atmospheric and Oceanic Physics, 3, No. 3, 297-333 (1967)

80. APPLEBY, J. F., OHMSTEDE, W. D., Proc. Army Sci. Conf. (1964), Washington, 1, 85-99 (1965)

81. ZILITINKEVICH, S. S., LAIKHTMAN, D. L., TSEITIN, G. H., Air-Ocean Interaction, 154-63 (Naukova Dumka, Kiev, 1966)

82. BOBYLEVA, I. M., ZILITINKEVICH, S. S., LAIKHTMAN, D. L., Atmospheric turbulence and radiowave propagation, 179-90 (Nauka Publishers, Moscow, 1967)

83. ZILITINKEVICH, S. S., CHALIKOV, D. V., Meteorol. Hydrol., No. 2, 11-26 (1968)

84. MONIN, A. S., Izv. Akad. Nauk SSSR, Atmospheric and Oceanic Physics, 1, No. 3, 258-65 (1965)

85. MONIN, A. S., Izv. Akad. Nauk SSSR, Atmospheric and Oceanic Physics, 1, No. 5, 490-500 (1965)

86. ESTOQUE, M. A., J. Geophys. Res., 868, No. 4 (1963)

87. SHARON, S. W., J. Geophys. Res., 70, No. 8, 1801-7 (1965)

88. KRISHNA, K., Monthly Weather Rev., 96, No. 5, 279-76 (1968)

89. VAGER, B. G., ZILITINKEVICH, S. S., Meteorol. Hydrol., No. 7, 3-18 (1968)

90. GOLITZYN, G. S., Izv. Ac. sci. USSR, Phys. Atm. Oceans, Vol. IV, No. 11, 1131-7 (1968)

91. GOLITZYN, G. S., Izv. Ac. Sci. USSR, Phys. Atm. Oceans, vol. V, No. 8, 775-81 (1969)

92. GOLITZYN, G. S., Doklady Ac. Sci. USSR, 190, No. 2, 523-6 (1970).

CONTENTS

SPACE AND TIME METEOROLOGICAL DATA ANALYSIS AND INITIALIZATION

P. MOREL

Laboratoire de Météorologie Dynamique
Ecole Normale Supérieure
75 - PARIS (France)

SPACE AND TIME METEOROLOGICAL DATA ANALYSIS AND INITIALIZATION

by

P. MOREL

Laboratoire de Météorologie Dynamique
Ecole Normale Supérieure
75 - PARIS (France)

1. - INTRODUCTION

The guideline of this Summer School on Dynamic Meteorology was to explore the problem of predicting the general circulation of the atmosphere from one specified state of motion at some initial time t_o. The previous speakers, particularly Professor PHILLIPS, showed that this initial value problem is approximately solved by replacing the continuous meteorological fields by a finite set of discrete values and integrating numerically the corresponding finite difference equations. Alternatively, one may choose to expand the continuous fields in series of orthogonal functions truncated at some finite order and solve numerically a set of algebraic interaction equations. In any case, the forecasting procedure starts from a set of initial values inferred from observations of the real atmosphere. The purpose of these talks is to review the methods used to infer these field values from the available experimental data. Several such methods, each claiming to be "optimal", have been proposed and some actually implemented in operational practice. It should be understood as the discussion progresses from straightforward to sophisticated approaches, that the best method from a mathematical or physical standpoint, may not be economical of computer processing time. But one must also keep in mind that the data assimilation and analysis process is an essential link between the world observing system and global extended range forecasting. It may be worth while then to embark upon expensive and involved data analysis methods if this allows a significant simplification of the observing system. In fact it is clear now that it would not be economically feasible to extend the present synoptic meteorological observation network to meet the requirements of future forecasting models. Consequently, further progresses of dynamic weather forecasting hinge on the utilization of mixed data from many different observing platforms, including non-synoptic satellite data. This is the reason why the design of the future global meteorological observing system cannot be separated from a fundamental understanding of observational data analysis and initialization.

2. - THE STATISTICAL APPROACH TO SYNOPTIC DATA ANALYSIS

We shall henceforth consider only the case of the discrete representation of a meteorological field $f(\vec{x},t)$ on a finite array of grid points distributed over the integration domain. Assume now that we know about the state of motion of the real atmosphere by simultaneous measurements $f(\vec{x}_i,t_o)$ at the synoptic time t_o and at fixed or mobile stations essentially distributed at random locations $\vec{x}_i$ over the Earth. There is naturally no reason why any observing station should coincide with one grid point. Accordingly, one particular grid point x_o is generally surrounded by a constellation of stations $\vec{x}_i$, each providing one measurement of the meteorological field under consideration (fig.1). The problem is to determine the most probable value of the field f at point $\vec{x}_o$, conditional upon the values $f_i = f(\vec{x}_i,t_o)$ observed at neighbouring stations.

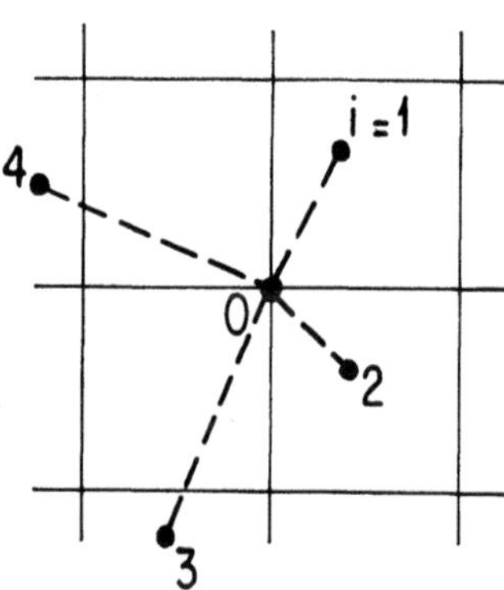

2.1. - Gandin's optimal interpolation method

Following GANDIN (1963), we first notice that the field value at location $\vec{x}_o$ is not completely unknown as the observation of many realizations of the meteorological field f or many successive meteorological situations has shown that $f(\vec{x},t)$ fluctuates about a climatological mean value, $\bar{f}(\vec{x},t)$. One can therefore separate each individual measurement f_i in two parts : the climatological average $\bar{f}_i$ and a random zero-average fluctuation f'_i which contains the significant information:

$$f_i = \bar{f}_i + f'_i \qquad (2.1)$$

similarly the (unknown) value of the meteorological field at $\vec{x}_o$ would be :

$$f_o = \bar{f}_o + f'_o \qquad (2.2)$$

How should we infer f'_o from the measured f'_i? In the absence of further knowledge of the random field $f'(\vec{x},t)$, the best approach consists in computing an estimate $\tilde{f}'_o$ for f'_o with a linear interpolation scheme :

$$\tilde{f}'_o = \sum_i p_i f'_i \qquad (2.3)$$

and try optimizing this estimate by a proper choice of the confidence weights p_i. Let then δ be the interpolation error :

$$\delta = f'_o - \tilde{f}'_o \qquad (2.4)$$

A reasonable optimization criterion is to ask that the mean square error $\overline{\delta^2}$ over an ensemble of independent realizations of the meteorological fields be minimum for a given geometrical configuration. Thus :

$$\overline{\delta^2} = \overline{f'^2_o} - 2 \sum_i p_i \overline{f'_o f'_i} + \sum_i \sum_j p_i p_j \overline{f'_i f'_j}$$

$$\frac{\partial}{\partial p_i} \overline{\delta^2} = -2 \overline{f'_o f'_i} + 2 \sum_j p_j \overline{f'_i f'_j} = 0 \qquad (2.5)$$

The system of linear equations (2.5) determines in general a unique set of values of the weights p_i corresponding to one particular configuration of grid points and stations. These values depend naturally upon the variance and spatial correlation $\overline{f'(x_i,t_o)\, f'(x_j,t_o)}$ of the random field f'. Considering that the variance $\overline{f'^2}$ does not vary much in the vicinity of $\vec{x}_o$, it is convenient (although not necessary) to divide equations (2.5) by $\overline{f'^2_o}$ and express the system in term of the correlation coefficients :

$$\mu_{ij} = \mu(\vec{x}_i,\vec{x}_j) = \frac{\overline{f'(\vec{x}_i,t_o)\, f'(\vec{x}_j,t_o)}}{\overline{f'^2_o}} \qquad (2.6)$$

The following system of n linear equations for n unknown weights p_i obtains :

$$\boxed{\sum_j \mu_{ij}\, p_j = \mu_{io}} \qquad (2.7)$$

The resulting set of p_i minimizes the variance of the true field value around the interpolated value $\overline{f_o} + \tilde{f}'_o$. It can therefore be said to be the optimal linear interpolation scheme from this statistical standpoint, for the given configuration of stations.

2.2. - Correlation properties of meteorological fields

Much depends then on the spatial correlation of the meteorological field under consideration. A first, simplified view of the matter will indicate that the fluctuating parts $f'(x,t)$ of all primary meteorological variables are essentially homogeneous isotropic random fields with a correlation range of the order of 1500 to 2000 Km. For obvious practical reasons, the subject has been given much attention by experimental investigators (see GANDIN, Objective Analysis of Meteorological Fields, 1968) from which the following charts are taken (fig. 2-4). The correlation coefficient is equal to 1 for zero separation, as it should from definition (2.6), and decreases to zero when the distance $|\vec{x}_i - \vec{x}_j|$ increases. The matter appears a bit more complicated on second sight. The correlation coefficient $\mu(\vec{x}_i,\vec{x}_j)$ is not truly isotropic nor homogeneous, and should actually be expressed as a symetrical tensor. Also, the coefficient depends upon the nature of the meteorological field (it is different for the X and Y components of the velocity, for example). But, the solution of system (2.7) is not really very sensitive to relatively gross approximations of the values of μ_{ij} so that it is quite adequate for our purpose to overlook the details and take μ_{ij} as an exponentially decreasing function of the distance r between $\vec{x}_i$ and $\vec{x}_j$ i.e :

$$\mu_{ij} \simeq \mu(r) = \exp\left(-\frac{r}{a}\right) \tag{2.8}$$

where a is a correlation range of the order of 1 500 Km.

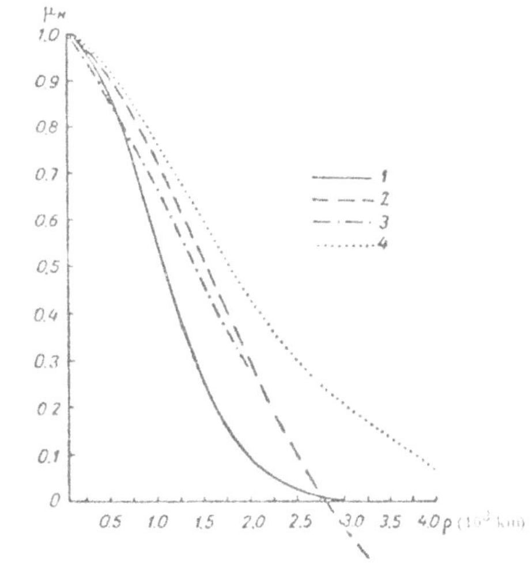

Fig.2 - Correlation coefficient of the 500 mb-geopotential

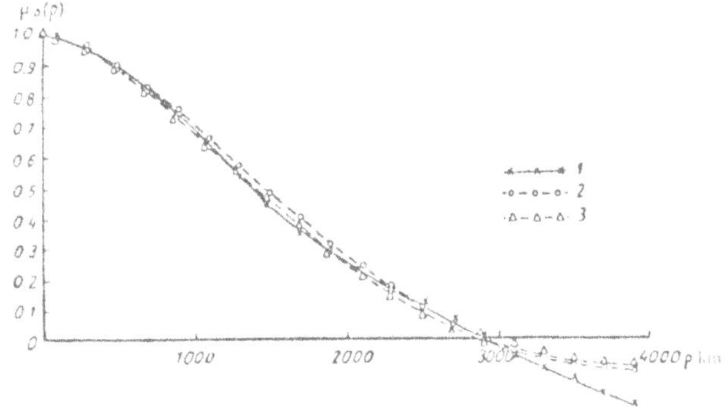

Fig.3 - Correlation coefficient of sea-level pressure

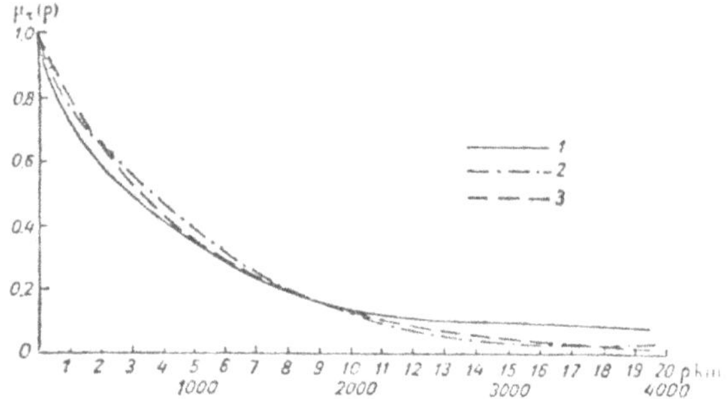

Fig.4 - Correlation coefficient of relative humidity at 850 mb

2.3. Four case-studies

We shall now illustrate what the method does for us by working out the values of the weights p_i for different geometrical configurations.

CASE I : Two symmetrical observing stations (fig. 5).

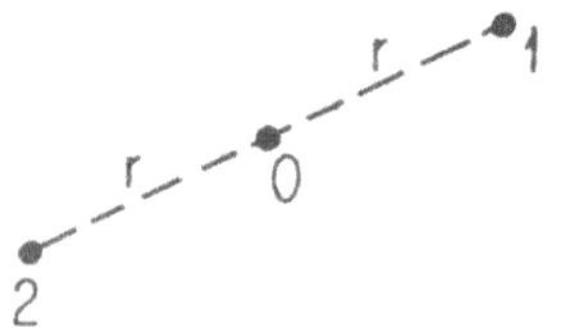

The linear system (2.7) can readily be written :

$$\begin{aligned} p_1 + \mu_{12}\, p_2 &= \mu_{01} \\ \mu_{21} p_1 + p_2 &= \mu_{02} \end{aligned} \qquad (2.9)$$

where :

$$\mu_{01} = \mu_{02} = \mu(r)$$

$$\mu_{21} = \mu_{12} = \mu(2r) \simeq \mu^2(r)$$

The solution is thus (fig. 6) :

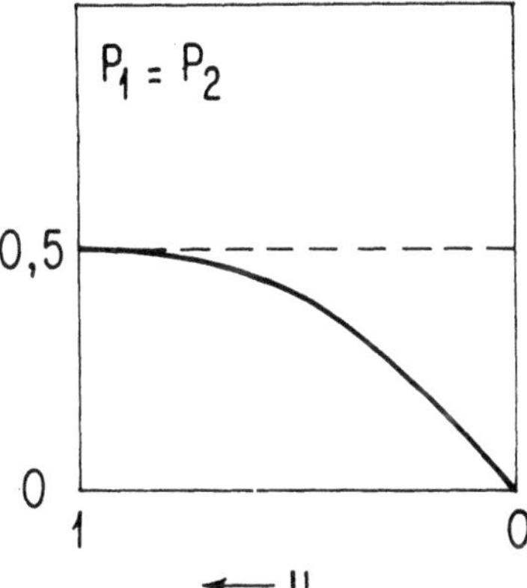

$$p_1 = p_2 = \frac{\mu}{1 + \mu^2}$$

If the distance between the grid point $\vec{x}_o$ and the stations $\vec{x}_1$, $\vec{x}_2$ is very small compared to the correlation range of the meteorological field, the most probable value is fully determined by the simultaneous measurements f'_1 and f'_2 , and is equal to the arithmetic average ($p_1 = p_2 = 0,5$). If, on the other hand, the distance is so large that the correlation coefficient is significantly smaller than 1, then the most probable value (from a climatological standpoint) is closer to the climatological average ($p_1 = p_2 < 0,5$). Thus , the interpolation automatically accounts for the proper relative weight of a priori information (the climatological mean value) and the new information afforded by observation.

CASE II : Two unsymmetrical stations.

Let us assume for example x_o is placed between x_1 and x_2 , at distances r and 2r respectively (fig. 7).

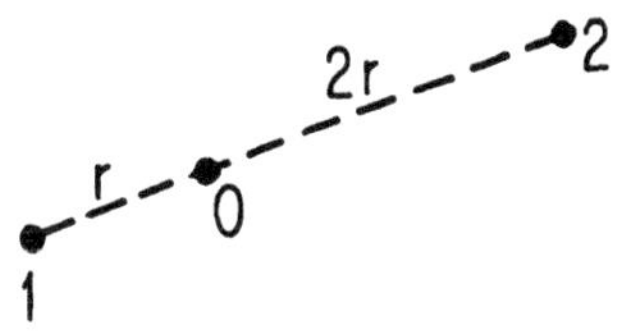

now :

$$\mu_{01} = \mu$$

$$\mu_{02} = \mu^2$$

$$\mu_{12} = \mu_{21} \simeq \mu^3$$

so that the optimal weights are (fig. 8) :

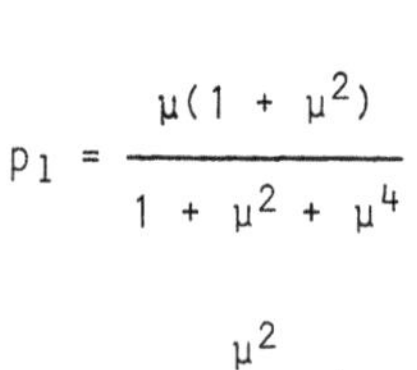

$$p_1 = \frac{\mu(1 + \mu^2)}{1 + \mu^2 + \mu^4}$$

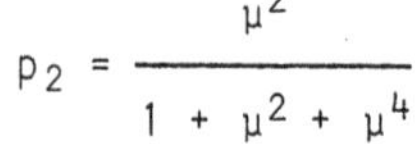

$$p_2 = \frac{\mu^2}{1 + \mu^2 + \mu^4}$$

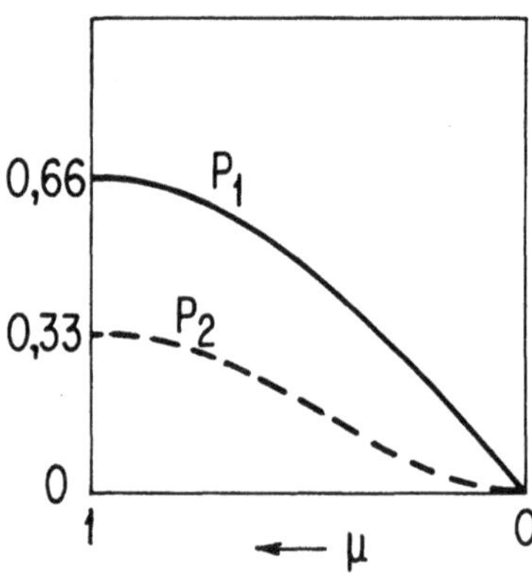

Again, for small distances, the best estimate is a weighted average of f'_1 and f'_2 with weights 0,66 and 0,33 inversely proportional to the distance. For large separations, the weights decrease to zero, thus making the optimal estimate closer to the climatological mean.

CASE III : Two redundant stations

Let us assume we have now three stations, two of which are so close together that they cannot be expected to provide independent information (fig. 9).

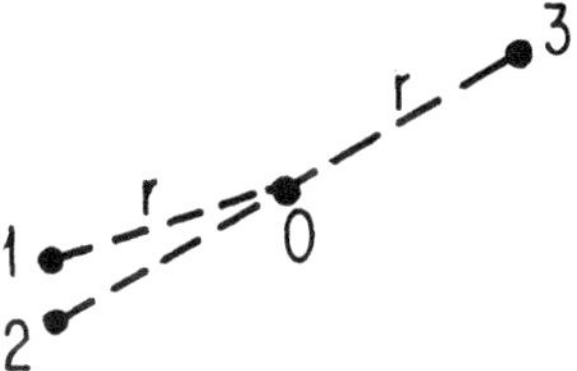

The observations from the three stations could nevertheless be included without extra care, as the optimal scheme accounts automatically for the redundancy. Indeed :

$$\mu_{01} = \mu_{02} = \mu_{03} = \mu$$

$$\mu_{12} = \mu_{21} \simeq 1$$

$$\mu_{13} = \mu_{23} = \mu_{31} = \mu_{32} \simeq \mu^2$$

so that the linear system (2.7) reads :

$$p_1 + p_2 + \mu^2 p_3 = \mu$$

$$p_1 + p_2 + \mu^2 p_3 = \mu$$

$$(p_1 + p_2)\mu^2 + p_3 = \mu$$

and is undetermined as p_1 and p_2 appear in the combination $(p_1 + p_2)$ only. The solution,is then :

$$p_1 + p_2 = p_3 = \frac{\mu}{1 + \mu^2}$$

giving equal weights to the two redundant stations on one hand, and to the third station on the other hand.

CASE IV : Four equally spaced stations.

The preceding case-studies gave us great confidence that this remarkable scheme will somehow produce the best possible interpolation under any geometrical configuration. One should however exercise some care as will be shown presently. Consider then a grid point $\vec{x}_o$ centrally located in a square array of observing stations (fig. 10).

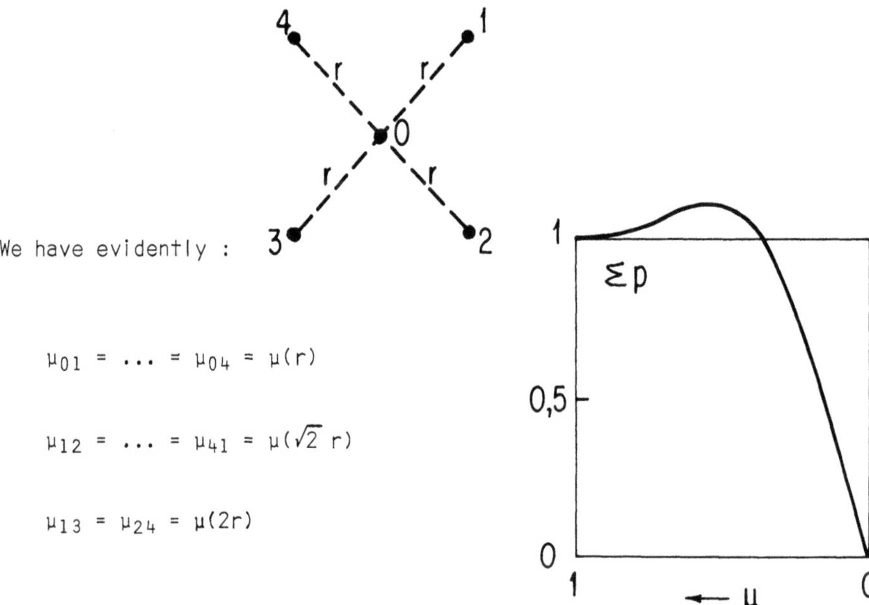

We have evidently :

$$\mu_{01} = \ldots = \mu_{04} = \mu(r)$$

$$\mu_{12} = \ldots = \mu_{41} = \mu(\sqrt{2}\, r)$$

$$\mu_{13} = \mu_{24} = \mu(2r)$$

The four p_i are equal and straightforward algebra yields :

$$p_1 = \ldots = p_4 = \frac{\mu(r)}{1 + 2\mu + (\sqrt{2}\, r) + \mu(2\, r)}$$

We have plotted (fig. 11) the sum $\sum_1^4 p_i$ versus different values of the distance or equivalently, the correlation coefficient μ estimated from (2.8). We find that the estimate is the arithmetic mean value when $\mu \simeq 1$, as expected. But we also find that the "optimal" estimate exceeds this mean value by 10 % when μ decreases. This certainly cannot be correct as no statistical consideration justifies the interpolated value to overshoot the surrounding observations. A cue to what may have gone wrong is given by trying a different approximation for the correlation coefficient, for example :

$$\mu(r) \approx \exp\left(-\frac{r^2}{a^2}\right) \qquad (2.10)$$

We find then an even more drastic overshooting and discover that the effect is very sensitive to the particular shape of the function (fig. 12).

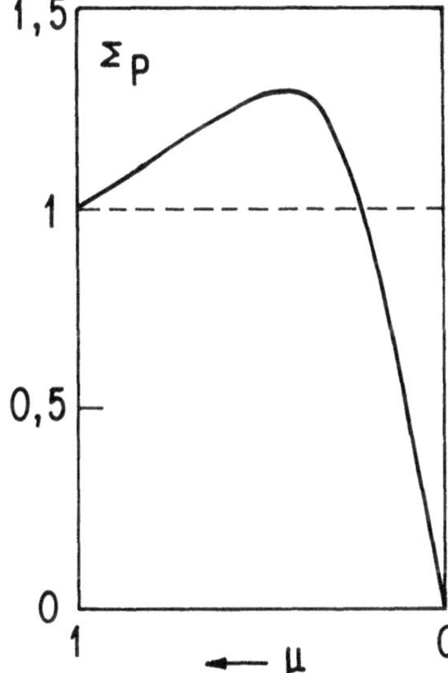

Considering that (2.8) is only a gross approximation of the actual correlation function, we feel justified to discard this difficulty as due to our oversimplifying the actual situation. One must however remember to renormalize the weights p_i by forcing the condition :

$$\sum_i p_i \leqslant 1 \qquad (2.11)$$

We notice further that condition (2.11) is better satisfied by model (2.8) than (2.10). One should therefore make sure that the particular approximation chosen for $\mu(r)$ decreases at least linearly for small r.

2.4. - Generalization to time like interpolation

There is no a priori reason why the same scheme could not be applied to measurements distributed not only in space around grid point $\vec{x}_o$ but also in time about the synoptic map time t_o. The same algebra applies of course, provided the following four dimensional correlation coefficients are known :

$$\mu_{ij} = \mu(\vec{x}_i - \vec{x}_j \ , \ t_i - t_j) \tag{2.12}$$

This linear interpolation procedure is certainly applicable when the time interval encompassed by the non-synoptic observations is short with respect to the typical time-constant of large meteorological perturbations (1-2 days typically) since the evolution of any field value can then be approximated fairly by a first order linear law. This approach should fail, on the other hand as soon as significant new meteorological developments appear. It is found experimentally that the time-correlation of dynamic variables remain remarkably high for periods of 12 to 24 hours (fig. 13), so high in fact as to suggest using persistence forecast rather than this moderately complicated interpolation. In other words, data spread over no more than 12 hours (possibly 24 hours) may just as well be considered as simultaneous for the purpose of applying Gandin's optimal interpolation scheme.

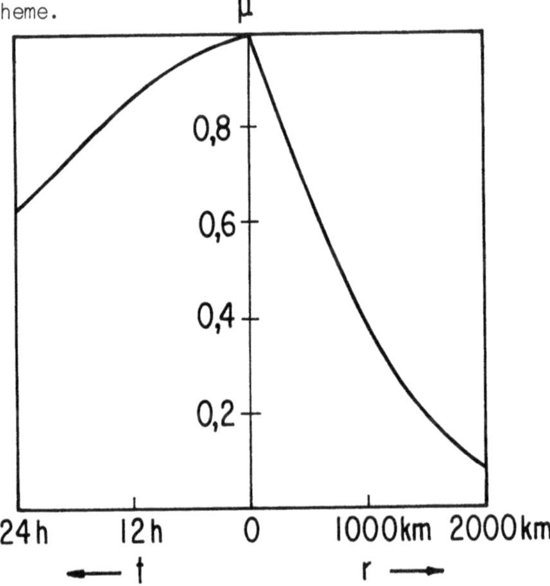

Fig. 13 - Comparison of spaco and time correlation of the zonal velocity at 200 mb.

This fact is a strong indication that we are not making the best possible use of available information : indeed we do have a much better pre-analysis knowledge of the meteorological fields than just climatic averages, if only taking the previous day estimate as a first guess for the current analysis. This is just what a human analyst would do naturally and this is indeed what we will consider in the next lecture.

3. - SEQUENTIAL ANALYSIS OF METEOROLOGICAL FIELDS

The obvious guiding idea here is that the latest observations are only supplemental information to update an already fair estimate of the meteorological fields given by previous analysis and forecasts. This idea was followed by Petersen (1968) who proposed an optimal sequential analysis procedure for linear random fields and to some extent, meteorological fields. We shall endeavour here to give the simplest account of Petersen's work without unessential mathematical developments.

3.1. - Field measurements as sequential differential corrections

We shall assume the existence of a reasonably accurate and unbiased model of the dynamic behaviour of the meteorological fields, meaning that the forecast value $f^*(x,t)$ of one variable at point $\vec{x}$ and time t is uncorrelated to the forecast error :

$$f'(x,t) = f(x,t) - f^*(x,t) \qquad (3.1)$$

This property must be understood as an application of the principle of orthogonal projection in the theory of linear stochastic processes. Defining then the stochastic average as the mean value over an ensemble of independent random realizations of the fields and forecasts, the following condition must be met for all $\vec{x}$ and $\vec{y}$:

$$\overline{f'(x,t) \quad f^*(y,t)} \equiv 0 \qquad (3.2)$$

or

$$\overline{\left[f(x,t) - f^*(x,t)\right] \quad f^*(y,t)} \equiv 0 \qquad (3.2')$$

Persistence forecast, although not very accurate, is indeed unbiased and would, for example, provide an acceptable first guess in the present approach. More sophisticated forecasting scheme would improve the matter provided they satisfy condition (3.2). Indeed, we shall see that Petersen's method **does** not actually depend upon faithful simulation of the real atmosphere dynamical response.

Given now a fair prior estimate of the fields by a good forecasting scheme, any new measurement $f(x_i,t)$ of a meteorological variable will appear as a <u>small</u> or differential correction (3.1) of the estimated value. This correction is not obtained in general at a regular grid point but instead the new piece of information should be incorporated as suitable corrections of the neighbouring grid point values at this time t. Assuming for generality that there are more than just one measurement at time t, the correction at one particular grid point $\vec{x}$ may be taken as a linear combination of the observed $f'(x_i,t)$ with due consideration of the correlation between the field value at $\vec{x}$ and $\vec{x}_i$. The corrected value will then be :

$$\tilde{f}(x,t) = f^{*}(x,t) + \sum_i p_i(x)\; f'(x_i,t) \qquad (3.3)$$

3.2. - <u>The optimization criterion</u>

On the basis of the theory of linear random systems and the principle of orthogonal projection applicable to such systems, the most probable corrected value (3.3) must be an <u>unbiased</u> estimate of the field, thus :

$$\overline{\left[f(x,t) - \tilde{f}(x,t)\right] \tilde{f}(y,t)} \equiv 0 \qquad (3.4)$$

Making use of condition (3.2) and the definition (3.3), this expression of the optimization criterion can be reduced to the following set of identities :

$$\sum_i p_i(y)\, \overline{f'(x)f'(x_i)} - \sum_{ij} p_i(y)\, p_j(x)\; \overline{f'(x_i)f'(x_j)} \equiv 0 \qquad (3.5)$$

where the argument t has been dropped for simplicity. Dividing by the essentially constant variance $\overline{f_o'^2}$ (in the neighbourhood of grid point $\vec{x}$) and introducing the correlation coefficients :

$$\mu_{oi}(x) = \frac{\overline{f'(x)\, f'(x_i)}}{\overline{f_o'^2}}$$

$$\mu_{ij} = \frac{\overline{f'(x_i)\, f(x_j)}}{\overline{f_o'^2}} , \qquad (3.6)$$

we obtain the familiar looking set of identities :

$$\sum_i p_i(y)\, \mu_{oi}(x) \equiv \sum_i \sum_j p_i(y)\, p_j(x)\, \mu_{ij} \qquad (3.7)$$

To complete the identification with Gandin's formulation, let us introduce the following vectors and matrices :

$$P(\mathbf{x}) = \begin{pmatrix} p_1(x) \\ " \\ " \\ p_n(x) \end{pmatrix}$$

$$\left[\mu_{0i}(x)\right] = \begin{pmatrix} \mu_{01}(x) \\ " \\ " \\ \mu_{0n}(x) \end{pmatrix} \qquad (3.8)$$

$$\left[\mu_{ij}\right] = \begin{pmatrix} \mu_{11} & \mu_{12} & \cdots \\ \mu_{21} & & \\ \vdots & & \\ \vdots & & \mu_{nn} \end{pmatrix}$$

and write (3.7) as one matrix identity :

$$[P(y)]^{+}[\mu_{oi}(x)] \equiv [P(y)]^{+}[\mu_{ij}]\ P(x) \tag{3.9}$$

equivalent to a single matrix equation :

$$[\mu_{oi}(x)] = [\mu_{ij}]\ P(x) \tag{3.10}$$

which is nothing but the linear equations system (2.7) arrived at by Gandin, with one modification however : the correlation coefficients now relate departures of the meteorological fields from their <u>forecast values</u> (instead of deviation from the climatological mean). The departures found in this sequential analysis scheme are truly <u>differential corrections</u> which can be taken in stride by the dynamic model. Accordingly, these corrections need not be applied to all dynamic variables at the same (synoptic) time. A partial set of differential corrections can instead be applied at some time t and carried forward (updated) by the forecasting scheme until it can be mixed with further information introduced at a later time t' , and so on.

3.3. - <u>Discussion</u>

There is much to be said in favor of the sequential analysis method which is a formal and mathematically consistent presentation of the empirical procedures used in current practice. It does incorporate the basic advantage of Gandin's method i.e. the capability to derive the proper weights of new information relative to "first guess" estimates based on previous data. Yet one should not expect more from the method than what is actually contained in it : a statistically optimal linear interpolation scheme, conditional upon the choice of a particular forecasting device.

The common shortcoming of the statistical approaches is that they refer only incidentally (if at all) to the real or model <u>atmosphere dynamics</u>. Truly, Petersen's sequential analysis scheme is based on the existence of a model of the dynamic behaviour of the atmosphere, but this model need not be accurate nor dynamically faithful, if it does not introduce a systematic bias. Indeed, persistence would be a perfectly valid scheme for this purpose. So, these optimal interpolation schemes do not depend upon the real dynamic response of the atmosphere to a sudden perturbation nor do they demand any faithful modelling of this dynamic response. No attention is paid to the dynamic balance or imbalance of the various meteorological fields updated independently from each other using whatever observations are available. Accordingly, missing data like wind measurements cannot be inferred from other data using a statistical approach alone.

Nor is the sequential method a truly objective analysis scheme either, because, unless the observations are very numerous and very evenly distributed indeed, the choice of one particular "first guess" forecasting model forces a significant bias onto the resulting meteorological fields. It is true that a memory of all previous observations is contained in the final product but only through the agency of one particular atmosphere model, determined by a subjective choice. In data sparse regions then, the resulting fields may reflect that particular model climatology more than anything else. There is little one can do about that, as the process of data assimilation is intimately linked to the dynamics of the forecasting model. It is best however to recognize this limitation and refrain from claiming a "universal objective" value for the resulting analysis.

Before leaving the subject, an attempt by Miyakoda and Talagrand (1970) to overcome this limitation, must be quoted. These authors try to free their method from the bias of one particular model by comparing the results of the numerical forecasting to real observations. The principle of the method consists in mixing (with optimal weights) the current data with the forecast values obtained from the best estimate at t_o - 1 day, as well as that obtained from t_o - 2 days, t_o - 3 days ... Unfortunately, the current general circulation models are not yet accurate enough to allow a reasonable utilization of previous data older than 1 or 2 days say. But research has not said its last word.

4. - THE DYNAMIC RESPONSE OF REAL AND SIMULATED ATMOSPHERES

4.1. - An analog linear model

Before tackling the practical problem of data assimilation, let us firstly feel our way and consider the response of a very simple one layer model of incompressible flow to an initial disturbance. Let then be u, v, the components of velocity, ϕ the geopotential of the free surface anf f_o the Coriolis parameter. Let us further restrict the motions to small perturbations of a basic uniform flow with velocity u_o, v_o, and geopotential ϕ. The Navier-Stokes equations and the continuity equation may then be written, to the first order of perturbation, as a set of simultaneous linear differential equations :

$$\frac{\partial u}{\partial t} + u_0 \frac{\partial u}{\partial x} + v_0 \frac{\partial u}{\partial y} + \frac{\partial \phi}{\partial x} - fv = 0$$

$$\frac{\partial v}{\partial t} + u_0 \frac{\partial v}{\partial x} + v_0 \frac{\partial v}{\partial y} + \frac{\partial \phi}{\partial y} + fu = 0 \qquad (4.1)$$

$$\frac{\partial \phi}{\partial t} + u_0 \frac{\partial \phi}{\partial x} + v_0 \frac{\partial \phi}{\partial y} + \phi_0 \left(\frac{\partial u}{\partial x} + \frac{\partial v}{\partial y} \right) = 0$$

It is convenient to collect these equations into one single matrix equation by introducing the vector :

$$U = \begin{pmatrix} u \\ v \\ \phi \end{pmatrix} \qquad (4.2)$$

and the constant matrices :

$$A = \begin{pmatrix} u_0 & 0 & 1 \\ 0 & u_0 & 0 \\ \phi_0 & 0 & u_0 \end{pmatrix} \qquad (4.3)$$

$$B = \begin{pmatrix} v_0 & 0 & 0 \\ 0 & v_0 & 1 \\ \phi_0 & 0 & v_0 \end{pmatrix}$$

$$C = \begin{pmatrix} 0 & -f & 0 \\ f & 0 & 0 \\ 0 & 0 & 0 \end{pmatrix} \tag{4.3}$$

(4.1) reads then :

$$\frac{\partial U}{\partial t} + A\frac{\partial U}{\partial x} + B\frac{\partial U}{\partial y} + CU = 0 \tag{4.4}$$

specializing now to plane wave solutions like :

$$U = U_{\ell m} \exp i\,[\,\ell x + my - \sigma t] \tag{4.5}$$

the following homogeneous algebraic equation obtains :

$$(\ell A + mB - iC - \sigma I)U = 0 \tag{4.6}$$

I is the unit 3 x 3 matrix. The secular equation derives immediately from this matrix relation and definition (4.3) and yields three eigenvalues corresponding to independent eigenvectors or "normal modes" of the fluid system :

$$\sigma^i = \ell u_0 + mv_0 + \varepsilon^i \sqrt{(\ell^2 + m^2)\,\phi_0 + f^2} \tag{4.7}$$

where $i = 0, 1, 2$ and $\varepsilon^i = 0, 1$ and -1, respectively.
We can condense this expression of the eigenvalues by introducing the wave-vector $\vec{K}(\ell, m)$ and the phase-velocity $c_0 = \sqrt{\phi_0}$ of the gravity waves which propagate in the system. Finally :

$$\sigma^i(\vec{K}) = \vec{K}.\vec{V}_0 + \varepsilon^i \sqrt{K^2 c_0^2 + f^2} \tag{4.7'}$$

It can be seen readily that the first eigenmode i = 0 of this very simple model flow, is not a travelling perturbation at all but a stationary disturbance, moving with the flow at the basic velocity $\vec{V}_0$. This is demonstrated best by taking the curl of the dynamic equation (4.1) to get the vorticity equation:

$$(\frac{\partial}{\partial t} + \vec{V}_0 \cdot \vec{\nabla})\zeta + f(\frac{\partial u}{\partial x} + \frac{\partial v}{\partial y}) = 0 \qquad (4.8)$$

$$\zeta = \frac{\partial v}{\partial x} - \frac{\partial u}{\partial y} \qquad (4.9)$$

A frozen-in wave obtains when the divergence $(\frac{\partial u}{\partial x} + \frac{\partial v}{\partial y})$ vanishes. Consequently, the first eigenmode i=0 is associated with zero divergence. This mode would then correspond, in a more realistic model of the atmospheric flow, to the quasi-bidimensional quasi-geostrophic waves generated by baroclinic instability.

The other two modes i=1 and 2, on the other hand, are associated with a large phase velocity, of the order of the speed of sound C_0, and correspondingly large vertical motions and horizontal flow divergence. These modes are thus gravity waves which are generated in the real atmospheric flow by meso-scale systems, thermal convection or passing over a mountain. External gravity waves similar to those allowed in the one-layer model flow, are associated with large vertical motions and large phase velocities. Internal gravity waves with an alternating vertical structure could also be generated: their phase velocity is not so large and possibly even close to that of geostrophic waves. But all gravity waves are characterized by significant vertical motions and correspondingly large horizontal divergence. An order of magnitude of the vertical velocity associated with both kinds of motion can be deduced from dimensional analysis of fluid dynamics equations for the general circulation of the atmosphere (see Charney, The Theory of Quasi-geostrophic

Flow in the same volume). It is found that inertia-gravity waves produce vertical velocities W of the order of VD/L (where V is the typical horizontal wind velocity, D the vertical scale height and L the horizontal scale of the motion). On the other hand, quasi-geostrophic waves produce only much smaller vertical velocities:

$$W \sim V \frac{D}{L} Ro$$

where Ro is the Rossby number, of the order of 0.1.

4.2. - Dynamic response to a perturbation

Coming back to our linearized one-layer model above, we note that the set of normal modes $U^i(\vec{K})$ constitutes a complete basis in the functional space of all possible motions of the model system. Thus, any initial perturbation of the fluid departing from the uniform basic flow velocity $\vec{V}_o$ can be expanded in series of these normal modes. The expansion will, in general, include gravity waves as well as quasi-geostrophic waves (in fact more gravity waves since they provide twice as many degrees of freedom as geostrophic waves).

A similar response can be expected from any mathematical analogue or model of the real atmosphere dynamics with a finite number of degrees of freedom and indeed, a similar response of the real flow would be excited if we were able to force the appropriate perturbation. Any small departure from one particular state of motion can be dealt with initially, in the frame of a linear perturbation theory described by equations formally similar to (4.4) and the dynamics of the fluid can be entirely described in term of its transient response to a percussive perturbation applied at one initial instant t=0. We may expect then that, in general, the initial motion of any numerical model flow simulating the general circulation of the atmosphere will include quasi-horizontal (bi-dimensional)quasi-geostrophic motions together with gravity waves associated with large vertical velocities.

4.3. - The condition of geostrophic balance

It is a fact that the circulation of the real atmosphere of the planet Earth includes large quasi two-dimensional quasi-geostrophic perturbations of all scales, which account for a large fraction of the total kinetic energy, and yet includes very little energy in the gravity wave modes. Why this should be so is not at all clear. It may be that gravity waves being essentially three-dimensional are coupled much more efficiently with three-dimensional clear-air turbulence, thereby dissipating their energy quickly while no such energy cascade develops in the essentially two-dimensional geostrophic perturbations. It may be also that gravity waves propagate upwards with increasing amplitudes as the density of air decreases, and eventually dissipate in upper atmosphere turbulence. In any case, we are led to conclude that a realistic model flow simulating the real atmosphere circulation, must be nearly in a state of "geostrophic balance" meaning that the amplitude of gravity oscillations sustained in the model must be vanishingly small.

It has not been proven that the best set of initial values of one particular mathematical analogue of the real atmosphere circulation, i.e. the initial values which would yield the best possible forecast, is that which corresponds to "geostrophic balance". But many unhappy experiments have indeed indicated time and time again that whenever the initial response of the mathematical analogue includes large inertia gravity oscillations, the forecast is immediately spoiled. It is also found, interestingly enough, that the forecast could recover to some extent from this "initialization shock" after 12 to 24 hours, when unphysical small scale gravity oscillations have dissipated through non-linear interaction.

We shall therefore conclude that the aim of a good initialization procedure is to infer from all information available, the one particular set of initial values of the forecast model parameters which:

- best resemble the real motion of the atmosphere at the same time,
- is consistent with the condition of "geostrophic balance", i.e. generates the least amount of gravity waves in the initial response of the model.

4.4. - How to restore the geostrophic balance

We can take it for granted that any set of raw measurements of meteorological parameters in the real atmosphere, will not be conducive to perfect geostrophic balance. For one thing, the measurements will be noisy from observational errors and from small scale fluctuations associated with transient turbulent motions or local gradients Also, the measurements will be incomplete so that a complete set of initial values can only be inferred by some more or less artificial interpolation method (see chapters 2 and 3 above). Finally, even if perfectly adequate space and time smoothed initial values were obtained everywhere the initial response of the mathematical analogue would still not be geostrophically balanced because the dynamics of the analogue cannot exactly mimick that of a real fluid with an infinite number of degrees of freedom. Consequently, initialization is indeed a matter of restoring the geostrophic balance lost when new observational data are introduced.

This problem was very neatly resolved in the early days of numerical forecasting, as meteorologists used filtered equations which could only describe the general circulation in term of a quasi-barotropic model flow (no gravity wave by design). But we are now using primitive equation models like (4.1), which do accommodate gravity wave solutions of any scale. So the simplest idea for eliminating the unwanted waves was to impose on the raw initial values the same dynamic conditions which would obtain in the previous filtered equations model, i.e., for example, various versions of the balance equation:

$$\nabla^2(\frac{1}{2}V^2 + \phi) + \vec{k}.\mathrm{curl}(f+\zeta)\,\vec{V} \simeq 0 \tag{4.10}$$

which can be derived from the primitive Navier-Stokes equations for two-dimensional flow, assuming that the divergence is small and nearly constant in time:

$$\frac{\partial u}{\partial x} + \frac{\partial v}{\partial y} \simeq 0 \tag{4.11}$$

One extra advantage of using a balance equation is that it allows computing the (non-divergent) wind velocity from the geopotential field only, a very convenient feature indeed when wind data are sparse and not reliable.

Now, one should realize that such a priori condition like the "balance equation" cannot be more than moderately successful because no two mathematical analogues of the general circulation show exactly the same dynamic response to one particular initial state of motion, let alone the same dynamic response as the real atmosphere. Thus, no single a priori "balance equation" could indeed produce a true geostrophic balance in all models. One must then convince oneself that the best set of initial values as defined in paragraph 4.3. is special to the one general circulation model under consideration. There is no "objectively" optimal set of initial values, only a variety of optimal sets, each being optimum for one particular mathematical analogue of the real flow. Indeed, the same initialization shock obtains when raw observational values are introduced in a forecast model or ready-made analysed fields produced with a different general circulation analogue. Values of the dynamic quantities which correspond to a balanced geostrophic flow in one general circulation model (or artificial data generated by this model) are not consistent with the geostrophic balance in another general circulation model. This is why the process of initialization or initial <u>data assimilation</u> must be taylor-made to fit the particular forecast model under consideration.

4.5. - <u>The damping of gravity-waves</u>

We have just seen that no single "filter" would do to eliminate that part of the initial values which will give rise to unwanted gravity oscillations. Another possible approach for restoring the geostrophic balance is thus to <u>test</u> the initial (essentially linear) dynamic response of the forecast model to the set of initial data and let the inertia-gravity modes develop to such an extent that they may be <u>discriminated</u> from the desired geostrophic modes. Note that discrimination is only possible after some time interval so that the initial data assimilation necessarily involves the time dimension: data assimilation for restoring the geostrophic balance is by nature a <u>four-dimensional</u> (time and space) process.

The problem could then be reduced to selectively eliminating or rather damping the inertia-gravity waves whenever they develop. The existence of such physical damping processes in the real atmosphere, has been alluded to in 4.3. but the fact is that such damping does not obtain naturally in numerical models probably because of heavy spectral truncation preventing the non-linear transfer of energy to sub-grid scale turbulent motions (not represented in the computation). It is thus appropriate to parameterize somehow these damping mechanisms and restore the proper selective dissipation of large scale inertia-gravity motions or even larger than nature dissipation when convenient.

The selective damping scheme proposed by Sadourny (1973) and used in all Laboratoire de Météorologie Dynamique assimilation experiments, consists in adding to the dynamic equation a modified eddy-viscosity dissipation term acting on the divergent part of the horizontal flow only:

$$\frac{\partial \vec{V}}{\partial t} + \ldots\ldots = K \operatorname{grad}(\operatorname{div} \vec{V}) \qquad (4.12)$$

Here K is an eddy-viscosity coefficient appropriate to mesoscale eddies, i.e. $K \simeq 10^5$ m^2 sec^{-1} or more if faster damping rates are needed. Since the divergence associated with quasi-geostrophic motions is about one order of magnitude smaller than the divergence associated with inertia-gravity oscillations of the same amplitude, it can be said that the right hand side of (4.12) produces effectively a selective damping of the latter. This has indeed been verified by several numerical experiments; it has been shown in particular, that the selective damping term (4.12) does not consume the geostrophic motion kinetic energy after dissipating the gravity wave energy (Sadourny, 1972).

Other selective damping schemes have been used for the same purpose, notably the Matsuno or Euler-backward time differencing scheme (Charney et al., 1969; Morel et al.,1971). It can be shown (Lilly, 1965) that step-wise time integration according to this finite difference approximation of the time derivative, induces an amplification factor slightly smaller than 1 (for each time step Δt) of all waves with frequencies smaller than $1/\Delta t$ (fig.14). Thus, this damping scheme is actually frequency dependent but not precisely mode-selective.

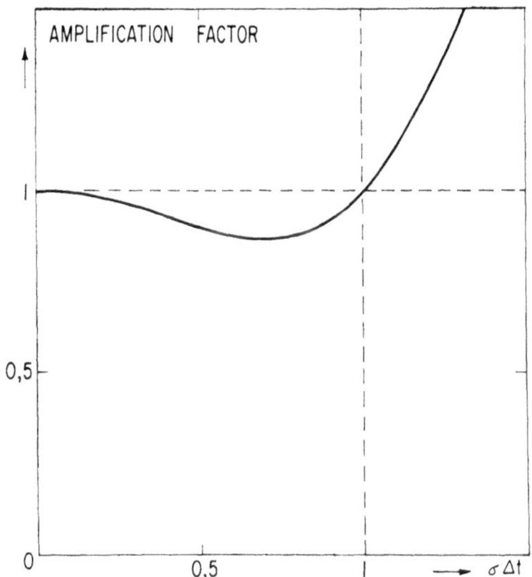

Indeed, low frequency internal gravity waves as well as "computational" modes akin to stationary gravity waves with 2 Δx wavelength, are not damped at all. Although this very simple device does provide some amount of damping, experiments with the truly selective damping scheme (4.12) have shown consistently better results.

5. - ASSIMILATION AND NON-SYNOPTIC DATA ANALYSIS

5.1. - Principle of four-dimensional analysis

We have seen that a numerical model of the atmospheric flow is necessarily a truncated version of the real fluid, with only a finite number of degrees of freedom. In the case of most general circulation models, the flow is described by 3N discrete field values (u, v, ϕ) at the nodes of a N points array or grid so that the model dynamic response is limited to 3N degrees of freedom. Any state of motion of this system is then specified by 3N parameters so that one particular dynamic history of the model may be represented by the trajectory of a point $X(t)$ in a 3N dimensional phase-space.

From a more mathematical standpoint, one may thus define one particular model of the atmospheric flow by specifying the nature of its degrees of freedom (so many points, so many layers ...) and the numerical scheme according to which the state of motion $X(t_1)$ at time t_1 is computed from the initial state $X(t_o)$. Let us write symbolically :

$$X(t_1) = f\left[X(t_0), t_1 - t_0\right]_{t_1>t_0} \tag{5.1}$$

In the real atmosphere, the dynamic evolution from t_o to t_1 involves many irreversible processes, both dynamic like molecular and eddy viscosity and thermodynamic, like condensation and rainfall. A special effort can be made to incorporate these irreversible processes (or at least their mean effect) as dependent functions in the forecasting scheme. But unless this special effort is done, a numerical model with a finite number of degrees of freedom includes basically only reversible exchanges of mass, momentum and energy between the different boxes or spectral components. This is indeed the very fundamental difference between the real and the simulated flows as dissipation occurs only if an exceedingly large (practically infinite) number of degrees of freedom is available to the system. Thus the numerical forecasting scheme (5.1) is reversible :

$$X(t_0) = f\left[X(t_1), t_0 - t_1\right]_{t_0<t_1} \tag{5.2}$$

and (5.1) together with (5.2) defines a one-to-one correspondence among all possible states of motion $X(t)$ of the system, at all times t. Indeed two states $X(t_o)$ and $X(t_1)$ can be said equivalent if they are related by (5.1) and (5.2) i.e., if they belong to the same possible history of the system. Hence, historical sequences of states $X(t)$ defined by the forecasting scheme (5.1) constitute classes of equivalence in the ensemble of all possible states of motion at all times and any state of the model atmosphere may be specified by the class (or historical sequence) X to which it belongs, and time t.

Conversely, one historical sequence is fully determined by specifying one particular state $X(t_o)$ i.e. the initial values of 3N parameters at time t_o or just as well, by any set of 3N independent conditions including non-simultaneous measurements of the dynamic variables and / or dynamic conditions restricting the number of degrees of freedom allowed to the system (absence of gravity wave, for example).

We conclude that, given an adequate number of observations in a short period (so short that a purely inertial model without source nor sink of energy is reasonably faithful), there is one and only one history of this model consistent with this set of data. The practical computational procedure by which this solution is found does not matter: inasmuch as it is unique, the solution is the limit of any convergent relaxation procedure. We shall describe one such relaxation scheme below.

5.2. - Assessment of data sets sufficiency

What is a sufficient data set, and what is the desirable amount of observation redundancy ? This question is of course, basic to the design of any global observing and weather forecasting system i.e. , the very core of GARP. But it is not a trivial question to ask with reference to such a complicated, non-linear fluid system as the atmosphere. We shall only attempt to discuss the problem here using analogy with the linearized system typified by equations (4.1).

Assume further that we are dealing with a finite array of N grid points and three independent variables (u, v, ϕ) per grid point so that (4.7) applies and that the system possesses 3N orthogonal normal modes which may serve to count the independent degrees of freedom. Consider now the restrictions we want to place on the dynamic response of the system, namely that there be no or very little excitation of the gravity waves. This leaves N amplitudes to be specified so that only one field value per grid point need be measured for determining the allowed quasi-geostrophic response to the input of new observations.

This straightforward property of a linear system governed by equations (4.1) must have its counterpart for non-linear systems, as can be guessed from equations (4.10) and (4.11). If the balance relation (4.10) is a prescribed condition, one dynamic variable, the geopotential Ø for example, can be computed from the other two (u,v,). If both relations (4.10) and (4.11) are prescribed, then two dynamic variables, the components of the velocity for example, may be computed from the third. Indeed, it is current practice in operational data analysis and initialization, to infer the wind velocity fields (u,v,) from the observed geopotential or mass field obtained from radiosondes. The reason for the success of this procedure is undoubtedly the fact that the desired quasi-geostrophic response of the forecasting model includes only N degrees of freedom.

Consider now a situation where the input data contain more information than strictly needed, considering the number of allowed degrees of freedom. It is clear then that the problem of finding the optimal response will be overdetermined and generally impossible. It is well known in operational practice than adding too many wind data to the analysis based on radiosonde (geopotential) data, can spoil the forecast. This is so, not so much because the wind measurements are objectively inaccurate but rather because they are not consistent with the geopotential field within the special dynamic balance demanded by the numerical forecasting scheme. Indeed, the analogy with a linear system indicates that more than N independent field values, taken from an external reference atmosphere, cannot be accomodated on N grid points without generating gravity waves. Consequently, redundant data generally do not improve the forecast but may rather add high frequency noise. This feature should be kept in mind when adding data in a running forecast : time for relaxation or shock damping must be allowed before new data are introduced.

There is one exception to the above conclusion, however : thè case of one general circulation model being checked against itself i.e. initialization experiments with artificial data generated by the same model. It is evident then that such artificial data reflect a priori the special dynamic balance required by the model and should not, if properly analysed, excite gravity oscillations. This is truly a very privileged situation, not likely to obtain with real data and one should accordingly be very careful not to draw overoptimistic conclusions from such experiments.

5.3. - Iterative relaxation of non-synoptic data input

From the formal standpoint developed in section (5.1), there is no basic difference between simultaneous (synoptic) and non-synoptic data for defining the most probable historical sequence fitting the available information. The practical analysis problem remains, however : one must solve a complicated set of non-linear equations to find the best fit in four dimensions (space and time). There can evidently be no straightforward analytic method to solve this problem, and we shall have to define the solution as the limit of a relaxation computation.

Let then $X^0(t_0)$ be a "first guess" of the initial state of the model atmosphere at the beginning t_0 of data acquisition and $X^0(t)$ the subsequent "first guess" trajectory. Assume furthermore that $X^0(t)$ is already a good approximation of the real or reference trajectory in phase-space. Then any measurement at time $\tau > t_0$ introduces a differential correction $\Delta X^0(\tau)$ with respect to the first-guess state $X^0(\tau)$ at that time. Our assumptions are equivalent to saying that ΔX^0 is a first-order small variation of the state of motion and can be carried forward as a differential correction along the basic trajectory $X^0(t)$, thus yielding :

$$X'(t) = f\left[X^0(\tau) + \Delta X^0(\tau),\ t - \tau\right] \tag{5.3}$$

and consequently an up-dated correction at the end t_1 of the data acquisition period (fig. 15).

$$\Delta X^0(t_1) = X'(t_1) - X^0(t_1) \tag{5.4}$$

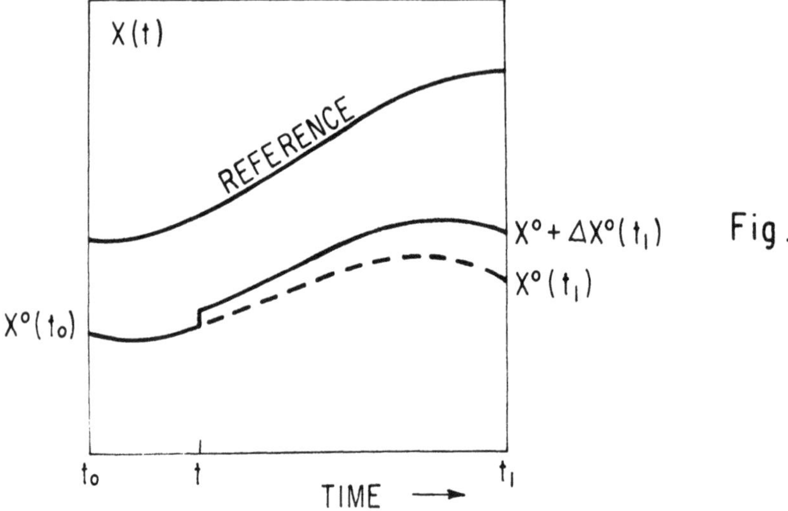

Fig.

This would appear a very cumbersome procedure to obtain one single up dated piece of information. But, it is not actually necessary to repeat this for all individual observation. Small (first-order) corrections can be considered to stay within the linear dynamic response range of the system, along the basic trajectory $X^0(t)$. Consequently, as many as 3N independent corrections can be carried along simultaneously with only second order interference between them. One can therefore follow a composite path (fig. 16) by introducing each new information at the proper time τ, τ', τ'' ... and pursue the integration as suggested by Charney, Halem and Jastrow (1969) for example. It is generally found then, that

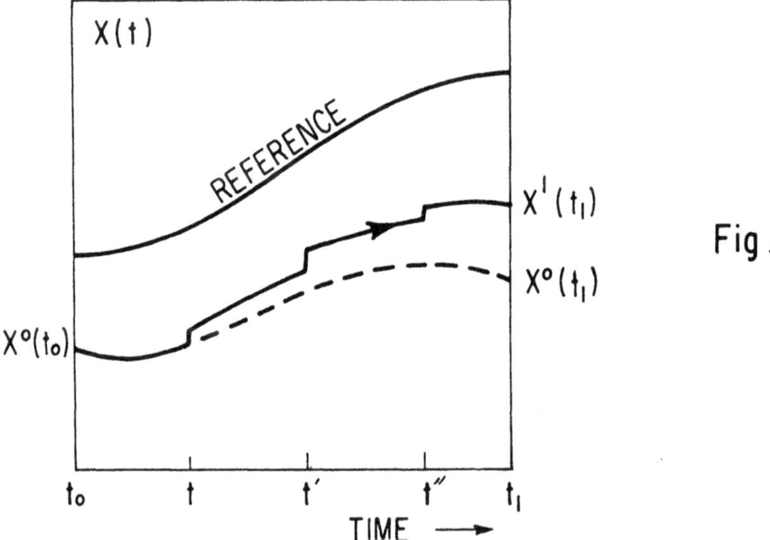

Fig.

each new piece of data makes the composite path approach closer to the reference trajectory. But this method, when unrestricted in time, involves adding many more information than the minimum required to determine one historical sequence X(t) : these redundant data may well then excite excessive gravity oscillations and never allow convergence toward the reference trajectory. This difficulty was not met by Charney, Halem and Jastrow, of course, because they used essentially perfect data i.e. an artificial data set generated by the model itself. But they did find that too many noisy information in too short a period could cause the computation to diverge.

In order to get away from these difficulties, and also, to get a better understanding of the basic data requirement for forecast initialization, Morel, Lefèvre and Rabreau (1971) developed an iterative relaxation scheme allowing one single complete set of data to be used over and over again until the best fit is finally arrived at. We have seen that the basic requirement is to provide enough time for selective damping of the unavoidable gravity waves and this means more time than the relatively short data acquisition period (12 or 24 hours). But extending the data assimilation run beyond this 12 or 24 hours period can only result from folding the trajectory backward (fig. 17) and retrace the finite time

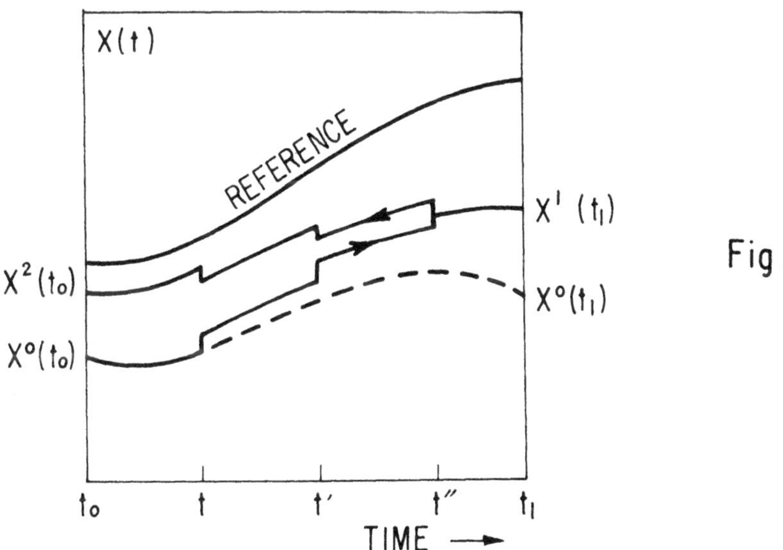

Fig.

step integration process backward from t_1 to t_o and complete one data assimilation cycle which yields a better "second guess" $X^2(t_o)$ initial state of motion. This is of course an iterative process which can be repeated over and over again using the same data set, as desired.

In actual practice, one should remember that real, thus incomplete and imperfect data (rather than perfect artificial data) are used so that four-dimensional data assimilation is applied for:

(a) extracting from the available data the best set of initial values, i.e. restore the geostrophic balance,

(b) reconstructing the missing information on the basis of dynamic consistency.

This dual role is normally tested by assessing how well and how fast a missing field (the geopotential field for example) can be reconstructed from the other field (horizontal wind). The result of one such experiment with a one-level primitive equation model is shown in Fig.18.

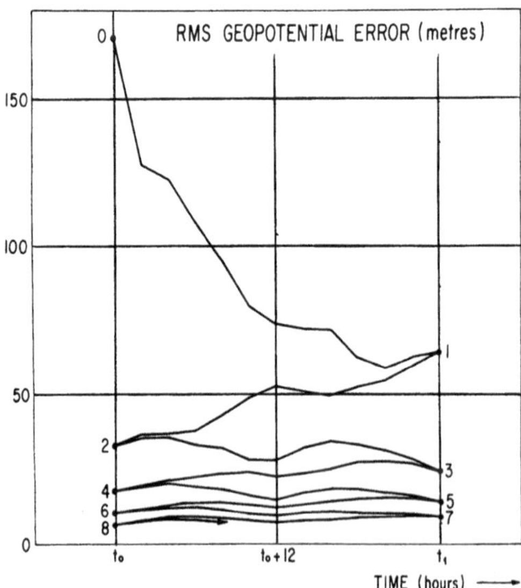

Here, a full historical sequence of complete and perfect artificial data was available but only one set of non-synchronous wind data covering the whole (hemispherical) domain within a 24 hour period, was used for input. The figure shows the residual RMS difference between the reconstructed geopotential field and the known geopotential of the reference history after several iterations of this forward and backward assimilation scheme. Figure 19 below shows the RMS geopotential error with one full cycle, i.e. 24 hours integration forward and 24 hours backward being indicated by two days assimilation time. An acceptable balanced state with residual geopotential errors less than 20 m (RMS) is arrived at after 3 cycles or 6 assimilation days for a particular value of the eddy-viscosity coefficient $K = 10^8 \text{ m}^2 \text{ sec}^{-1}$.

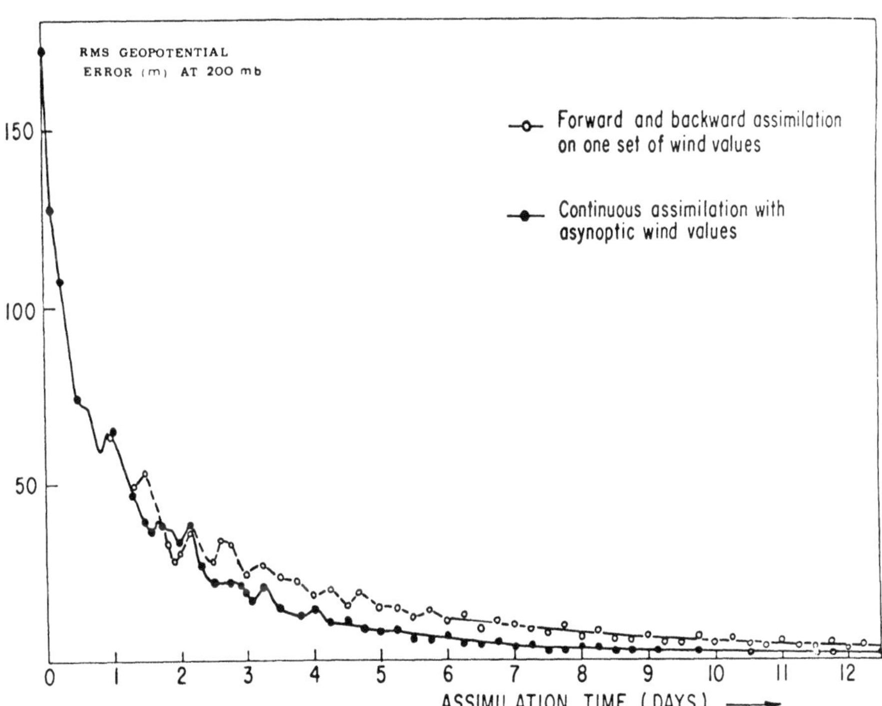

5.4. - Continuous forward assimilation

Similar results have been first reported by Charney, Halem and Jastrow (1969) using continuous assimilation of temperature data taken from one (artificial) historical sequence produced by the same general circulation model. One continuous sequence does not, in principle, contain more information than a single complete set of field values when no error is involved (because the sequence can always be reconstructed by running the model). This is not true however when random errors are added (if only round-off errors of the machine). Accordingly, it is expected that continuous assimilation of artificial data taken from a reference historical sequence should converge slightly better than backward and forward assimilation. For evidence supporting this view, we offer the second curve on Fig.19 showing the residual RMS error of the reconstructed geopotential field after several days of continuous assimilation. The data base here consisted of non-synchronous wind data introduced by successive sectors to make up one complete hemispheric field every 24 hours (Fig. 20).

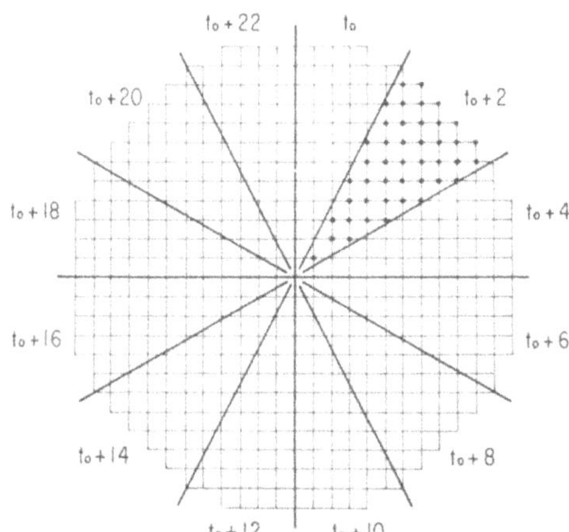

Very similar (practically undistinguishable) results are obtained if synoptic wind data are introduced instead of non-synchronous data; an assimilation experiment run under the same conditions above, only introducing one complete synoptic wind field every 24 hours instead of the corresponding amount of non-synchronous data yields the same geopotentia error curve as shown in Fig. 19. Thus,from the standpoint of four-dimensional assimilation, there is no difference between synchronous (synoptic) data and non-synchronous (asynoptic) data: only the quality and quantity of information matters.

5.5. - The effect of prediction error growth

Now we have only examined the results of four-dimensional data assimilation under most favorable circumstances, i.e. when matching a numerical forecast with an hypothetical, artificial historical record generated by the same general circulation model. Andyet, we know that the whole problem of restoring the geostrophic balance bears upon the dynamical inconsistencies of forecast models.

Up to now, we have considered only situations in which given a correct initial state, prediction errors would remain zero forever or rather, would grow slowly to meet the finite predictability limit of turbulent flow (see Lilly, same volume). In practice, prediction errors grow much faster mainly on account of numerical truncation (discrepancy between differential operators and finite difference approximations) and also spectral truncation modifying the non-linear exchanges of momentum and energy. This has been excellently demonstrated in a series of experiments designed by Williamson (1973) using the NCAR global circulation model in two versions I and II having different basic spatial resolution ,i.e. 5° x 5° and 2.5° x 2.5° respectively, everything else being the same. Fig. 21 taken from Williamson's paper, shows that the prediction error growth

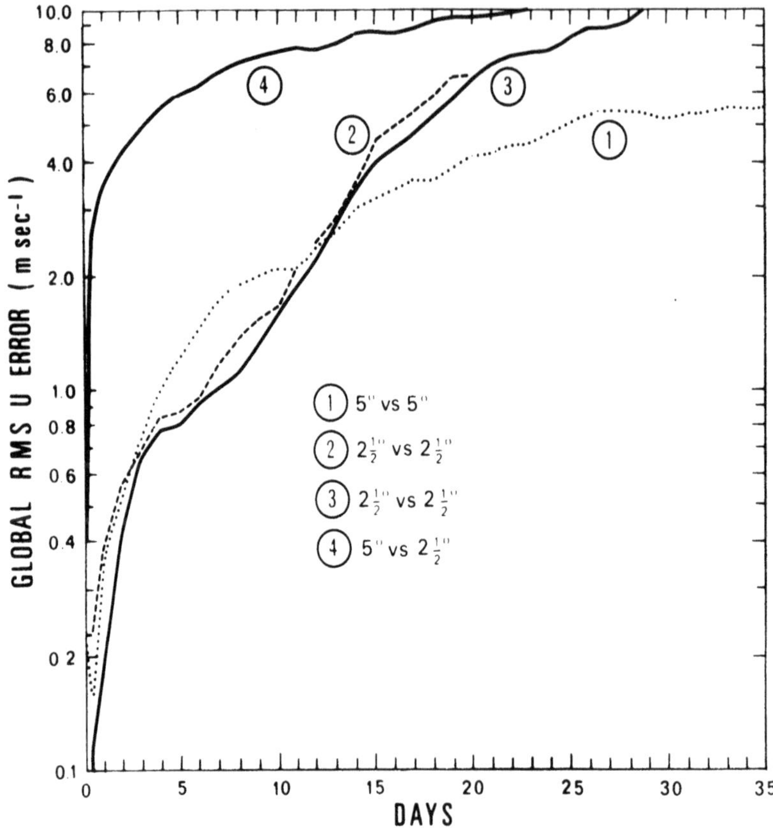

rate of version I with respect to an artificial history generated by itself, is quite similar to the prediction error growth of version II also tested against itself. The prediction error grows much faster however when a forecast by version I is tested against the reference history generated by version II with four times as many degrees of freedom and correspondingly, better accuracy.

Version I and II of the NCAR global circulation model were thereafter used in various continuous four-dimensional assimilation experiments in which the wind field was reconstructed from artificial temperature and pressure data only. Now the ultimate RMS error obtained with version I operating on data generated by version II is considerably worse than the error obtained with either version I or II operating on data generated by itself (Fig.22).

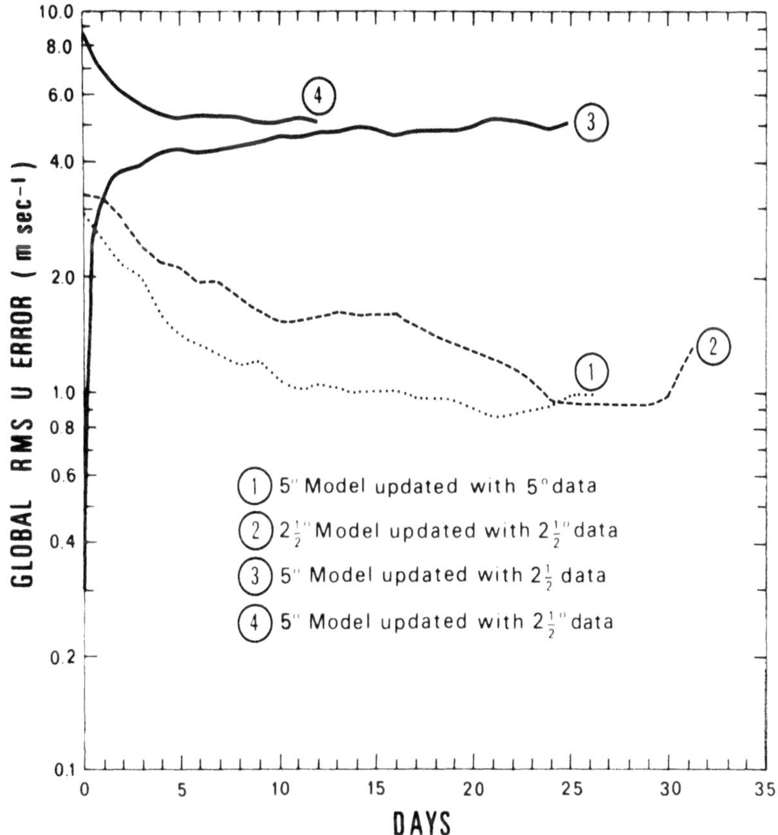

Since the gist of four-dimensional assimilation is to establish the proper dynamical relation between sets of incomplete data introduced at different times, one expects this process is quite sensitive to error growth between successive inputs. In the limit where errors will grow extremely fast, no memory of previous information would remain when new data become available. Given then a prediction error growth time constant E for one particular forecast model compared to one particular reference history, four-dimensional assimilation succeeds inasmuch as a sufficient amount of data is made available. Morel and Talagrand (1973) have estimated that the data introduction rate is measured by the time interval A needed to acquire a complete description of one of the three dynamical fields u, v or ϕ, or an equivalent amount of information. Four-dimensional data assimilation to work satisfactoryly, A must be shorter than E

On the other hand, the more often new data are introduced in the forecast, the more gravity waves are generated. An inordinate amount of kinetic energy in these unwanted modes is not conducive to a good forecast because gravity waves eventually produce large gradients which may lead to large truncation errors and numerically unstable conditions. It is thus imperative that gravity wave energy be kept down to a manageable level by suitable damping. Thus, the gravity wave damping time constant D must be adjusted to the data acquisition time interval A so that $D \leqslant A$, say.But then, if A is rather short like a full set of new observations every 6 hours, a very fast (unphysical) damping of gravity waves must be provided. This would in turn deteriorate the forecast and further reduce the error growth time constant ! One can see now that the condition $D < A < E$ for a successful assimilation of synoptic or asynoptic meteorological data may not be easy to achieve with current high spatial resolution operational forecasting models, let alone simple barotropic models or similar laboratory devices. But one can rest assured that the art of numerical weather forecasting will rise to the point where fairly infrequent (once a day) observations of the <u>full</u> global domain from Earth satellites and other automatic stations will provide an adequate data base for extended predictions. Four-dimensional data assimilation and initialization schemes will be an essential link of this logical chain.

REFERENCES

CHARNEY J., M. HALEM and R. JASTROW (1969) : Use of incomplete historical data to infer the present state of the atmosphere. J. Atmos Sciences 26, 1160-1163

GANDIN (1963) : Objective analysis of meteorological fields, Leningrad

LILLY D.K. (1965) : On the computational stability of numerical solutions of time-dependent non-linear geophysical fluid dynamics problems.
Monthly Weather Review 93, 11-26

MESINGER F. (1972) : Computation of the wind by forced adjustment to the height field.
J. Applied Meteo. 11, 60-71

MIYAKODA K. and O. TALAGRAND (1971) : The assimilation of passed data in the dynamical analysis. I and II
Tellus 23, 310-327

MOREL P., G. LEFEVRE and G. RABREAU (1971) : On initialization and non-synoptic data assimilation.
Tellus 23, 197-206

MOREL P. and O. TALAGRAND (1973) : The dynamic approach to meteorological data assimilation.
(submitted to Tellus)

PETERSEN D.P. (1968) : On the concept and implementation of sequential analysis for linear random fields.
Tellus 20, 673-686

SADOURNY R. (1972) : Approximations en différences finies des équations de Navier-Stokes appliquées à un écoulement géophysique.
Thèse de Doctorat, University of Paris VI

SADOURNY R. (1973): Forced geostrophic adjustment in large scale flow.
Laboratoire de Météorologie Dynamique, C.N.R.S., Paris

TALAGRAND O.(1972) : A note on the damping of inertia-gravity waves in four-dimensional assimilation of meteorological data.
J. Atmos. Sciences 29, 1571-1574

WILLIAMSON D.L. (1972) : The effect of forecast error accumulation on four-dimensional data assimilation.
National Center for Atmospheric Research, NCAR/LAS 2120-72-4

WILLIAMSON D.L. and R.E. DICKINSON (1972) : Periodic updating of meteorological variables.
J. Atmos. Sciences 29, 190-193

WILLIAMSON D.L. and A. KASAHARA (1971) : Adaptation of meteorological variables forced by updating.
J. Atmos. Sciences 28, 1313-1324

TRANSFER AND DISSIPATION OF ENERGY BY MOUNTAIN WAVES

P. QUENEY

University of Paris
Laboratoire de Météorologie
75 - PARIS (France)

CONTENTS

TRANSFER AND DISSIPATION OF ENERGY BY MOUNTAIN WAVES

by

P. QUENEY

UNIVERSITY OF PARIS
Laboratoire de Météorologie
75 - PARIS (France)

ABSTRACT

The main purpose of this paper is to show that, when the perturbation of an atmospheric flow by a mountain range has a well defined wavy structure, there may be a very large transfer of energy from the flow to the perturbation, the greatest part of this energy then being dissipated in the stratosphere or mesosphere, so that the global result is equivalent to a damping of the flow by friction. This requires a proper structure of the flow for a given mountain, and eventually the effect may be so important that the flow is quite unable to cross the mountain range, and passes round. The models used are such that a linear theory can be applied.

1. GENERAL PROPERTIES OF MOUNTAIN PERTURBATIONS.

Any ground corrugation necessarily produces a deformation in the air flow aloft, and as it is the case for all atmospheric motions in general the properties of this orographic perturbation largely depend on its horizontal scale, itself mainly determined by the horizontal extent of the corrugation in the direction of the wind (longitudinal extent). Roughly speaking, one may adopt a classification in three groups : small-scale, meso-scale, and synoptic-scale perturbations.

a) Small-scale orographic perturbations.- If the longitudinal extent of the corrugation is of the order of 1 km at most, the resulting wind deformation is usually rapidly damped upwards, and becomes negligible beyond a height which is comparable to this horizontal extent. Accordingly it is a small-scale perturbation, which may be considered as a part of the general ground-layer turbulence.

b) Meso-scale orographic perturbations.- In the case of a mountain range with a width of 1 to 20 km, overflowed by a transversal wind, what is usually observed is a system of lee-waves which are approximately stationary, as shown on fig. 1. Any air particle crossing the mountain range performs one or several vertical oscillations before going back to its initial level, and this wave motion may be observed up to 20 km or more, without any appreciable damping in the absence of turbulence in the stratosphere : this is the well-known phenomenon of mountain-waves, often associated with standing lenticular clouds with a smooth contour and regularly spaced. In the troposphere the wave length is usually between 5 and 25 km, and in some cases more than six successive waves may be observed. With a high mountain the waves often change, in the vicinity of the ground, into vortices with horizontal axes, called rotors, associated with highly turbulent cumulus clouds. The frequent occurrence of stationary clouds in the lee of mountains is well-known, but the mountain waves themselves were not really known before 1935, when glider pilots first used them in Germany. In 1939 one of these pilots, KÜTTNER, who was at the same time a meteorologist, noted the close analogy between them and the standing barrier-waves observed on the free surface of a river flowing over a corrugation at the bed, which are also lee-waves and have to be explained as a consequence of the static stability of the flow. Indeed the first theories of mountain waves attributed them to the only static stability of the atmosphere, but the later observations led to the conclusion that the vertical distribution of the wind also plays an important role, in other words the waves must be explained as a result of the combined effects of static stability and dynamic stability of the air flow, as first suggested by SCORER in 1949.

In the following years several field experiments were undertaken in order to determine the exact structure of mountain waves, mainly for aeronautic purposes, first in U.S.A. (Sierra-Wave Project), then in France, and the results obtained were completed by theoretical researches.

In the case of a mountain range with a width of the order of 20 to 100 Km, there are no lee-waves in general, but the theory suggests that above the mountain the deformation of the air flow must still have an oscillatory vertical variation, and that the perturbation may extend to a very high level, eventually up to 100 Km if the wind keeps an approximately constant direction. This may explain, in particular, the stratospheric clouds (nacreous clouds) sometimes observed in winter, between 20 and 30 Km, above Scandinavia, Scotland or Alaska ; these clouds have the same lenticular shape, with smooth contours, as those associated with mountain waves.

Finally, again according to theoretical results, most of meso-scale orographic perturbations are likely to operate an important upward transfer of mechanical energy in the atmosphere, as a consequence of their wavy structure.

c) _Synoptic-scale orographic perturbations_.- With a mountain range the width of which exceeds 100 Km, the perturbation of the air flow, when considered at the synoptic scale, is controlled not only by the static stability and the vertical distribution of the wind, but also by the geostrophic force if the vicinity of the equator is excluded, and accordingly there is a predominant dissymmetric horizontal deformation of the streamlines associated with their vertical deformation. Fig. 2 shows a typical aspect of this horizontal deformation, at some level of the lower troposphere, in the case of a straight isolated mountain range located in the northern hemisphere (see chapter 6 for the theory) :there is an anticyclonic deformation of streamlines and horizontal isobars above the mountain, followed by a cyclonic one, and as a result the air flow is concentrated at the left of the mountain range and reduced in speed at the right (in the southern hemisphere the concentration is at the right). Eventually a cyclonic vortex appears at the lee of the mountain range. In addition the deformation of the air flow may be an oscillatory function of altitude if the mountain range is not too wide, in which case an important upward energy transfer may again result.

This description explains most of the well-known regional winds associated with high mountains, such as the Mistral, due to a concentration of a NW-wind by the Pyrenees (fig.3), and the strong westerlies sometimes observed along the eastern coast of North Africa, due to a similar concentration of a NW-wind by the Atlas (fig.4). The high speed of these winds is therefore essentially an effect of earth's rotation.

2. METHODS USED FOR THE THEORY OF ATMOSPHERIC PERTURBATIONS IN GENERAL

It often happens that a fluid motion appears as a perturbation of small amplitude superimposed on a simple motion of the same fluid, or eventually on a state of equilibrium. Accordingly it is possible, as a first approximation, to consider the amplitude of this perturbation as an infinitesimal quantity, and this simplifies considerably the theoretical study of the disturbed motion or equilibrium since the system of hydrodynamical equations and associated boundary conditions may be replaced with another system of linear equations and conditions. This substitution is usually referred to as a linearization, and the linearized equations are called perturbation equations. In particular it is possible to apply the principle of superposition, according to which any solution of a system of linear equations and conditions can be obtained by adding a finite or infinite number of particular solutions. The simplification in question explains why in most of the theoretical models of mountain perturbations which have been used, the height of the mountain is assumed to be infinitesimal, and more generally it has been usual to try to explain the properties of the various atmospheric waves of synoptic or planetary scale by treating these waves as infinitesimal-amplitude perturbations. In the following we shall always adopt the same infinitesimal assumption.

Disturbances of the hydrodynamical parameters.

Let us consider in general a first simple motion or basic motion (it may be eventually a state of equilibrium), and a second motion or disturbed motion, deduced from the first one by some infinitesimal deformation. For the description of these motions one may use either eulerian or lagrangian variables, and accordingly we shall have to consider two sorts of disturbances for the various hydrodynamical parameters, namely the "eulerian disturbances" and the "lagrangian disturbances". By hydrodynamical parameter we mean a quantity such as the density, the velocity, the viscous stress, etc..., so that it may be a scalar, a vector, or a tensor.

Let us first assume that the motions are described in terms of eulerian variables $\mathbf{x}$, $\mathbf{y}$, $\mathbf{z}$, $\mathbf{t}$, ($\mathbf{x}$, $\mathbf{y}$, $\mathbf{z}$, = space coordinates ; $\mathbf{t}$ = time). If $\mathbf{A}$ is any hydrodynamical parameter, its value in the basic motion is then given by an expression such as

(2.1) $A = \bar{A}(x, y, z, t)$,

and in the disturbed motion we may write

(2.2) $A = \bar{A} + \delta A$,

where δA is another function of x, y, z, t , which we assume to be infinitesimal. This quantity δA defines the <u>eulerian disturbance</u> of A : It is the variation of A , when one proceeds from the basic to the disturbed motion, at a fixed point and time.

The lagrangian disturbance of A is similarly defined by assuming that the motions are described in terms of lagrangian variables a, b, c, t . In this case the value of A in the basic motion is given by an expression such as

(2.3) $A = \bar{A}'(a, b, c, t)$,

and in the disturbed motion we may write

(2.4) $A = \bar{A}' + \Delta A$,

where ΔA is another function of a, b, c, t , which we again assume to be infinitesimal and in general differs from δA . This quantity ΔA defines the <u>lagrangian disturbance</u> of A : It is the variation of A , when one proceeds from the basic to the disturbed motion, at a fixed fluid particle and time.

Of course this definition assumes that a correspondence has been established between the fluid particles of the two motions, such that the position M of a fluid particle in the basic motion at time t remains infinitely close to the position M' of the corresponding fluid particle in the disturbed motion, at the same time t (this condition only needs to be satisfied at one particular time).

If $\vec{M}$ denotes the position vector of M, the infinitesimal vector $\overrightarrow{MM'}$, which we shall refer to as the displacement of the fluid at M and t, is nothing but the lagrangian disturbance of $\vec{M}$, so that we may write

(2.5) $$\Delta\vec{M} = \overrightarrow{MM'} .$$

If cartesian coordinates x, y, z, are used, the components of $\Delta\vec{M}$ are also the lagrangian disturbances of these coordinates, say $\Delta x, \Delta y, \Delta z$, and for any hydrodynamical parameter A we have the following relationship between δA and ΔA :

(2.6) $$\Delta A = \delta A + \frac{\partial \bar{A}}{\partial x}\Delta x + \frac{\partial \bar{A}}{\partial y}\Delta y + \frac{\partial \bar{A}}{\partial z}\Delta z ,$$

that is

(2.7) $$\Delta A = \delta A + (\overrightarrow{\text{grad}}\,\bar{A}).\Delta\vec{M} .$$

In these formulae the bar above A may be omitted, and more generally (2.6) is valid with any system of eulerian variables.

Let us also note the analogy between (2.7) and the classical formula relating the lagrangian derivative dA/dt to the eulerian derivative $\partial A/\partial t$, namely

(2.8) $$\frac{dA}{dt} = \frac{\partial A}{\partial t} + (\overrightarrow{\text{grad}}\,A).\frac{d\vec{M}}{dt} ,$$

where $d\vec{M}/dt$ is nothing but the velocity vector.

Computation rules for disturbances.

In general all the disturbances of the various hydrodynamical parameters $A_1, A_2, \ldots, A_n$ are infinitesimal quantities of the same order, and accordingly the methods used for the computation of differentials can be applied to them. We shall retain the following rules, where δ and Δ are considered as operators, and the sign (=) means an equality in the physical sense :

$$\delta[f(A)] = [f'(A)]\,\delta A \;; \tag{2.9}$$

$$\delta[f(A_1, A_2, \ldots, A_n)] = \sum_{i=1}^{n} \frac{\partial f}{\partial A_i}\,\delta A_i \;. \tag{2.10}$$

In particular

$$\delta(\Sigma A_i) = \Sigma\,\delta A_i \;. \tag{2.11}$$

For two scalars :

$$\delta(A_1 A_2) = A_1 \delta A_2 + A_2 \delta A_1 \;. \tag{2.12}$$

For two vectors :

$$\delta(\vec{A}_1 . \vec{A}_2) = (\vec{A}_1 . \delta\vec{A}_2) + (\vec{A}_2 . \delta\vec{A}_1) \;; \tag{2.13}$$

$$\delta(\vec{A}_1 \times \vec{A}_2) = (\delta\vec{A}_1 \times \vec{A}_2) + (\vec{A}_1 \times \delta\vec{A}_2) \;, \tag{2.14}$$

and we have exactly the same rules for lagrangian disturbances (δ everywhere replaced with Δ).
In addition we have the following important permutation rules :

(a) the operator δ is permutable with any operator of eulerian derivation, so that

$$\delta\left(\frac{\partial A}{\partial x_i}\right) = \frac{\partial}{\partial x_i}(\partial A) \;; \quad \delta\left(\frac{\partial A}{\partial t}\right) = \frac{\partial}{\partial t}(\delta A) \;; \tag{2.15}$$

$$\delta(\overrightarrow{\mathrm{grad}}\,A) = \overrightarrow{\mathrm{grad}}\,(\delta A) \;; \quad \delta(\mathrm{div}\,\vec{A}) = \mathrm{div}\,(\delta\vec{A}) \;; \tag{2.16}$$

(b) similarly the operator Δ is permutable with any operator of lagrangian derivation. In particular :

$$\Delta\left(\frac{dA}{dt}\right) = \frac{d}{dt}(\Delta A) \;. \tag{2.17}$$

But the operators δ and d/dt are not permutable in general. For instance if A is a scalar we have, denoting by $\vec{U}$ the velocity vector :

$$\delta\left(\frac{dA}{dt}\right) = \frac{d}{dt}(\delta A) + (\overrightarrow{\mathrm{grad}}\,A) . \delta\vec{U} \;. \tag{2.18}$$

By applying these various rules it is easy to linearize each one of the hydrodynamical equations, in other words to deduce from it a linear equation, and since one is free to introduce either eulerian or lagrangian disturbances, many different forms can thus be obtained for the system of perturbation equations, but of course they are all equivalent. In particular if the continuity equation is written in the form

$$\operatorname{div} \vec{U} + \frac{1}{\rho}\frac{d\rho}{dt} = 0 \ , \tag{2.19}$$

one may use for the corresponding perturbation equation the analogous form

$$\frac{d}{dt}\left(\operatorname{div} \Delta\vec{M} + \frac{\Delta\rho}{\rho}\right) = 0 \ , \tag{2.20}$$

which can be established directly in the same manner as (2.19), and since the application of the operator Δ to this latter equation gives

$$\Delta\left(\operatorname{div}\vec{U}\right) + \frac{d}{dt}\left(\frac{\Delta\rho}{\rho}\right) = 0 \ , \tag{2.21}$$

by comparing to (2.20) we get the important identity

$$\Delta\left(\operatorname{div}\vec{U}\right) = \frac{d}{dt}\left(\operatorname{div}\Delta\vec{M}\right) . \tag{2.22}$$

Proof of (2.19).- The mass of a fluid element of volume dv is dm = ρdv, and since the continuity equation must express the convervation of mass, one of its forms is $\frac{d}{dt}(\rho\, dv) = 0$, that is $\rho\frac{d}{dt}dv + \frac{d\rho}{dt}dv = 0$.
But $\frac{d}{dt}dv$ is the flux of $\vec{U}$ going out through the contour of the fluid element, therefore $\frac{d}{dt}dv = (\operatorname{div}\vec{U})dv$ by Gauss theorem, whence $\rho\operatorname{div}\vec{U} + \frac{d\rho}{dt} = 0$.

Proof of (2.20).- By applying the operator Δ to $\frac{d}{dt}(\rho dv) = 0$ we get $\frac{d}{dt}[\Delta(\rho dv)] = 0$, that is $\frac{d}{dt}[\rho\Delta(dv) + (\Delta\rho)dv] = 0$.
But $\Delta(dv)$ is the flux of $\Delta\vec{M}$ going out through the contour of the fluid element, therefore $\Delta(dv) = (\operatorname{div}\Delta\vec{M})dv$, whence $\frac{d}{dt}[(\rho\operatorname{div}\Delta\vec{M} + \Delta\rho)dv] = 0$, that is $\frac{d}{dt}[(\operatorname{div}\Delta\vec{M} + \frac{\Delta\rho}{\rho})dm] = 0$, whence $\frac{d}{dt}(\operatorname{div}\Delta\vec{M} + \frac{\Delta\rho}{\rho}) = 0$.

It is to be noted that :

(i) the identity (2.22) can be proved directly, but this is rather cumbersome ;

(ii) the equation (2.20) means that $\operatorname{div} \Delta\vec{M} + \frac{\Delta\rho}{\rho}$ is a conservative quantity. Therefore if the correspondence between the fluid elements, in the basic and disturbed motion, is such that this quantity is equal to zero at a particular time, the equation can be replaced with

$$(2.23) \quad \operatorname{div} \Delta\vec{M} + \frac{\Delta\rho}{\rho} = 0 \ .$$

Results concerning Fourier integrals.

a) If $F(x)$ is a real or complex function of the real variable x, supposed to be continuous from $x = -\infty$ to $x = +\infty$ and vanishing at these limits, it can be represented by the Fourier integral (i denotes the imaginary unit)

$$(2.24) \quad F(x) = \int_{-\infty}^{\infty} f(k)\, e^{ikx}\, dx \ ,$$

$f(k)$ being itself given by the Fourier integral

$$(2.25) \quad f(k) = \frac{1}{2\pi} \int_{-\infty}^{\infty} F(x)\, e^{-ikx}\, dx \ ,$$

where the path of integration in the plane of the complex variable k is the real axis. This is one form of a classical Fourier theorem, according to which any continuous function such as $F(x)$ can be considered as the superposition of sinusoidal elements. If $F(x)$ is a real function, it is more convenient to write

$$(2.26) \quad F(x) = Re \int_{0}^{\infty} 2 f(k)\, e^{ikx}\, dk \ ,$$

in order to have to consider the only positive values of k.

For instance, if we take successively

$$(2.27) \quad F(x) = \frac{h}{1 + (\frac{x}{a})^2} \ ; \qquad F(x) = h e^{-(\frac{x}{2a})^2} \ ,$$

where a and h are two real positive constants, we get

(2.28) $$\frac{h}{1+(\frac{x}{a})^2} = Re\int_0^\infty ah\,e^{-ak}\,e^{ikx}\,dk \quad ;$$

(2.29) $$h\,e^{-(\frac{x}{2a})^2} = Re\int_0^\infty \frac{2}{\sqrt{\pi}}\,ah\,e^{-a^2k^2}\,e^{ikx}\,dk \quad ,$$

giving the Fourier representation of a bell-shape profile, which is a Lorentz profile in the case of (2.28), a Gauss profile in the case of (2.29). In the theory of barrier waves, it will be very convenient to use one of these particular profiles for the corrugation of the ground surface.

b) Riemann theorem.- The two examples just given show that the superposition of an infinite number of sinusoidal elements may result in a practically complete vanishing by interference, as it is actually the case for $|x| \gg a$. More generally we have such a result for any integral of the form

(2.30) $$F(X) = \int_\alpha^\beta f(k)\,e^{ikX}\,dk \quad ,$$

where X is real, α and β are two real constants (eventually $\alpha = -\infty$, or $\beta = +\infty$, or both), the integration path is the real axis, and $f(k)$ is such that along this path its real and imaginary parts are not oscillating many times and nowhere large in magnitude. Then a partial integration proves that if $|X|$ is large, $F(X)$ is of the order X^{-1} at most, and this important result may be referred to as "Riemann theorem" for Fourier integrals :

(2.31) $$F(X) \lessapprox X^{-1} \quad , \text{ for } |X| \text{ large} .$$

c) Stationary-phase principle and formula.- Let us now consider the more general integral

(2.32) $$F(X) = \int_\alpha^\beta f(k)\,e^{iX\varphi(k)}\,dk \quad ,$$

where, in addition to the conditions just specified, the function $\varphi(k)$ is assumed to be real along the integration path. If along this path the derivative $\varphi'(k)$ keeps a constant sign and is nowhere small in magnitude, we can take $\varphi(k)$ as new variable of integration, and then by applying Riemann theorem we can conclude that, when $|X|$ is large, $F(X)$ is again of the order X^{-1} at most. But if $\varphi'(k)$ is vanishing for one or several values of k, either along the path of integration or outside of it but close to α or β, this result is no longer valid, and we shall first assume that this is occurring for only one value of k, say $k = k_1$.
In this case the phase $X\varphi(k)$ is stationary for $k = k_1$, therefore when $|X|$ is large its variation with k is rapid only in the vicinity of $k = k_1$, and accordingly the asymptotic value of the integral can be obtained by except retaining the contribution of this vicinity, a result often referred to as the "principle of stationary-phase". Now since $\varphi'(k_1) = 0$, the Taylor expansion of $\varphi(k)$ at k_1 starts as follows :

$$\varphi(k) = \varphi_1 + \frac{1}{2}\varphi_1''(k-k_1)^2 + \frac{1}{6}\varphi_1'''(k-k_1)^3 + \cdots \quad , \tag{2.33}$$

where φ_1, φ_1'', φ_1''' are written for $\varphi(k_1), \varphi''(k_1), \varphi'''(k_1)$, respectively. Therefore if $|\varphi_1''|$ is not too small it is to be expected that the asymptotic value of the integral can be obtained by replacing $\varphi(k)$ by the first two terms of this expansion, at the same time as $f(k)$ with $f_1 = f(k_1)$ if this value is also not too small in magnitude. It can be proved that this procedure is actually correct, and this gives

$$F(x) \approx f_1 e^{iX\varphi_1} \int_\alpha^\beta e^{\frac{iX\varphi_1''}{2}(k-k_1)^2} dk \quad , \tag{2.34}$$

or, after a change of variable

$$F(x) \approx f_1 e^{iX\varphi_1} \int_{-\lambda(k_1-\alpha)}^{\lambda(\beta-k_1)} e^{i\varepsilon_1\theta^2} d\theta \quad , \tag{2.35}$$

with

$$(2.36)\quad \lambda = \sqrt{\tfrac{1}{2}|X\varphi_1''|}\ ; \quad \varepsilon_1 = \operatorname{sgn}(X\varphi_1'') .$$

But since λ is positive and large at the same time as $|X|$, in the general case when $|\varphi'(\alpha)|$ and $|\varphi'(\beta)|$ are not small (k_1 not close to α or β when these limits are finite), we may replace the lower and upper limit of the integral in (2.35) with $-\infty$ and $+\infty$ respectively, and this gives a classical Fresnel integral the value of which is $\sqrt{\pi}\, e^{\varepsilon_1 i \frac{\pi}{4}}$. Finally we thus get the following asymptotic formula, referred to as the "stationary-phase formula" :

$$(2.37)\quad F(X) \approx f_1 \sqrt{\frac{2\pi}{|X\varphi_1''|}}\ e^{i(X\varphi_1 + \varepsilon_1 \frac{\pi}{4})}, \quad \text{for } |X| \text{ large},$$

valid provided $f_1, \varphi_1'', \varphi'(\alpha)$ and $\varphi'(\beta)$ are not too small in magnitude. Therefore $F(X)$ is of the order $X^{-\frac{1}{2}}$ instead of X^{-1}, and the formula also shows that the real part of $F(X)$ is asymptotically an oscillating function of X. If $|\varphi'(\alpha)|$ or $|\varphi'(\beta)|$ is small, only the upper or lower limit of the integral in (2.35) may be replaced with $\pm\infty$, so that we get an asymptotic formula involving a Fresnel function, but $F(X)$ is still of the order $X^{-\frac{1}{2}}$. On the other hand if $|\varphi_1''|$ is small but $|\varphi_1'''|$ not small, the same method leads to an asymptotic formula involving an Airy function (the classical one is $Ai(\xi) = \int_0^\infty e^{i(\frac{\lambda^3}{3} + \xi\lambda)}\, d\lambda$), and in this case $F(X)$ is of the order $X^{-\frac{1}{3}}$.

If $\varphi'(k)$ is vanishing for several well distinct values $k_1, k_2, \ldots, k_n$ all located between α and β, except eventually one of them which may be outside of this interval but close to one of its limits, the asymptotic value of $F(X)$ when $|X|$ is large is obtained by simply adding their contributions, so that we have in general

$$(2.38)\quad F(X) \approx \sum_{j=1}^{\infty} f_j \sqrt{\frac{2\pi}{|X\varphi_j''|}}\ e^{i(X\varphi_j + \varepsilon_j \frac{\pi}{4})}, \quad \text{for } |X| \text{ large},$$

where $f_j, \varphi_j, \varphi_j''$ are written for $f(k_j), \varphi(k_j), \varphi''(k_j)$, respectively, and $\varepsilon_j = \operatorname{sgn}(X\varphi_j'')$. Therefore in general $F(X)$ is again of the order $X^{-\frac{1}{2}}$.

d) Generalization for double Fourier-integrals.- If $F(x,z)$ is a function of the two real variables x, z, supposed to be continuous in the whole x,z-plane and vanishing at infinity, it can be represented by the double Fourier-integral

$$(2.39)\quad F(x,z) = \int_{-\infty}^{\infty}\int_{-\infty}^{\infty} f(k,m)\, e^{i(kx+mz)}\, dk\, dm\,,$$

$f(k,m)$ being itself given by the double Fourier-integral

$$(2.40)\quad f(k,m) = \frac{1}{(2\pi)^2}\int_{-\infty}^{\infty}\int_{-\infty}^{\infty} F(x,z)\, e^{-i(kx+mz)}\, dk\, dm\,,$$

where k and m are supposed to be real, and this generalization of Fourier integral theorem can be extended to a function of more than two variables.

For instance we have, if a and b are two real constants,

$$(2.41)\quad h e^{-\frac{1}{4}\left(\frac{x^2}{a^2}+\frac{z^2}{b^2}\right)} = \int_{-\infty}^{\infty}\int_{-\infty}^{\infty} \frac{1}{\pi}\, abh\, e^{-(a^2k^2+b^2m^2)}\, e^{i(kx+mz)}\, dk\, dm\,,$$

a formula which gives the Fourier representation of a Gauss surface.

Let us finally consider the function of the real variable X defined by

$$(2.42)\quad F(X) = \iint_A f(k,m)\, e^{iX\varphi(k,m)}\, dk\, dm$$

where k and m are real, A denotes some domain of the k,m-plane, $\varphi(k,m)$ is a real function in this domain, and $f(k,m)$ is such that in this domain its real and imaginary parts are not oscillating many times when considered as functions of either k or m, and nowhere large in magnitude. If in the domain, or in the vicinity of its boundary, no point exists where $\frac{\partial\varphi}{\partial k}$ and $\frac{\partial\varphi}{\partial m}$ are both vanishing, then when $|X|$ is large the function $F(X)$ is of the order X^{-2} at most, and this generalizes Riemann theorem. But if

such a point (k_1, m_1) exists, $F(X)$ is in general of the order X^{-1}, its asymptotic value being obtained by only retaining the contribution of the vicinity of this point (generalization of the stationary-phase principle).
Using the subscript 1 for a value at (k_1, m_1), and putting

$$D_1 = \left(\frac{\partial^2 \varphi_1}{\partial k \partial m}\right)^2 - \frac{\partial^2 \varphi_1}{\partial k^2} \frac{\partial^2 \varphi_1}{\partial m^2} \quad , \tag{2.43}$$

if this quantity is not vanishing or very small and if neither $\left|\frac{\partial \varphi}{\partial k}\right|$ nor $\left|\frac{\partial \varphi}{\partial m}\right|$ is small somewhere along the boundary of A (this is the case if the point (k_1, m_1) is not close to this boundary supposed to be at finite distance), then the asymptotic value of $F(X)$ is found to be given by

$$F(X) \approx \varepsilon_1 \frac{2\pi f_1}{X\sqrt{|D_1|}} e^{iX\varphi_1} \quad , \quad \text{for} \quad |X| \text{ large} \quad , \tag{2.44}$$

with

$$\begin{cases} \varepsilon_1 = 1 \text{ , for } D_1 > 0 \text{ ;} \\ \varepsilon_1 = i \operatorname{sgn}\left(\frac{\partial^2 \varphi_1}{\partial k^2}\right) = i \operatorname{sgn}\left(\frac{\partial^2 \varphi_1}{\partial m^2}\right), \text{ for } D_1 < 0 \text{ .} \end{cases} \tag{2.45}$$

This is the generalized stationary-phase formula for a double integral.
If there are in the domain A several points such as $(k_1 m_1)$, the asymptotic value of $F(X)$ is obtained, as a rule, by adding their contributions.

3. GRAVITY-WAVES IN A WATER LAYER.

The purpose of this chapter is to give a theory of the barrier waves observed at the free surface of a water flow downstream from a corrugation at the bed, the study of which is a good introduction to the theory of mountain waves. If the corrugation is not too high, the waves are nearly stationary, and they are essentially a consequence of the static stability of the flow, which in this case is all concentrated at the free surface of the water.
For this reason the waves belong to the category of *surface waves*, characterized by the fact that their energy

is concentrated in the vicinity of a surface. The phenomenon was first studied in 1880 by RAYLEIGH, who tried to explain it by using a simple model : a water flow with a uniform speed, an infinitesimal amplitude for the perturbation due to the corrugation, both capillarity and viscosity neglected. Accordingly the perturbation, in such a model, is a pure gravitational one, and we are led to consider the gravitational waves (or gravity-waves) in general.

Free perturbations in a water layer at rest.

Here the basic motion is a water layer with a uniform density ρ , at rest upon a fixed horizontal plane, and the thickness of which has the constant value H (fig.5). We neglect compressibility, capillarity and viscosity, assume for gravity the constant value g , and that the water is surmounted by air also at rest but with a negligible density. In such conditions any infinitesimal perturbation of the system is a purely gravitational one, but here we first consider the free perturbations, such that the bottom is undeformed. We use as reference frame a cartesian system of axes $Oxyz$, with Oz vertical upwards and $z=0$ along the free surface undisturbed, so that the equation of the bottom is $z=-H$. We only consider the motions independent of y and taking place in planes parallel to Oxz, so that they may be treated as two-dimensional in this plane, and the vorticity reduces to its y-component, say $q=\frac{\partial u}{\partial z}-\frac{\partial w}{\partial x}$, where as usual u and w denote the velocity components along Ox and Oz respectively.

The hydrodynamical equations reduce to the following three ones (since the assumption $\rho=\mathrm{const}$ takes place of the thermal equation) :

$$\frac{du}{dt}+\frac{1}{\rho}\frac{\partial p}{\partial x}=0;\ \frac{dw}{dt}+\frac{1}{\rho}\frac{\partial p}{\partial z}=-g;\ \frac{\partial u}{\partial x}+\frac{\partial w}{\partial z}=0, \tag{3.1}$$

where p denotes the pressure. For the basic state they

reduce to the hydrostatic equation $\frac{\partial \bar{p}}{\partial z} = -g$, whence (since $g = \text{const}$ and $\rho = \text{const}$)

(3.2) $\bar{p} = p_0 - g\rho z$,

where p_0 is the pressure above the water, equal to a constant since the air density is neglected.
The eulerian perturbation equations deduced from (3.1) are simply

(3.3) $\frac{\partial}{\partial t}\delta u + \frac{\partial}{\partial x}\frac{\delta p}{\rho} = 0 ; \frac{\partial}{\partial t}\delta w + \frac{\partial}{\partial z}\frac{\delta p}{\rho} = 0 ; \frac{\partial}{\partial x}\delta u + \frac{\partial}{\partial z}\delta w = 0$,

and between the components of the velocity and those of the fluid displacement we have the relations

(3.4) $\delta u = \frac{\partial}{\partial t}\Delta x \quad ; \quad \delta w = \frac{\partial}{\partial t}\Delta z$.

The elimination of δp from the first two equations (3.3) gives the vorticity equation $\frac{\partial}{\partial t}\delta q = 0$,according to which the most general perturbation may be considered as the superposition of a stationary rotational perturbation and some irrotational one, such that $\delta q = 0$ and for which we may therefore adopt, instead of (3.3), the system

(3.5) $\frac{\partial}{\partial z}\delta u - \frac{\partial}{\partial x}\delta w = 0 ; \frac{\partial}{\partial x}\delta u + \frac{\partial}{\partial z}\delta w = 0 ; \frac{\partial}{\partial t}\delta u + \frac{\partial}{\partial x}\frac{\delta p}{\rho} = 0$.

Altogether (3.4) and (3.5) form a complete set of five homogeneous equations with constant coefficients for the five disturbances δu ,δw , δp, Δx, Δz, so that any one of these disturbances must satisfy the same differential equation, which is nothing but the Laplacian equation ($\nabla^2 \delta u = 0$; $\nabla^2 \delta w = 0$; etc..).

In addition we have the following boundary conditions :

a) at the bottom, the kinematic condition

(3.6.) $\Delta z = 0$ (or $\delta w = 0$), for $z = -H$;

b) at the free surface, a kinematic condition and a dynamic one, but they may be combined in one unique condition. If Δz_0 denotes the value of Δz for $z=0$, the kinematic condition is that the equation of the disturbed free-surface must be $z=\Delta z_0$. On the other hand, since capillarity and air density are neglected, the dynamic condition is that the pressure on this surface must be equal to p_0, and according to the kinematic condition this gives $\Delta p = 0$. But we have $\Delta p = \delta p + \frac{\partial \bar{p}}{\partial z}\Delta z = \delta p - g\rho\,\Delta z$, whence finally

(3.7) $\frac{\delta p}{\rho} = g\,\Delta z_0$, for $z=0$.

Fondamental waves.- As particular irrotational perturbations we have the "fundamental waves", such that for each of them all the disturbances figuring in (3.4) or (3.5) depend on x and t by the exponential factor

(3.8) $$e^{i(kx-\omega t)} = e^{ik(x-ct)},$$

where k is a real positive constant (wave number), and ω, c two other constants real or complex (in fact we shall find that they are real), with $\omega = kc$. In particular we have

(3.9) $$\Delta z_0 = A_0\, e^{i(kx-\omega t)} = A_0\, e^{ik(x-ct)},$$

where A_0 is some infinitesimal constant, real or complex, and the perturbation equations (3.5) completed with the condition (3.6) enable to express all the elements of the wave in terms of Δz_0. In particular we get

(3.10) $$\Delta z\ \frac{\sinh k(z+H)}{\sinh kH}\,\Delta z_0\,;\quad \Delta x = i\,\frac{\cosh k(z+H)}{\sinh kH}\,\Delta z_0;\quad \frac{\delta p}{\rho} = \frac{\omega^2}{k}\,\frac{\cosh k(z+H)}{\sinh kH}\,\Delta z_0.$$

Then the condition (3.7), the only remaining to be satisfied, gives between k and ω the relation

(3.11) $$\omega^2 = gk\tanh kH,$$

giving for ω, therefore also for c, two opposite real values for every value of k, namely

(3.12) $\omega = \pm\, \Omega(k)\,;\ c = \pm\, C(k)\,,$

with

(3.13) $\Omega(k) = \sqrt{gk \tanh kH}\ ;\ C(k) = \sqrt{\frac{g}{k} \tanh kH}\,.$

By taking the real part of this complex perturbation, we again obtain a free perturbation (according to the principle of superposition) which is an irrotational sinusoidal wave progressing with a constant shape and with the constant phase-velocity C. The wave length is $L = \frac{2\pi}{k}$, and ω is the angular frequency. We have two families of waves : the "positive waves", corresponding to $C = C(k)$, and the negative waves", corresponding to $c = -C(k)$. If $H = \infty$ we get

(3.14) $C(k) = \sqrt{\frac{g}{k}} = \sqrt{\frac{g}{2\pi} L}\,,$

So that the dispersion curve of the waves (giving $\pm C$ as a function of L) is a parabola. More generally it is a curve symmetrical about the L-axis (fig.6), with the asymptotes $c = \pm\sqrt{gH}$, and C is a monotic increasing function of L, so that $0 \leqslant C \leqslant \sqrt{gH}$, the limit values being reached for $L=0$ and $L=\infty$, respectively.
Any irrotational free perturbation of the system may be obtained by superposing a finite or infinite number of fundamental waves. In particular the superposition of an infinite number of either positive or negative waves with a wave number very close to a given value k, and the amplitude of which is a continuous function of their wave number, gives a train of modulated quasi-fundamental waves (fig.7), therefore propagating with a phase velocity very close to the value of c corresponding to k, but the modulation of which propagates with another velocity γ, or "group velocity", given by

(3.15) $\gamma = \frac{d\omega}{dk} = \frac{d}{dk}(kc) = c + k\frac{dc}{dk} = c - L\frac{dc}{dL}\,,$

representing the velocity of propagation of the energy of the waves (this result applies, more generally, to any family of waves defined by some relationship between k and ω). According to (3.15), γ is the ordinate of the point where the tangent to the dispersion curve intersects the c-axis, and its variation with L is also given by fig.7.

We have $\gamma = G(k)$ for positive waves, $\gamma = -G(k)$ for negative waves, with

$$G(k) = \frac{d\Omega}{dk}, \tag{3.16}$$

and just as c this group velocity increases with L from 0 for $L = 0$, to $\sqrt{gH}$ for $L = \infty$, but except for these limit values we have

$$\frac{c}{2} \leqslant G < c, \tag{3.17}$$

the limit case $G = \frac{c}{2}$ corresponding to $H = \infty$.

Dispersion of an initial local perturbation (dispersion waves).

Because of the existence of two opposite values of ω (or c) for each value of k, the initial conditions necessary to determine any irrotational free perturbation of the water layer must consist of two x-distributions, for instance those of Δz_0 and δw_0 (value of δw for $z = 0$). This result is easily proved if the perturbation depends on by the factor e^{ikx}, where k is a given real constant, and then can be generalized by Fourier theorem. It is thus found that, if the initial values of Δz_0 and δw_0 (for $t = 0$), say Δz_{00} and δw_{00}, are given by the Fourier integrals

$$\Delta z_{00} = \int_{-\infty}^{\infty} f_1(k) e^{ikx} dk; \quad \delta w_{00} = \int_{-\infty}^{\infty} f_2(k) e^{ikx} dk, \tag{3.18}$$

where $f_1(k)$ and $f_2(k)$ are two infinitesimal functions, then at any other time t the value of Δz_0 for the corresponding free perturbation is given by

(3.19) $$\Delta z_0 = \int_{-\infty}^{\infty} g_1(k)\, e^{i(kx-\Omega t)}\, dk + \int_{-\infty}^{\infty} g_2(k)\, e^{i(kx+\Omega t)}\, dk ,$$

where Ω is the function of k defined by (3.13), and with

(3.20) $$g_1(k) = \frac{1}{2} f_1(k) + \frac{i}{2\Omega} f_2(k) ; \quad g_2(k) = \frac{1}{2} f_1(k) - \frac{i}{2\Omega} f_2(k) ,$$

so that the perturbation appears, as a rule, as a superposition of positive and negative waves.

For instance if we assume a Lorentz profile for Δz_{00}, and no initial vertical velocity at the free surface, that is

(3.21) $$\Delta z_{00} = \frac{h}{1+\left(\frac{x}{a}\right)^2} ; \quad \delta w_{00} = 0 ,$$

where h is an infinitesimal constant, and a a positive constant, we have

(3.22) $$f_1(k) = \frac{ah}{2} e^{-|ak|} ; \quad g_1(k) = g_2(k) = \frac{1}{2} f_1(k) ,$$

whence, according to (3.19),

(3.23) $$\Delta z_0 = F(x,t) + F(-x,t) ,$$

with

(3.24) $$F(x,t) = \frac{ah}{2} \; \mathrm{Re} \int_0^{\infty} e^{-ak} e^{i(kx-\Omega t)}\, dk .$$

Here Δz_0 is obviously an even function of x, in agreement with the symmetry of the initial conditions, and its asymptotic value when t is large can be obtained by the stationary-phase principle. For this purpose we note that the integral in (3.24) can be reduced to the form (2.32) by putting

(3.25) $$X = t ; \quad \varphi(k) = \frac{x}{t} k - \Omega(k) ,$$

whence

(3.26) $$\varphi'(k) = \frac{x}{t} - G(k) ; \quad \varphi''(k) = -G'(k) > 0 ,$$

and since $0 < G(k) < \sqrt{gH}$ for $k > 0$, there is a positive wave number k_1 such that $\varphi'(k_1) = 0$ in the only case where $\frac{x}{t}$ is between the same limits, and it is then given by

$$G(k_1) = \frac{x}{t} . \tag{3.27}$$

Accordingly, if x is not close to 0 or $(\sqrt{gH})t$, we have when t is large

$$F(x,t) \sim t^{-1}, \text{ if } x < 0 \text{ or } x > (\sqrt{gH})t ; \tag{3.28}$$

$$F(x,t) \approx \frac{ah}{2} e^{-ak_1} \sqrt{\frac{2\pi}{|G_1'|t}} \cos\left(k_1 x - \Omega_1 t + \frac{\pi}{4}\right), \text{if } 0 < x < (\sqrt{gH})t, \tag{3.29}$$

where $\Omega_1 = \Omega(k_1)$ and $G_1' = G'(k_1)$.

In the vicinity of $x = (\sqrt{GH})t$, since $\varphi''(0) = 0$ we get for $F(x,t)$ an asymptotic value involving an Airy integral, so that the function is of the order $t^{-1/3}$, and if $\left|\frac{x}{t}\right|$ is small it can be established that the function has ultimately a negligible value.

Finally we can conclude that $F(x,t)$ represents ultimately a train of modulated waves located in the interval $0 < x < (\sqrt{gH})t$, with a fringe around $x = (\sqrt{gH})t$, and we have to complete this system by the symmetrical one with respect to the z-axis in order to get the asymptotic profile of Δz_0 when t is large (more precisely, when $t \gg \sqrt{H/g}$). The aspect to this profile is shown on fig.8 in the case $a = 0{,}4H$, and except for the x-variation of the amplitude of the waves a similar aspect is obtained for any other value of a provided it is at most comparable to H : two symmetrical trains of modulated waves, or "dispersion waves", separated by a central region where the free surface is practically undeformed, and the fronts of which are progressing with the constant speed $\sqrt{gH}$, equal to the maximum group-velocity. The length of the frontal fringes (around $x = \pm(\sqrt{gH})t$) is proportional to $t^{\frac{1}{3}}$, so that it is more and more negligible in comparison with the total length of the trains which is proportional to t.

According to (3.27) the wave length at x is such that the corresponding group velocity is $\frac{x}{t}$, in other words each wave length propagates with a constant speed equal to the corresponding group velocity, and in each train the wave length is, at any time, increasing from the central region to the front. The evolution consists in a continual apparition of new waves at the limit of the central region, these waves then progressing towards the fronts at the same time as their wave length increases, so that the total number of waves becomes larger and larger, the minimum wave length being shorter and shorter, and the frontal waves being longer and longer.

We have the same type of evolution for the irrotational part of any perturbation the horizontal extent of which is initially at most comparable to H (except for a phase difference between the two trains if the initial conditions are not symmetrical about the z-axis), and in any case the irrotational part of any initially limited perturbation spreads over a larger and larger portion of the water layer, so that its amplitude becomes ultimately negligible owing to the fact that its total energy must remain constant in the absence of viscosity : it is essentially this conclusion that we shall retain (a similar dispersion occurs in any system where the group velocity of the fundamental waves varies with the wave length, and in reality the viscosity of the fluid is an additional damping factor). On the other hand, we have seen that the eventual rotational part of the perturbation remains stationary.

Free waves in a water layer flowing with a uniform velocity.

We now assume that the same water layer as before, instead of being at rest, is moving horizontally with a constant and uniform positive velocity U . Obviously the structure of any free perturbation is unchanged, and in the case of a fundamental wave we only have to replace c, γ and ω with $c+U, \gamma+U$ and $\omega+kU$, respectively, in order that the disturbances still depend on x and t by the factor (3.8). We shall again use the names "positive waves" and "negative waves" for the two families of fundamental waves, and the new formulae to be used for them are now

$$(3.30) \quad c = U \pm C(k); \quad \gamma = U \pm \Omega'(k); \quad \omega = kU \pm \Omega(k),$$

where $C(k)$ and $\Omega(k)$ are again given by (3.13). The new dispersion curve is deduced from the diagram of fig.6 by a U-translation parallel to the c-axis, and this gives the aspects shown on fig.9 according as $U > \sqrt{gH}$ or $U < \sqrt{gH}$: in the first case we have $c > 0$ for all the waves ; on the contrary in the second case the dispersion curve intersects the L-axis, and if L_s denotes the wave length of the intersection point we have $c = 0$ for the corresponding negative waves, in other words these waves are stationary. The wave number k_s corresponding to L_s is given by the equation

$$(3.31) \quad C(k_s) = U, \text{ that is } \quad g \tanh k_s H = k_s U^2 .$$

It is obvious on fig.9 that the group velocity γ_s of the stationary waves is always positive, and we shall see that this is the reason why the barrier waves due to a corrugation at the bottom are always located on the lee side of this corrugation (the fact that γ_s is positive is itself a consequence of the fact that $C(k)$ is a monotonic decreasing function of k, whence also $G(k) < C(k)$). Just as for a water layer at rest, the irrotational part of any initially limited perturbation spreads over a larger and larger portion of the water flow and therefore becomes ultimately negligible, while its eventual rotational part is carried away with the flow without any deformation.

<u>Forced perturbation in a water layer flowing with a uniform velocity over a corrugated bed (resonance waves, or barrier waves).</u>

We finally assume that, in the foregoing model, the boundary $z=-H$ is replaced with a cylindrical surface infinitely close to this plane, the equation of which is

(3.32) $$z=-H+F(x) \quad ,$$

where $F(x)$ is a given infinitesimal function, and we shall again only consider the perturbations which are independent of y.

If we can find a stationary perturbation as a particular solution, the most general perturbation will be obtained by superposing on it an arbitrary free perturbation, and since the irrotational part of this latter becomes ultimately negligible by dispersion, while its rotational part is carried away with the flow, only will remain the stationary perturbation after a long time : accordingly we shall say that, in this case, the stationary perturbation is the "forced perturbation" due to the bottom corrugation. Again in virtue of the principle or superposition we shall first take the case of a sinusoidal corrugation, then generalize by Fourier integrals.

Case of a sinusoidal corrugation. - We assume that

(3.33) $$F(x) = A_1 e^{ikx} ,$$

where k is a real constant, and A_1 an infinitesimal constant, real or complex, and seek for a stationary perturbation. Such a perturbation necessarily depends on x by the factor e^{ikx}, so that we must have in particular

(3.34) $$\Delta z_0 = \lambda A_1 e^{ikx} ,$$

where λ is a constant to be determined. The perturbation equations to be used differ from (3.4) and (3.5) by changing $\frac{\partial}{\partial t}$ into $U\frac{\partial}{\partial x}=ikU$, but the condition (3.7) relative to the free surface has not to be changed, whereas the new condition at the bottom is

(3.35) $$\Delta z = A_1 e^{ikx} , \quad \text{for } z=-H .$$

If we only use the perturbation equations completed by this condition (3.35) and the relation (3.34), we can express all the disturbances in terms of λ and Δz_0.
In particular we get

(3.36) $$\Delta z = \frac{\lambda \sinh k(z+H) - \sinh kz}{\lambda \sinh kH} \Delta z_0 ; \quad \frac{\delta p}{\rho} = kU^2 \frac{\lambda \cosh k(z+H) - \cosh kz}{\lambda \sinh kH} \Delta z_0 ,$$

and then the condition (3.7) gives

(3.37) $$kU^2 \frac{\lambda \cosh kH - 1}{\lambda \sinh kH} = g ,$$

whence

(3.38) $$\lambda = \frac{1}{\cosh kH} \frac{U^2}{U^2 - C^2(k)} .$$

Accordingly :

a) If $U \neq C(k)$, this formula gives for λ one finite real value, so that we get one forced perturbation, which is a stationary wave. Taking its real part, we get a sinusoidal wave, the streamlines of which have one of the two aspects shown on fig.10 : if $U > C(k)$, all the streamlines have the same phase along any vertical ; on the contrary if $U < C(k)$ the deformations of the free surface and of the bottom are out of phase, and there is one disturbed streamline at some intermediate level.

b) If $U = C(k)$, in other words if the stationary free waves have precisely the wave number $k\ (k = k_s)$, an occurence which is possible only if $U < \sqrt{gH}$, the formula (3.37) would give an infinite value for λ, therefore a perturbation with an amplitude infinitely large as compared with A_1, so that we are dealing with a resonance effect. In this case one gets as a particular solution the perturbation defined by

(3.39) $$\Delta z_0 = i A_1 \frac{kU}{2\cosh kH} t e^{ikx} ,$$

the amplitude of which increases proportionally to time, and which therefore ultimately predominates when one starts from arbitrarily given initial conditions.

If one assumes a small viscosity in the fluid (molecular viscosity), with a uniform value of the kinematic coefficient ν (this assumption requires the absence of viscous stress at the bottom $z=-H$ in the basic motion, otherwise this motion could not be a uniform flow), one gets instead of (3.37)

$$\lambda = \frac{1}{\cos kH} \frac{(U-2ik\nu)^2}{(U-2ik\nu)^2 - C^2(k)} , \tag{3.40}$$

so that λ is no longer real but always finite for k real, and accordingly in any case there is one stationary perturbation.
In particular for $U = C(k)$ this formula gives, if ν is very small,

$$\lambda = \frac{iU}{4k\nu \cosh kH} . \tag{3.41}$$

The corresponding perturbation has a very large amplitude, which increases indefinitely as $\nu \to 0$, and as it was to be expected the phase of Δz_0 for this perturbation is the same as for the perturbation defined by (3.39).

In any case, if one starts from arbitrary initial conditions, there is, as a rule, a transfer of energy from the flow to the perturbation, but in the absence of viscosity the transfer stops when the forced perturbation is established, except for $U = C(k)$.
With a viscous fluid, the transfer never stops, and once the forced perturbation is established the energy transferred to the perturbation is exactly balanced by the energy dissipated by viscosity during the same time ; when ν is very small this energy is also very small except if U is close to $C(k)$, and if $\nu \to 0$ we get at the limit the same result as obtained in the absence of viscosity, namely no transfer except for $U = C(k)$.

Case of an arbitrary corrugation with a limited horizontal extent. -
In this case we can assume that $F(x)$ is given by a Fourier integral, say

(3.42) $$F(x) = \mathrm{Re}\int_0^\infty f(k)\, e^{ikx}\, dk \, .$$

Two sub-cases must be considered, according as U is larger or smaller than $\sqrt{gH}$.

a) $U > \sqrt{gH}$.- For any positive value of k we then have $U > C(k)$; therefore any sinusoidal corrugation of the bottom produces one well-defined stationary forced wave, and by superposing the forced waves corresponding to the various elements of the integrand in (3.42) we get a well-defined stationary perturbation, which is therefore the forced perturbation due to the corrugation $F(x)$. For this perturbation we have, in particular ,

(3.43) $$\Delta z_0 = \mathrm{Re}\int_0^\infty \lambda(k) f(k)\, e^{ikx}\, dk \, ,$$

where $\lambda(k)$ is the function defined by (3.37), real and positive in this case. Since the integrand has no pole the integral has a well-defined value, and according to Riemann theorem Δz_0 represents a deformation with a limited horizontal extent. If the profile of the bottom corrugation is a Lorentz profile, the streamlines of the forced perturbation have the aspect shown on fig.11 (left part).

b) $U < \sqrt{gH}$.- In this case there is a particular wave number k_s such that the corresponding negative free waves are stationary, so that $\lambda(k_s)$ is infinite, and the integral (3.43) is undetermined. Accordingly we are led to examine what ultimately happens if one starts from some initial local deformation (that is, the horizontal extent which is finite) of the flow, which may be chosen arbitrarily since with any other local deformation the resulting perturbation will differ from the first one by a free perturbation, and we know that this latter will ultimately become negligible by dispersion, or disappear at large distance.

As a particular perturbation, let us take the irrotational perturbation obtained by superposing upon the forced wave corresponding to each element of the integral (3.43),

the free negative wave with an opposite value of $\Delta \zeta_0$
For this perturbation we have

$$\Delta \zeta_0 = Re \int_0^\infty \lambda(k) f(k) \left\{ e^{ikx} - e^{i[kx - \bar{\Omega}(k)t]} \right\} dk , \tag{3.44}$$

where $\bar{\Omega}(k)$ denotes the angular frequency of the free negative waves expressed in terms of k , that is

$$\bar{\Omega}(k) = kU - \Omega(k) = k[U - C(k)] . \tag{3.45}$$

From (3.43) we deduce

$$\delta w_0 = \left(\frac{\partial}{\partial t} + U \frac{\partial}{\partial x} \right) \Delta \zeta_0 = Re \int_0^\infty i \lambda(k) f(k) \left\{ kUe^{ikx} - \Omega(k) e^{i[kx - \bar{\Omega}(k)t]} \right\} dk , \tag{3.46}$$

and since $\bar{\Omega}(k_s) = 0$, in contrast to (3.43) the integrals in (3.44) and (3.46) have not $k = k_s$ as a pole, so that they have well-defined values, and the perturbation in question is actually a local one at the initial time $t = 0$ according to Riemann theorem.

Now, instead of taking the real axis as the path of integration in (3.44) for instance (in the plane of the complex variable k), we may as well take an indented path such as shown on fig.12, the identation being small and avoiding the point k_s .
Then we may split the integral into two parts, and write

$$\Delta \zeta_0 = (\Delta \zeta_0)_1 - (\Delta \zeta_0)_2 , \tag{3.47}$$

with

$$(\Delta \zeta_0)_1 = Re \int_0^{\infty(-)} \lambda(k) f(k) e^{ikx} dk ; \tag{3.48}$$

$$(\Delta \zeta_0)_2 = Re \int_0^{\infty(-)} \lambda(k) f(k) e^{i[kx - \bar{\Omega}(k)t]} dk , \tag{3.49}$$

where $\int_0^{\infty(-)}$ means an integration along the negatively indented path, leaving k_s at its left.
Obviously $(\Delta \zeta_0)_1$ is a stationary deformation, and it is to be expected that the vicinity of $k = k_s$ must give a predominant contribution to the integral (3.48).

But in this vicinity we have

(3.50) $$\lambda(k)f(k) \sim \frac{r_s f(k_s)}{k-k_s},$$

where r_s is the residue of $\lambda(k)$ for $k=k_s$, given by

(3.51) $$r_s = \frac{k_s U}{2\gamma_s \cosh k_s H}.$$

(γ_s is the group velocity of the stationary free waves, and we have seen that it is positive, so that r_s is also positive). We are thus led to write

(3.52) $$(\Delta z_0)_1 = Re\left\{r_s f(k_s)\int_{-\infty}^{\infty(-)} \frac{e^{ikx}}{k-k_s}dk + r_s f(k_s)\int_0^{-\infty}\frac{e^{ikx}}{k-k_s}dk + \int_0^{\infty}\left[\lambda(k)f(k) - \frac{r_s f(k_s)}{k-k_s}\right]e^{ikx}dk\right\}.$$

The first integral is equal to $2i\pi e^{ik_s x}$ for $x>0$, equal to zero for $x<0$, and the other two integrals have no pole, so that, according to Riemann theorem, they represent altogether a local deformation about $x=0$. Therefore $(\Delta z_0)_1$ represents a half-infinite train of sinusoidal waves of wave number k_s, defined by

(3.53) $$\varepsilon\, Re\left[2i\pi r_s f(k_s)e^{ik_s x}\right], \text{ with } \begin{cases}\varepsilon=1, \text{ for } x>0;\\ \varepsilon=0, \text{ for } x<0,\end{cases}$$

with a fringe near $x=0$, as shown on fig.13 (where a Lorentz profile is assumed for the bottom corrugation, so that $f(k_s)$ is real).

A similar procedure can be applied to $(\Delta z_0)_2$, representing the dispersion of $(\Delta z_0)_1$. Since $\bar{\Omega}(k_s)=0$ and $\bar{\Omega}'(k_s)=\gamma_s$, the Taylor expansion of $kx-\bar{\Omega}(k)t$ near $k=k_s$ starts as follows :

(3.54) $$kx-\bar{\Omega}(k)t = k_s x + (x-\gamma_s t)(k-k_s)+\ldots.$$

Therefore, if x is not close to $\gamma_s t$, we may retain the only terms written in order to obtain the contribution of the vicinity of $k = k_s$, and we are thus led to write

$$(3.55)\quad (\Delta z_0)_2 = Re\left\{ r_s f(k_s) e^{ik_s x} \int_{-\infty}^{\infty(-)} \frac{e^{i(x-\gamma_s t)(k-k_s)}}{k-k_s}\, dk \right.$$

$$+ r_s f(k_s) e^{ik_s x} \int_0^{-\infty} \frac{e^{i(x-\gamma_s t)(k-k_s)}}{k-k_s}\, dk$$

$$\left. + \int_0^{\infty} \left[\lambda(k) f(k) e^{i[kx - \bar{\Omega}(k)t]} - \frac{r_s f(k_s)}{k-k_s} e^{i(x-\gamma_s t)(k-k_s)} \right] dk \right\}.$$

The first integral is equal to $2i\pi$ or zero according as $x > \gamma_s t$ or $x < \gamma_s t$, and it can be shown (see ref. 6) that the other two integrals, which have no pole, represent altogether a system of dispersion waves which are ultimately negligible, except in a fringe region around $x = \gamma_s t$, the length of which is proportional to $\sqrt{t}$, where they appear as modulated waves with a wave number close to k_s, the modulation being given by a Fresnel function. Therefore $(\Delta z_0)_2$ represents ultimately a half-infinite train of the same sinusoidal waves as $(\Delta z_0)_1$, but located at the right of $x = \gamma_s t$, given by

$$(3.56)\quad \varepsilon' \, Re\left[2i\pi r_s f(k_s) e^{ik_s x} \right], \text{ with } \begin{cases} \varepsilon' = 1, \text{ for } x > \gamma_s t\,; \\ \varepsilon' = 0, \text{ for } x < \gamma_s t\,, \end{cases}$$

with a fringe around $x = \gamma_s t$, representing the dispersion of these sinusoidal waves, and the length of which increases proportionally to $\sqrt{t}$ (fig. 13).

Finally, by subtracting $(\Delta z_0)_2$ from $(\Delta z_0)_1$, we get for Δz_0 a train of sinusoidal stationary waves, or "resonance waves", inside of the interval $0 < x < \gamma_s t$, with a fixed fringe at the left and a moving fringe at the right (fig. 13). Accordingly, if we ignore what happens at large distance when t is itself large, we can say that the perturbation tends towards a stationary one such that the free-surface deformation is nothing but $(\Delta z_0)_1$, and may therefore be called the "forced perturbation due to the bottom corrugation", the streamlines of which have the aspect shown on fig.11 (right part). We see that this perturbation is located on the lee of the corrugation because the group velocity γ_s is positive : the resonance waves develop in the direction of the propagation of their energy, which is continually extracted from the energy of the flow in the vicinity of the corrugation at a constant rate, in agreement with the fact that the length of the train increases proportionally to time (in the case of capillary waves, for instance, γ_s is negative, and accordingly the forced waves develop on the upstream side of the corrugation).

In reality the viscosity of the fluid puts an end to the extension of the train, so that the perturbation actually tends towards a stationary limit, and once this limit is reached the energy transferred from the flow to the waves is exactly equal to the energy dissipated by viscosity during the same time.

In the case of a small molecular viscosity with a uniform value of the kinematic coefficient ν, this limit is given by the integral (3.43) where $\lambda(k)$ is the function defined by (3.40). If ν is very small the pole of the integrand close to k_s is practically

$$k_s' = k_s + \frac{2ik_s^2\nu}{\gamma_s} , \tag{3.57}$$

and this gives for the modified train of resonance waves

$$Re\left[r_s f(k_s)\int_{-\infty}^{\infty}\frac{e^{ikx}}{k-k_s'}dk\right] = \varepsilon Re\left[2i\pi r_s f(k_s) e^{-\frac{2k_s^2\nu}{\gamma_s}x} e^{ik_s x}\right], \tag{3.58}$$

with $\varepsilon = 1$, for $x > 0$; $\varepsilon = 0$, for $x < 0$.

Therefore the forced waves are characterized by an exponential leeward decrease of their amplitude, at the free surface, as shown on fig.14. The smaller the viscosity, the slower is this exponential decrease, and when $\nu \to 0$ we get at the limit the stationary perturbation $(\Delta z_0)_1$, which can therefore be obtained in this manner. It is important to note that this result is a consequence of the fact that the pole k_s' is located on the positive side of the real axis (in the plane of the complex variable k), and this because γ_s is positive, according to (3.57) : at the limit the integration path must still be such that the pole of the integrand be located on its left, and this is actually the case for the indented path used in (3.48). However it is clear that it is not the viscosity of the fluid which is responsible of the fact that the forced waves are on the lee side of the corrugation rather than on the other side, since we get the same result with a quite unviscid fluid.

If we assume for the corrugation the Lorentz profile (2.28), according to (3.53) the ratio of the amplitude A of the resonance waves to the height h of the corrugation is

$$\text{(3.59)} \qquad \frac{A}{h} = 2\pi r_s a e^{-k_s a} .$$

For a given basic flow the value of this ratio is maximum for $k_s a = 1$, that is, for $2\pi a = L_s$, and it is negligible if the practical width $10a$ is either small or large as compared to L_s. We get a similar result with the Gauss profile (2.29), and more generally <u>the amplitude of the resonance waves is significant only if the practical width of the corrugation is comparable to L_s</u>; otherwise the forced perturbation has the same aspect as for $U > \sqrt{gH}$, given by fig.11 (left part).

4. SHEAR-WAVES IN A SYSTEM OF TWO BAROTROPIC COUETTE FLOWS

In the foregoing chapter the basic motion was characterized by a purely gravitational stability concentrated along the unique plane $z=0$, and accordingly the wave-like perturbations of this motion were referred to as particular "gravity-waves", belonging to the category of surface waves. Of course the water may be replaced with any incompressible fluid with a constant density ρ , and the theory presented can be easily generalized if the air above the plane $z=0$ is also replaced with another incompressible fluid with a smaller constant density ρ' and unlimited upwards, provided this fluid is moving horizontally with the same uniform velocity as the lower one : this gives a two-layer uniform flow with a density discontinuity at the interface $z=0$, therefore again characterized by a purely gravitational stability concentrated along this plane (the model considered in the preceding chapter is nothing but the limit case where $\rho'=0$, and the generalized theory leads to a quite similar dispersion curve for the fundamental waves, according to which the condition for the occurrence of stationary lee-waves is $U<\sqrt{gH\frac{\rho-\rho'}{\rho}}$; in the lower fluid the structure of the waves is unchanged, whereas in the upper fluid there is an upward exponential decrease of the amplitude, given by the factor e^{-kz}).

Now it was shown in 1880 by RAYLEIGH that quite similar waves might again be observed, at least theoretically, if in this two-layer model we assume $\rho'=\rho$, and that instead of being uniform the flow consists of two Couette flows with different values of the vertical shear, and such that the velocity itself be continuous on the interface $z=0$. In other words the density discontinuity is replaced with a vorticity discontinuity, and instead of a uniform value of the velocity we have a uniform value of the density. Accordingly instead of a purely gravitational stability we have now a purely dynamic stability, and the waves of the system may be referred to as "shear-waves" ; in addition since the vorticity is constant and uniform in each Couette flow we have again a stability concentrated along the unique plane $z = 0$, so that the waves again belong to the category of surface waves. Since we know that mountain waves are to be explained by the combined effects of gravitational and dynamic stability of the air flow, it is useful to consider now such a model of a two-layer barotropic flow.

The frame of reference being unchanged, our model is defined as follows :

a) The fluid is incompressible and unviscid, with the same density ρ everywhere.

b) For the basic motion we have

$$\bar{v} = \bar{w} = 0 \; ; \; \bar{u} = U_0 + \bar{q} z \; , \tag{4.1}$$

with

$$\begin{cases} \bar{q} = Q \; , \text{ for } \; -H \leqslant z < 0 & \text{(lower flow) ;} \\ \bar{q} = Q' \; , \text{ For } \; z > 0 & \text{(upper flow),} \end{cases} \tag{4.2.}$$

where U_0, Q and Q' are three constants, and thanks to a proper choice of the x -direction we assume

$$Q' < Q \; . \tag{4.3}$$

The velocity profile ($\bar{u}$ versus z) may have any one of the several aspects shown on fig.15 (upper part), but in any case it is made of two straight lines with a kink pointing towards the x -direction.

The velocity U_1 at the bottom is given by

$$U_1 = U_0 - QH \; . \tag{4.4}$$

c) The disturbed motion is again two-dimensional in the xz-plane.

Free perturbations.

The system (3.1) is again valid, and since ρ is constant it leads to the equation $\frac{dq}{dt} = 0$, giving $\frac{d}{dt}\Delta q = 0$, and therefore $\frac{d}{dt}\delta q = 0$ since $\bar{q}$ is constant in each Couette flow. Accordingly any perturbation is again the superposition of a rotational one carried along with the basic flow, and some irrotational perturbation (such that $\delta q = 0$) for which the system (3.5) has to be replaced with the following one :

(4.5) $$\frac{\partial}{\partial z}\delta u - \frac{\partial}{\partial x}\delta w = 0;\ \frac{\partial}{\partial x}\delta u + \frac{\partial}{\partial z}\delta w = 0;\ \frac{d}{dt}\delta u + \bar{q}\,\delta w + \frac{\partial}{\partial x}\frac{\delta p}{\rho} = 0,$$

where

(4.6) $$\frac{d}{dt} = \frac{\partial}{\partial t} + \bar{u}\frac{\partial}{\partial x} \quad ,$$

and instead of (3.4) we must write

(4.7) $$\delta u = \frac{d}{dt}\Delta x;\quad \delta w = \frac{d}{dt}\Delta z \ .$$

For a free perturbation the boundary conditions are :

a) the Kinematic condition for $z = -H$: $\Delta z = 0$, equivalent to $\delta w = 0$;

b) the Kinematic condition for $z = +\infty$: all the disturbances must remain at most comparable to their values at $z = 0$) ;

c) the Kinematic condition for $z = 0$: Δz must be continuous (the same in both Couette flows) and equal to the vertical displacement of the interface ; the continuity of Δz is equivalent to that of δw ;

d) the dynamic condition for $z = 0$: the pressure must be continuous. In virtue of the Kinematic condition, this requires the continuity of Δp, equivalent to that of δp, itself equivalent to the continuity of $\Delta u = \delta u + \bar{q}\,\Delta z$ in virtue of (4.7) and the third equation (4.5).

Therefore there is no sliding along the interface, contrarily to the case of gravity-waves.

Fundamental waves.- As particular irrotational free perturbations we have again the "fundamental waves", depending on x and t by an exponential factor of the form (3.8). If we only use (4.5) and (4.7) completed by the three kinematic conditions, we get the following expressions for δu and δw in terms of Δz_0, value of Δz for $z=0$ (we have again $\nabla^2 \delta u = 0$ and $\nabla^2 \delta w = 0$):

(4.8) In the lower flow :

$$\delta u = -k(U_0 - c)\frac{\cosh k(z+H)}{\sinh kH}\Delta z_0 ; \quad \delta w = ik(U_0 - c)\frac{\sinh k(z+H)}{\sinh kH}\Delta z_0 .$$

In the upper flow :

$$\delta u = k(U_0 - c)\, e^{-kz}\Delta z_0 ; \quad \delta w = ik(U_0 - c)\, e^{-kz}\Delta z_0 ,$$

and then the dynamic condition for $z=0$ gives c in terms of k, namely

$$c = U_0 - \frac{\mu}{2k}\left(1 - e^{-2kH}\right), \quad \text{with } \mu = Q - Q' . \tag{4.9}$$

We thus get only one family of fundamental waves (one value of c for each value of k), for which c is a real decreasing function of L, its limit values being : $c = U_0$ for $L = 0$, and $c = U_0 - \mu H$ for $L = \infty$. The group velocity is given by

$$\gamma = U_0 - \mu H e^{-2kH} , \tag{4.10}$$

and has the same properties. We see that three cases must be considered, according to the value of U_0 as compared to 0 and μH :

a) If $U_0 > \mu H$, the dispersion diagram (fig.15) is entirely above the L-axis, therefore all the waves have a positive phase velocity.

b) If $0 < U_0 < \mu H$, the dispersion diagram intersects the L-axis, therefore there is one wave-length L_s such that the corresponding waves are stationary. The corresponding group velocity γ_s is again positive.

c) If $U_0 < 0$, we have $c < 0$ for all the waves.

In each case there are four possible aspects of the velocity profile of the basic motion, shown on fig.15, and it is noteworthy that in the case where L_s exists $(0<U_0<\mu H)$, the basic velocity $\bar{u}$ is necessarily vanishing at least at one level (eventually at two levels, one in each Couette flow), whereas in the other two cases this can happen at one level at most. Now if we consider the streamlines of the perturbation defined by the real part of a fundamental wave, they have the aspect given by fig. 16 or fig.17 according as $\bar{u}$ has a constant sign in the whole basic motion, or is vanishing at some level $z=z_0$ in the upper flow : in the first case all the streamlines are sinusoidal with the same phase along any vertical, their amplitude decreasing exponentially upwards in the upper flow ; in the second case there is a "cat's eye pattern" in the vicinity of $z=z_0$, in other words a system of vortices limited by a double streamline and forming the transition between sinusoidal streamlines which are out of phase (the use of formulae (4.7) would give infinite values for $\frac{\Delta x}{\delta w}$ and $\frac{\Delta z}{\delta w}$ at the level $z=z_0$; however the streamlines in the cat's eye region can be obtained from the expression of streamfunction, the disturbance of which is proportional to δw). There is such a system of vortices in the vicinity of any level where $\bar{u}=0$, and therefore in the case of a stationary wave there is always at least one of them.

Dispersion of an initial local perturbation.-Since there is only one family of fundamental waves, any irrotational perturbation is completely defined by one initial x-distribution, for instance that of Δz_0 (vertical deformation of the interface), and accordingly the method used in the preceding chapter gives for the asymptotic profile of the interface, when t is large, only one train of dispersion waves. But the properties of these waves are quite similar to those of the left train of dispersion gravity-waves described in this preceding chapter : the train is ultimately limited at the points $x=(U_0-\mu H)t$ and $x=U_0 t$, progressing with the extreme values of the group velocity, the wave length decreases regularly from the left to the right, each wave length propagating with the corresponding group velocity, and there is a continual apparition of shorter and shorter waves at the right end of the train.

Finally we can again conclude that any initially limited perturbation becomes ultimately negligible, its irrotational part being more and more dispersed, and its rotational part being carried away with the flow.

Forced perturbation due to a corrugation at the bottom.

By using again the same method as in the preceding chapter, it is found that in any case it is possible to define one unique forced perturbation due to a small-amplitude corrugation at the bottom, and that in the case where k_s exists ($0 < U_0 < \mu H$) this perturbation involves a train of stationary waves (barrier waves) with the wave length L_s, the amplitude of which is significant provided the width of the corrugation is comparable to this wave length. If the magnitude of the basic velocity U_1 at the bottom is not too small, it is also possible to use the same general formula as for gravitational perturbations, assuming that the profile $z = F(x)$ of the bottom corrugation is given by the Fourier integral

$$F(x) = Re \int_0^\infty f(k) e^{ikx} dk . \tag{4.11}$$

it is thus found that :

a) If $U_0 > \mu H$ or $U_0 < 0$, for the forced perturbation the interface deformation is given by

$$\Delta z_0 = Re \int_0^\infty \lambda(k) f(k) e^{ikx} dk , \tag{4.12}$$

where

$$\lambda(k) = \frac{U_1 e^{-kH}}{c(k)} , \tag{4.13}$$

$c(k)$ being the function defined by (4.9), and the integration path being the real axis. Since the integrand has no pole along this path, this gives a forced perturbation the practical horizontal extent of which is comparable to that of the bottom corrugation, and the aspect of its streamlines is shown on fig.18 and 19 in the case of a Lorentz profile for $F(x)$, and for two typical velocity profiles of the basic flow (in fig.19 we have $\bar{u}=0$ at some level $z=z_0$, and accordingly there are two open vortices in the vicinity of that level, limited by a double streamline ; we know that when k_s does not exist, $\bar{u}$ cannot vanish at more than one level).

b) If $0<U_0<\mu H$, so that k_s exists and is a pole of $\lambda(k)$, the interface deformation is given by

$$\Delta z_0 = Re\int_0^{\infty(-)} \lambda(k) f(k) e^{ikx} dk \,, \tag{4.14}$$

where the integration path is the indented one shown on fig.12, leaving the pole k_s at its left. This gives a half-infinite train of resonance waves, extending from the vicinity of $x=0$ to $x=\infty$, with a fringe around $x=0$. The resonance waves themselves are given by

$$Re\left[2i\pi r_s f(k_s) e^{ik_s x}\right] \,, \tag{4.15}$$

where r_s is the residue of $\lambda(k)$ for $k=k_s$, given by the formula

$$r_s = \frac{k_s U_1}{\gamma_s} e^{-k_s H} \,. \tag{4.16}$$

In this case $\bar{u}$ is vanishing at one or two levels, and in the vicinity of such a level the aspect of the streamlines on the positive side of the corrugation is a cat's eye pattern, as shown on fig.20 where $\bar{u}$ is assumed to vanish in the upper flow.

The fact that the resonance waves are located on the positive side of the corrugation (which is the lee side relatively to the basic flow at the interface, but not necessarily relatively to the basic flow at the bottom) is again a consequence of the fact that γ_s is positive.

In any case when $|U_1|$ is small the determination of the forced perturbation requires another method, involving the use of the expression of the streamfunction (in particular the resonance waves change into a system of vortices in the vicinity of the bottom, so that Δz is no longer comparable with δw).

5. ADIABATIC MESO-SCALE PERTURBATIONS IN A STRAIGHT ATMOSPHERIC FLOW.

In view of a theory of meso-scale orographic perturbations it is necessary to study further models in which the stability of the basic motion, instead of being concentrated at one level, is distributed continuously throughout the fluid. We shall first consider the case of a purely gravitational stability uniformly distributed, indeed the only one such that the methods used in the preceding chapters can be applied without any important difficulty. Then we shall examine the more general cases of a straight atmospheric flow where both static stability and vertical shear are changing with height.

Gravitational perturbations in a motionless incompressible fluid with a static stability uniformly distributed.

The frame of reference being the same as before (Oz vertical upwards), the static stability of an incompressible fluid at rest is expressed quantitatively by the coefficient

$$(5.1) \quad \sigma^2 = -\frac{g}{\rho}\frac{\partial \rho}{\partial z} \quad ,$$

representing the restoring force applied to a fluid element, per unit mass and unit vertical displacement, after this element has received an infinitesimal adiabatic displacement from its equilibrium level, without any appreciable perturbation of the surrounding fluid.

Here we assume, not only that the fluid is incompressible and at rest in the basic state, but in addition that both g and σ are independent of height, and according to (5.1) this requires an exponential upward decrease of $\bar{\rho}$, say

$$(5.2) \quad \bar{\rho} = \rho_0 e^{-2sz} ,$$

where ρ_0 and s are two positive constants (the factor 2 is introduced for sake of convenience). Accordingly we have, in the basic state,

$$\sigma = \sqrt{2gs}\,. \tag{5.3}$$

We also assume that earth's rotation may be neglected.

Thanks to the incompressible-motionless assumption the perturbations of this basic state are purely gravitational, and in addition we assume that the fluid is unbounded in all directions, an assumption which is indeed unrealistic since it requires an infinite density for $z=-\infty$, but may be physically accepted as far as the perturbations are concerned. Again we also assume no viscosity, at least up to a certain level, and that the perturbations are adiabatic and two-dimensional in the xz-plane.

The mechanical equations used in the preceding chapters must be completed by the equation $\frac{d\rho}{dt}=0$ taking place of the thermal equation (in virtue of the adiabatic assumption), and this gives

$$\rho\frac{du}{dt}+\frac{\partial p}{\partial x}=0;\ \rho\frac{dw}{dt}+\frac{\partial p}{dz}+g\rho=0;\ \frac{\partial u}{\partial x}+\frac{\partial w}{\partial z}=0;\ \frac{d\rho}{dt}=0\,. \tag{5.4}$$

From this system the following perturbation equations are deduced, after use is made of (5.2) :

$$\begin{cases}\bar\rho\frac{\partial}{\partial t}\delta u+\frac{\partial}{\partial x}\delta p=0; & \bar\rho\frac{\partial}{\partial t}\delta w+\frac{\partial}{\partial z}\delta p+g\,\delta\rho=0\,;\\ \frac{\partial}{\partial x}\delta u+\frac{\partial}{\partial z}\delta w=0; & \frac{\partial}{\partial t}\left(\delta\rho-2s\bar\rho\,\Delta z\right)=0\,,\end{cases} \tag{5.5}$$

and if we complete them by

$$\delta u=\frac{\partial}{\partial t}\Delta x\ ;\quad \delta w=\frac{\partial}{\partial t}\Delta z\,, \tag{5.6}$$

we get a system of six equations for the six disturbances involved. Furthermore, again in virtue of (5.2), this system can be reduced to one with constant coefficients by putting

$$\begin{cases}\delta u=e^{sz}\delta u';\ \delta w=e^{sz}\delta w';\ \Delta x=e^{sz}\Delta x';\ \Delta z=e^{sz}\Delta z';\\ \delta p=e^{-sz}\delta p';\ \delta\rho=e^{-sz}\delta\rho',\end{cases} \tag{5.7}$$

and using the "modified disturbances " $\delta u'$, $\delta w'$, ... as new unknowns. After eliminating $\delta\rho'$, and performing some substitutions, we thus get the new system

$$\text{(5.8)} \quad \begin{cases} \frac{\partial^2}{\partial t^2}\Delta x' + \frac{\partial}{\partial x}\frac{\delta p'}{\rho_0} = 0 ; \quad \frac{\partial}{\partial t}\left[\left(\frac{\partial^2}{\partial t^2} + \sigma^2\right)\Delta z' + \left(\frac{\partial}{\partial z} - s\right)\frac{\delta p'}{\rho_0}\right] = 0 ; \\ \frac{\partial}{\partial t}\left[\frac{\partial}{\partial x}\Delta x' + \left(\frac{\partial}{\partial z} + s\right)\Delta z'\right] = 0 ; \quad \delta u' = \frac{\partial}{\partial t}\Delta x' ; \quad \delta w' = \frac{\partial}{\partial t}\Delta z' . \end{cases}$$

Fundamental waves.- If we impose no boundary condition we have as particular solutions what we shall again refer to as "fundamental waves", such that any one of the modified disturbances figuring in (5.7) depends on $x, z, t,$ by the exponential factor

$$\text{(5.9)} \quad e^{i(kx + mz - \omega t)} = e^{i(\vec{K}.\vec{M} - \omega t)},$$

where k, m are two real constants, ω is another constant, $\vec{K}$ is the "wave vector", the components of which are k and m (so that its magnitude is $K = \sqrt{k^2 + m^2}$) , and $\vec{M}$ is the position vector of the point $M(x, z)$.
It will be convenient to assume

$$\text{(5.10)} \quad k \geqslant 0 ,$$

and to denote by α the polar angle of $\vec{K}$ such that $|\alpha| \leqslant \frac{\pi}{2}$. We thus have

$$\text{(5.11)} \quad -\frac{\pi}{2} \leqslant \alpha \leqslant \frac{\pi}{2} ; \quad k = K\cos\alpha ; \quad m = K\sin\alpha .$$

Now according to (5.8) we must have, if we disregard the stationary perturbations (such that $\omega = 0$),

$$\text{(5.12)} \quad \left(k^2 + m^2 + s^2\right)\omega^2 - \sigma^2 k^2 = 0 ,$$

and this gives two opposite real values for ω when $\vec{K}$ is given, namely

$$\text{(5.13)} \quad \omega = \pm\, \Omega(k, m),$$

where

$$\text{(5.14)} \quad \Omega(k, m) = \frac{\sigma k}{K'} , \quad \text{with } K' = \sqrt{K^2 + s^2} = \sqrt{k^2 + m^2 + s^2} .$$

Therefore we have again two families of waves : the "positive waves", such that $\omega=\Omega$, and the "negative waves", such that $\omega=-\Omega$. Obviously we are no longer dealing with surface waves, but with perturbations of the same type as the classical sound waves or electromagnetic waves propagating in a continuous medium : the surfaces of constant phase or "wave surfaces" are the planes perpendicular to $\vec{K}$, the wave length is $L=2\pi/K$, and the phase velocity in the direction of $\vec{K}$ is $c=\omega/K$, so that it may be represented by the vector $\vec{c}=\pm\vec{C}$ (upper sign for positive waves, lower sign for negative waves), with

(5.15) $$\vec{C}=\frac{\Omega}{K}\vec{k}=\frac{\sigma k}{KK'}\vec{k},$$

$\vec{k}$ denoting the unit vector of the $\vec{K}$ direction.
From the expression obtained for ω we can also derive that of the group velocity $\vec{\gamma}$, representing the propagation velocity of the wave energy (in the case of the sinusoidal waves defined by the real part of the fundamental waves) and the components of which are $\gamma_x=\partial\omega/\partial k$ and $\gamma_z=\partial\omega/\partial m$ (generalization of the result used in chapter 3).
Here we have $\vec{\gamma}=\pm\vec{G}$, the components of $\vec{G}$ being

(5.16) $$G_x=\frac{\partial\Omega}{\partial k}=\sigma\left(\frac{1}{K'}-\frac{k^2}{K'^3}\right)=\sigma\frac{m^2+s^2}{K'^3};\quad G_z=\frac{\partial\Omega}{\partial m}=-\sigma\frac{km}{K'^3}.$$

Therefore G_x and C_x have the same sign, but G_z and C_z have opposite signs.
From (5.15) and (5.16) the following vectorial formulae are deduced :

(5.17) $$\frac{\vec{C}}{C_0}=\lambda\vec{k}\cos\alpha;\quad \frac{\vec{G}}{C_0}=\lambda\vec{x}-(\lambda-\lambda^3)\vec{k}\cos\alpha,$$

where

(5.18) $$C_0=\frac{\sigma}{s};\quad \lambda=\frac{s}{K'},$$

and $\vec{x}$ denotes the unit vector of Ox. The undimensional parameter λ is a function of the only wave length (or of K), increasing from 0 to 1 as L varies from

0 to ∞ , and the formulae (5.17) lead to the simple construction shown on fig. 21. We see that, if $\vec{C}$ and $\vec{G}$ are drawn with O as origin, their K-hodographs (hodographs for K = const) are two tangent circles, and the angle between them is acute (we have $\cot\beta = \lambda^2 \cot\alpha$). The $\vec{C}$-domain (area described by $\vec{C}$ as $\vec{K}$ varies arbitrarily) is bounded by the circle (C) with $OO' = C_0$ as diameter, and the $\vec{G}$ -domain is the area enclosed by the envelope (G) of the K -hodograph of $\vec{G}$: it is a leaf-shaped curve intersecting the x-axis at the same points O and O' as does (C) , and which are respectively a cusp point and a double point with a $60°$-kink.

The point where the K-hodograph of $\vec{G}$ is tangent to its envelope is such that $\tan\alpha = \lambda\sqrt{3}$, and this leads to the following two systems of parametic equations for (G) :

$$(5.19) \quad x = C_0 \frac{4\lambda^3}{1+3\lambda^2} \; ; \quad z = \pm C_0 \frac{\sqrt{3}\,\lambda^2(1-\lambda^2)}{1+3\lambda^2} \; ;$$

$$(5.20) \quad x = C_0 \frac{4}{3\sqrt{3}} \sin^2\alpha \tan\alpha \; ; \; z = \pm C_0 \frac{1}{3\sqrt{3}} \sin^2\alpha \left(3 - \tan^2\alpha\right).$$

In view of atmospheric applications it is interesting to consider the particular case where $K \gg s$ (in the lower atmosphere we have $2s \sim 10^{-6}\,\mathrm{cm}^{-1}$, and for a meso-scale wave $L < 30\,\mathrm{km}$, whence $K > 2 \times 10^{-6}\,\mathrm{cm}^{-1}$). In this limit case we have $\lambda \ll 1$, therefore $\vec{C}$ and $\vec{G}$ have practically the same K-hodograph, as shown on fig. 21 (right part), and $\vec{G}$ is practically perpendicular to $\vec{K}$: in other words the group velocity is nearly transversal. Furthermore in the same case the third equation (5.8) may be replaced with $\frac{\partial}{\partial x}\Delta x' + \frac{\partial}{\partial z}\Delta z' = 0$, whence $k\Delta x + m\Delta z = 0$, and this means that the vibration of the fluid is also transversal, just as that of the electric or magnetic vector in electromagnetic waves.

It is important to note that, whereas the modified disturbances $\delta u', \delta w', \Delta x', \Delta z'$ keep the same amplitude at all levels, according to (5.7) the amplitudes of the corresponding disturbances themselves $(\delta u, \delta w, \Delta x, \Delta z)$

are increasing indefinitely with z. This result is in agreement with the fact that, in any sinusoidal wave defined by the real part of a fundamental wave, the energy per unit volume must remain constant owing to the unviscid assumption : if A denotes the amplitude of the fluid displacement for instance, this energy is proportional to $\bar{\rho}A^2$, and since $\bar{\rho}$ is decreasing upwards as e^{-2sz}, A must increase as e^{sz}. Therefore we have to admit that beyond some height the infinitesimal assumption is no longer valid, unless the waves are properly damped by viscosity or some other factor of dissipation, and the same conclusion is to be expected for the actual atmospheric waves as a consequence of the unlimited upward decrease of air density.

Dispersion of an initial local perturbation (dispersion waves). - The method used in chapter 3 can be readily generalized, leading to the following results :

a) Because of the existence ot two opposite values of ω for each wave vector, the initial conditions necessary to determine a perturbation of our basic state must consist of two xz-distributions, for instance those of $\Delta z'$ and $\delta w'$, and the perturbation can be obtained as the superposition of fundamental waves.

b) If these initial distributions are given by the Fourier integrals

$$(5.21)\quad \begin{cases} \Delta z'_{(t=0)} = \int_{-\infty}^{\infty}\int_{-\infty}^{\infty} f_1(k,m)\, e^{i(kx+mz)}\, dk\, dm\,; \\ \delta w'_{(t=0)} = \int_{-\infty}^{\infty}\int_{-\infty}^{\infty} f_2(k,m)\, e^{i(kx+mz)}\, dk\, dm\,, \end{cases}$$

where f_1 and f_2 are two infinitesimal functions, then at any other time t the value of $\Delta z'$ is given by

$$(5.22)\quad \Delta z' = \Delta z'_1 + \Delta z'_2\,,$$

with

$$(5.23)\quad \Delta z'_1 = \int_{-\infty}^{\infty}\int_{-\infty}^{\infty} g_1(k,m)\, e^{i(kx+mz-\Omega t)}\, dk\, dm\,;$$

(5.24) $$\Delta z_2' = \int_{-\infty}^{\infty}\int_{-\infty}^{\infty} g_2(k,m)\, e^{i(kx+mz+\Omega t)}\, dk\, dm,$$

where Ω is written for $\Omega(k,m)$, given by (5.14), and

(5.25) $$g_1(k,m) = \frac{1}{2} f_1(k,m) + \frac{i}{2\Omega} f_2(k,m);\; g_2(k,m) = \frac{1}{2} f_1(k,m) - \frac{i}{2\Omega} f_2(k,m).$$

c) The integral in (5.23), representing the superposition of positive waves, can be reduced to the form (2.42) by putting

(5.26) $$X = t;\quad \varphi(k,m) = k\frac{x}{t} + m\frac{z}{t} - \Omega(k,m),$$

so that its asymptotic value, when t is large, can be obtained by applying the generalized stationary-phase principle. The values of k and m such that $\partial\varphi/\partial k$ and $\partial\varphi/\partial m$ are both vanishing are given by the system

(5.27) $$G_x(k,m) = \frac{x}{t};\quad G_z(k,m) = \frac{z}{t}.$$

If the point $M\left(\frac{x}{t}, \frac{z}{t}\right)$ is outside the boundary (G) of the $\vec{G}$-domain represented on fig.21, this system has no real solution, and therefore if M is not close to (G) the asymptotic value of $\Delta z_1'$ is of the order of t^{-2} at most. But if M is inside the area enclosed by (G) the system (5.26) has two solutions such that $k>0$, say k_1, m_1 and k_2, m_2, and the corresponding wave vectors $\vec{K}_1$ and $\vec{K}_2$ can be obtained by drawing through M the two circles tangent to (G), and then using the relationships shown on fig.21. In this case, if M is again not close to (G), the asymptotic value of $\Delta z_1'$ is obtained by adding two expressions of the form (2.44), and this gives a system of two families of dispersion waves with an amplitude of the order t^{-1}, located inside the curve (G) and forming the main part of the perturbation defined by $\Delta z_1'$ when t is large. If M is in the vicinity of (G) the asymptotic value of $\Delta z_1'$ involves the product of a Fresnel function by an Airy function (or the product of two Airy functions if M is close to the kink of (G)), and this gives a complex system of waves forming a fringe around the dispersion-waves domain.

The same method applied to $\Delta z_2'$ (representing the superposition of negative waves) leads to a quite similar result : the perturbation defined by $\Delta z_2'$ is ultimately made of a system of dispersion waves inside the area bounded by the curve (G') symmetrical of (G) with respect to O, this system being bordered by a fringe region, and the constant-phase curves of the waves (profiles of the wave surfaces) are also symmetrical of those of the first system with respect to O.

The whole pattern of constant-phase curves for the dispersion waves themselves (as they are defined by the stationary-phase formula) is represented on fig.22, where these curves are drawn in full line for equidistant values of the phase. It reminds the gravity-waves pattern produced at the surface of a quiet river for instance by a slowly-moving ship, also characterized by two families of wave lines, with cusp points along their common boundary.

d) According to (5.27) and the similar formulae concerning $\Delta z_2'$, the wave vector $\vec{K}$, in each family of dispersion waves, is at any point P such that the corresponding group velocity $\pm \vec{G}$ is equal to $\vec{OP}/t$, and this generalizes the result obtained for the dispersion waves at the free surface of a water layer, namely that the wave length (or the wave number) propagates with the wave energy, at a constant speed equal to the corresponding group velocity : here the wave vector $\vec{K}$ also propagates with the wave energy, along OP and with the corresponding group velocity $\pm \vec{G}$.

e) Finally we shall conclude that, as a result of its dispersion, any local perturbation of the fluid becomes ultimately negligible if what happens at large distance is unknown, and it is essentially this conclusion that we shall retain (the energy of the perturbation is ultimately spread over an area which is increasing proportionally to t^2, and this agrees with the fact that the amplitude is decreasing as t^{-1}).

Free perturbations in a half-bounded incompressible fluid with uniform static stability and uniform velocity.

We now assume that the same fluid as before, still unbounded upwards, is bounded downwards by the fixed plane $z=0$, and that in addition it may be moving in the x-direction with the uniform velocity U.
The introduction of the boundary requires the kinematic condition that $\Delta z=0$ (or $\delta w=0$) at $z=0$, obviously not satisfied by the fundamental waves as we have defined them, except if their wave vector is vertical. But the condition is satisfied by the perturbation obtained by superposing two fundamental waves of the same family, with opposite values of m and of the complex amplitude of Δz (or δw). For such a free perturbation we thus have

$$(5.28)\quad \Delta z' = A\left[e^{i(kx+mz-\omega t)} - e^{i(kx-mz-\omega t)}\right],$$

where k and m are two real constants, with $k>0$, A is an infinitesimal constant, and $\omega=\pm\Omega$. Any free perturbation can be obtained by superposing such elementary perturbations, and this gives for $\Delta z'$ (or $\delta w'$) a distribution represented at any time by an odd function of z, therefore by a double Fourier integral of the form (2.39) where $f(k,m)$ is an odd function of m.
In particular this must be the case for the Fourier integrals representing the initial distributions of $\Delta z'$ and $\delta w'$, but excepting this condition nothing has to be changed in the theory just given for the dispersion of a local perturbation if the fluid is supposed at rest in the basic state, and the constant-phase lines of the dispersion waves are obtained by simply taking the upper half of the pattern shown on fig.22.

If the fluid is moving in the x-direction with the uniform velocity U, all the results obtained with $U=0$ can be used provided x is changed into $x-Ut$, and if the fundamental waves are still defined by the condition that the modified disturbances depend on x, z, t by the factor $e^{i(kx+mz-\omega t)}$ we have to replace ω with $\omega+kU$. We thus get $\omega = kU \pm \Omega$ (upper sign for positive waves), and for the group velocity we have $\vec{\gamma} = U\vec{x} \pm \vec{G}$, where as before $\vec{x}$ denotes the unit vector of Ox.

Stationary waves.- The condition for a wave to be stationary is $\omega=0$, and this may happen for a negative wave only : we must have $\Omega = kU$, and if we substitute for Ω its expression given by (5.14), we get $K' = \frac{\sigma}{U}$, which is possible only if $U < C_0$ (with $C_0 = \frac{\sigma}{s}$). This is the condition for the existence of stationary waves, and if it is satisfied these waves are such that

$$(5.29) \quad K = k_s, \text{ with } k_s = \sqrt{\frac{\sigma^2}{U^2} - s^2} = \frac{\sigma}{U}\sqrt{1 - \frac{U^2}{C_0^2}},$$

so that they all have the same wave length, namely $L_s = 2\pi/k_s$.

Therefore the value of $\vec{G}$ for a stationary wave is a function of the only parameter α (polar angle of $\vec{K}$), say $\vec{G} = \vec{G}_s(\alpha)$, the expression of which can be obtained by using (5.17) : since $K' = \frac{\sigma}{U}$ we have $\lambda = \frac{U}{C_0}$, whence

$$(5.30) \quad \vec{G}_s = U\vec{x} - U\left(1 - \frac{U^2}{C_0^2}\right)\vec{k}\cos\alpha,$$

and this gives for the group velocity $\vec{\gamma} = \vec{\gamma}_s(\alpha)$ of a stationary wave

$$(5.31) \quad \vec{\gamma}_s = U\vec{x} - \vec{G}_s = U\left(1 - \frac{U^2}{C_0^2}\right)\vec{k}\cos\alpha.$$

We thus get this important result that <u>the group velocity of a stationary wave has the same direction as $\vec{K}$</u>, therefore it is longitudinal (with respect to the fluid the group velocity of the wave is $-\vec{G}_s$, and we have seen that it is practically transversal if $K \gg s$, as it is the case if $U \ll C_0$; but the addition of $U\vec{x}$ makes it longitudinal, as shown on fig. 23). According to (5.31) when C_0 is given the hodograph of $\vec{\gamma}_s$ relative to the origin 0 is the circle of diameter $U\left(1-\frac{U^2}{C_0^2}\right)$ tangent to $0z$ at 0 (also shown on fig. 23), and the maximum of γ_s (for $\alpha = 0$) is vanishing for $U=0$ and $U=C_0$, reaching its largest value $\frac{2}{3\sqrt{3}}C_0$ for $U=\frac{1}{\sqrt{3}}C_0$.

Forced perturbations in a half-bounded incompressible fluid with uniform static stability and uniform velocity.

Let us now assume that the plane boundary $z=0$ is replaced with the cylindrical surface $z=F(x)$, where $F(x)$ is an infinitesimal function, and seek for what may be considered as the"forced perturbation"due to the corrugation defined by this function. We shall use the same method as in the preceding chapters : first take the case of a sinusoidal corrugation, then generalize by Fourier integrals. In view of atmospheric applications we shall assume

$$U < C_0 \tag{5.32}$$

(if we adopt for $2s$ and σ the values observed in the troposphere, namely $2s \sim 10^{-6}\text{cm}^{-1}$ and $\sigma \sim 10^{-2}\text{sec}^{-1}$, we get $C_0 \sim 200\text{m/s}$ so that the above condition is largely satisfied by usual winds).

Case of a sinusoidal corrugation. - We assume that

$$F(x) = A e^{ikx}, \tag{5.33}$$

where k is a real positive constant, A an infinitesimal constant, and first seek for a stationary perturbation. Then we have necessarily

$$\Delta z' = h(z)\, e^{ikx}, \tag{5.34}$$

where $h(z)$ is a function to be determined. The condition at $z=0$ being $\Delta z' = F(x)$, we must have

(5.35) $$h(0) = A.$$

On the other hand, in the perturbation equation **(5.8)** we have to replace $\frac{\partial}{\partial t}$ with $\frac{d}{dt} = U\frac{\partial}{\partial x} = ikU$, and the partial differential equation to be satisfied by any one of the modified disturbances can be directly deduced from (5.12) by the substitutions

(5.36) $$\omega^2 = k^2 U^2 ; \quad m^2 = -\frac{\partial^2}{\partial z^2} .$$

This gives for $h(z)$ the differential equation with constant coefficients

(5.37) $$h''(z) + (k_s^2 - k^2)\, h(z) = 0 ,$$

where k_s is the particular wave number defined by (5.29), the existence of which results from the assumption (5.32). Two cases must be considered in succession :

a) If $k > k_s$, the general stationary solution for $\Delta z'$ is a linear combination of the two particular ones satisfying the condition (5.35) :

(5.38) $$\Delta z_1' = A e^{-\sqrt{k^2 - k_s^2}\, z} e^{ikx} ; \quad \Delta z_2' = A e^{\sqrt{k^2 - k_s^2}\, z} e^{ikx} ,$$

and obviously the perturbation P_1 defined by $\Delta z_1'$ is the only stationary perturbation such that $\Delta z'$ is negligible when z is large. Now any perturbation P can be obtained as the superposition of P_1 and some free perturbation P', and therefore if P is such that $\Delta z'$ and $\delta w'$ are initially negligible when z is large the same holds for P', whence it follows that when t is large this free perturbation is negligible at any level as a result of its dispersion, provided what happens at large distance upwards is ignored.

We can say accordingly that the perturbation P tends towards the stationary perturbation P_1, which we shall for this reason adopt as the forced perturbation due to the sinusoidal corrugation.

It is to be noted that, whereas $\Delta z_1'$ is vanishing for $z=\infty$, the corresponding vertical deformation Δz_1 of the fluid has not necessarily the same property : this is true only if $\sqrt{k^2-k_s^2}>s$, whence $k>\frac{\sigma}{U}$. Therefore if $k_s<k<\frac{\sigma}{U}$ the amplitude of the fluid displacement is increasing indefinitely upwards, so that we are facing with the same difficulty as in the case of fundamental waves, and the conclusions concerning the behaviour of these waves at large distance upwards also hold for P_1. This means that this forced perturbation cannot be considered as determined by some kinematic upper-boundary condition, and we shall be led to the same statement for atmospheric perturbations in general.

The aspect of the "modified streamlines" for the perturbation P_1 (curves $z+\Delta z'=\text{const}$) is shown on fig.24 (left part).

b) If $k<k_s$, the general stationary solution for $\Delta z'$ is a linear combination of the two particular ones

$$(5.39)\quad \Delta z_1' = A e^{i(kx+\sqrt{k_s^2-k^2}\,z)} \;;\; \Delta z_2' = A e^{i(kx-\sqrt{k_s^2-k^2}\,z)},$$

again satisfying the condition at $z=0$. These are nothing but two stationary fundamental waves, with $m=\pm\sqrt{k_s^2-k^2}$ ($K=k_s$; $\cos\alpha=k/k_s$), for both of which Δz is increasing indefinitely upwards as already mentioned. But if we note that the group velocity $\vec{\gamma_s}$ is directed upwards for the first one, and downwards for the second one (see fig.23), we may expect that if we start from a deformation of the fluid with a limited vertical extent, it is the stationary wave P_1 defined by $\Delta z_1'$ which will ultimately predominate up to the height $z=\gamma_{sz}t$, where γ_{sz} is the vertical component of $\vec{\gamma_s}$.

This result can actually be confirmed by the same method as used in chapter 3 for the theory of resonance waves in a water layer flowing over a corrugated bed, the appropriate perturbation P_0 to be considered for this purpose being obtained by superposing upon P_1 the free perturbation resulting itself of the superposition of negative fundamental waves and such that the initial value of $\Delta z'$ is opposite to $\Delta z'_1$. Since we have

(5.40) $$\Delta z'_1 = A e^{ikx} e^{im_0 z} = A e^{ikx} \frac{1}{2i\pi} \int_{-\infty}^{\infty(-)} \frac{1}{m - m_0} e^{imz} dm ,$$

where $m_0 = \sqrt{k_s^2 - k^2}$, and the path of integration is the real axis with a negative indentation around the pole $m = m_0$ (such as shown on fig.12), this perturbation P_0 is defined by

(5.41) $$\Delta z' = A e^{ikx} \left[e^{im_0 z} - \frac{1}{2i\pi} \int_{-\infty}^{\infty(-)} \frac{1}{m - m_0} \left(e^{i(mz - \bar{\Omega} t)} - e^{i(-mz - \bar{\Omega} t)} \right) dm \right],$$

where $\bar{\Omega} = kU - \Omega$ is the angular frequency of negative fundamental waves, considered as a function of m, and vanishing for $m = m_0$. This gives

(5.42) $$(\Delta z')_{t=0} = 0 \; ; \; (\delta w')_{t=0} = A e^{ikx} \frac{1}{2\pi} \int_{-\infty}^{\infty} \frac{\bar{\Omega}}{m - m_0} \left(e^{imz} - e^{-imz} \right) dm ,$$

and since the latter integral has no real pole, both $\Delta z'$ and $\delta w'$ are initially negligible when z is large, whence it follows that the whole perturbation P_0 has initially a limited vertical extent.

Accordingly any other perturbation P satisfying the same condition differs from P_0 by a free perturbation with again a limited initial vertical extent, and we know that it is ultimately negligible as a result of its dispersion : therefore for the perturbation P the distribution of $\Delta z'$ is ultimately the same as for the particular perturbation P_0.

Now the integral in(5.41) is one of the same type as the

integral (3.49) used in the theory of resonance gravity-waves, and its asymptotic value when t is large can be obtained by the same method : a first approximation is obtained by replacing the phase $\pm m z - \bar{\Omega} t$ with the first two terms of its Taylor expansion in the vicinity of $m = m_0$, namely

(5.43) $$\pm m z - \bar{\Omega} t = \pm m_0 z + (\pm z - \gamma_{sz} t)(m - m_0) + \cdots ,$$

and this gives

(5.44) $$\Delta z' = \Delta z'_1 , \text{ for } 0 < z < \gamma_{sz} t \; ; \; \Delta z' = 0 , \text{ for } z > \gamma_{sz} t ,$$

in other words the wave P_1 limited to the layer $0 < z < \gamma_{sz} t$, the upper boundary of which is progressing upwards with the constant velocity γ_{sz}. This train of waves has to be modified in a fringe layer around $z = \gamma_{sz} t$, the thickness of which is proportional to $\sqrt{t}$ and where the z-profile of $\Delta z'$ along each vertical has exactly the same aspect as shown on fig.13 for the x-profile of Δz_0 in the vicinity of $x = \gamma_s t$.

Finally we can say that, if we ignore what happens at large distance upwards, the perturbation P defined by any initial deformation of the fluid with a limited vertical extent always tends towards the stationary wave P_1 defined by $\Delta z'_1$, which may therefore be referred to as the forced perturbation due to the sinusoidal corrugation. The aspect of the modified streamlines of P_1 is shown on fig.24 (right part) : the wave surfaces are the planes making the angle α with the verticals, and sloping in the direction of the basic flow.

Just as for the resonance gravity-waves considered in chapter 3, the train of waves representing the ultimate stage of the evolution of P develops in the direction of the corresponding group velocity $\vec{\gamma_s}$ (this is one form of the so-called "radiation condition"), its front progressing with the velocity γ_{sz}, and its energy being continually extracted from the energy of the basic flow at a constant rate (contrary to the case $k > k_s$, where there is no transfer of energy once the forced perturbation is set up). According to (5.31) we have

$$\gamma_{sz} = U\left(1-\frac{U^2}{C_0^2}\right)\cos\alpha\,\sin\alpha = U\left(1-\frac{U^2}{C_0^2}\right)\frac{k}{k_s^2}\sqrt{k_s^2-k^2}, \tag{5.45}$$

So that γ_{sz} is vanishing for $\alpha=0$ and $\alpha=\pi/2$, corresponding to $k=k_s$ and $k=0$ respectively, its maximum value being reached for $\alpha=\pi/4$, corresponding to $k=k_s/\sqrt{2}$. Therefore in the limit cases $k=k_s$ and $k=0$ the setting up of the forced perturbation would require an infinite time.

In reality the upward extension of the train is necessarily stopped either by viscosity or by the fact that beyond some level the amplitude of the fluid displacement is too large for the infinitesimal assumption to remain valid: if the viscosity is small enough the waves must change into a system of vortices, forming some sort of an organized turbulence which may eventually produce a partial downward reflection of the ascending flux of energy, and therefore some modification in the stationary perturbation established in the lower part of the fluid. This means that a proper generalization of the above theory would be necessary and the best we can do here is to admit that the perturbation P_1 can actually be established up to some level and then remains stationary, this assumption being apparently confirmed by the existence of stationary mountain-waves in the lower atmosphere.

Case of an arbitrary corrugation with a profile of limited horizontal extent. -Such a corrugation can be represented by a Fourier integral, say

$$F(x) = Re\int_0^\infty f(k)\,e^{ikx}\,dk, \tag{5.46}$$

in other words by a superposition of sinusoidal elementary corrugations, and it is easy to deduce from the preceding results that, when t is large, the limit of any perturbation defined by some "local" initial deformation of the fluid (negligible beyond a certain distance in all directions of the xz-plane) is obtained by superposing the various elementary forced perturbations due to these sinusoidal corrugations, provided what happens at large distance is ignored. One thus get a stationary perturbation which may be called the "forced perturbation" due to the corrugation defined by (5.46), and such that $\Delta z'$ is given by

$$\Delta z' = \Delta z'_a + \Delta z'_b \, , \tag{5.47}$$

with

$$\begin{cases} \Delta z'_a = Re \int_0^{k_s} f(k)\, e^{i(kx + \sqrt{k_s^2 - k^2}\, z)}\, dk ; \\ \Delta z'_b = Re \int_{k_s}^{\infty} f(k)\, e^{ikx - \sqrt{k^2 - k_s^2}\, z}\, dk , \end{cases} \tag{5.48}$$

the paths of integration being along the real axis. In the first integral we may use as new variable of integration the polar angle α already introduced, such that

$$k = k_s \cos\alpha \; ; \; \sqrt{k_s^2 - k^2} = k_s \sin\alpha \; ; \; 0 \leqslant \alpha \leqslant \frac{\pi}{2} \, , \tag{5.49}$$

and if we also introduce the polar coordinates r, θ defined by

$$x = r\cos\theta \; ; \; z = r\sin\theta \; ; \; 0 \leqslant \theta \leqslant \pi \, , \tag{5.50}$$

we get the formula

$$\Delta z'_a = Re \int_0^{\frac{\pi}{2}} (k_s \sin\alpha) f(k_s \cos\alpha)\, e^{ik_s r \cos(\alpha - \theta)}\, d\alpha , \tag{5.51}$$

whence an symptotic value of $\Delta z'_a$, when $k_s r$ is large, can be directly obtained by the stationary-phase method.

If $x>0$ the phase $k_s r\cos(\alpha-\theta)$ is stationary for $\alpha=\theta$, and this leads to an asymptotic value of the order $r^{-\frac{1}{2}}$ if z is not too small, whereas in the other cases ($x<0$, or z small) we have $\Delta z'_a \sim r^{-1}$ (the phase is monotonic if $x<0$), except in the vicinity of $\theta = \pi/2$ where $\Delta z'_a$ is again of the order $r^{-\frac{1}{2}}$. On the other hand if f_M denotes the maximum of $|f(k)|$ along the path of integration in the second integral (5.48), we can write

$$(5.52)\quad |\Delta z'_b| < f_M \int_{k_s}^{\infty} e^{-kz}\, dk = \frac{f_M}{z}\, e^{-k_s z},$$

whence we deduce that $\Delta z'_b \sim z^{-1}$ at most when z is large, and since by Riemann theorem we know that $\Delta z'_b \sim x^{-1}$ at most when $|x|$ is large, we see that in any case we have $\Delta z'_b \sim r^{-1}$ at most, so that the asymptotic value or order of magnitude of $\Delta z'$ is always the same as that of $\Delta z'_a$:

a) If we except the vicinity of $\theta=\frac{\pi}{2}$ (such that $|k_s x| \sim \sqrt{k_s z}$ at most), and also the vicinity of $\theta=0$ (such that $k_s z \sim \sqrt{k_s x}$ at most), we have for $k_s r \gg 1$,

$$(5.53)\quad \Delta z' \approx Re\left[(k_s \sin\theta)\, f(k_s\cos\theta)\sqrt{\frac{2\pi}{k_s r}}\, e^{i(k_s r-\frac{\pi}{4})}\right], \text{ for } x>0;$$

$$(5.54)\quad \Delta z' \sim r^{-1}, \text{ for } x<0,$$

the expression (5.53) being given by the stationary-phase formula.

b) In the vicinity of $\theta=\frac{\pi}{2}$, we have

$$(5.55)\quad \Delta z' \approx Re\left[k_s f(0)\sqrt{\frac{2}{k_s z}}\, e^{ik_s z}\int_{-\infty}^{\xi} e^{-i(\lambda^2-\xi^2)}\, d\lambda\right], \text{ where } \xi=\frac{k_s x}{2\sqrt{k_s z}},$$

provided the width of the bottom corrugation is not large compared to $L_s = 2\pi/k_s$, and in the vicinity of $\theta=0$ we again have $\Delta z' \sim r^{-1}$.

We thus get a system of lee waves, consisting in quasi-fundamental stationary waves with the wave length L_s, bordered by a Fresnel fringe in the vicinity of Oz, and a comparatively negligible deformation of the basic flow on the upstream side of the corrugation, the aspect of the "modified streamlines" being such as shown on fig. 25. The lee-waves themselves are given by formula (5.53), according to which the constant-phase curves are the circles centered at O , and the amplitude of these waves decreases as $t^{-\frac{1}{2}}$ along each radius. Therefore at any point M , the group velocity $\vec{\gamma_s}$ is directed radially upwards from the origin O , in agreement with the radiation condition, its magnitude being constant along OM (for the wave system deduced from the present one by a symmetry about Oz,and which also defines a possible stationary perturbation, the group velocity is directed downwards, so that the radiation condition can be used to decide which solution must be adopted as the forced perturbation).

By generalizing the theory given in the case of a sinusoidal corrugation, it is readily proved that if one starts from a local deformation of the basic flow, after a long enough time t the lee-waves of the forced perturbation are practically set up in the domain bounded by the hodograph of the vector $\vec{\gamma_s}t$ relative to the origin,and according to the properties of $\vec{\gamma_s}$ it is the half-circle of diameter $U\left(1-\frac{U^2}{C_0^2}\right)t$ tangent to Oz at the origin (shown on fig.26 in the case $U \ll C_0$) : we see that the radial velocity of the progression of this front is maximum in the horizontal direction (equal to U if $U \ll C_0$), and very small in the vicinity of the vertical direction (in the vertical direction itself we have $\gamma_s = 0$, but since the asymptotic formula (5.53) is no longer valid the half-circle in question does not represent the actual front of the progressing perturbation). Of course we encounter the same difficulty as with a sinusoidal corrugation, namely that the amplitude of the fluid displacement increases indefinitely upwards beyond some level.

In the particular case of the Lorentz-profile corrugation defined by

(5.56) $$F(x) = \frac{h}{1+\left(\frac{x}{a}\right)^2}, \quad \text{whence} \quad f(k) = h a e^{-ak},$$

the asymptotic formula (5.55) gives for the lee-waves,

$$(5.57)\quad \Delta z' = h(k_s a)e^{-(k_s a)\cos\theta}(\sin\theta)\sqrt{\frac{2\pi}{k_s r}}\ \cos\left(k_s r - \frac{\pi}{4}\right),$$

whence it follows that, in any direction defined by some value of θ, the successive maxima of $|\Delta z'|$ are located on the circles $k_s r = \frac{\pi}{4} + n\pi$, where n is a positive integer, and the maximum $h_n(\theta)$ corresponding to a fixed value of n is given by

$$(5.58)\quad \frac{h_n(\theta)}{h} = (k_s a)\, e^{-(k_s a)\cos\theta}(\sin\theta)\sqrt{\frac{8}{4n+1}}\ ,$$

provided n is large enough for the formula (5.57) to give practically the correct value of $\Delta z'$ (if $k_s a = 1$, as it is the case in fig.25, the formula is practically valid for $k_s r > \pi$).

Therefore the ratio h_n/h is maximum for $k_s a \sim 1$, and negligible for $k_s a \ll 1$ as well as for $k_s a \gg 1$, so that we get the same result as obtained for the barrier gravity-waves : the amplitude of the lee waves is appreciable, when compared to the height of the corrugation, only if the width of this corrugation is comparable with the wave length L_s of the waves, and it can be shown that the same conclusion holds with any shape of the corrugation profile, in other words that there are practically no lee waves if the width b of the corrugation is either small compared to L_s (case of a "narrow barrier") or large compared to L_s (case of a "broad barrier"). In both of these extreme cases the forced perturbation can be easily computed for a lorentz-profile corrugation.

Case of a narrow barrier ($b \ll L_s$).- By taking bk as new variable of integration we can see that in the integrals (5.48) we may everywhere replace k_s with 0. Therefore $\Delta z'_a$ is negligible, and we may write

$$(5.59)\quad \Delta z' = Re\int_0^\infty f(k)\, e^{(ix-z)k}\, dk,$$

whence it follows that $\Delta z'$ is of the order r^{-1} at most when r is large, and each modified streamline has the same general shape as the bottom-corrugation profile. With the Lorentz-profile (5.56) we get

$$\Delta z' = Re\left(\frac{ah}{a+z-ix}\right) = h\,\frac{a(a+z)}{(a+z)^2+x^2}, \tag{5.60}$$

and the corresponding modified streamlines are shown on fig.27 : the width of the disturbed portion of a streamline is increasing with z at the same time as $\Delta z'$ decreases, the total area above the undisturbed streamline being independent of z in any case.

The forced perturbation is negligible beyond a height which is comparable to the width of the barrier.

Case of a broad barrier ($b \gg L_s$).- In this case it is inversely $\Delta z'_b$ which may be neglected, and in the integral giving $\Delta z'_a$ we may replace the upper limit with $+\infty$,and neglect k^2 beside k_s^2. We thus get

$$\Delta z' = Re\left[e^{ik_s z}\int_0^\infty f(k)\,e^{ikx}\,dk\right], \tag{5.61}$$

So that $\Delta z'$ is a sinusoidal function of z with the wave length L_s , and the same result holds for all the modified disturbances such as $\Delta x'$, or $\delta p'$. With the Lorentz-profile (5.56) we have

$$\Delta z' = ah\,Re\left(\frac{e^{ik_s z}}{a-ix}\right) = h\,\frac{\cos k_s z - \frac{x}{a}\sin k_s z}{1+\left(\frac{x}{a}\right)^2}, \tag{5.62}$$

and the corresponding modified streamlines are shown on fig.28.

Therefore in spite of the absence of lee waves in this case, the forced perturbation still possesses a wavelike structure.

Comparison with the case of a stability concentrated at one level. - The principal difference appears in the aspect of the lee waves.

If the stability is concentrated at one level, the wave length is the same along all the streamlines, it is equal to that of the free stationary fundamental waves, and there is no leeward damping in the absence of viscosity, the flow of energy from the basic motion to the forced perturbation being asymptotically horizontal at large distance.

On the contrary in the case of a static stability uniformly distributed there is an important leeward damping of the lee waves when they are not negligible, the flow of energy being dispersed in all directions on the lee side of the corrugation, and in addition the wave length is not constant, these properties corresponding to the fact that there are no free stationary fundamental waves.

Perturbations in a straight air flow.

As a generalization of the foregoing model, we assume in this section that :

a) The incompressible fluid is replaced with dry air of standard composition.

b) The basic flow is still parallel to $0x$ and independent of t and y , but both $\bar{u}$ and $\bar{\rho}$ are continuous functions of z , except for eventual discontinuities at one or several levels, their vertical variations being consistent with what may be observed in the atmosphere at aerological scale, and the boundary $z=0$ representing the ground surface.

c) We again assume $g = \text{const}$, no geostrophic force and no viscosity, and that the perturbations are adiabatic and two-dimensional in the xz-plane.

First of all we have to take account of the compressibility of the fluid, by introducing the coefficient of adiabatic compressibility α such that

(5.63) $d\rho = \alpha\, dp$,

if $d\rho$ and dp represent infinitesimal variations for a fluid element in adiabatic motion. In the case of dry air we have

(5.64) $$\alpha = \frac{1}{\gamma RT} \quad ; \quad \frac{1}{\alpha} = C_s^2 \ ,$$

where $\gamma = C_p/C_v$, ratio of specific heats, R is the specific gas constant, T is Kelvin temperature, and C_s denotes sound velocity. The first relation is valid for any perfect gas, and the second one for any compressible fluid. In addition we may take $\gamma = \text{const}(= 1.40)$.
According to (5.63) the expression (5.1) for the coefficient of static stability must be replaced with

(5.65) $$\sigma^2 = -g\left(\frac{1}{\bar{\rho}}\frac{d\bar{\rho}}{dz} + g\bar{\alpha}\right) ,$$

the additional term $g\bar{\alpha}$ representing the value of $-\frac{1}{\rho}\frac{d\rho}{dz}$ for a fluid element displaced adiabatically with respect to the surrounding fluid supposed to remain undisturbed and to satisfy the hydrostatic equation. Using (5.64) and this latter equation, we get the classical formula

(5.66) $$\sigma^2 = \frac{g}{\bar{T}}\left(\frac{d\bar{T}}{dz} + B\right), \quad \text{with} \quad B = \frac{\gamma - 1}{\gamma}\frac{g}{R} ,$$

this quantity B being the so-called dry-adiabatic gradient, equal to a constant since we assume $\gamma = \text{const}\ (B = 9{,}8\,°K/km)$.
In the following it will be convenient, for any parameter $\bar{A}$ of the basic flow, to write $\bar{A}'$ and $\bar{A}''$ for $\frac{d\bar{A}}{dz}$ and $\frac{d^2\bar{A}}{dz^2}$, respectively.
The system of hydrodynamical equations to be used instead of (5.4) is now

(5.67) $$\rho\frac{du}{dt} + \frac{\partial p}{\partial x} = 0; \ \rho\frac{dw}{dt} + \frac{\partial p}{\partial z} + g\rho = 0; \ \rho\left(\frac{\partial u}{\partial x} + \frac{\partial w}{\partial z}\right) + \frac{d\rho}{dt} = 0; \ \frac{d\rho}{dt} = \alpha\frac{dp}{dt} .$$

After replacing $\frac{d\rho}{dt}$ with $\alpha\frac{dp}{dt}$ in the third equation, and using hydrostatic equation, we derive from this system the following complete set of perturbation equations:

(5.68) $$\begin{cases} \bar{\rho}\frac{d^2}{dt^2}\Delta x + \frac{\partial}{\partial x}\delta p = 0; \ \bar{\rho}\frac{d^2}{dt^2}\Delta z + \frac{\partial}{\partial z}\delta p + g\delta p = 0; \\ \bar{\rho}\frac{\partial}{\partial x}\Delta x + \bar{\rho}\left(\frac{\partial}{\partial z} - g\bar{\alpha}\right)\Delta z + \bar{\alpha}\delta p = 0; \ g\delta\rho = \bar{\rho}\sigma^2\Delta z + g\bar{\alpha}\delta p. \end{cases}$$

where $\frac{d}{dt} = \frac{\partial}{\partial t} + \bar{u}\frac{\partial}{\partial x}$. In addition we have

(5.69) $$\delta u = \frac{d}{dt}\Delta x - \bar{u}'\Delta z; \ \delta w = \frac{d}{dt}\Delta z .$$

<u>Stationary perturbations</u>.- For a stationary perturbation we have $\frac{d}{dt} = \bar{u}\frac{\partial}{\partial x}$, and after the easy elimination of Δx and $\delta\rho$ we deduce from (5.68) the following system for Δz and δp:

(5.70) $$\bar{\rho}\left(\bar{u}^2\frac{\partial^2}{\partial x^2} + \sigma^2\right)\Delta z + \left(\frac{\partial}{\partial z} + g\bar{\alpha}\right)\delta p = 0; \bar{\rho}\bar{u}^2\left(\frac{\partial}{\partial z} - g\bar{\alpha}\right)\Delta z = (1-\eta^2)\delta p,$$

where

(5.71) $$\eta = \bar{u}\sqrt{\bar{\alpha}} = \frac{\bar{u}}{\bar{C}_s}$$

is the Mach number of the basic flow, always small in the atmosphere.

Let us now introduce the "virtual density" ρ_v and the "virtual static stability" σ_v, defined by

(5.72) $$\bar{\rho} = (1 - \eta^2)\rho_v \ ; \ \sigma_v^2 = -g\left(\frac{\rho_v'}{\rho_v} + g\bar{\alpha}\right),$$

at the same time as the "modified" disturbances $\Delta z'$ and $\delta w'$ such that

(5.73) $$\Delta z = \sqrt{\frac{\rho}{\rho_v}}\frac{u_1}{\bar{u}}\Delta z'; \ \delta w = \sqrt{\frac{\rho_1}{\rho_v}}\delta w',$$

whence

(5.74) $$\delta w' = u_1\frac{\partial}{\partial x}\Delta z',$$

where ρ_1 is a fixed density, and u_1 is a fixed velocity.

Then the second equation (5.70) gives;

$$\delta p = \rho_v \bar{u}^2 \left(\frac{\partial}{\partial z} - g\bar{\alpha}\right)\Delta z = \left[\rho_v \bar{u}^2 \frac{\partial}{\partial z} - g(\rho_v - \bar{\rho})\right]\Delta z, \tag{5.75}$$

whence, after substituting in the first equation (5.70),

$$\left[(1-\eta^2)\frac{\partial^2}{\partial x^2} + \frac{\partial^2}{\partial z^2} + \frac{(\rho_v \bar{u}^2)'}{\rho_v \bar{u}^2}\frac{\partial}{\partial z} + \frac{\sigma_v^2}{\bar{u}^2}\right]\Delta z = 0, \tag{5.76}$$

and this gives for $\Delta z'$ a partial-differential equation without term in $\frac{\partial}{\partial z}$, namely

$$\left[(1-\eta^2)\frac{\partial^2}{\partial x^2} + \frac{\partial^2}{\partial z^2} + S_v\right]\Delta z' = 0, \tag{5.77}$$

where

$$S_v = \frac{\sigma_v^2}{\bar{u}^2} - \frac{(\sqrt{\rho_v}\,\bar{u})''}{\sqrt{\rho_v}\,\bar{u}}. \tag{5.78}$$

According to (5.74) we have exactly the same equation for $\delta w'$.

The quantity S_v, which may be referred to as the "stability parameter", is in general variable with z, and may be negative as well as positive. In the atmosphere it is usually not very different from the so-called "Scorer's parameter" S, defined by

$$S = \frac{\sigma^2}{\bar{u}^2} - \frac{\bar{u}''}{\bar{u}}, \tag{5.79}$$

where the first term is the contribution of static stability, and the second one that of dynamic stability.

We note that if one uses $x/\sqrt{1-\eta^2}$ instead of x as horizontal coordinate, one gets for $\Delta z'$ an equation where the compressibility coefficient does not figure explicitly, so that the fluid may as well be thought of as an incompressible liquid with the density ρ_v and the static stability σ_v, as far as the space distribution of $\Delta z'$ is concerned. But this is not necessarily true for the distribution of $\delta p'$ or $\Delta x'$, involving the operator $\frac{\partial}{\partial z} - g\bar{\alpha}$

where $g\bar{\alpha}$ may be comparable to $\frac{\partial}{\partial z}$. Therefore it is not correct to say that the only practical effect of air compressibility is a modification of the static stability, even when dealing with nearly stationary perturbations. Since all the coefficients of the perturbation equations (5.68) are independent of x, we may adopt as fundamental solutions the stationary perturbations depending on x by the factor e^{ikx}, where k is a real positive constant. For such a perturbation we have in particular

$$(5.80)\qquad \Delta z' = h(z)\, e^{ikx},$$

where according to (5.77) the function $h(z)$ must satisfy the differential equation

$$(5.81)\qquad h''(z) + \left[S_v - (1-\eta^2)k^2\right] h(z) = 0.$$

This is the equation to be used, together with the boundary condition (5.36) at $z=0$, in order to obtain the forced perturbation due to the sinusoidal ground corrugation defined by (5.33), and once this problem is solved for any real positive value of k the forced perturbation due to any meso-scale local ground corrugation can be obtained by Fourier integrals (technique used in the preceding sections). But since the equation (5.81) is one of the elliptic type at all levels if k is smaller than the maximum of S_v, which in the atmosphere is always positive, the determination of the forced perturbation in question appears as a really difficult problem, except for the particular models that we shall now consider.

Case of a uniform airflow with uniform temperature.- We begin with the simplest case, that of an isothermal uniform flow : $\bar{u} = \text{const}\ (=U)$; $\bar{T} = \text{const}\ (=T)$.
Then we have also $\bar{\alpha} = \text{const}$, $\eta = \text{const}$, and

$$(5.82)\qquad \sigma_v^2 = \sigma^2 = \frac{gB}{T} = \frac{\gamma-1}{\gamma}\frac{g^2}{RT} = \text{const}\,;\ \frac{\rho_v'}{\rho_v} = \frac{\bar{\rho}'}{\bar{\rho}} = \frac{g}{RT} = \text{const}.$$

Therefore we have a uniform static stability together with an exponential upward decrease of $\bar{\rho}$,say

(5.83) $\bar{\rho} = \rho_0 e^{-2sz}$, with $s = \frac{g}{2RT}$

and all the results of the preceding section can be easily generalized. In particular :

a) The system (5.68) of perturbation equations is reduced to one with constant coefficients by taking as new unknowns the "modified disturbances defined by (5.7), and we note that the definitions of $\Delta z'$ and $\delta w'$ agree with (5.73) if we take $\rho_1 = \frac{\rho_0}{1-\eta^2}$, value of ρ_v for $z=0$. Accordingly we can again use as fundamental solutions the particular ones depending on x, z, t by an exponential factor of the form (5.9), consisting of plane waves.

b) The condition for such a wave to be stationary is

(5.84) $(1-\eta^2)(k^2-k_s^2)+m^2=0$,

where k_s^2 is now defined by

(5.85) $(1-\eta^2)k_s^2 = S_v = \frac{\sigma^2}{U^2} - s^2 = \frac{\sigma^2}{U^2}\left(1-\frac{U^2}{C_0^2}\right)$,

and we have

(5.86) $C_0^2 = \frac{4(\gamma-1)}{\gamma^2} C_s^2 = (0,82)C_s^2$, whence $\frac{U^2}{C_0^2} = \frac{\gamma^2}{4(\gamma-1)}\eta^2$,

so that the condition $U^2 \ll C_0^2$ is practically equivalent to $\eta^2 \ll 1$, generally satisfied in the actual atmosphere.

c) If $U < C_0$, the group velocity $\vec{\gamma}_s$ of a stationary wave is again directed leewards, and upwards or downwards according as m is positive or negative (as shown on Fig.23). Therefore the radiation condition requires that one takes $m > 0$.

d) If $\eta^2 \ll 1$, all the formulae given in the preceding section for the group velocity can be used without any change (with $U_0^2 \ll C_0^2$), and the same holds for the streamlines and setting up of the forced perturbations, as shown on fig.24 to 28. More generally these results concerning the forced perturbations can be applied provided that x is replaced with $x/\sqrt{1-\eta^2}$, and in any case the theory for meso-scale perturbations fails at high levels because of the upward increase of the amplitude resulting of the unlimited decrease of air density.

The case of an isothermal uniform air flow is the only one where the system of perturbation equations can be reduced to one with constant coefficients, which is the condition for the existence of plane waves progressing without changing their orientation and at a constant speed with respect to the basic flow. In the other cases one may use as fundamental perturbations those which are defined by some particular solutions of equation (5.81), but their progression involves a diffraction effect (the wave surfaces are curved and changing in shape as they progress), so that the above theory based on the concept of group velocity cannot be easily generalized, and actually such a generalization has not yet been undertaken. For this reason in most of the models used by various authors the basic flow is topped by a uniform isothermal current which is supposed to be unlimited upwards, so that the radiation condition can be applied without difficulty and then takes place of the necessary upper-boundary condition.

Case of a uniform air flow with constant temperature-gradient.- If instead of $\bar{T} = \text{const}$ we assume $\bar{T}' = \text{const}$, this vertical gradient being at most of the order of $B(10°\,K/km)$, the parameters $\bar{\alpha}$, σ_v^2 and $\bar{\rho}'/\bar{\rho}$ are no longer independent of z, but their vertical variations are relatively slow, so that the results obtained for $\bar{T} = \text{const}$ can be applied as a first approximation.

Therefore the fig.25 for instance may be thought of as representing the meso-scale mountain waves in a theoretical troposphere (such that $\bar{\bar{T}}' = \text{const}$, of the order of $6° K/km$) supposed to be in uniform motion and unlimited upwards. Assuming $U < 40\ m/sec$, we may take practically, in this case

$$(5.87) \quad \sigma^2 = \frac{g}{T_m}(\bar{\bar{T}}' + B)\,; \quad k_s = \frac{\sigma}{U}, \quad \text{whence } L_s = \frac{2\pi U}{\sigma},$$

where T_m is the mean value of $\bar{\bar{T}}$ between $z = 0$ and $z = 10\ km$ for instance. Indeed several features shown on fig.25 give a good representation of what is actually often observed when mountain waves are set up in a straight air flow with a weak vertical shear, and the same remark applies to fig. 28 when dealing with a broad mountain range (in particular the mother-of-pearl clouds observed in the lower stratosphere can be explained by a similar pattern).

Models involving a discontinuous stratification of the air flow.- The solutions of the equation (5.81) being very simple if S_v is equal to a constant, it is natural to consider the particular models in which the air flow consists of a superposition of currents having each a constant but different value of S_v : this subdivision of the flow may actually be observed in the atmosphere, and if not, it gives a first approach to the actual stratification.

Models of this sort were first used by SCORER in 1949, then by several other authors, the uppermost current being uniform and isothermal in all models for the reason already given.

In general $\bar{\bar{T}}'$ is taken constant in each current, with the condition that $\bar{\bar{T}}$ must be continuous at the interface levels, and then the vertical distribution of $\bar{u}$ is obtained, if $\bar{u}$ is not also assumed constant, by solving numerically the equation expressing that S is equal to another given constant (all the authors made the approximation $S_v = S$, which is actually quite justified), with the condition that $\bar{u}$, just as $\bar{T}$, must be continuous at all levels. Then the function $h(z)$, giving the amplitude of $\Delta z'$ in the case of a sinusoidal ground corrugation, is obtained in each current by solving the equation (5.81) which for the purpose is simplified by replacing the coefficient of $h(z)$ with $S - k^2$, and since this coefficient is a constant the general solution is a linear combination of two exponential functions, therefore involving two arbitrary constants, except for the uppermost current where only one arbitrary constant is involved (the solution must either vanish for $z = \infty$, or satisfy the radiation condition, according as k^2 is larger or smaller than S). Therefore if the flow is made of n currents, there are $2n - 1$ arbitrary constants, and also $2n - 1$ boundary conditions to be satisfied (one at ground level, and two at each interface, namely that Δz and δp must be continuous), so that all the constants can be expressed in terms of A , the amplitude of the ground corrugation. If the values obtained are finite for all values of k, the procedure gives a well-defined forced perturbation for any local ground corrugation, which may include a system of lee waves such as shown on fig. 25, characterized by a leeward damping along each streamline. But it may also happen that the computations give an infinite amplitude for the forced perturbation corresponding to one or several values of k , in which case we are dealing with a resonance effect (the air flow allows for free perturbations depending on x by the factor e^{ikx} , for each one of these resonance wave numbers), and then the forced perturbation has to be computed by the same method as used in chapter 3, leading to resonance waves of the category of surface waves, superposed upon a perturbation of the type given of fig. 25 or 26 : in this case the deformation of the streamlines does not vanish at large distance downstream, it tends towards a sinusoidal deformation, or a superposition of sinusoidal deformations.

Now it is important to note that most of these resonance waves are the result of the introduction of discontinuities in the vertical variation of S_v , which in most of the models is a purely mathematical artifice, used for the only purpose of making the computations more feasible. With modern computers at hand, it would be better to only use models with a continuous S_v-distribution, except for the discontinuities possessing a real physical meaning. Actually the use of a model with artificial discontinuities cannot give better than a rather crude approach to the structure of mountain perturbation at large distance downstream, and it is unfortunate that most of the authors have been mainly interested in the determination of the resonance waves corresponding to the models they have used.

6. ADIABATIC SYNOPTIC-SCALE PERTURBATIONS IN A STRAIGHT UNIFORM ATMOSPHERIC FLOW.

In this chapter we proceed to another generalization of the models studied in the first part of the foregoing one : the fluid is again supposed to be incompressible with a uniform static stability, either at rest or in uniform motion, but the earth's rotation is taken into account by the introduction of geostrophic force, this generalization being necessary for an application to synoptic-scale orographic perturbations. However we assume that the earth's curvature may be neglected, so that the three components of the earth's angular velocity may be considered as constant parameters, thanks to what we shall obtain a system of perturbation equations with constant coefficients and be able to use the same method as in the preceeding chapters for the determination of forced perturbations.

Because of the deflecting effect of the geostrophic force the perturbations cannot be two-dimensional in the xz-plane, but we may assume that the fluid displacement is independent of y $\left(\frac{\partial}{\partial y}\Delta\vec{M}=0\right)$ in order that the theory can be used in the case of a cylindrical ground corrugation parallel to the y-axis.

Case of an incompressible fluid at rest. - We first assume that the fluid is at rest in the basic state, with $g=$ const and an exponential upward decrease of the density given by(5.2), so that σ^2 has the constant value $2gs$.

If $\vec{\Omega}_t$ is the vector representating the earth's angular velocity, and Ω_t its magnitude, the three components of $2\vec{\Omega}_t$ are (we assume that the frame of reference is a direct cartesien one) :

$$(6.1)\quad f_x = 2\Omega_t \cos\varphi\cos\theta\,;\; f_y = 2\Omega_t\cos\varphi\sin\theta;\; f = 2\Omega_t\sin\varphi,$$

where φ is latitude, θ is the polar angle of the east-direction in the xy-plane, and f is the so-called Coriolis parameter, positive in the northern hemisphere and negative in the southern hemisphere. These three components are of the order of $10^{-4}\,\text{sec}^{-1}$ at most, and we assume that they can be considered as three constants. In addition we assume for σ the order of magnitude usually observed in the lower atmosphere, namely $\sigma \sim 10^{-2}\,\text{sec}^{-1}$. Then the hydrodynamical equations to be used are :

$$(6.2)\quad \begin{cases} \rho\dfrac{du}{dt} - f\rho v + f_y\rho w + \dfrac{\partial p}{\partial x} = 0;\ \rho\dfrac{dv}{dt} + f\rho u - f_x\rho w + \dfrac{\partial p}{\partial y} = 0; \\ \rho\dfrac{dw}{dt} - f_y\rho u + f_x\rho v + \dfrac{\partial p}{\partial z} + g\rho = 0;\ \dfrac{\partial u}{\partial x} + \dfrac{\partial v}{\partial y} + \dfrac{\partial w}{\partial z} = 0;\ \dfrac{d\rho}{dt} = 0, \end{cases}$$

and from them we can deduce the following system of perturbation equations with constant coefficients, generalizing (5.8) and applicable to the perturbations independent of y :

$$(6.3)\quad \begin{cases} \dfrac{\partial^2}{\partial t^2}\Delta x' - f\dfrac{\partial}{\partial t}\Delta y' + f_y\dfrac{\partial}{\partial t}\Delta z' + \dfrac{\partial}{\partial x}\dfrac{\delta p'}{\rho_0} = 0; \\ \dfrac{\partial^2}{\partial t^2}\Delta y' + f\dfrac{\partial}{\partial t}\Delta x' - f_x\dfrac{\partial}{\partial t}\Delta z' = 0; \\ \dfrac{\partial}{\partial t}\left[\left(\dfrac{\partial^2}{\partial t^2} + \sigma^2\right)\Delta z' - f_y\dfrac{\partial}{\partial t}\Delta x' + f_x\dfrac{\partial}{\partial t}\Delta y' + \left(\dfrac{\partial}{\partial z} - s\right)\dfrac{\delta p'}{\rho_0}\right] = 0; \\ \dfrac{\partial}{\partial t}\left[\dfrac{\partial}{\partial x}\Delta x' + \left(\dfrac{\partial}{\partial z} + s\right)\Delta z'\right] = 0;\ \delta u' = \dfrac{\partial}{\partial t}\Delta x';\ \delta v' = \dfrac{\partial}{\partial t}\Delta y';\ \delta w' = \dfrac{\partial}{\partial t}\Delta z', \end{cases}$$

where the unknowns are the "modified" disturbances defined by the formulae (5.7) completed with

$$(6.4)\quad \delta v = e^{sz}\,\delta v';\quad \Delta y = e^{sz}\,\Delta y'.$$

Therefore we have again as particular solutions the "fundamental waves" depending on x, z, t by a factor of the form (5.9), and if we disregard the stationary perturbations we get from (6.3) the following dispersion equation for these waves :

(6.5) $$\left(k^2+m^2+s^2\right)\omega^2+\left(2f_y sk\right)\omega-\left[\left(\sigma^2+f_x^2\right)k^2+f^2\left(m^2+s^2\right)-2ff_x km\right]=0.$$

But in virtue of the order of magnitude adopted for σ we have $f_x^2 \ll \sigma^2$ and $f_y \ll \sigma$, whence it follows (it is easy to prove it) that all the terms containing f_x or f_y may be omitted in this equation, in conformity with the well-known statement that the only vertical component of $\vec{\Omega}_t$ has to be retained in the dynamics of the lower atmosphere. We shall accordingly adopt the simplified equation

(6.6) $$\left(k^2+m^2+s^2\right)\omega^2-\left[\sigma^2 k^2+f^2\left(m^2+s^2\right)\right]=0,$$

giving two opposite real values for ω when $\vec{K}$ is given, namely $\omega = \pm\Omega$, where

(6.7) $$\Omega = \frac{1}{K'}\sqrt{\sigma^2 k^2+f^2\left(m^2+s^2\right)}\,,\ \text{with}\ K'=\sqrt{K^2+s^2}=\sqrt{k^2+m^2+s^2},$$

and we have again a family of positive waves $(\omega=\Omega)$ and a family of negative waves $(\omega=-\Omega)$.

From (6.7) we deduce that $f \leqslant \Omega \leqslant \sigma$, the lower limit $\Omega = f$ being reached for $k=0$ (horizontal waves), and the upper limit $\Omega=\sigma$ for $m=0$ with $K \gg s$ (vertical short waves). The fluid vibration is in general elliptic and anticyclonic, it is circular if $k=0$, and rectilinear in the xz-plane if $m=0$. For $K \gg s$ (short waves) the vibration is transversal.

For the positive waves the components of the group velocity $\vec{G}$ are given by

(6.8) $$G_x=\frac{\partial\Omega}{\partial k}=\frac{\sigma^2-f^2}{\Omega K'^4}\,k\left(m^2+s^2\right);\ G_z=\frac{\partial\Omega}{\partial m}=-\frac{\sigma^2-f^2}{\Omega K'^4}\,k^2 m.$$

Therefore we have the same result as for $f=0$: G_x is always positive, and the sign of G_z is opposite to that of m. In virtue of the fact that $f \ll \sigma$ the maximum of G (magnitude of $\vec{G}$) is also practically the same as for $f=0$, it is very close to $C_0 = \frac{\sigma}{s}$, and reached with the vertical waves ($m=0$) such that K is practically equal to $3^{-\frac{1}{4}} s\sqrt{\frac{f}{\sigma}}$, corresponding to a wave length of 1500 to 2000 km. Accordingly if $\vec{G}$ is given the point O as origin, the boundary of its domain has roughly the same shape as the curve (G) of fig.21 or 22, except that the apex O' on Ox is no longer a double point. Another difference is that the dispersion waves resulting from a local deformation of the fluid near the origin, and which have the same boundary completed by its symmetrical with respect to O, are now consisting of four families instead of two, so that the pattern of these dispersion waves is still more intricate than the one shown on fig.22. It is also to be noted that now the phase velocity C is infinite for $K=0$, so that the domain of the vector $\vec{C}$ is unbounded.

Case of a uniform flow.- We now assume that in the basic state the fluid is moving in the x-direction with the uniform velocity U, of the order of 100 m/s at most, We have first to verify that, just as in the preceding models involving a uniform basic motion, the perturbations when referred to a frame moving with the flow can be practically identified with those obtained in the case of a fluid at rest, if it is specified that the fluid displacement is independent of y.
For the basic motion the system (6.2) reduces to the following two equations :

$$(6.9)\quad \bar{\rho}(fU)+\frac{\partial\bar{p}}{\partial y}=0;\quad \bar{\rho}(g-f_y U)+\frac{\partial\bar{p}}{\partial z}=0,$$

whence we deduce

$$(6.10)\quad \frac{\partial\bar{p}}{\partial z}-\varepsilon\frac{\partial\bar{p}}{\partial y}=0\ ;\ \frac{\partial\bar{\rho}}{\partial z}-\varepsilon\frac{\partial\bar{\rho}}{\partial y}=0,\ \text{with } \varepsilon=\frac{fU}{g-f_y U}\sim 10^{-3}\ \text{at most},$$

and this result requires that the isosteric surfaces are the same as the isobaric ones (a classical result for a uniform wind), consisting of parallel planes with the small uniform slope ε. Therefore if we assume that the vertical distribution of $\bar{\rho}$ is unchanged, we have now

$$(6.11)\quad \bar{\rho} = \rho_0 \, e^{-2s(z+\varepsilon y)},$$

and in formulae (5.7) and (6.4) the factor $e^{\pm sz}$ must be changed into $e^{\pm s(z+\varepsilon y)}$. Accordingly, in order to have $\frac{\partial}{\partial y}=0$ when this operator applies to the fluid displacement, we must take $\frac{\partial}{\partial y}=-\varepsilon s$ when the same operator applies to any "modified" disturbance. Taking account of this condition at the same time as of the modifications in the basic state, we thus get the following system of perturbation equations with constant coefficients to be used instead of (6.3), where $\frac{d}{dt}=\frac{\partial}{\partial t}+U\frac{\partial}{\partial x}$, and $\sigma'^2=\sigma^2\left(1-\frac{f_y U}{g}\right)$:

$$(6.12)\quad \begin{cases} \frac{d^2}{dt^2}\Delta x' - f\frac{d}{dt}\Delta y' + f_y\frac{d}{dt}\Delta z' + \frac{\partial}{\partial x}\frac{\delta p'}{\rho_0} = 0 \,; \\ \left(\frac{d^2}{dt^2}+\varepsilon^2\sigma'^2\right)\Delta y' + f\frac{d}{dt}\Delta x' + \left(\varepsilon\sigma'^2 - f\frac{d}{dt}\right)\Delta z' - 2\varepsilon s\frac{\delta p'}{\rho_0} = 0 \,; \\ \frac{d}{dt}\left[\left(\frac{d^2}{dt^2}+\sigma'^2\right)\Delta z' + f_y\frac{d}{dt}\Delta x' + \left(\varepsilon\sigma'^2 + f_x\frac{d}{dt}\right)\Delta y' + \left(\frac{\partial}{\partial z}-s\right)\frac{\delta p'}{\rho_0}\right] = 0 \,; \\ \frac{d}{dt}\left[\frac{\partial}{\partial x}\Delta x' + \left(\frac{\partial}{\partial z}+s\right)\Delta z'\right] = 0 \,;\ \delta u' = \frac{d}{dt}\Delta x' \,;\ \delta v' = \frac{d}{dt}\Delta y' \,; \delta w' = \frac{d}{dt}\Delta z'. \end{cases}$$

If the motion is referred to a frame moving with the basic flow we have to take $\frac{d}{dt}=\frac{\partial}{\partial t}$, and this gives for the dispersion equation of the fundamental waves, after omitting the terms in f_x or f_y which we know to be negligible :

$$(6.13)\quad \left(k^2+m^2+s^2\right)\omega^2 - \left(2\varepsilon f s k\right)\omega - \left[(1+\varepsilon^2)\sigma^2 k^2 + \varepsilon^2\sigma^2(m-is)^2 + f^2(m^2+s^2)\right] = 0$$

But, since $\varepsilon \sim 10^{-3}$ at most (this value corresponding to $U \sim 100\,m/s$) it is easy to prove that the omission of the terms containing ε or ε^2 results in a quite negligible error, much smaller indeed than the error resulting of the omission of the terms in f_x or f_y, so that the simplified equation (6.6) may again be adopted, and for the same reason we may also use the formulae (5.7) and (6.4) unmodified.

If the frame of reference is no longer moving with the basic flow, the stationary fundamental waves (negative waves such that $\Omega = kU$) are obtained by putting $\omega^2 = k^2 U^2$ in (6.6), and this gives a condition leading to the formulae

(6.14) $$K'^2 = \frac{k^2(\sigma^2 - f^2)}{k^2U^2 - f^2} \; ; \quad m^2 + s^2 = \frac{k^2(\sigma^2 - k^2U^2)}{k^2U^2 - f^2} \, ,$$

whence, practically

(6.15) $$m^2 = \frac{k^2(k_s^2 - k^2)}{k^2 - k_f^2} \, ,$$

with

(6.16) $$k_s = \sqrt{\frac{\sigma^2}{U^2} - s^2} \; ; \quad k_f = \frac{|f|}{U} \, .$$

Therefore k_s is the same parameter as already introduced in chapter 5, and since U^2 is much smaller than $C_0'^2 = \sigma^2/s^2$ we have $k_s \sim \frac{\sigma}{U}$, whence

(617) $$k_f < k_s \; ; \quad \frac{k_f}{k_s} \sim 10^{-2} \text{ at most} \, .$$

Obviously there are always some stationary waves, for which we have necessarily $k_f \leqslant k \leqslant k_s$ whereas m may take any positive or negative value (m^2 increases from 0 to $+\infty$ as k decreases from k_s to k_f), and for such a stationary wave the components of the group velocity are found to be given by

(6.18) $$\gamma_{sx} = U \frac{k^2(k^2 - k_f^2) + k_f^2(k_s^2 + s^2 - k^2)}{k^2(k_s^2 + s^2 - k_f^2)} \; ; \quad \gamma_{sz} = U \frac{m(k^2 - k_f^2)^2}{k^2(k_s^2 + s^2 - k_f^2)} \, .$$

Therefore γ_{sx} is always positive, and γ_{sz} has the same sign as m , just as without geostrophic force (as shown on fig. 23), whence it follows that the radiation condition is again satisfied by the only waves such that m is positive. If γ_s is the magnitude of $\vec{\gamma_s}$, we have also $\gamma_s \leqslant U$, this maximum being reached for $k = k_f$ (horizontal waves with $m^2 = +\infty$). If $k \ll k_s$ (long waves) we have practically

(6.19) $m^2 = \dfrac{k_s^2 k^2}{k^2 - k_f^2}$; $\gamma_{sx} = U \dfrac{k_f^2}{k^2}$; $\gamma_{sz} = U \dfrac{k^2 - k_f^2}{km}$, for $k \ll k_s$.

Mountain perturbations. - By using the above results the theory of the forced waves given in chapter 5 is easily generalized, as follows :

a) In the case of the sinusoidal corrugation defined by (5.33), we must have for $\Delta z'$ an expression of the form (5.34), where the function $h(z)$ must satisfy the differential equation

$$h''(z) + \frac{k^2(k_s^2 - k^2)}{k^2 - k_f^2} h(z) = 0, \tag{6.20}$$

deduced from (6.6) by the substitutions (5.36). This equation being hyperbolic if $k < k_f$ or $k > k_s$ and elliptic if $k_f < k < k_s$, the solution to be adopted for the forced perturbation is

$$\begin{cases} \Delta z' = A e^{ikx - \mu z}, & \text{if } k < k_f \text{ or } k > k_s ; \\ \Delta z' = A e^{i(kx + mz)}, & \text{if } k_f < k < k_s , \end{cases} \tag{6.21}$$

where

$$m = k\sqrt{\frac{k_s^2 - k^2}{k^2 - k_f^2}} \; ; \quad \mu = k\sqrt{\frac{k^2 - k_s^2}{k^2 - k_f^2}} . \tag{6.22}$$

In the first case the modified streamlines have the aspect shown on the left part of fig.24, characterized by an upward damping.

In the second case the forced perturbation is a stationary fundamental wave satisfying the radiation condition, and the modified streamlines have the aspect shown on the right part of fig.24.

b) In the case of a "local" ground corrugation independent of y, the profile of which is defined by the Fourier integral (5.46), the forced perturbation is obtained by superposing the contributions of the various sinusoidal elements of the integral, and this gives

$$(6.23)\quad \Delta z' = Re\left[\int_0^{k_f} f(k)e^{ikx-\mu z}dk + \int_{k_f}^{k_s} f(k)e^{i(kx+mz)}dk + \int_{k_s}^{\infty} f(k)e^{ikx-\mu z}dk\right]$$

But, thanks to the fact that $k_f \ll k_s$, this formula can be largely simplified in any case, the simplification depending on the order of magnitude of the width b of the corrugation as compared to the characteristic wave lengths

$$(6.24)\quad L_s = \frac{2\pi}{k_s} \sim \frac{2\pi U}{\sigma}\ ;\ L_f = \frac{2\pi}{k_f} = \frac{2\pi U}{|f|},$$

which in the average case $U \sim 15\,m/s$ are of the order of 10 Km and 1000 Km respectively if $f \sim 10^{-4}\,sec^{-1}$, and in any case we have $(L_f/L_s) \sim 100$ at least.

The three cases $b \ll L_s$, $b \sim L_s$ and $L_s \ll b \ll L_f$ have already been considered in chapter 5 : the first integral in (6.23) is negligible, and in the other ones k_f may be replaced with 0, so that we get the same formula (5.47) as without geostrophic force. These three cases are corresponding respectively to a "narrow barrier" (fig.27), a normal meso-scale barrier (fig.25), and a "broad barrier" (fig.28)

Here we only have to consider the remaining case of a synoptic-scale orographic perturbation, such that $b \sim L_f$, or eventually $b \gg L_f$. The main contribution being given by the sinusoidal elements with a wave number comparable to k_f, the third integral in (6.23) is now negligible, and in the other ones we may neglect k^2 beside k_s^2, and replace the upper limit k_s with $+\infty$. We thus get the practically correct formula

(6.25) $$\Delta\zeta' = Re\int_0^{k_f} f(k)e^{ikx - \frac{k_s k}{\sqrt{k_f^2-k^2}}z} + Re\int_{k_f}^{\infty} f(k)e^{i(kx+\frac{k_s k}{\sqrt{k^2-k_f^2}}z)}\,dk ,$$

and similar formulae could be written for the other modified disturbances such as $\Delta y'$, $\delta p'$ etc...
At large distance the main contribution to the asymptotic value of $\Delta\zeta'$ is that of the second integral, and it can be obtained by the stationary-phase method. For this purpose it is convenient to introduce the modified cartesian coordinates X, Z and the corresponding polar coordinates R, ψ defined by

(6.26) $$k_f x = X = R\cos\psi;\ k_s z = Z = R\sin\psi;\ 0 \leqslant \psi \leqslant \pi ,$$

and to use as new variable of integration the angle τ defined by

(6.27) $$k_f = k\cos\tau;\ \sqrt{k^2-k_f^2} = k\sin\tau;\ 0 \leqslant \tau \leqslant \frac{\pi}{2} .$$

This gives, for the second integral,

(6.28) $$\Delta\zeta_2' = Re\int_0^{\frac{\pi}{2}} \frac{\sin\tau}{\cos^2\tau} f\left(\frac{k_f}{\cos\tau}\right) e^{iR\left(\frac{\cos\psi}{\cos\tau}+\frac{\sin\psi}{\sin\tau}\right)}\,d\tau ,$$

and we must consider the phase function

(6.29) $$\varphi(\tau) = R\left(\frac{\cos\psi}{\cos\tau} + \frac{\sin\psi}{\sin\tau}\right) .$$

This phase can be stationary only if $X > 0$, and in this case the stationary value τ_0 is given by

(6.30) $$\tan\tau_0 = (\tan\psi)^{\frac{1}{3}} ,$$

Then we have

(6.31) $$\varphi(\tau_0) = R\frac{\sin\psi}{\sin^3\tau_0} = R\frac{\cos\psi}{\cos^3\tau_0};\ \varphi''(\tau_0) = 3\varphi(\tau_0) .$$

So that the stationary-phase formula gives

$$(6.32)\quad \Delta z' \approx Re\left[k_f (\tan^2 \tau_0) f\left(\frac{k_f}{\cos \tau_0}\right) \sqrt{\frac{2\pi}{3R}\,\frac{\sin \tau_0}{\sin \psi}}\; e^{i\left(R\frac{\sin \psi}{\sin^3 \tau_0}+\frac{\pi}{4}\right)}\right] \text{for } R \gg 1,$$

and this formula is valid for $0 < \psi < \frac{\pi}{2}$ with ψ not close to 0 or $\frac{\pi}{2}$. This gives a system of lee waves, shown on fig.29 in the case of a Lorentz-profile corrugation, the coordinates used being proportional to X and Z. The constant-phase curves are given by the equation

$$(6.33)\quad X^{\frac{2}{3}} + Z^{\frac{2}{3}} = \text{const},$$

and we have again this important result that at any point M in the lee-waves domain, the group velocity $\vec{\gamma}_s$ of the waves is along OM (it is easy to prove that this property belongs to any asymptotic system of barrier waves produced in a straight flow the perturbations of which can be obtained by the superposition of fundamental plane waves depending on x, z, t by a factor of the form (5.9), whatever the dispersion equation of these waves may be). Therefore the energy supplied by the basic flow to the waves is again propagating radially from the mountain in all leeward directions, with the speed γ_s reaching its maximum value U in the horizontal direction (its minimum value, of the order of $U\sqrt{\frac{k_f}{k_s}} \sim \frac{U}{10}$, is in the vicinity of the vertical direction). If one starts from a local perturbation near the mountain range, the front of the lee-waves domain after the time t is practically the hodograph of the vector $\vec{\gamma}_s t$ relative to the origin, shown on fig.30 and given by the equation

$$(6.34)\quad XZ^2 = (k_f Ut - X)^3,$$

deduced from formulae (6.19) which are actually valid if the directions close to Ox or Oz are excluded.

It is important to note that in the vertical direction the forced perturbation is again periodical with the wave length L_s , but now the wave length of the lee waves tends towards L_f as the direction approaches the horizontal, and accordingly the amplitude of these waves has a significant value only if $b \sim L_f$, a result which can be easily verified, in the case of Lorentz-profile mountain, by using formula (6.32). If $b \gg L_f$ the second integral in (6.25) is negligible, and in the first one we may replace the upper limit with $+\infty$ and neglect k^2 beside k_f^2 , so that we get an integral of the same type as (5.59), namely

$$(6.35) \quad \Delta z' = Re \int_0^\infty f(k) e^{(ix - \frac{k_s}{k_f} z)k} \, dk ,$$

giving for the streamlines a pattern such as shown on fig.27.

Finally it is interesting to consider the horizontal deformation of the streamlines and horizontal isobars, represented on fig.29 (lower part) at the ground level. In agreement with the typical pattern of fig.2, based on observational evidence, there is an anticyclonic deformation just above the mountain range, followed by lee waves the amplitude of which, in the case of the streamlines, is much larger than that of the vertical deformation. This anticyclonic deformation is associated with a positive δp, this excess of pressure being explainable as a consequence of the upward displacement of the lower air layers crossing the mountain, making this part of the flow colder than the surrounding air at each level (because of the static stability of the air flow) ; however it is necessary to take account of the upward decrease of the δp-amplitude in order to make this explanation acceptable, since the downward deformation of the air flow at various levels tends to cancel the excess of pressure in question, actually found at ground level.

7. COMPUTATION OF THE DISSIPATION OF MECHANICAL ENERGY RESULTING FROM A MOUNTAIN PERTURBATION.

In the case of the model just considered, including all scales of mountain perturbations, we have found that, except for a very narrow or very broad mountain range ($b \ll L_s$ or $b \gg L_f$), the forced perturbation shows a wavelike structure corresponding to the intervention of the second integral in (6.23) : its vertical extent has no upper limit as far as the infinitesimal assumption is valid, the periodicity along the vertical above the mountain being characterized by the wave length L_s, and this requires a constant upward transfer of mechanical energy supplied by the basic flow. Beyond some level the waves necessarily change into a system of eddies (as a consequence of the upward decrease of air density requiring an increasing amplitude of the fluid displacement) where this energy is dissipated, whereas in the lower part of the flow the forced perturbation is likely to remain practically stationary. In the actual atmosphere the mechanical energy of large-scale motions is probably dissipated in the same way by mountain perturbations, and therefore it is important to compute the dissipation rate in the case of the model in question in spite of its rather unrealistic character. We shall see that this computation requires the only knowledge of the structure of the perturbation at ground level, so that it is relatively easy.

If the motion is referred to a frame moving with the uniform basic flow, the corrugated bed appears as moving with the opposite velocity, but since nothing is changed in the structure of the perturbation, the energy dissipation is the same, the only difference being that the energy is now supplied by the moving ground surface (and not by the fluid which is at rest) in the form of the work done by this surface in order to maintain its motion. Actually we have the same situation as in the case of a boat moving uniformly on the sea : there is also a system of forced waves moving solidly with the boat, and the energy dissipated by these waves is equal to the work done by the boat during the same time if the viscous forces acting on it may be neglected. We are thus led to compute the work W done par unit time by the ground surface supposed to move with the horizontal velocity $u = -U$,and to adopt the result as giving the mechanical energy D dissipated per unit time by the mountain perturbation.

It is to be noted that, in order to have the same basic velocity at large distance downstream as at large distance upstream, as we have assumed it, we must also assume that some internal factors are supplying to the flow an amount of energy equal to D , and which in the actual atmosphere are nothing but those which are producing the winds. The fact that these factors had not to be considered in the theory of the forced perturbation is easy to understand, it is because we have assumed an infinitesimal amplitude for the perturbation ; then the energy $W = D$ is also infinitesimal but of the second order, whence the same follows for the factors supplying this energy, and accordingly they cannot appear in the theory where all the second-order quantities have been omitted. Actually we may as well assume that the air flow is not quite re-established after crossing the mountain range, as it is probably often the case in the atmosphere, but the equality $W = D$ is likely to remain valid. In the following we shall be interested in the relative depletion $r = \frac{W}{P}$,where P denotes the total kinetic energy carried by the basic flow, per unit time, through any vertical plane perpendicular to it.

In the case of the model used in the preceding chapter, since the fluid as well as the ground corrugation are unlimited in the y-direction the values of W and P are actually infinite, and therefore what we have to do is to consider the portion of the model between two planes parallel to the xz -plane, which we may take at a unit-distance interval. Then P is a finite quantity, and r is an infinitesimal one of the second order with respect to the height of the mountain, just as W. For P we have

$$P = U\int_0^\infty \tfrac{1}{2}\bar{\rho}\, U^2 dz = \tfrac{1}{2}\rho_0 U^3 \int_0^\infty e^{-2sz} dz = \frac{\rho_0 U^3}{4s}. \tag{7.1}$$

On the other hand since there is no viscosity in the lower part of the flow, the force exerted on the fluid by each element of the ground surface is simply the upward pressure force, so that it is easy to compute its work when this surface is moving in the x-direction with the velocity $u = -U$, and then W is obtained by an integration along the surface. Assuming that the equation of this surface is $z = F(x)$, and noting that it is also a fluid surface, we thus get, since $F(\pm\infty) = 0$,

$$W = U\int_{-\infty}^{\infty} (\bar{p}_0 + \Delta p_0)\, F'(x)\, dx = U\int_{-\infty}^{\infty} (\Delta p_0)\, F'(x)\, dx, \tag{7.2}$$

the subscript 0 being used, in general, for a value at $z = 0$.

Now we have, without any approximation,

$$\Delta p = \delta p + \frac{\partial \bar{p}}{\partial y}\Delta y + \frac{\partial \bar{p}}{\partial z}\Delta z = \delta p - \bar{\rho}(fU)\Delta y - \bar{\rho}(g - f_y U)\Delta z, \tag{7.3}$$

and by using only the first and the third one of equations (6.2), we get

$$\delta p = \bar{\rho} U^2 \frac{\partial}{\partial z}\Delta z - \bar{\rho}(fU)\Delta y - \bar{\rho}(f_y U)\Delta z, \tag{7.4}$$

whence

$$\Delta p = \bar{\rho} U^2\left(\frac{\partial}{\partial z} - \frac{g}{U^2}\right)\Delta z = \bar{\rho} U^2\left(\frac{\partial}{\partial z} - \frac{g}{U^2} + s\right)\Delta z'. \tag{7.5}$$

Therefore, since $\Delta z'_0 = F(x)$,

$$\Delta p_0 = \rho_0 U^2\left(\frac{\partial}{\partial z}\Delta z'\right)_0 - \rho_0 U^2\left(\frac{g}{U^2} - s\right)F(x), \tag{7.6}$$

whence, according to (7.1) and (7.2),

$$(7.7)\quad r = \frac{W}{P} = 4s\int_{-\infty}^{\infty}\left(\frac{\partial}{\partial z}\Delta z'\right)_0 F'(x)\,dx - 4s\left(\frac{g}{U^2}-s\right)\int_{-\infty}^{\infty} F(x)F'(x)\,dx .$$

But the second integral is equal to zero, and if we transform the first one by a partial integration we thus get the following formula :

$$(7.8)\quad r = 4s\int_{-\infty}^{\infty}\left(-\frac{\partial^2\Delta z'}{\partial x\,\partial z}\right)_0 F(x)\,dx .$$

Now by using the expression (6.23) obtained for $\Delta z'$ we get

$$(7.9)\quad \left(-\frac{\partial^2\Delta z'}{\partial x\,\partial z}\right)_0 = \mathrm{Re}\left[\int_0^{k_f} ik\mu f(k)e^{ikx}dk + \int_{k_f}^{k_s} km f(k)e^{ikx}dk + \int_{k_s}^{\infty} ik\mu f(k)e^{ikx}dk\right],$$

and we know that the second integral, representing the contribution of the stationary fundamental waves, is for this reason the only one which can contribute to the value of r . Actually the contribution of the first integral for instance is

$$(7.10)\quad r_1 = 4s\int_{-\infty}^{\infty}\left[\mathrm{Re}\int_0^{k_f} ik\mu f(k)e^{ikx}dk\right]F(x)\,dx ,$$

or, after reversing the order of the integrations,

$$(7.11)\quad r_1 = 4s\ \mathrm{Re}\int_0^{k_f} ik\mu f(k)\left[\int_{-\infty}^{\infty}F(x)e^{ikx}dx\right]dk .$$

But we know by Fourier theorem that the integral between brackets is equal to $\pi f^*(k)$, where $f^*(k)$ is the imaginary conjugate of $f(k)$, whence

$$(7.12)\quad r_1 = 4\pi s\ \mathrm{Re}\int_0^{k_f} ik\mu|f(k)|^2 dk = 0,$$

as expected. The same computation gives $r_3 = 0$ for the contribution r_3 of the third integral, so that only remains the contribution r_2 of the second integral, and finally we get, again by the same computation,

$$(7.13)\quad r = 4\pi s\int_{k_f}^{k_s} k\, m\, |f(k)|^2\, dk = 4\pi s\int_{k_f}^{k_s} k^2\sqrt{\frac{k_s^2-k^2}{k^2-k_f^2}}\,|f(k)|^2\, dk .$$

This result requires the same condition as required for the validity of formula (6.15), namely that $U < C_0 \left(= \frac{\sigma}{s}\right)$ with U not close to C_0, and therefore it may be preferable to use the following more general formula, obtained by taking for m the value given by (6.14) instead of (6.15) :

$$(7.14)\quad r = 4\pi s\int_{k_f}^{k_s'} k\sqrt{k^2+\frac{s^2 k_f^2}{k_s'^2}}\;\sqrt{\frac{k_s'^2-k^2}{k^2-k_f^2}}\;|f(k)|^2\, dk,$$

where

$$(7.15)\quad k_s' = \sqrt{\frac{1}{2}\left(k_s^2+\sqrt{k_s^4+4s^2k_f^2}\right)} \quad .$$

For a numerical computation this formula is not practicable because of the existence of the pole k_f, but this difficulty is easily removed by taking $\theta = \sqrt{k^2-k_f^2}/k_s'$ as new variable of integration, and after neglecting k_f^2 beside $k_s'^2$ we thus get

$$(7.16)\quad r = 4\pi s\, k_s'^2\int_0^1 \sqrt{k_s'^2\theta^2+k_f^2\left(1+\frac{s^2}{k_s'^2}\right)}\,\sqrt{1-\theta^2}\,\left|f\left(\sqrt{k_s'^2\theta^2+k_f^2}\right)\right|^2 d\theta .$$

Let us now apply these general results to the typical Gauss-profile mountain range defined by

$$(7.17)\quad F(x) = h\,e^{-\frac{8x^2}{b^2}}, \quad \text{whence} \quad f(k) = \frac{hb}{\sqrt{8\pi}}\,e^{-\frac{b^2}{32}k^2},$$

where h and b represent respectively the height and the practical width of this mountain range.

If we assume $U^2 \ll C_0^2$, so that (7.13) is valid, we deduce from this formula

$$(7.18)\quad r = \frac{s}{2}\,h^2 b^2\int_{k_f}^{k_s} k^2\sqrt{\frac{k_s^2-k^2}{k^2-k_f^2}}\;e^{-\frac{b^2}{16}k^2}\, dk ,$$

whence, by putting $\theta = \sqrt{k^2-k_f^2}/k_s$ and then replacing

k_s and k_f with σ/U and f/U respectively,

(7.19) $r = s\frac{h^2}{b}R(\lambda)$, with $\lambda = \frac{\sigma}{4}\frac{b}{U} = \frac{\pi}{2\varepsilon}\frac{b}{L_f}$ $\quad\left(L_f = \frac{2\pi U}{f}\right)$,

the function $R(\lambda)$ being defined by

(7.20) $R(\lambda) = 32\,\lambda^3\int_0^1\sqrt{(\theta^2+\varepsilon^2)(1-\theta^2)}\,e^{-\lambda^2(\theta^2+\varepsilon^2)}\,d\theta$, with $\varepsilon\,\frac{f}{\sigma}$.

Therefore if σ, f and s are given, the graphs of r versus U for various values of b and h are all reducible to the unique graph of this function if we only consider the portions of these graphs such that $U^2 \ll C_0^2$. Such graphs, computed by formula (7.16), are represented on fig.31 with a logarithmic U-scale, the values adopted for σ, f and s being those reported on the same figure ($\sigma = 10^{-2}\,sec^{-1}$; $f = 10^{-4}\,sec^{-1}$; $s = 10^{-6}\,cm^{-1}$; whence $\varepsilon = 10^{-2}$ and $C_0 \doteq 100\,m/s$) and it is quite apparent that, except for the small portions where U is comparable with C_0, all these curves can actually be deduced from any one of them by a translation parallel to the U-axis, combined with a vertical expansion or contraction (the relative variation of r is a function of the only ratio U/b), the salient feature being a sharp maximum preceded by a rapid increase of r. Of course since the computation is based on the infinitesimal assumption, the value obtained for r is acceptable as a first approximation only if it is small enough, and for this reason the portions of the curves beyond $r = 0{,}8$ are drawn in dotted line (it is physically impossible to have $r > 1$). According to (7.19) the dimensionless quantity λ may be used as the scale parameter of the mountain perturbation (it is a function of the ratio b/L_f, or b/L_s), and we may adopt for the scale ranges the limits shown on the following table (for the present case where $\varepsilon = 10^{-2}$ and $\sigma = 10^{-2}\,sec^{-1}$), also giving some corresponding values of λ, U/b and $R(\lambda)$, those relative to the maximum of r being underlined :

	synoptic-scale range: synoptic-scale lee-waves (fig.29)				meso-scale range: no lee-waves (fig.28)			meso-scale range: meso-scale lee-waves (fig.25)				
λ →	500	200	<u>91.80</u>	50	20	10	5	2	1	0.5	0.2	0.1
U/ℓ → (m/s per km)	0.005	0.0125	<u>0.0272</u>	0.05	0.125	0.25	0.5	1.25	2.5	5	12.5	25
$R(\lambda)$ →	10^{-6}	220	<u>1267</u>	886	338	163	79	27	7.4	1.2	0.08	0.01

Thanks to the smallness of ε, the function $R(\lambda)$ can be expressed in terms of known functions, as follows :

a) If $\lambda < 3$ ($\ell < 2L_s$, approximately), $R(\lambda)$ is given within less than $0,3\ \%$ by a formula involving Dawson's integral, namely

$$(7.21)\quad R(\lambda) = 16\left(\lambda - \int_0^\lambda e^{t^2-\lambda^2}dt\right), \quad \text{whence} \quad R'(\lambda) = 32\lambda\int_0^\lambda e^{t^2-\lambda^2}dt,$$

deduced from (7.20) by omitting ε^2 (earth's rotation neglected), putting $t = \lambda\sqrt{1-\theta^2}$, and then performing a partial integration. For $\lambda < 2$ this formula gives $R(\lambda)$ within less than $0,1\ \%$.

b) If $\lambda > 25$ ($\ell > 16\ L_s$, approximately), the same approximation of $0,1\%$ is obtained by using the following formula involving the classical Bessel functions $K_0(\zeta)$ and $K_1(\zeta)$:

$$(7.22)\quad R(\lambda) = 16\lambda\zeta e^{-\zeta}\left[K_0(\zeta) + K_1(\zeta)\right] = \frac{16\sqrt{2}}{\varepsilon}\zeta^{\frac{3}{2}}e^{\zeta}\left[K_0(\zeta)+K_1(\zeta)\right],$$

where

$$(7.23)\quad \zeta = \frac{1}{2}(\varepsilon\lambda)^2 = \frac{1}{2}\left(\frac{f}{4}\frac{\ell}{U}\right)^2 = \frac{1}{2}\left(\frac{\pi}{2}\frac{\ell}{L_f}\right)^2,$$

this formula being deduced from (7.18) by neglecting k^2 beside k_s^2, replacing the upper limit of the integral with $+\infty$, and then putting $k = k_f\sqrt{\frac{t+1}{2}}$. For $\lambda > 70$ the approximation is better than $0,01\%$, and in particular the formula can be used to obtain the maximum of $R(\lambda)$: the corresponding value of ζ is given by the equation

$$H(\zeta) \equiv (3-4\zeta)K_0(\zeta) - (4\zeta-1)K_1(\zeta) = 0, \tag{7.24}$$

the first member of which is a solution of

$$\zeta H'' + 2H' + (3-\zeta)H = 0. \tag{7.25}$$

This gives $\zeta = 0,4214$, whence the underlined values of the above table have been deduced.

c) If $3 < \lambda < 25$ the value of $R(\lambda)$ is obtained within less than $0,3\%$ by the formula

$$R(\lambda) = 16\lambda\zeta e^{-\zeta}\left[K_0(\zeta) + K_1(\zeta) - \frac{\varepsilon^2}{4\zeta^2}\right], \tag{7.26}$$

deduced from (7.18) in the same way as (7.23), except that $\sqrt{k_s^2 - k^2}$ is replaced with $k_s\left(1 - \frac{k^2}{2k_s^2}\right)$ instead of k_s , and an approximate value is adopted for the resulting additional term.

The location of the maximum of r in the synoptic-scale range is to be related to the fact that, in the corresponding mountain perturbations, the upward flux of energy is mainly vertical, which is not the case for the meso-scale perturbations. Actually the value of U corresponding to this maximum is not far from the value giving the largest lee-waves amplitude for the same mountain range and the same values of σ and s , and the rapid decrease of r as U gets smaller values also corresponds to a rapid decrease of this amplitude : we are dealing with some sort of a resonance effect, existing more generally in any barrier perturbation produced in a steady air flow possessing a static or dynamic stability. For instance if $b = 100\,km$ and $h = 700\,m$, the relative depletion r

of the wind energy is less than 2% for $U = 1\,m/s$, reaches the maximum 62% for $U = 2{,}7\,m/s$, then falls to 20% for $U = 10\,m/s$, and to 4% for $U = 50\,m/s$; with $b = 1000\,Km$ we have the same values of r, if $h = 2200\,m$, for $U = 10\,m/s$, $27\,m/s$, $76\,m/s$ and $108\,m/s$, respectively. The importance of this friction effect due to the ground orography is also quite remarkable : we have seen that, with $b = 100\,Km$, the relative depletion r is exceeding 60% for $h = 700\,m$, this height representing less than 1% of the width of the mountain range; for $h > 1000\,m$ the theory gives $r > 1$, this result meaning that the air flow is probably unable to cross the mountain range (this conclusion is confirmed by a computation which no longer assumes and infinitesimal amplitude for the perturbation), and since any actual mountain range has necessarily a length comparable to its width it is to be expected that in such a case the air flow will turn around the mountain range by its cyclonic side (by the left in the northern hemisphere), in conformity with the pattern of fig. 2 but with a much larger horizontal deflection.

It would be very important to take account of these results in the models of atmospheric motions used for numerical forecasting, in order to get a realistic simulation of the friction forces due to the various mountains, and in particular it would be quite incorrect to assume a friction force represented by some increasing monotonic function of the wind speed measured at a fixed small height, as suggested by the case of a small-scale orographic perturbation

8. CONCLUSION

In spite of the fact that the results obtained in this paper are essentially based on the infinitesimal assumption, it is very likely that they give at least a first approximation of what actually happens when a steady air flow is crossing a mountain range with a smooth profile and moderate height ; this statement is well confirmed by the observed structure of mountain perturbations in the lower atmosphere, which in agreement with the theory may vary considerably with the vertical distribution of the parameters controlling the static or dynamic stability of the air flow. Furthermore what we have inferred concerning the behaviour of the perturbations at higher levels is also confirmed by observations, in particular the existence on several occasions of strong vortices or clear-air turbulence above mountainous areas, brought to evidence by aeronautical events or recent field experiments.

The conclusions concerning the dissipation of mechanical energy due to mountain perturbations are perhaps not so well corroborated, but they are probably still more important for the understanding of the general evolution of large-scale atmospheric motions, and above all the fact that, in the case of a nearly uniform air flow, the dissipation by a given mountain range is practically restricted to a narrow range of wind speed. Another consequence of the theory also merits a special attention, namely the fact that the vertical transfer of mechanical energy between the troposphere and higher levels is operated mainly by meso-scale and short synoptic-scale perturbations thanks to their wavelike structure : a significant progress is to be expected in weather forecasting once this transfer is properly taken into account, but of course the problem to be solved is a very difficult one since the transfer in question depends on the structure of the perturbations, itself controlled by the existing vertical distribution of many atmospheric parameters up to a high level and eventually by the evolution of this distribution.

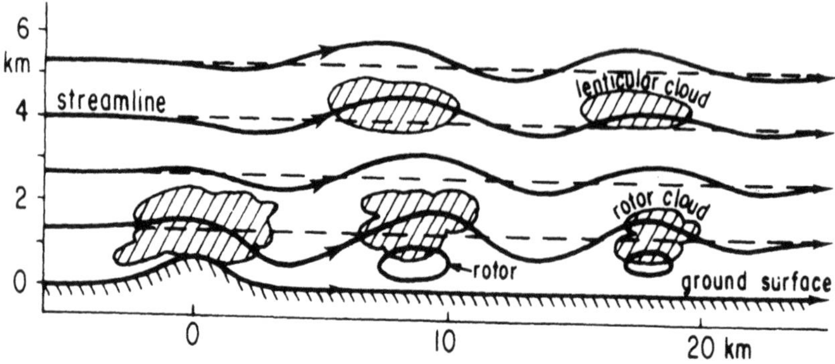

Fig. 1 - Typical mountain-waves (streamlines and clouds in a vertical plane).

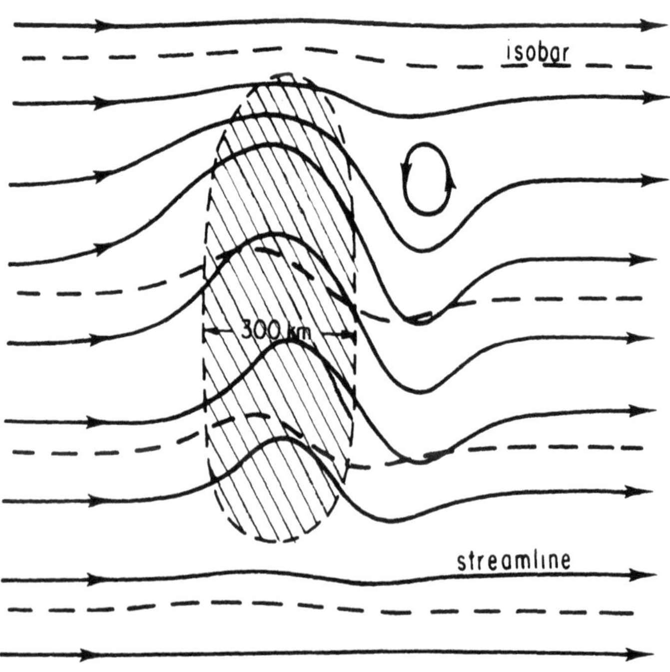

Fig. 2 - Typical synoptic scale orographic perturbation in northern hemisphere (horizontal projection of streamlines and isobars; contour of mountain range at 1 km)

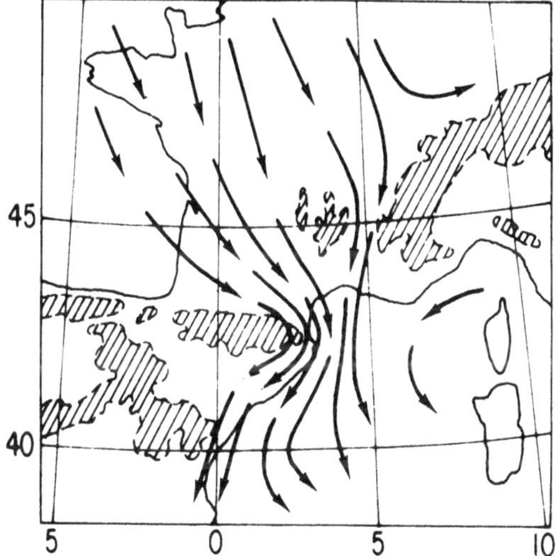

Fig.3 - Formation of the Mistral
(air flow at 1,5 km; isohypses at 1 km)

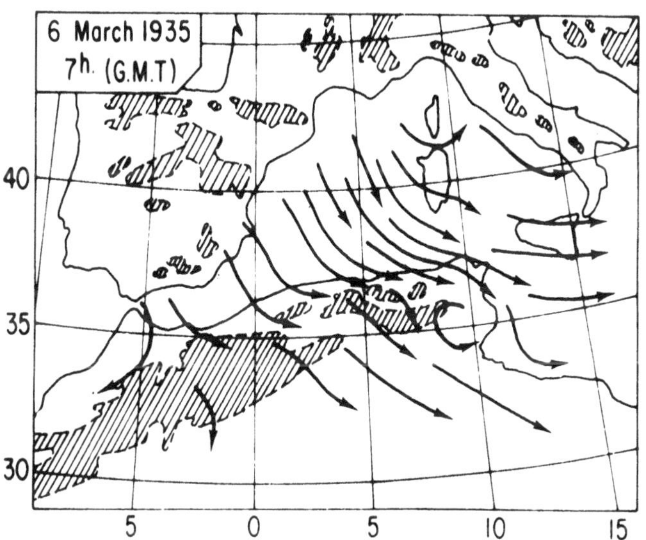

Fig. 4 - Perturbation of a NW-air flow by the Atlas
(air flow at 1,5 km; isohypses at 1 km)

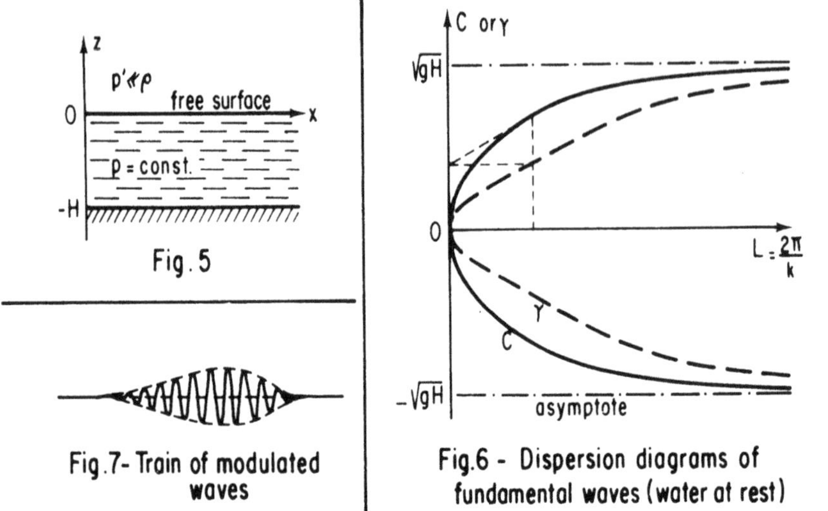

Fig. 5

Fig. 7 - Train of modulated waves

Fig. 6 - Dispersion diagrams of fundamental waves (water at rest)

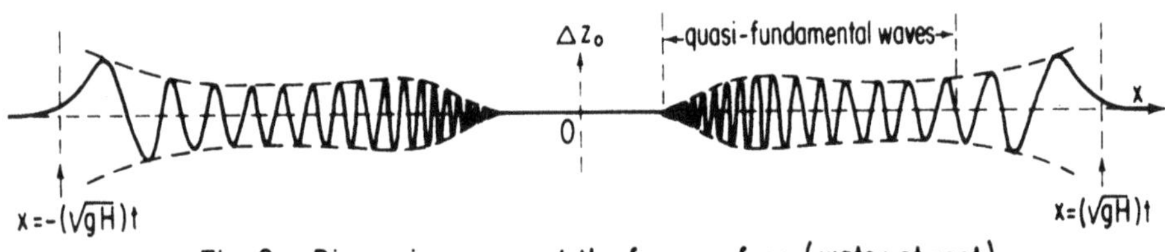

Fig. 8 - Dispersion waves at the free surface (water at rest)

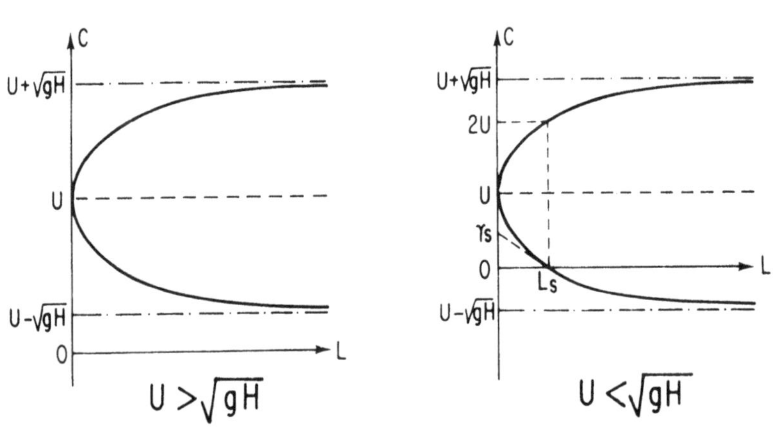

Fig. 9 - Dispersion diagrams of fundamental waves (uniform water-flow)

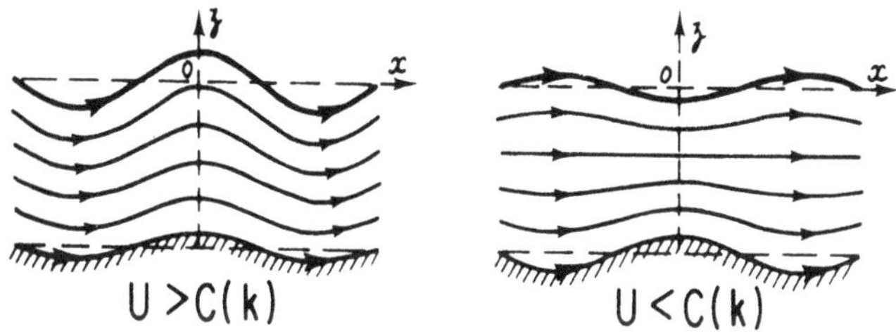

Fig.10 - Forced waves in a water layer flowing over a sinusoidal corrugation at the bed (stationary streamlines)

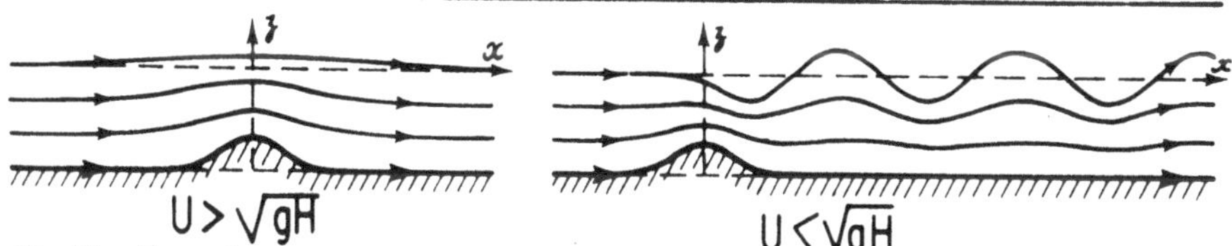

Fig.11 - Forced perturbation in a water layer flowing over a local corrugation at the bed (stationary streamlines)

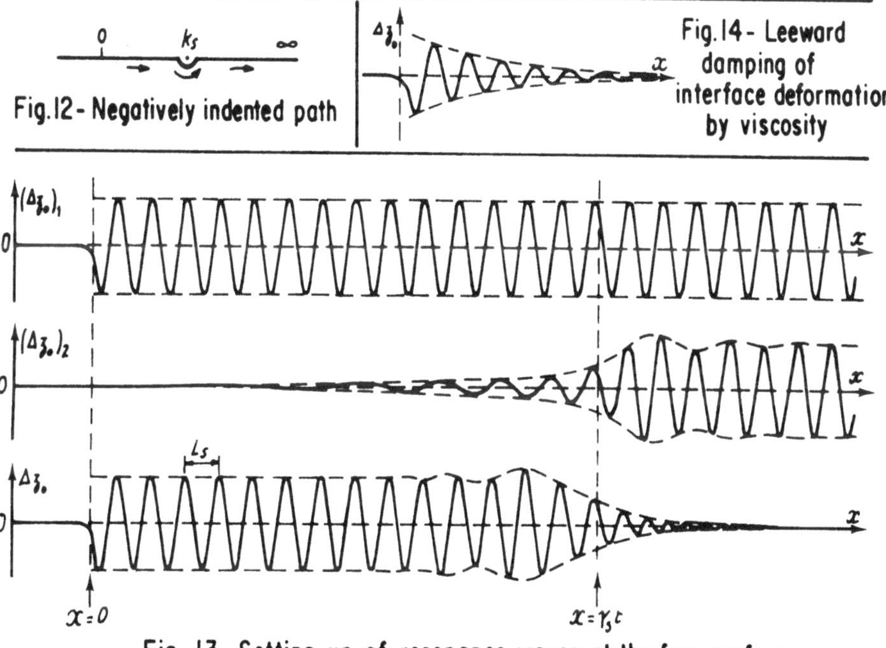

Fig.12 - Negatively indented path

Fig.14 - Leeward damping of interface deformation by viscosity

Fig. 13 - Setting up of resonance waves at the free surface of a water layer flowing over a local corrugation at the bed

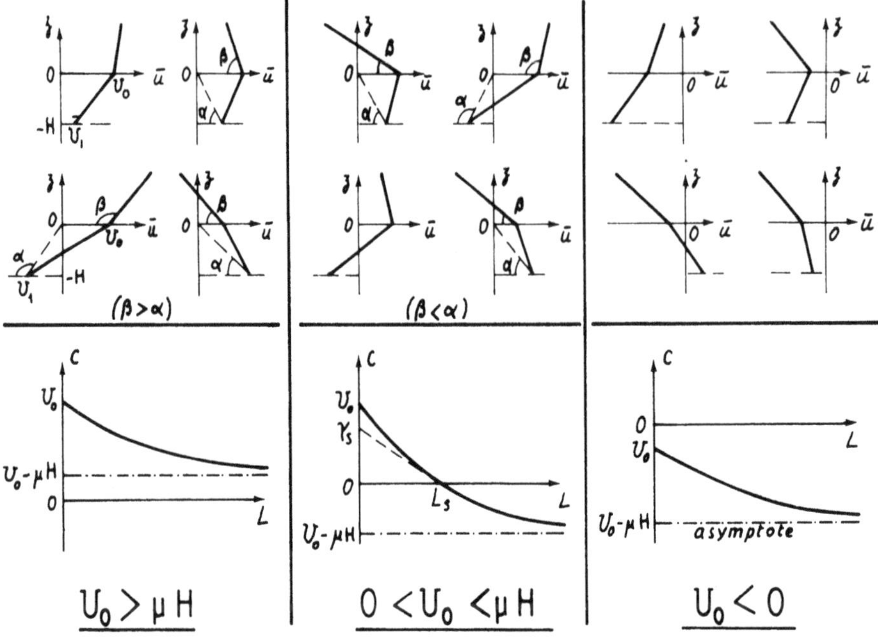

Fig. 15 - Dispersion diagrams of fundamental shear-waves and corresponding velocity profiles

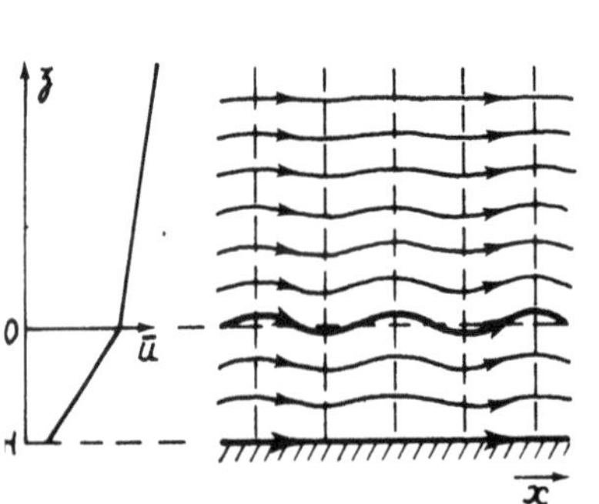

Fig. 16 - Streamlines of fundamental shear-waves

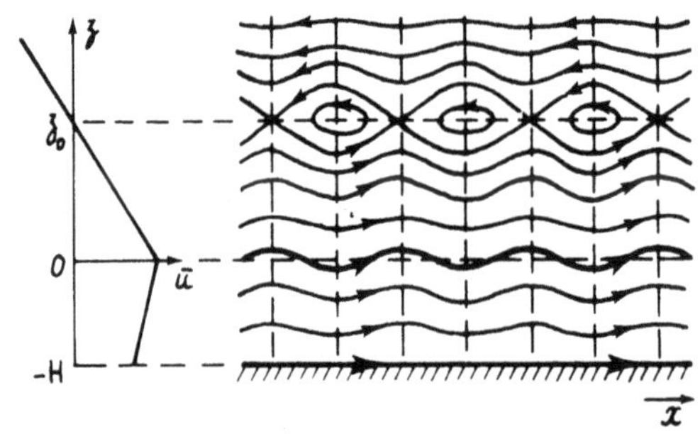

Fig. 17 - Streamlines of fundamental shear-waves

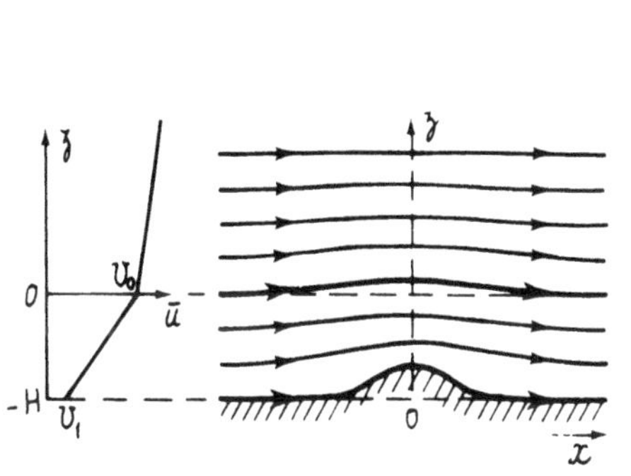

Fig.18 - Streamlines of a forced perturbation

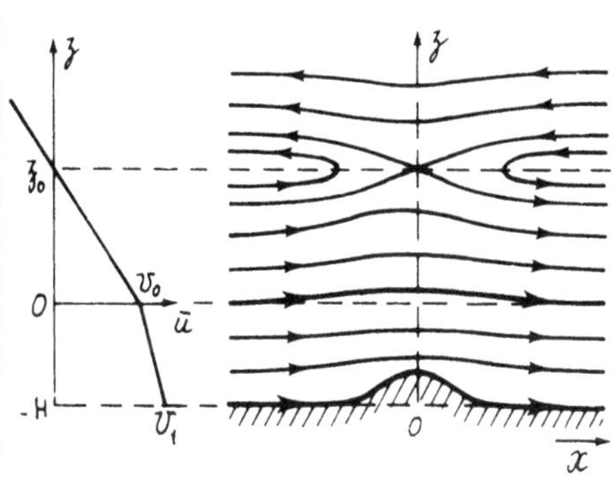

Fig.19 - Streamlines of a forced perturbation

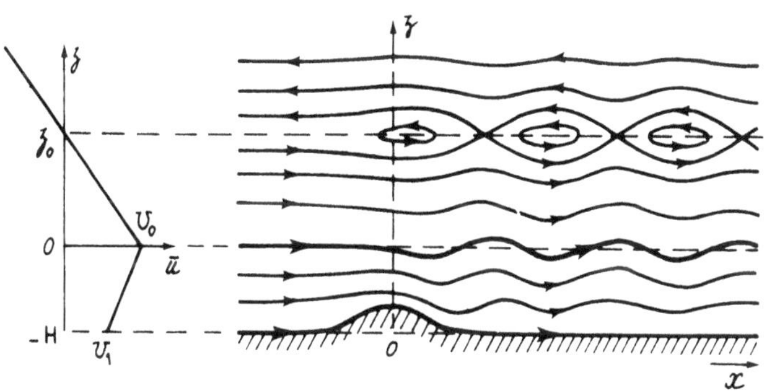

Fig.20 - Streamlines of a forced perturbation involving resonance waves (barrier waves)

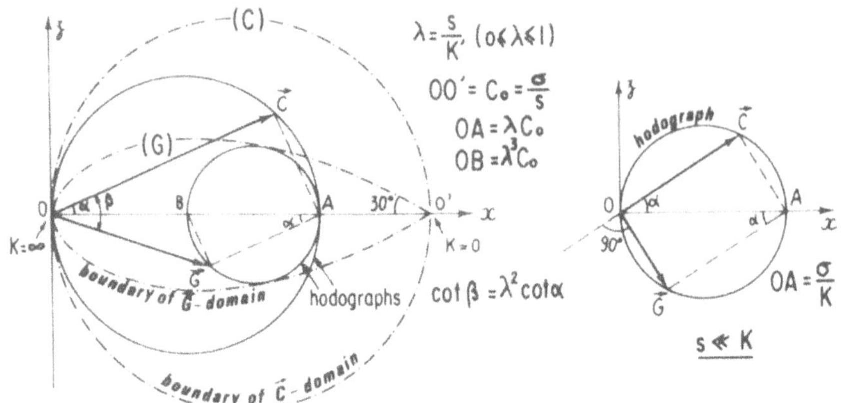

Fig. 21 - Construction of $\vec{C}$ and $\vec{G}$; hodographs for K = const. (gravity - waves for σ = const; $\bar{u}$ = 0)

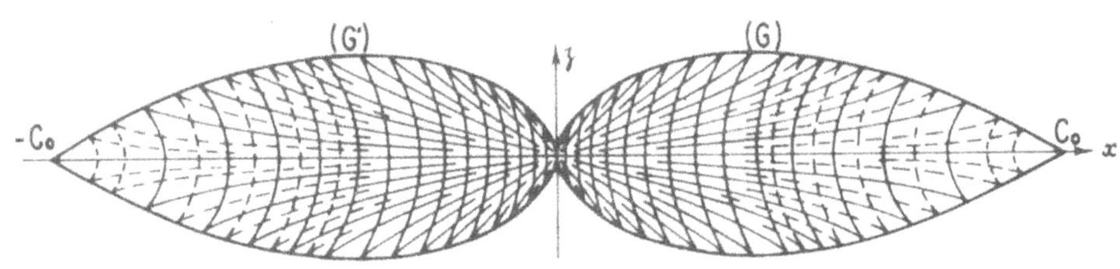

Fig .22 - Dispersion gravity-waves; boundary and profiles of wave - surfaces (σ = const; $\bar{u}$ = 0)

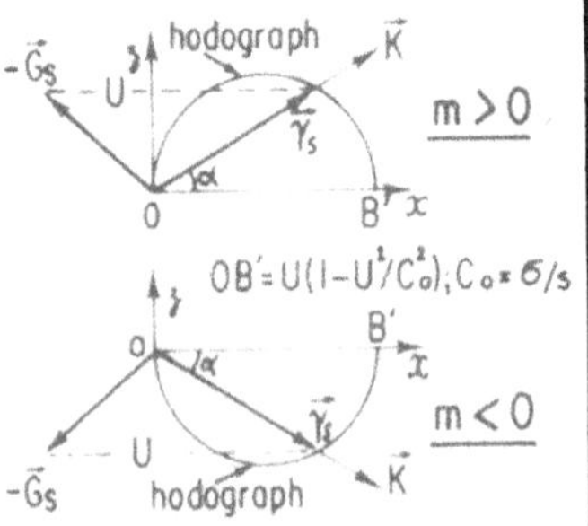

Fig.23- Group velocity of stationary waves (σ = const; $\bar{u}$ = const = U)

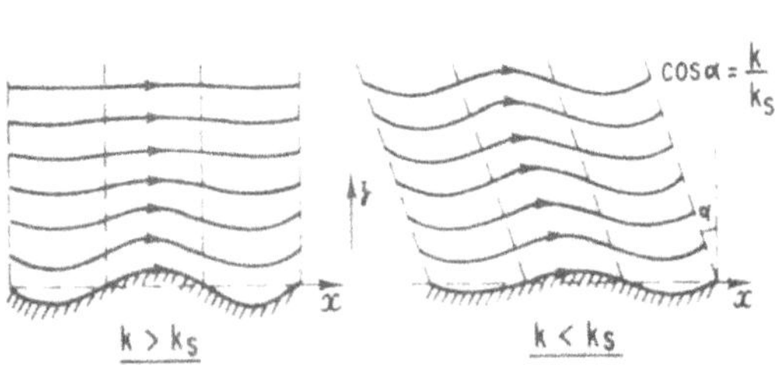

Fig .24 - Forced waves over a sinusoidal bottom corrugation ($\sigma, \bar{u}$ = const.)

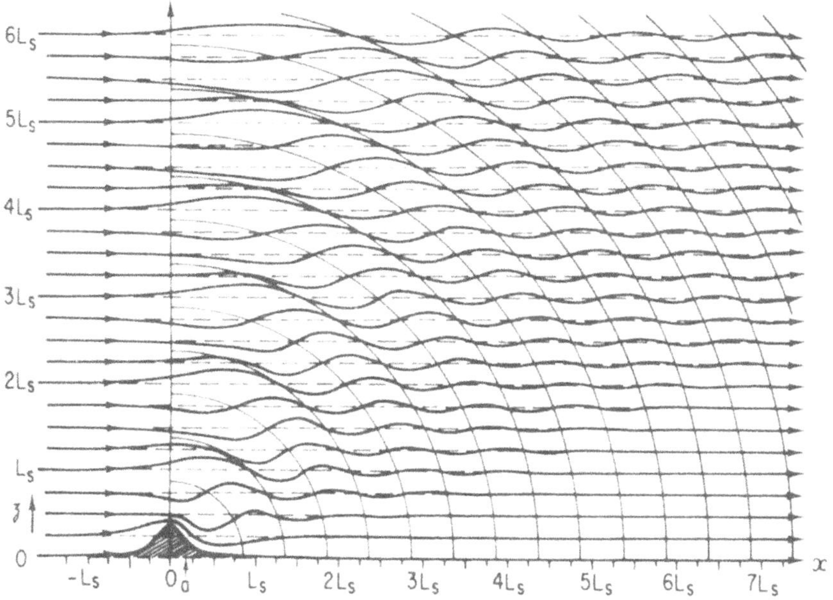

Fig.25 - Barrier perturbation involving significant lee-waves ($\sigma, \bar{u}$ = const; Lorentz-profile barrier with $k_S a = 1$)

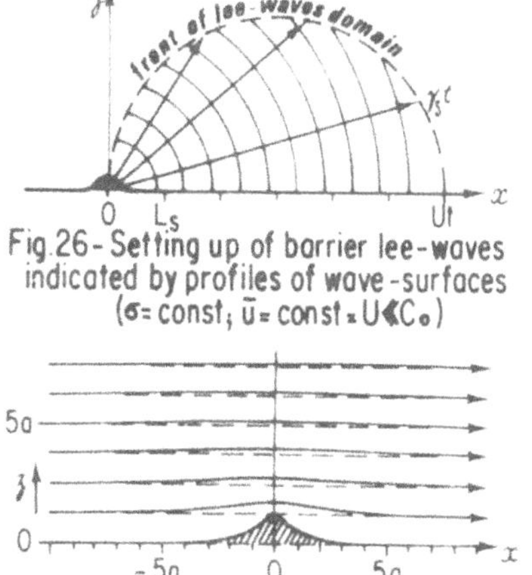

Fig.26 - Setting up of barrier lee-waves indicated by profiles of wave-surfaces (σ = const; $\bar{u}$ = const = $U \ll C_0$)

Fig.27 - Forced perturbation by a narrow barrier ($\sigma, \bar{u}$=const, Lorentz-profile barrier with $k_S a \ll 1$)

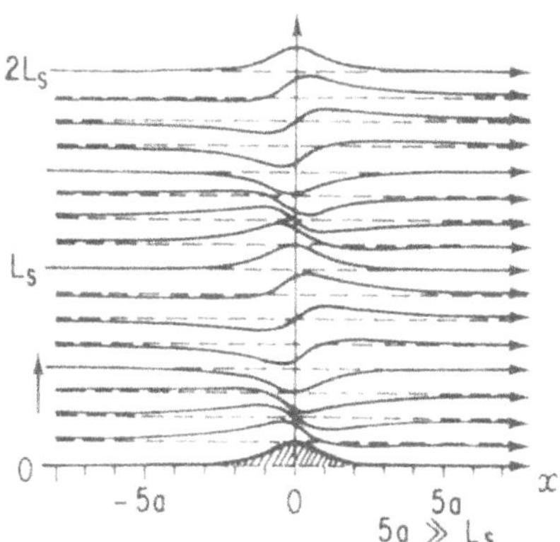

Fig.28 - Forced perturbation by a broad barrier ($\sigma, \bar{u}$ = const.; Lorentz-profile barrier with $k_S a \gg 1$)

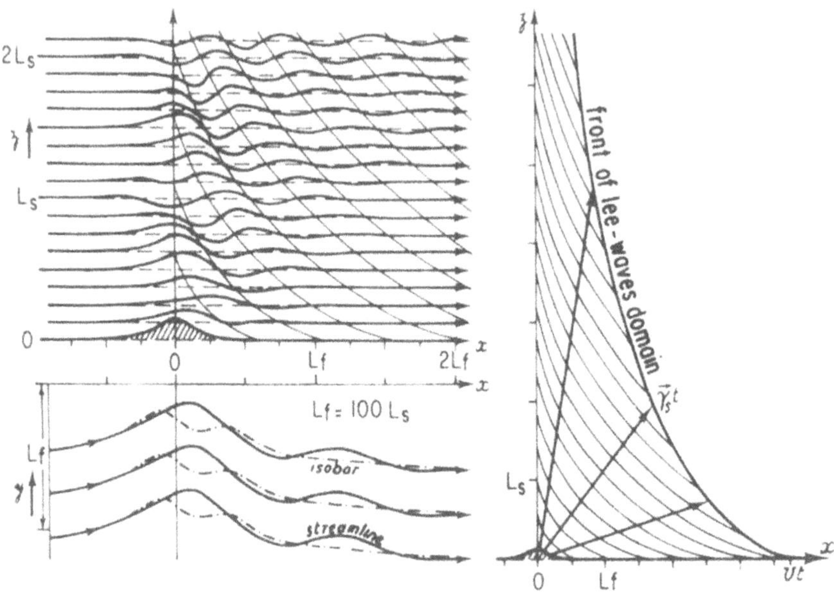

Fig. 29 - Synoptic-scale barrier perturbation ($\sigma, \bar{u}, f$ = const; northern hemisphere)

Fig. 30 - Setting up of synoptic-scale barrier lee-waves, indicated by profiles of wave surfaces ($\sigma, \bar{u}, f$ = const.)

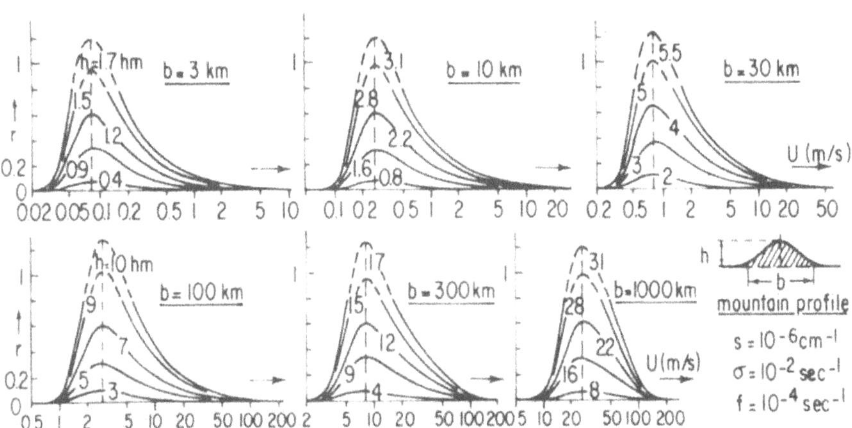

Fig. 31 - Depletion of the mechanical energy of a steady uniform wind crossing a mountain range (r = dissipated energy per unit wind energy)

BIBLIOGRAPHY.

1. HOLMBOE, J., and KLIEFORTH, H.,1957.- Investigations on mountain lee waves and the airflow over the Sierra Nevada.- Dept of Met., Univ. of Cal. at Los Angeles, Final Report- (Sponsored by the Geophysical Research Directorate, U.S. Air Force).

2. QUENEY, P., 1947.- Theory of perturbations in stratified currents with applications to airflow over mountains barriers.- Dept of Met., Univ. of Chicago, Report N° 23, Univ. of Chicago Press.

3. QUENEY, P., 1948.- The problem of airflow over mountains : a summary of theoretical studies.- Bull, Amer, Met, Soc., Vol. 29, N° 1.

4. QUENEY, P., 1954.- Initial value problems in a double Couette flow.- Dept of Met., Univ. of Cal. at Los Angeles, Autobarotropic Flow Project, Sc. Rep. N°1.

5. QUENEY, P., and coll., 1960.- The airflow over mountains.- Techn. Note N° 34, OMM, Genève.
(there is a revised edition).

6. QUENEY, P., 1956.- Dispersion and resonance phenomena in a simple baratropic flow.- Dept of Met., Univ. of Cal. at Los Angeles.

INDEX

GPSR Compliance
The European Union's (EU) General Product Safety Regulation (GPSR) is a set of rules that requires consumer products to be safe and our obligations to ensure this.

If you have any concerns about our products, you can contact us on

ProductSafety@springernature.com

In case Publisher is established outside the EU, the EU authorized representative is:

Springer Nature Customer Service Center GmbH
Europaplatz 3
69115 Heidelberg, Germany

www.ingramcontent.com/pod-product-compliance
Ingram Content Group UK Ltd.
Pitfield, Milton Keynes, MK11 3LW, UK
UKHW061655190726
13853UKWH00008B/2223

* 9 7 8 9 4 0 1 0 2 6 0 0 0 *